Management Science in Fisheries

A key goal of fisheries management is to regulate extractive pressure on a resource so as to ensure social, economic and ecological sustainability. This text provides an accessible entry point for students and professionals to management science as developed in fisheries, in order to facilitate uptake of the latest ideas and methods.

Traditional management approaches have relied upon a stock assessment based on existing understanding of resource status and dynamics, and a prediction of the likely future response to a static management proposal. However all such predictions include an inherent degree of uncertainty, and the last few decades have seen the emergence of an adaptive approach that uses feedback control to account for unknown future behaviour. Feedback is achieved via a control rule, which defines a relationship between perceived status of the resource and a management action. Evaluations of such rules commonly include computer simulation testing across a broad range of uncertainties, so that an appropriate and robust rule can be selected by stakeholders and managers. The book focuses on this approach, which is usually referred to as Management Strategy Evaluation.

The book is enriched by case study examples from different parts of the world, as well as insights into the theory and practice from those actively involved in the science of fisheries management.

Charles T.T. Edwards is a Fisheries Scientist in the Fisheries Modelling Group at NIWA Ltd, Wellington, New Zealand. He has previously worked as a fisheries consultant in the UK and South Africa, and held academic research positions at the University of Cape Town, South Africa, and Imperial College London, UK.

Dorothy J. Dankel is a Researcher at the University of Bergen, Norway, Centre for the Study of the Sciences and the Humanities, and previously at the Institute of Marine Research, Bergen, Norway. She is also a board member of the Nordic Marine Think Tank and served two terms as Chair of the Working Group Marine Systems (WGMARS) in the International Council for the Exploration of the Sea.

Management Science
in Fisheries

An introduction to simulation-based
methods

Edited by Charles T. T. Edwards
and Dorothy J. Dankel

Routledge
Taylor & Francis Group

LONDON AND NEW YORK

earthscan
from Routledge

First published 2016
by Routledge

2 Park Square, Milton Park, Abingdon, Oxfordshire OX14 4RN
711 Third Avenue, New York, NY 10017

Routledge is an imprint of the Taylor & Francis Group, an informa business

First issued in paperback 2018

British Library Cataloguing-in-Publication Data
A catalogue record for this book is available from the British Library

Library of Congress Cataloging in Publication Data
Names: Edwards, Charles T.T., editor. | Dankel, Dorothy J., editor.
Title: Management science in fisheries : an introduction to simulation-
  based methods / edited by Charles T.T. Edwards and Dorothy J.
  Dankel.
Description: New York : Routledge, 2016. | Includes bibliographical
  references and index.
Identifiers: LCCN 2015036002 | ISBN 9781138806801 (hbk) |
  ISBN 9781315751443 (ebk)
Subjects: LCSH: Fishery management. | Fishery management—Simulation
  methods.
Classification: LCC SH328 .M353 2016 | DDC 639.2068—dc23
LC record available at http://lccn.loc.gov/2015036002

ISBN: 978-1-138-80680-1 (hbk)
ISBN: 978-1-138-36420-2 (pbk)

Typeset in Bembo
by Apex CoVantage, LLC

Contents

Notes on authors

Nokome Bentley, Trophia, Kaikoura, New Zealand.

Aaron M. Berger, Northwest Fisheries Science Center, National Marine Fisheries Service, NOAA, Newport, OR, USA.

Trevor A. Branch, School of Aquatic and Fishery Sciences, University of Washington, Seattle, WA, USA.

Paul A. Breen, Breen Consulting, Porirua, New Zealand.

Rik C. Buckworth, CSIRO Oceans and Atmosphere, St Lucia, QLD, Australia.

Douglas S. Butterworth, Marine Resource Assessment and Management Group (MARAM), Department of Mathematics and Applied Mathematics, University of Cape Town, South Africa.

Steven X. Cadrin, School for Marine Science & Technology, University of Massachusetts Dartmouth, New Bedford, MA, USA.

Matthew J. Catalano, School of Fisheries, Aquaculture, and Aquatic Sciences, Auburn University, Auburn, AB, USA.

Anthony Charles, School of Business & School of the Environment, Saint Mary's University, Halifax, Canada.

Sean M. P. Cox, School of Resource and Environmental Management, Simon Fraser University, Burnaby, BC, Canada.

K. Alexandra Curtis, Ocean Associates Inc., under contract to Marine Mammal and Turtle Division, Southwest Fisheries Science Center, National Marine Fisheries Service, NOAA, La Jolla, CA, USA.

Dorothy J. Dankel, Center for the Study of the Sciences and the Humanities, University of Bergen, Norway and the Nordic Marine Think Tank, Copenhagen, Denmark.

Campbell R. Davies, CSIRO Oceans and Atmosphere, Hobart, TAS, Australia.

Roy A. Deng, CSIRO Oceans and Atmosphere, Brisbane, QLD, Australia.

Catherine M. Dichmont, CSIRO Oceans and Atmosphere, Brisbane, QLD, Australia.

Mark Dickey-Collas, International Council for the Exploration of the Sea, Copenhagen, Denmark and DTU- Aqua (National Institute for Aquatic Resources, Technical University of Denmark).

Charles T. T. Edwards, Fisheries Modelling Group, NIWA, Wellington, New Zealand.

Daniel R. Goethel, Southeast Fisheries Science Center, National Marine Fisheries Service, NOAA, Miami, FL, USA.

Chris Grandin, Department of Fisheries and Oceans Canada, Pacific Biological Station, Nanaimo, Canada.

Vivian Haist, Haist Consultancy, Nanoose Bay, Canada.

Katell G. Hamon, LEI, Wageningen University & Research Centre, Netherlands.

Shelton Harley, Secretariat of the Pacific Community, Oceanic Fisheries Programme, Noumea, New Caledonia.

Allan C. Hicks, Northwest Fisheries Science Center, National Marine Fisheries Service, NOAA, Seattle, WA, USA.

Richard M. Hillary, CSIRO Oceans and Atmosphere, Hobart, TAS, Australia.

James N. Ianelli, Alaska Fisheries Science Center, National Marine Fisheries Service, NOAA, Seattle, WA, USA.

Tomoyuki Itoh, National Research Institute of Far Seas Fisheries, Orido, Shimizu, Japan.

Michael L. Jones, Quantitative Fisheries Center, Department of Fisheries and Wildlife, Michigan State University, East Lansing, MI, USA.

Laurence T. Kell, ICCAT Secretariat, Corazón de María, Madrid, SPAIN.

Lisa A. Kerr, Gulf of Maine Research Institute, Portland, ME, USA.

Dale S. Kolody, CSIRO Oceans and Atmosphere, Hobart, TAS, Australia.

Sarah B. M. Kraak, Thünen-Institute of Baltic Sea Fisheries (TI-OF), Rostock, Germany.

A. Robert Kronlund, Department of Fisheries and Oceans Canada, Pacific Biological Station, Nanaimo, Canada.

Hiroyuki Kurota, Seikai National Fisheries Research Institute, Tairamachi, Nagasaki, Japan, and National Research Institute of Far Seas Fisheries, Orido, Shimizu, Japan.

Polina Levontin, Centre for Environmental Policy, Imperial College London, Ascot, Berkshire, United Kingdom.

Mark N. Maunder, Inter-American Tropical Tuna Commission, La Jolla, CA, USA.

Carryn L. de Moor, Marine Resource Assessment and Management Group (MARAM), Department of Mathematics and Applied Mathematics, University of Cape Town, South Africa.

Jeffrey E. Moore, Marine Mammal and Turtle Division, Southwest Fisheries Science Center, National Marine Fisheries Service, NOAA, La Jolla, CA, USA.

Iago Mosqueira, European Commission, Joint Research Center (EC JRC), Institute for the Protection and Security of the Citizen (IPSC), Maritime Affairs Unit G03, Ispra VA, Italy.

Coby L. Needle, Marine Scotland - Science, Marine Laboratory, Aberdeen, Scotland.

Jackie Odell, Northeast Seafood Coalition, Gloucester, MA, USA.

Ana M. Parma, Centro Nacional Patagonico, Centro para el Estudio de Sistemas Marinos, Puerto Madryn, Argentina.

Sean Pascoe, CSIRO Oceans and Atmosphere, Brisbane, QLD, Australia.

Martin A. Pastoors, Pelagic Freezer-trawler Association (PFA), Zoetermeer, Netherlands.

Lisa K. Peterson, Quantitative Fisheries Center, Department of Fisheries and Wildlife, Michigan State University, East Lansing, MI, USA.

Graham M. Pilling, Secretariat of the Pacific Community, Oceanic Fisheries Programme, Noumea, New Caledonia.

Éva E. Plagányi, CSIRO Oceans and Atmosphere, Brisbane, QLD, Australia.

Jan-Jaap Poos, IMARES, Wageningen University & Research Centre, Netherlands.

Ann L. Preece, CSIRO Oceans and Atmosphere, Hobart, TAS, Australia.

André E. Punt, CSIRO Oceans and Atmosphere, Hobart, TAS, Australia and School of Aquatic and Fishery Sciences, University of Washington, Seattle, WA, USA.

Jake Rice, Department of Fisheries and Oceans Canada, Ottawa, Canada.

Marie-Joëlle Rochet, Ifremer, Nantes, France.

Osamu Sakai, National Research Institute of Far Seas Fisheries, Orido, Shimizu, Japan

Rishi Sharma, IOTC Secretariat, Victoria, Mahé, Seychelles.

Anthony D. M. Smith, CSIRO Oceans and Atmosphere, Hobart, TAS, Australia.

Sarah Lindley Smith, Environmental Defense Fund, Boston, MA, USA.

Paul J. Starr, Starifish, Wellington, New Zealand.

Daryl R. Sykes, NZ Rock Lobster Industry Council Ltd., Wellington, New Zealand.

Cody S. Szuwalski, Bren School of Environmental Science and Management and the Marine Science Institute, University of California, Santa Barbara, CA, USA.

Ian G. Taylor, Northwest Fisheries Science Center, National Marine Fisheries Service, NOAA, Seattle, WA, USA.

Nathan Taylor, Department of Fisheries and Oceans Canada, Pacific Biological Station, Nanaimo, Canada.

Verena M. Trenkel, Ifremer, Nantes, France.

Foreword

Anthony Charles
Director, School of the Environment, and Professor,
School of Business, Saint Mary's University, Halifax, Canada

Management science – essentially the 'science of management' – involves applying a systematic, structured and scientific approach to decision-making. What could be more crucial for informed choices, whether in fisheries or in another field? I have a certain bias, admittedly, as my university position lies in both management science (in our School of Business) and fishery research (in our School of the Environment). It is thus a great pleasure to be invited to write this foreword for *Management Science in Fisheries*. I have seen the value, over the years, of bringing management science and fisheries together, and this book meets the challenge of making that connection in a comprehensive manner. The book will undoubtedly help many more fishery people to become familiar with the ideas, methods and results of management science.

This foreword places the book in a historical and thematic context. For many decades management science has been applied in business, public utilities, healthcare and similar sectors, and yes, in natural resource management. Its essence lies typically in determining the 'best' or a 'good enough' or an 'acceptable' decision, given a specific interacting 'system' of key components, together with that system's dynamics, a set of objectives being pursued, and a set of constraints. For example, consider a nonfishery scenario: the challenge an electricity utility faces in determining a desirable (or acceptable) supply and distribution scheme for providing electricity to a diverse set of consumers in a certain jurisdiction (all part of the 'system'). This requires incorporating dynamics in the form of changing demographic and economic conditions over the coming decades, in order to meet a set of objectives (such as reducing greenhouse gases and providing a reliable electrical service), subject to a set of economic, social and environmental constraints. Therein lies a classic management science problem.

Fisheries and management science

In fisheries, while applications of the corresponding techniques have a long history, explicit references to management science, or the largely equivalent term 'operations research', have appeared only over the past few decades. Early in that time period, Rodrigues (1990) provided a major compilation of practical applications to fisheries, while the broad role of 'fishery management science' was highlighted soon after (by Lane and Stephenson, 1995), with an emphasis on identifying a set of explicit objectives as the key initial prerequisite. Every application of management science, in fisheries as elsewhere, must also define the system of interest: a set of interacting components (species, fleets, gear types, management arrangements, etc.) with accompanying dynamic relationships. This reflects a 'fishery systems' perspective in fishery science and management (Charles, 1995b, 2001) and can draw on modern frameworks that support this perspective (Garcia and Charles, 2007).

The scope of each management science application depends on the management issue being addressed. For example, while it is common for a management strategy evaluation to provide advice on the setting of a total allowable catch, under uncertainty, this may be entirely biologically based or may also involve economic and/or social considerations. For the latter, a bioeconomic (or biosocioeconomic) modelling approach can be used (e.g. Clark and Munro, 1975), drawing on over 40 years of experience. These approaches have strong links to management science, with both having a focus on (1) building models of systems of interacting biological and human components and (2) determining the harvesting strategy to maximize a certain objective function.

Whether biologically focused or bioeconomic in nature, management science applications in fisheries also draw on a range of analytical techniques. Many emphasized closed-form optimization techniques, which led to major conceptual insights. However, these typically had to be restricted to models of 'low dimensionality', that is involving only a small number of variables, and to models with a relatively simple objective function. Over time, a broader application of the toolkit of management science has included greater emphasis on simulation models and scenario analyses, which correspondingly provide greater flexibility with model structure and with the set of objectives to be considered.

The history of management science in fisheries is brought to the present day within this book, which applies the analytical toolkit conceptually, thematically and in a significant set of real-world case studies. Each of these case studies involves specifying – as described initially in Chapters 1 and 2 – (1) the components and the boundaries of the fishery system, (2) the fishery objectives and constraints, (3) the model and incorporated dynamics, (4) the sequence of stages of decision-making and (5) the role of uncertainty and risk in decision-making.

Uncertainty

The analysis of risk and uncertainty in fisheries has an interesting history. In earlier days of my career, I was part of a wave of efforts to incorporate uncertainty into fishery models. Gradually, those efforts became mainstream, and indeed this book reflects a further extension and consolidation of that work, adopting more sophisticated ways, such as systematic management strategy evaluations. What has remained constant, however, is the usefulness of the standard typology of uncertainty, ranging from 'simple' random fluctuations to parameter uncertainty to the most challenging, structural uncertainty.

Paralleling the quantitative analysis of uncertainty in fisheries has been an increasing focus on developing management and policy approaches that support fishery sustainability under uncertainty. While quantitative modellers and policy analysts are most often different people with different skill sets, a few of us modellers have shifted (or expanded) into the management and policy realm. In my case, that came as a result of a profound event in the early 1990s, just outside my door – the infamous 'cod collapse' on the Atlantic coast of Canada. My assessment is that this dramatic collapse revolved around a range of uncertainties, and the failure of decision-makers and institutions to properly allow for the accompanying risks (Charles, 1995a, 1998, 2002). The resulting messages, I argue, are (1) a need for humility in how we analyze and manage fisheries, given the wide-ranging uncertainties involved and (2) the need for an effective precautionary approach, that includes not only quantitative analysis of uncertainty and risk, but also qualitative methods of robust management (that are more likely to give decent results even under severe uncertainty) and adaptive policy (providing built-in flexibilities).

I like to think these messages have become widespread (e.g. Charles, 2001), though this is not always as consistent as it should be, and there remains a need for broader implementation. Modern management science applications in fisheries can contribute to a strong analysis of uncertainty, through management strategy evaluations and similar approaches. To that end, *Management Science in Fisheries* – while not about fisheries management and policy as such – provides important analytical insights into how management might be shifted to become more robust, and how policy could be more adaptive.

There is, however, an enduring lesson from the past for those engaged in or considering the use of management science in fisheries: the need for humility. As time passes, we may think we know now (in retrospect) why *past* errors were made in scientific support for fishery decision-making (and specifically in stock assessment). We may imagine that we are now able to handle uncertainty in fisheries. This would be a very unfortunate attitude. At some future point, a major error will surface in a model-based fishery analysis, such as a management strategy evaluation. This is particularly true since – as the editors of this book note – models are never correct. Humility, again, is an important companion in decision-making and decision support. There is always room for improvement.

That said, it is the presence of uncertainties that makes it so essential to adopt systematic decision-making, within the context of a suitable precautionary approach, in order to properly take into account the corresponding risks. Models are needed, and management science provides the framework within which the models are applied. Within the book's analyses, uncertainty is introduced early on and recurs throughout. The analysis of uncertainty appears in many ways, most notably in simulation methods but also with such tools as scenario analyses, parameter and model estimation, and development of harvest control rules that allow for random fluctuations. It is useful, while reading the book's case studies, to ponder the performance of each method, in its capability to guide the fishery system toward 'living with uncertainty', within a world of ubiquitous uncertainties.

Human dimensions

While some models can be physical (such as scale models of aircraft) or put on paper (blueprints and maps), models in management science are mathematical and computer based. The heavily quantitative nature of management science keeps it from being bedtime reading for very many people. However, there have been remarkable efforts over the years to add 'human dimensions' into fishery models. This began with basic harvesting economics (revenues and costs of fishing), followed by broader economics (e.g. the postharvest sector and ecosystem valuation) and to some extent social aspects (e.g. labour dynamics, and social goals such as employment and community stability).

It is important that human dimensions be prominent in fishery management science applications (Charles, 1995b, Fulton et al., 2011). In several cases in this volume, human-oriented considerations are built into the performance criteria with which management strategies are assessed, and in some cases also into the fishery dynamics. This is positive, although as discussed in the first and last chapters of this book (see e.g. the section 'Steps to meet the human system in fisheries' in Chapter 22), there remains much room to grow. This is an emerging horizon in fishery management science. Progress is already occurring – for example, in the inclusion of human behaviour and social aspects in the Atlantis ecosystem model.

What is notable widely in fisheries is the shift to involving fishers and other stakeholders in the decision support, and perhaps also the decision-making process. This broadening of participation in governance (Garcia et al., 2014) is certainly supported in the case studies presented in this book, as well as being highlighted in the final chapters in Part 4. Indeed, within some fishery systems, the book shows how stakeholders themselves are closely involved in conducting management strategy evaluations. If such moves translate into sharing of real decision-making power, then the management science applications are serving as vehicles to make true co-management and cooperative governance a reality – a crucial factor in building sustainability of the world's fisheries.

Conclusion

I began this short piece by noting that there is a need for more fishery scientists, managers and fishing sector participants to know something about management science. This book is certainly directly aimed at those who may want to delve fully into the field by learning how to carry out real-world management science analyses in fisheries. It succeeds in serving that audience, but will also be useful for those who are looking not for technical details, but rather for an understanding of how models and quantitative analyses can help in supporting decisions in fishery management, and particularly what 'management strategy evaluations' are all about. This book provides a solid basis for that nontechnical audience too.

For me, it is important to pay close attention, in fishery management science, to key fundamentals: the components of the fishery being considered, together with their dynamics, objectives, constraints and uncertainties. It is crucial to limit what is included in an analysis since not everything can be considered. However, it is equally important to pay attention to what is omitted, since omitting the wrong elements can lead to misleading results and misguided advice. A true interdisciplinary and 'integrated' approach is needed, applying a 'fishery systems' perspective, to figure out this balance of inclusion and exclusion, with adequate coverage of both ecological and human considerations. This, I might argue, is one of the crucial lessons I have learned over the past several decades: the need to think broadly.

My hope would be that the next generation of fishery scientists, analysts and decision-makers will further broaden the scope of fishery models used in management strategy evaluations, so as to better cover the suite of social, economic, ecological and biological objectives and constraints, and the dynamic relationships arising across the fishery system (i.e. in the ecosystem, human system and management system). I expect that the right degree of 'broadening' will lead to increases in the explanatory power and the usefulness of management science analyses – a worthy goal for the next generation of interdisciplinary management science practitioners in fisheries.

References

Charles, A. 1995a. The Atlantic Canadian groundfishery: Roots of a collapse. *Dalhousie Law Journal*, 18, 65–83.

Charles, A. 1995b. Fishery science: The study of fishery systems. *Aquatic Living Resources*, 8, 233–239.

Charles, A. 1998. Living with uncertainty in fisheries: Analytical methods, management priorities and the Canadian groundfishery experience. *Fisheries Research*, 37, 37–50.

Charles, A. 2001. *Sustainable Fishery Systems*. Wiley-Blackwell, Oxford, UK.

Charles, A. 2002. The precautionary approach and 'burden of proof' challenges in fishery management. *Bulletin of Marine Science*, 70, 683–694.

Clark, C.W., and Munro, G.R. 1975. Economic of fishing and modern capital theory: A simplified approach. *Journal of Environmental Economics and Management,* 2, 92–106.

Fulton, E.A., Smith, A.D.M., Smith, D.C. and van Putten, I.E. 2011. Human behavior: The key source of uncertainty in fisheries management. *Fish and Fisheries,* 12, 2–17.

Garcia, S.M. and Charles, A. 2007. Fishery systems and linkages: From clockwork to soft watches. *ICES Journal of Marine Science,* 64, 580–587.

Garcia, S.M., Rice, J. and Charles, A. 2014. *Governance of Marine Fisheries and Biodiversity Conservation: Interaction and Coevolution,* Wiley-Blackwell, Oxford, UK.

Lane, D.E. and Stephenson, R.L. 1995. Fisheries management science: The framework to link biological, economic, and social objectives in fisheries management. *Aquatic Living Resources,* 8, 215–221.

Rodrigues, A.G. (ed.) 1990. *Operations Research and Management in Fishing,* Kluwer Academic Publishers, Dordrecht, Netherlands.

Preface

Natural resource management continues to expand as a field of scientific study, becoming increasingly complex and encompassing a broader range of systems within its remit. This book, "Management Science in Fisheries" describes how it practiced in the marine environment, where our exploitation of the resident populations is probably one of the most direct and significant anthropogenic impacts. Science has looked for ways to mitigate these impacts and provide the technical solutions needed for sustainable and healthy oceans. We have sought in this volume to bring together practitioners that have been working at the forefront of this still young scientific field. By providing an introduction that is grounded in their real world experience, we hope that the techniques they have developed can be disseminated to meet the increasingly pressing need for sustainable fisheries management.

Fisheries management science currently places a strong emphasis on computer-based simulation, in which the relevant components of the system are represented by one or more mathematical-statistical models that allow its behaviour to be predicted. The performance of different management approaches, or strategies, which describe how future decisions might be made concerning exploitation of the fishery, can then be evaluated against the simulation model. This simulation based approach forms an important part of what is now conventionally called management strategy evaluation. This book is designed to give an introduction to this simulation based approach – where it came from, how it is currently practiced, and how it is being developed.

The book begins by placing fisheries management science in the context it is required to operate, along with an overview of the theoretical background (Part I). In Part II, the book then offers case studies that cover a wide range of ecosystems and jurisdictions. The case studies are diverse, from formidable species such as bluefin tuna to the more humble haddock, and differ in the scientific approach taken. To accommodate a reader new to the field we have listed them in order of increasing scientific complexity. In Part III, contributing authors have addressed some of the major topics in current scientific practice, namely perspectives on the construction of decision making strategies, spatial and ecosystem complexities, how to evaluate and communicate performance of particular management strategies, and what to do when a simulation based approach is not suitable. Finally, the book ends with a more encompassing view (Part IV), acknowledging that science in a management context must be performed in collaboration with those that will ultimately be affected by its outcome. The reader will note, particularly from the case studies in Part II, that this inclusivity is a recurrent feature of best practices of fisheries management science.

We have sought to make this book accessible, whilst retaining enough detail for it to be a comprehensive and independent resource for both graduate students and practitioners. Although some of the material is quite technical, it provides the detail needed for the reader to understand this exciting scientific field in terms of how it is actually practiced. Furthermore, it will give insight into the science-policy dialogue that not only provides context, but forms an integral part of the work.

All contributions in this book have been peer-reviewed for scientific rigor and edited for ease of access, including an attempt to ensure consistent terminology throughout. It has been our aim that anyone interested in fisheries management science will find this book a useful resource, and we hope that it will support their own endeavors in applying the approaches described.

Good and relevant science for policy is dependent on strong networks of scientists who are attuned to the real problems and challenges around them. We are grateful to our research institutions, NIWA and the University of Bergen, and our networks there within who have supported us in this work, as well as the many external colleagues with whom we are privileged to interact. We also thank the following colleagues who have helped guide our own thinking in a way that has shaped the content of this book: Anthony Smith, Jake Rice, Richard Hillary, Éva Plagányi and Doug Butterworth.

Charles T. T. Edwards, Wellington, New Zealand

Dorothy J. Dankel, Bergen, Norway

Part I

An introduction to fisheries management science

Fisheries management science operates at the interface with policy, meaning that it is a practical exercise designed to meet the needs of society, as outlined in policy-orientated management objectives. Context is therefore very important, and Part 1 of the book aims to introduce some of the background that will be required by the reader for the remaining chapters.

Chapter 1 begins by outlining the concept of a fisheries system, which includes interacting parts that will all influence how the science is conducted and what it is intended to achieve. The actual process of designing a functioning mechanism for managing the fishery is then outlined in broad terms, emphasising how the science fits into this broader management cycle. Given the contextual background, Chapter 2 provides an introduction to the scientific process itself. Management science in general is concerned with the construction of decision-making mechanisms that are designed to meet explicit objectives for the system being managed. It is an extremely broad scientific field, and this chapter describes how it has been developed and applied in fisheries. Starting from early theoretical approaches the chapter brings the reader up to how it is currently practiced, which places a heavy emphasis on computer-based simulation experiments. Since this is technically the most challenging part of the work, Chapter 3 is dedicated to the practical aspects of a simulation-based approach, including the computer code needed to execute an actual application. Familiarity with these three chapters, detailing the context, scientific background and practical aspects, will equip the reader for the remaining chapters of the book, particularly those in Part 2, which includes a range of interesting case study examples.

## Chapter 1

Fishery systems and the role of management science

Dorothy J. Dankel and Charles T. T. Edwards

Introduction

The driving force behind fisheries science is to produce advice relevant to the exploitation and conservation of marine resources and ecosystems. It therefore has a very practical motive. The collapse and closure of fish stocks is well known and fisheries scientists are primarily tasked with quantifying and describing ways to promote "sustainability" in fisheries. Sustainability is an internationally recognised objective used to guide fishery management decisions, stipulating that fishing should not detriment the stock more than it can naturally continue to renew itself. Sustainable fisheries are dependent not only on healthy fish stocks in a healthy ecosystem, but on functioning and self-repairing systems of politics, management, fishing communities and consumers. This is due to the strong interconnectivity between anthropogenic forces and the ecosystems on which they depend. So how do we understand and manage a sustainable fishery, and how does science accommodate the human dimension whilst simultaneously providing advice on a natural resource that is often poorly understood?

The fishery system

Fisheries management is a very broad concept, but fundamentally includes a decision-making process intended to result in a set of actions that will allow the fishery to meet explicit, policy-driven management objectives. It requires that decisions are made (often on an annual basis) regarding, for example, the total catch, season length or equipment regulations. Fisheries management science is primarily concerned with how we make these decisions – or more generally, how we select a particular decision-making strategy from the suite of alternatives (Stephenson and Lane, 1995). Management science, the managers that actually make decisions and the guiding policy principles together constitute the management system, with the term *system* referring to entities composed of interacting parts. These parts can be dynamic and have a positive, negative or neutral effect on other interconnected components. We can think of a fishery as a system in which management is just one part.[1] But there are other essential components that are equally important. To understand how the management system works, it needs to be placed in the context of other system components, namely the natural and human systems (Figure 1.1; Charles, 2001).

Within the three components of the natural ecosystem, the human system and the management system (Charles, 2001, ICES, 2006), there are numerous internal dynamics. Temperature change, ocean current changes, algal blooms and resulting stochastic fish reproduction and recruitment are some examples of events that can occur in the natural system. The human system includes fishers, conservationists and other stakeholders and their associated dynamics including employment and economic variations, interactions and conflicts. The management system includes the process of applying knowledge of the natural and human systems towards actions (like fishing regulations and quotas) that best fulfil policy objectives.

Fisheries managers are professionals who make decisions regarding regulations in order to administer local, regional, national or sometimes international laws governing a fishery. For example, a fishery manager may be tasked with ensuring that local fishermen use the regulated mesh size (a gear regulation), fish within restricted spaces (an area regulation), fish during certain days or seasons (a temporal restriction), fish only species allowed in accordance with a fishing license (a bycatch regulation) and/or fish an amount that is allocated to them (a quota regulation). Regulations change according to the "best available science" or other information that supports them, thus there is an inherent

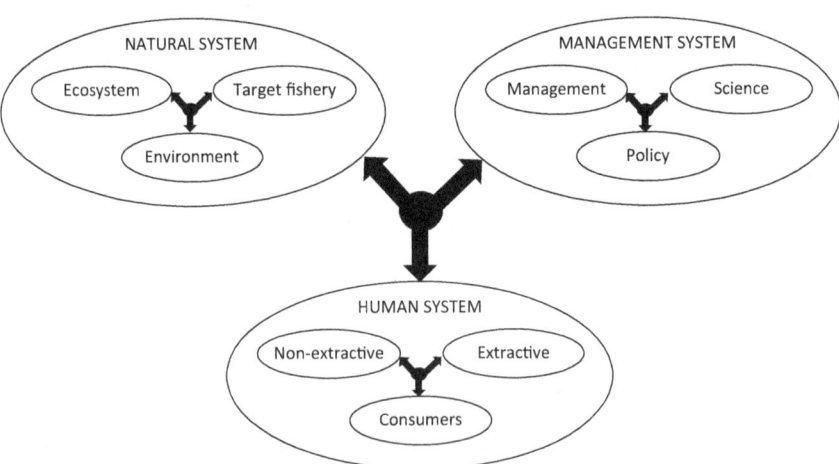

Figure 1.1 A simplified view of the fisheries system, adapted from Charles (2001). This conceptual framework facilitates understanding of what factors and influences can affect management, and in turn the science that is used to inform decision-making within the management system.

connection between science and management, especially pronounced for well-studied and commercially important species around the world.

Scientific advice to support fisheries management is produced routinely by national marine institutes in developed countries. Internationally, institutions and organisations like the International Council for the Exploration of the Sea (www.ices.dk) or the International Commission for the Conservation of Atlantic Tunas (www.iccat.int) pool and integrate scientific data and resources in order to collectively produce science and advice to inform decision-makers on sustainable practices of fishery and ecosystem exploitation.

Objectives in fisheries management

Successes and failures in fisheries management are assessed according to whether the underlying objectives are achieved. A regime considered a biological success may also be judged an economic failure (Cunningham and Bostock, 2005, Hilborn, 2007b, Dankel et al., 2008), and vice versa. Objectives are often elicited on different levels and can be broad or specific, explicit or implicit. International agreements often state broad objectives like "long term conservation and sustainable use of fisheries resources" (FAO, 1995) but may not be legally binding. The United Nations Convention of the Law of the Sea sets standards for marine conservation and protection of resources and is legally binding (UN, 1982). More specific objectives, like maintaining catch stability at a specified level, can be set for a fishery according to stakeholders' desires. Political objectives are often kept implicit. One example of an implicit objective is to have the "minimum sustainable whinge" (Pope, 1983, Hilborn, 2007a) or to curtail stakeholder discontent as much as possible within a management regime.

There are many different objectives for resource use, like maximum sustainable yield, maximum profit, tourism (Caddy and Mahon, 1995) or a compatible combination of these or others. The resource's stakeholders include anyone with an interest in the fishery and can be broadly partitioned into groups according to their objectives or preferences. For example, commercial fishers are usually stakeholders who may forgo biological risks for high and stable annual profit, whereas conservationists are those who are vocal towards habitat preservation and biodiversity issues. Managers, on the other hand, are guided by national policy that would usually stipulate long-term social or economic objectives for the fishery, such as job creation. Thus a resource's management objectives should aim to strike a balance among different, possibly conflicting, objectives. However, a politician may not want to take a stance to lay down explicit biological, economic, and social objectives. Doing so runs the risk of unnecessary stakeholder conflict and discontent aimed at the decision-maker, and also the risk of being held to account for not meeting objectives. Even though clear objectives in management are imperative, they are not always the norm (Dankel et al., 2008).

It is important, but not easy, to strike a multidimensional balance in fisheries objectives that can reflect the biological and socioeconomic components related to the resource and its stakeholders (Cunningham and Bostock, 2005). The lack of clear and explicit objectives can act as a barrier blocking sustainable management of marine fisheries. However, setting objectives does not necessarily mean that one must endure the wrath of conflict. On the contrary, conflict arising from stakeholder diversity can be utilised (Follett, 1955; this idea is further developed by Dankel, Chapter 22, this volume).

The role of fisheries scientists in marine fisheries management has traditionally been to collect and disseminate data from scientific surveys and catch reports from the fishing industry to managers in the form of a recommended total allowable catch (TAC; Reeves and Pastoors, 2007). But when stocks face crises, the scientists' role in routine stock assessment and quota setting (the so-called "TAC machine"; Holm and Nielsen, 2004, Nielsen and Holm, 2007) can become less relevant (Schwach et al., 2007). This is especially true if the crisis was created by science-based management regulations being ignored, such as when TACs are routinely overshot (Cardinale and Svedäng, 2008). Nevertheless, fisheries scientists can become easily overworked by political pressure to come up with the "answer," when in most instances the questions asked by fisheries managers are beyond the boundaries of natural science. These questions could be better confronted across scientific disciplines (Schwach et al., 2007), in a post-normal scientific framework (Funtowicz and Ravetz, 2008). This is a theme that will be revisited, along with the pioneering work on an "integrated solution" by Mary Parker Follett, in the concluding chapter of this volume (Dankel, Chapter 22).

The fisheries management decision cycle

To understand the role of fisheries science within the fisheries management system, it is necessary to describe in more detail the process by which decisions are made. This is most easily represented as a *decision cycle* (Figure 1.2) containing various decision points or nodes. In contrast to the conceptual idea of a fishery system, which is useful for exploring *what* we need to consider to manage a fishery well, the decision cycle should be understood as a directive framework for making decisions regarding *how* the fishery should be managed.

Errors in any part of the decision cycle may lead to unsustainability (ecological, social and/or economic), but even taken individually, many of the decisions are not straightforward, such as choosing an optimal harvesting rate for a stock with variable recruitment or allocating catch among several fleet sectors with different ecosystem impacts. However, the challenge of good fisheries management is actually even harder, because as explained later, many of the decisions can be made independently, yet the outcome of good fisheries management requires that the independent decisions are integrated to be collectively coherent. There is ample evidence that once one aspect of a fishery is performing unsustainably, problems are likely to spread to other aspects of the same or other fisheries (Rice, 2006, Walters and Maguire, 2006).

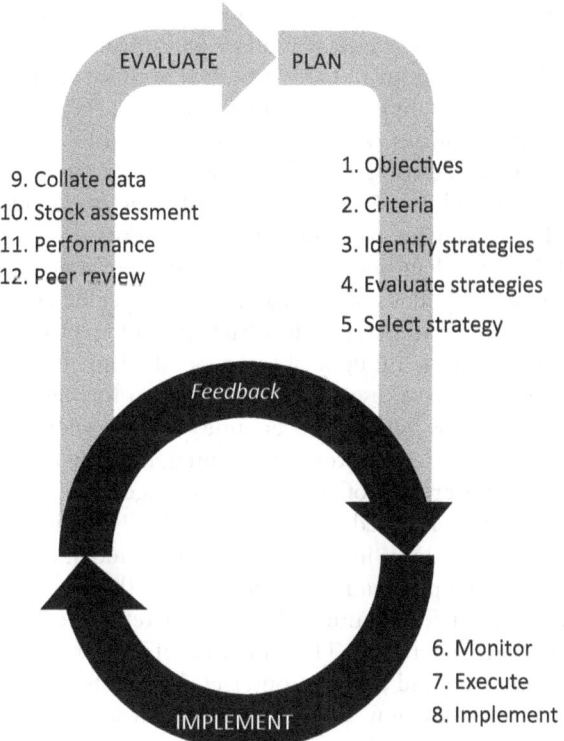

EVALUATE   PLAN

9. Collate data

10. Stock assessment

11. Performance

12. Peer review

1. Objectives

2. Criteria

3. Identify strategies

4. Evaluate strategies

5. Select strategy

Feedback

6. Monitor

7. Execute

8. Implement

IMPLEMENT

Figure 1.2 A schematic representation of the fisheries management decision cycle, derived from a similar double-loop cycle in Jones (2005), with 12 different nodes of action. The inner loop represents the implementation of a feedback control, management strategy, typically on an annual cycle. These management strategies are designed during the planning phase, which usually occurs with less frequency and forms part of the outer loop. The cycle of planning, implementation and evaluation, whereby our understanding of the system improves, is referred to here as *adaptive management*.

The decision cycle contains a double-loop that is best understood using the concepts of *feedback control* and *adaptive management*, which we define here as corresponding to the inner and outer loops respectively (Figure 1.2).

Feedback control means that a current action is dependent on the perceived system state at the current time (see also Edwards, Chapter 2, and Kell et al., Chapter 17, this volume). In fisheries the link between state and action is usually a simple algorithm commonly known as a *harvest control rule* (HCR) or *decision rule*. When inputs to the HCR are also defined, including the data and data processing steps (the *assessment*), these components are together known as a *management procedure* (MP; Rademeyer et al., 2007). A management procedure is a type of management (or harvest) strategy, with the latter being less strictly defined. For example, a management strategy will usually contain a HCR but may allow the data and assessment to be modified or updated during the period in which the strategy is implemented. In some cases, the term may also be applied to describe situations where an HCR is not specified, which makes it useful for discussing more complicated systems where the strategies themselves are still being developed (e.g. Fulton et al., 2014). The important point is that both MPs and management strategies are mechanisms for decision-making. The design of these mechanisms is a central objective of fisheries management science, and the primary focus of this volume, with specific emphasis given to management strategies that include a feedback control mechanism (HCR).

Feedback control can be thought of as a type of adaptive management, but to avoid ambiguity we have reserved this latter term for the learning process by which management is improved over time (the outer loop in Figure 1.2). The adaptive cycle includes the steps of planning, implementation and evaluation. Planning is the process of selecting a management strategy, which in the current context is assumed to include an HCR. This strategy is then implemented (the inner loop in Figure 1.2), which includes a feedback step that would usually take place annually. Finally, after the management strategy has been implemented for a number of years, a retrospective evaluation of performance can be undertaken. The outcome of the evaluation phase is an improved planning step, and thus the outer cycle repeats (Figure 1.2). Monitoring is key to this paradigm, since it allows both feedback control in the inner loop and subsequent evaluation of the management process in the outer loop (i.e. adaptation).

Under the adaptive management headings, we consider further subdivisions of the management process (based on personal communication on February 27, 2015, with J. Rice and A.D.M. Smith):

Plan

1   Elicit and clarify management objectives and constraints: Objectives.
2   Define appropriate performance criteria: Criteria.
3   Identify candidate management strategies: Identify strategies.
4   Conduct scientific evaluation: Evaluate strategies.
5   Select most appropriate strategy: Select strategy.

Implement

6   Collect monitoring data: Monitor.
7   Execute management strategy: Execute.
8   Implement decision: Implement.

Evaluate

9 Collate monitoring data: Collate data.
10 Perform comprehensive stock assessment : Stock assessment.
11 Evaluate performance of management strategy: Performance.
12 Solicit input from peer community: Peer review.

The planning phase begins with setting the conceptual and operational objectives for managing the fishery, including acceptable social, economic and ecological outcomes (Step 1). It is then necessary to define how these can be measured (Step 2). Given the objectives and constraints (and also information from evaluation of previous cycles in the adaptive strategy at Step 11), candidate management strategies can be identified (Step 3). These alternative strategies are then scientifically evaluated using the performance criteria specified in Step 2. This evaluation process (Step 4) can be either qualitative or quantitative, but must accommodate uncertainty in what we know about the system, highlight the trade-offs across management objectives and should be transparent, so as to be understood by the wider stakeholder group. This is because the final strategy is usually selected (Step 5) following a period of dialogue between the managers, scientists and other interested groups, most notably the fishing industry.

Following selection of a management strategy, it is then implemented for a defined number of years, involving the sequential application of the strategy in Steps 6 to 8. During this implementation, monitoring data are collected to iteratively execute the management strategy, which involves the collection of monitoring data and an assessment. The assessment can take the form of a stock assessment, whereby a process-based model is fitted to the data (see Hamon and Poos, Chapter 3, this volume, for an example). In this case it is known as a *model-based* management strategy (Rademeyer et al., 2007). Alternatively, the assessment step can simply represent the extraction of a summary statistic from the raw data, such as the mean length of fish caught or the abundance index trend. In this case the strategy is referred to as *empirical* (Rademeyer et al., 2007).

Finally, following a period of implementation, usually between 5 and 10 years, the monitoring data collected at Step 6, together with other information reflecting experience with implementation, and possibly new research findings, can be collated (Step 9). At this point a more comprehensive stock assessment can be performed, which may include both the target stock and other ecosystem components (Step 10). Although execution of a management strategy often involves undertaking an annual stock assessment, periodic stock assessments conducted during the slower evaluation cycle are usually to a higher standard, involving a reappraisal of the model design and assumptions on which it is based. These are sometimes referred to as benchmark or comprehensive assessments. This importantly allows measurement of how well the previously implemented management strategy has performed against the predefined objectives, since these objectives would usually include achieving some measure of stock status. The outcome of the assessment, and other measures corresponding to the criteria specified in Steps 1 and 2, are then used to evaluate management strategy performance retrospectively (Step 11) in a manner that can inform their future design in the next loop of the adaptive management cycle. Specifically, if a strategy has performed well, it is more likely to be included in the range of strategies considered during the next planning phase.

Before the next planning phase is initiated, input from the extended peer community should be solicited (Step 12). The "extended peer community" (Funtowicz and Ravetz, 1997, Douguet et al., 2009) is a term used to describe who, with their associated knowledge and perspectives, should be involved in decision-making for the fishery. It is used in the current context to describe a review process that extends beyond the stakeholder group that was involved in the previous round of decision-making. An inevitable consequence is that this review process can lead to a revision of who should be involved in the next round, which can in turn lead to a revision of the management objectives. In the case studies in this book, it is typical that scientists have a continuing dialogue with the wider stakeholder group, such as fishermen or environmental organisations (see also Pastoors, Chapter 20, and Odell and Smith, Chapter 21, this volume). Many have argued that fisheries and ecosystem science is inherently interdisciplinary (Degnbol et al., 2006, Nielsen and Holm, 2007, Steffen, 2009, Haapasaari et al., 2012), and therefore disciplines such as social science and economics should routinely integrate with fisheries scientists. The concept of the extended peer community goes even further and recommends that laypeople with relevant knowledge should also be involved in processes that will affect them (Mikalsen and Jentoft, 2008, Dreyer et al., 2009, Haapasaari et al., 2009, Hegland and Wilson, 2009, Marschke and Sinclair, 2009, Mackinson et al., 2011, Röckmann et al., 2012). These points are revisited by Dankel (Chapter 22, this volume).

Arguably, all the steps in the adaptive cycle described here are part of overall "governance," referring to formal and informal modes of decision-making. An additional factor that needs to be considered is that of formal, high-level governance, which ultimately decides the policy that directs management. This is something that we don't include in the adaptive management cycle because it is largely outside the control of fisheries scientists involved. However it will have a large influence on the process and can create or remove difficulties for the effective long-term planning and implementation of management strategies. For example, during a retrospective review of how a certain strategy has performed, the scientific team may be presented with a revised set of policy objectives or constraints on how these objectives should be achieved, meaning that past performance of the management strategy may not be relevant. Social, political or institutional change of this type can sometimes be accommodated, but may otherwise threaten continuity of the management cycle or even viability of the fishery itself (e.g. Isaacs, 2006). The important point here is that it is almost impossible to predict.

Social, political and institutional uncertainties are consequences of the continuously changing relationship between the management and human systems (Figure 1.1). This is why the concept of a fishery system, which explicitly recognises these interacting parts, is so important for success (see Dankel, Chapter 22, this volume).

The role of simulation

This book is focused on the design of feedback control strategies for fisheries management, which takes place during the planning phase of the adaptive management cycle in Figure 1.2. Data, theory and the scientific method come together during this phase, which roughly corresponds to a paradigm known as *management strategy evaluation* (MSE; Smith, 1993, Smith et al., 1999).

Management strategy evaluation begins and ends with stakeholder engagement, but at its core it involves a scientific evaluation of alternative candidate management strategies to assist the selection of one that is appropriate for actual management. Evaluation usually involves computer models that can replicate dynamics of the fishery system in silico, allowing exploration of the consequences of adopting each strategy (see Edwards, Chapter 2, this volume). Fisheries management science now places a strong emphasis on the design of management strategies using MSE, and Part 2 of this book provides a range of examples. The approach is widely seen as best practice in fisheries management, and is specifically advocated by policy in the United States (Restrepo et al., 1998), Europe (as multiannual plans; European Commission, 2013), Australia (DAFF, 2007) and New Zealand (MFISH, 2011). Not only is MSE scientifically sound, but it has also been credited with facilitating dialogue and increasing transparency in the decision-making process (Butterworth and Punt, 1999, Butterworth, 2007).

Scientific evaluation is only a tool, and to be used effectively it must be placed in the correct context (Dankel, Chapter 22, this volume). This includes a scientific acknowledgement, as far as is possible, of the fishery system, and in particular an engagement with other stakeholders involved in the decision-making process. Because of this, more complexity is not necessarily better (Rochet and Rice, 2009), and there are circumstances where simple, qualitative analyses are sufficient to support good decision-making (Trenkel et al., Chapter 18, this volume). Although beyond the scope of the current volume, fisheries management science should ultimately support good decision-making at all nodes of the decision cycle, which are all necessary for long-term ecological, economic, and socially sustainable fisheries

Outlook of the themes in this book

Management strategy evaluation and related exercises are often highly technical. But despite their prominent role within fisheries management science, the science itself is only part of the broader fisheries system (Figure 1.1); human, political and institutional factors will all influence how management science is conducted. Yet whilst acknowledging this point, the remainder of this book deals exclusively with the evaluation process itself, beginning with the definition of objectives, and ending with a decision on how to best manage the fishery in the immediate future (Figure 1.2).

Reflecting on the role of MSE and its increasing prominence in practice, the collection of case studies (Part 2) and perspectives (Part 3) on fisheries management science in this book is a timely contribution to the growing literature base. The remaining chapters in Part 1 provide much of the background necessary for an understanding of the subsequent material. Edwards (Chapter 2) first describes the evaluation process in more detail, with a particular emphasis on the simulation process itself. Hamon and Poos (Chapter 3) then give an actual example of how a simulation exercise can be performed, given that this is the most technically demanding aspect of MSE. These two chapters will prepare the reader for the case studies in Part 2.

Careful selection of the case studies in Part 2 ensures a geographic and species diversity to showcase the current state-of-the-art. Through discussions with the contributing authors, we have made an effort to preserve consistent definitions and technical vocabulary across these chapters, despite the diverse application. For further reference and clarity, Rademeyer et al. (2007) defines most of the necessary technical semantics.

Following the case study examples, Part 3 provides a glimpse of the research horizons currently being explored by scientific practitioners in the field, including aspects of model structure (Hillary, Chapter 13), ecosystem considerations (Moore and Curtis, Chapter 14, and Plagányi, Chapter 15), spatial complexity (Goethel et al., Chapter 16) and risk definition and communication (Kell et al., Chapter 17). Finally, Part 3 addresses the utility of alternative, sometimes qualitative methods in MSE, which may both augment the evaluation process and provide a framework for reviewing the MSE process itself (Trenkel et al., Chapter 18).

Part 4 presents some longer term and nonscientific stakeholder perspectives on MSE from Europe (Dickey-Collas, Chapter 19, and Pastoors, Chapter 20) and the east coast of the United States (Odell and Smith, Chapter 21). These are regions where MSE has been applied only in more recent years but is currently gaining momentum, partly based on the success of the MSE framework in the Southern Hemisphere and due to local policy needs for multiannual plans with stakeholder dialogue. Part 4 concludes with one more reminder of the place of MSE in the larger fisheries system (Dankel, Chapter 22), and the need to continually improve the human dimensions working in parallel with the technical simulation modelling work and data collection. Ultimately MSE is performed at the science-policy interface, and new conceptual perspectives are put forward as a prospectus to further advance how MSE is conducted, with a view to ensuring economic, social and biologically sustainable fisheries.

Acknowledgements

We thank our colleagues Jake Rice (Department of Fisheries and Oceans, Ottawa, Canada), Anthony D.M. Smith (CSIRO, Hobart, Australia), Anthony Charles (Saint Mary's University Halifax, Canada), Verena M. Trenkel (Ifremer, Nantes, France) and Doug S. Butterworth (University of Cape Town, South Africa) for discussions that helped us unify the literature on management science and management strategy evaluation in fisheries. We also acknowledge Jake Rice, Anthony D.M. Smith and Verena M. Trenkel for proposing the idea of a decision cycle now represented in Figure 1.2.

Note

1 One of the most renowned systems intellectuals is Ludvig von Bertalanffy (1901–1972), whose work in general systems theory is now widely applied in biology, sociology, education, psychiatry, cybernetics and other fields. Von Bertalanffy, L. 1950. An Outline of General System Theory. *British Journal for the Philosophy of Science*, 1, 114–129.

References

Butterworth, D.S. 2007. Why a management procedure approach? Some positives and negatives. *ICES Journal of Marine Science*, 64, 613–617.

Butterworth, D.S. & Punt, A.E. 1999. Experiences in the evaluation and implementation of management procedures. *ICES Journal of Marine Science*, 56, 985–998.

Caddy, J. & Mahon, R. 1995. Reference points for fisheries management. FAO technical paper No. 347. Rome, FAO.

Cardinale, M. & Svedäng, H. 2008. Mismanagement of fisheries: Policy or science? *Fisheries Research*, 93, 244–247.

Charles, A.T. 2001. *Sustainable Fishery Systems*, Blackwell Science, Oxford.

Cunningham, S. & Bostock, T. (eds.) 2005. *Successful Fisheries Management: Issues, Case Studies and Perspectives*, Delft, Eburon Academic Publishers.

DAFF 2007. *Commonwealth Fisheries Harvest Strategy: Policy and Guidelines*, Department of Agriculture Fisheries and Forestry, Australia.

Dankel, D.J., Skagen, D.W. & Ulltang, Ø. 2008. Fisheries management in practice: A review of 13 commercially-important fish stocks. *Reviews in Fish Biology and Fisheries*, 18, 201–233.

Degnbol, P., Gislason, H., Hanna, S., Jentoft, S., Raakjær Nielsen, J., Sverdrup-Jensen, S. & Clyde Wilson, D. 2006. Painting the floor with a hammer: Technical fixes in fisheries management. *Marine Policy*, 30, 534–543.

Douguet, J.-M., O'connor, M., Petersen, A., Janssen, P.H.M. & van der Sluijs, J. 2009. Uncertainty assessment in a deliberative perspective, In: Pereira Guimaraes, A. & Funtowicz, S., (eds.) *Science for Policy*, pp. 15–47. Delhi: Oxford University Press.

Dreyer, M., Renn, O., Drakeford, B. & Borodzicz, E. 2009. JAKFISH Deliverable 2.3: Review of literature about participatory modelling in fisheries management with a focus on the Invest in Fish South West project and the PRONE project. JAKFISH (Judgement and Knowledge in Fisheries Involving StakeHolders).

European Commission 2013. Regulation (Eu) No 1380/2013 of the European Parliament and of the Council of 11 December 2013, on the Common Fisheries Policy, amending

Council Regulations (EC) No 1954/2003 and (EC) No 1224/2009 and repealing Council Regulations (EC) No 2371/2002 and (EC) No 639/2004 and Council Decision 2004/585/EC. *In:* Parliment, E. (ed.) *Official Journal of the European Union*, pp. 1–40.

FAO 1995. Precautionary approach to fisheries. Part 1: Guidelines on the precautionary approach to capture fisheries and species introductions. *FAO Fisheries Technical Paper. No. 350, Part 1.* Rome, FAO, Lysekil, Sweden, 6–13 June 1995 (A scientific meeting organized by the Government of Sweden in cooperation with FAO).

Follett, M.P. 1955. *Dynamic Administration: The Collected Papers of Mary Parker Follett,* New York, Harper & Row Publishers.

Fulton, E.A., Smith, A.D.M., Smith, D.C. & Johnson, P. 2014. An integrated approach is needed for ecosystem based fisheries management: Insights from ecosystem-level management strategy evaluation. PLoS ONE 9(1): e84242. DOI: 10.1371/journal.pone.0084242.

Funtowicz, S. & Ravetz, J. 1997. Environmental problems, post-normal science, and extended peer communities. *Études et Recherches sur les Systèmes Agraires et le Développement,* 30, 169–175.

Funtowicz, S. & Ravetz, J. 2008. Post-normal science. *In:* Cleveland, C.J. (ed.) *Encyclopedia of Earth,* Environmental Information Coalition, National Council for Science and the Environment, Washington, DC.

Haapasaari, P., Kulmala, S. & Kuikka, S. 2012. Growing into interdisciplinarity: How to converge biology, economics, and social science in fisheries research? *Ecology and Society,* 17, http://dx.doi.org/10.5751/ES-04503-170106.

Haapasaari, P., Mäntyniemi, S. & Kuikka, S. 2009. Participatory modeling to enhance understanding and consensus within fisheries management: the Baltic herring case. *In:* ICES (ed.) *ICES CM 2009/O:13,* p. 12. Copenhagen, DK: ICES.

Hegland, T.J. & Wilson, D.C. 2009. Participatory modelling in EU fisheries management: Western Horse Mackerel and the Pelagic RAC. Maritime Studies, 8(1) 75–96.

Hilborn, R. 2007a. Defining success in fisheries and conflicts in objectives. *Marine Policy,* 31, 153–158.

Hilborn, R. 2007b. Moving to sustainability by learning from successful fisheries. *Ambio,* 34, 296–303.

Holm, P. & Nielsen, K.N. 2004. The TAC machine. Appendix B, Working document 1. *In: ICES 2004 report of the Working Group for Fisheries Systems (WGFS). Annual Report.* Copenhagen, Denmark.

ICES 2006. Report of the Working Group on Fishery Systems (WGFS), 26–28 April, 2006, Charlottenlund, Denmark. ICES CM 2006/RMC:05, Ref. ACFM.

Isaacs, M. 2006. Small-scale fisheries reform: Expectations, hopes and dreams of "a better life for all". *Marine Policy,* 30, 51–59.

Jones, G. 2005. Is the management plan achieving its objectives? pp. 555-567 *In:* Worboys, G., De Lacy, T. & Lockwood, M. (eds.) *Protected Area Management. Principles and Practices.* Second ed. Oxford University Press, Oxford, UK.

Mackinson, S., Wilson, D.C., Galiay, P. & Deas, B. 2011. Engaging stakeholders in fisheries and marine research. *Marine Policy,* 35, 18–24.

Marschke, M. & Sinclair, A.J. 2009. Learning for sustainability: Participatory resource management in Cambodian fishing villages. *Journal of Environmental Management,* 90, 206–216.

MFISH 2011. *Operational Guidelines for New Zealand's Harvest Strategy Standard,* Ministry of Fisheries New Zealand.

Mikalsen, K.H. & Jentoft, S. 2008. Participatory practices in fisheries across Europe: Making stakeholders more responsible. *Marine Policy,* 32, 169–177.

Nielsen, K.N. & Holm, P. 2007. A brief catalogue of failures: Framing evaluation and learning in fisheries resource management. *Marine Policy,* 31, 669–680.

Pope, J. G. 1983. Fisheries resource management theory and practice. In: Taylor, J. L. and Baird, G. G. (Eds.) New Zealand finfish fisheries: the resources and their management. Trade Publications, Auckland. p. 56–62.

Rademeyer, R.A., Plagányi, E.É. & Butterworth, D.S. 2007. Tips and tricks in designing management procedures. *ICES Journal of Marine Science,* 64, 618–625.

Reeves, S.A. & Pastoors, M.A. 2007. Evaluating the science behind the management advice for North Sea cod. *ICES Journal of Marine Science,* 64(4): 671–678.

Restrepo, V.R., Thompson, G.G., Mace, P.M., Gabriel, W.L., Low, L.L., Maccall, A.D., Methot, R.D., Powers, J.E., Taylor, B.L., Wade, P.R. & Witzig, J.F. 1998. *Technical guidance on the use of precautionary approaches to implementing national standard 1 of the Magnuson-Stevens Fishery Conservation and Management act,* National Marine Fisheries Service, National Oceanic and Atmospheric Administration of the US Department of Commerce.

Rice, J.C. 2006. Every which way but up: The story of cod in the North Atlantic. *Bulletin of Marine Science,* 78, 429–465.

Rochet, M.-J. & Rice, J.C. 2009. Simulation-based management strategy evaluation: Ignorance disguised as mathematics? *ICES Journal of Marine Science,* 66, 754–762.

Röckmann, C., Ulrich, C., Dreyer, M., Bell, E., Borodzicz, E., Haapasaari, P., Hauge, K.H., Howell, D., Mäntyniemi, S., Miller, D., Tserpes, G. & Pastoors, M. 2012. The added value of participatory modelling in fisheries management – what has been learnt? *Marine Policy,* 36, 1072–1085.

Schwach, V., Bailly, D., Christensen, A.-S., Delaney, A.E., Degnbol, P., Van Densen, W.L.T., Holm, P., Mclay, H.A., Nielsen, K.N., Pastoors, M.A., Reeves, S.A. & Wilson, D.C. 2007. Policy and knowledge in fisheries management: A policy brief. *ICES Journal of Marine Science,* 64(4): 798–803.

Smith, A.D.M. 1993. Risk assessment or management strategy evaluation: What do managers need and want? *ICES CM,* D:18, 1–6.

Smith, A.D.M., Sainsbury, K.J. & Stevens, R.A. 1999. Implementing effective fisheries-management systems – management strategy evaluation and the Australian partnership approach. *ICES Journal of Marine Science,* 56, 967–979.

Steffen, W. 2009. Interdisciplinary research for managing ecosystem services. *Proceedings of the National Academy of Sciences,* 106, 1301–1302.

Stephenson, R.L. & Lane, D.E. 1995. Fisheries management science: A plea for conceptual change. *Canadian Journal of Fisheries and Aquatic Sciences,* 52, 2051–2056.

UN 1982. United Nations Convention on the Law of the Sea (UNCLOS). 10 December 1982. Available at www.un.org/Depts/los/convention_agreements/texts/unclos/closindx.htm.

von Bertalanffy, L. 1950. An outline of general system theory. *British Journal for the Philosophy of Science,* 1, 114–129.

Walters, C. & Maguire, J.-J. 2006. Lessons for stock assessment from the northern cod collapse. *Reviews in Fisheries Biology and Fisheries,* 6, 125–137.

Chapter 2

Feedback control and adaptive management in fisheries

Charles T. T. Edwards

Introduction

To meet the key objective of sustainable exploitation, fisheries management requires that decisions are made regarding, for example, the total catch, season length or equipment regulations. Fisheries management science is primarily concerned with how we make these decisions – or more generally, how we select a particular decision-making strategy from the suite of alternatives. The decision-making strategy itself is usually referred to as a *management* (or *harvest*) *strategy* (Dankel and Edwards, Chapter 1, this volume). Selection of a management strategy using scientific methods requires that they are well defined, and must be achieved despite a pervasive uncertainty that is characteristic of natural (and human) systems.

Uncertainty means that neither the current system state nor its future trajectory can ever be completely known, and it is a concept of fundamental philosophical importance to resource management science. First, if the future dynamics of a system cannot be predicted with certainty, then the most appropriate management action is one that is state-dependent, because otherwise management will not be responsive to unforeseen changes. A control strategy requiring sequential interrogation of the current system state as the strategy is being applied is known as *feedback control* (Walters and Hilborn, 1978). Second, if the system state or dynamics are poorly understood, then the manager can pursue one of two options. Either management action is deferred under the assumption that acting with incomplete knowledge is too risky, or management proceeds with the understanding that management itself can lead to improved knowledge of the system, allowing better-informed management in the future. This latter option is known as *adaptive resource management* (Holling, 1978).

Notwithstanding that delayed corrective action may precipitate a negative outcome, Walters (1986a) articulated the flawed reasoning behind the idea that further understanding of the system should precede management action, stating that because the system is continually changing, it is impossible to know exactly which of its components should be studied to provide the understanding necessary for management at any given future time. Moreover, there is no reason to believe that accumulated knowledge of the system under the current set of management rules will be useful for identifying other rules or system states associated with more desirable management outcomes. Deferred action will therefore not necessarily lead either to more informed action or a more desirable outcome. The requirement for action despite uncertainty has been articulated through the precautionary approach to natural resource management (see later), and the dominant paradigm has subsequently become one of "learning by doing" (Walters and Hilborn, 1978, Walters and Holling, 1990).

If management is to proceed despite uncertainty, then feedback control is the most appropriate means of ensuring sustainability whilst information on the system is collected. Collection of this information is necessary both for the implementation of a feedback mechanism and as a means to improve our understanding of the resource (Williams, 2011). Because of these requirements, a characteristic of management strategies is that they typically include specification of a feedback mechanism, expressed as an algorithm that is usually referred to as a *decision rule* or *harvest control rule* (HCR), alongside a monitoring program and an assessment method for providing the control rule input. The design and evaluation of these types of approaches provides a focus for the current volume.

Because management strategies are intended to facilitate action despite uncertainty, it is natural that the definition and quantification of uncertainty is central to how they should be designed. This chapter therefore begins with an introduction to the types of uncertainty usually considered in fisheries management science, followed by sections on HCR design and adaptive management. It attempts to place fisheries within the broader context of natural resource management, but includes relevant fisheries examples and concludes with an exposition of how fisheries management science is currently practised.

Models and the classification of uncertainty

Our understanding of a system is typically represented by a model, which in the broadest sense is an abstraction designed to facilitate our interaction with the real world. In the current fisheries context we are concerned only with mathematical models. These are necessarily simplifications of reality, but are designed in such a way as to be useful.[1] The modeller should not necessarily aspire to an accurate description of reality but should instead create a working hypothesis, to be evaluated against whether or not it is able to provide information that is useful for management. A *model* therefore refers to the mathematical framework used by the fisheries scientist to interpret information collected from the system under study. Estimation of model parameters involves a statistical fitting procedure by which the model output is tuned to match the observed data. This fitting procedure is a science in itself. An example is given by Hamon and Poos (Chapter 3, this volume) and the interested reader is referred to Hilborn and Walters (1992c) and Quinn and Deriso (1999a) for well-established introductions to the field.

Table 2.1 Types of uncertainty.

Irreducible (aleatoric)
— Process
— Implementation
Reducible (epistemic)
— Statistical
— Structural

Since our knowledge of a system is coalesced into a model or suite of models, uncertainty should be understood as a measure of the difference between a given model and reality in terms of the model formulation, component parameter values and outputs, and more importantly how useful it is in predicting behaviour of the system in response to different management prescriptions. Within this context, uncertainty is most usefully defined as "incompleteness of knowledge about a state or process (past, present or future) of nature" (Francis and Shotton, 1997). Thus uncertainty includes both error and ignorance, but also encompasses apparently random components such as environmentally driven variations in stock productivity. The following description attempts to summarise various uncertainty typologies that have appeared in the literature, and should not be considered definitive (e.g. an alternative description is given by Kell et al., Chapter 17, this volume).

Within fisheries, at least four types of uncertainty are now routinely acknowledged; namely *process*, *implementation*, *statistical*, and *structural* uncertainties (Table 2.1). These can be usefully categorised according to whether or not they appear to be *reducible* (Fogarty et al., 1996). If uncertainty is reducible then it can be lessened with improved knowledge or effort. Reducible uncertainty in a fisheries context typically refers to *epistemic uncertainty*, which reflects incomplete knowledge of the state of the system. Statistical and structural uncertainties fall into this category. Irreducible uncertainty, on the other hand, includes those processes considered to be *aleatoric* (i.e. random) – although very few biological processes are truly random, it is acceptable to adopt a working definition that reflects the boundaries of what can be practically understood (Regan et al., 2002). This type of uncertainty includes process and implementation uncertainties.

Process uncertainty

Aleatoric uncertainty (from the Latin word *alea* meaning "dice") in a fisheries system consists primarily of inherent background variation, driven largely by environmental change and apparently stochastic birth, death and growth of individuals in an ecosystem. This type of uncertainty represents the combined effects of various unknown system components and has been summarised as random fluctuation (Charles, 1998), natural variation (Regan et al., 2002) or simply noise (Walters, 1986a, Hilborn, 1987). However it is more often referred to as *process uncertainty* (Rosenberg and Restrepo, 1994, Francis and Shotton, 1997, Hilborn and Mangel, 1997).

Implementation uncertainty

A secondary form of irreducible uncertainty is implementation uncertainty (Rosenberg and Brault, 1993) or *partial controllability* (Williams, 2011). This may have both deterministic and stochastic components but is usually considered a random, and possibly directional, deviation around the management prescription. For example, the catch output from a management strategy, the quota set by managers and the amount caught may differ to some extent for a variety of reasons, in part concerning political motivations and human responses to management regulations (Fulton et al., 2011). This becomes important for predicting the consequence of alternative management actions.

A characteristic of aleatoric uncertainty is that it can be described mathematically, usually with a probability distribution, but never predicted with complete accuracy. It also has a multiplicative effect within any model. Process uncertainties will interact via the different modelled components, so that the inclusion of additional noise will compound to inflate the uncertainty around model predictions. This is an argument for model simplicity, since complicated models, if they include process uncertainty in all their components, eventually become unusable.

Statistical uncertainty

Also known as *estimation uncertainty* (Rosenberg and Restrepo, 1994, Francis and Shotton, 1997), statistical uncertainty surrounds particular parameter values within a model, but more generally reflects our inability to exactly define the "state of nature" (Hilborn, 1987, Charles, 1998), which in this case is represented by a derived model quantity. A model may be used for example to produce an estimate of current biomass or fishing mortality rates, but compared to the true value this will have some error associated with it. This type of uncertainty is therefore specific to the context of a given model, since the model defines the parameters being estimated.

Statistical uncertainty is a consequence of process uncertainty, *observation error* and *structural uncertainty* (see later). Observation error, also referred to as *partial observability* (Williams, 2011), includes sampling error (since we only observe a small fraction of the population), measurement error and any systematic bias in the data collection procedure. Given the data, a subtle form of statistical uncertainty relates to the error structure assumed during model fitting and the extent to which process uncertainty is accommodated, both of which may influence model-based inference. Estimators based on minimising the difference between observed and model-predicted values often conflate observation error with process and structural uncertainties (e.g. Polacheck et al., 1993). State-space formulations, on the other hand, attempt to separate observation error from the other two forms of uncertainty (e.g. Punt, 2003, Newman and Lindley, 2006).

Statistical uncertainty is epistemic and characteristically reducible, at least in part, since error could be lessened through the collection of additional data and improvements to the model or estimation procedure. However reduction of the uncertainty is limited. Process uncertainty will always place an upper bound on the accuracy of model prediction, as will structural uncertainty, since if the model is a poor representation of reality then more data will not necessarily help.

Structural uncertainty

Because any model is necessarily an abstraction, one that defines the questions we can use it to answer, the most fundamental form of uncertainty is structural uncertainty. This includes *model uncertainty* (Francis and Shotton, 1997), which refers to the fact that the model formulation itself may take on a variety of forms. However it more broadly describes our limited understanding of the system and which parts are important to consider, reflecting "basic ignorance" (Charles, 1998) of what we are trying to represent. Structural uncertainty can have a profound influence on what we can hope to achieve with any given model, with our ignorance of system properties often apparent through "surprise" events (Holling, 1973, Hilborn, 1987) – dramatic changes that contradict model predictions.

During model construction it is not always clear which processes should be included or even considered for inclusion. Those processes selected may also be described in a variety of ways. For example, the recruitment (R) of new individuals to the population is an important predictor of productivity. It is usually, but not always, thought of as a function of spawning biomass (B) and represented as either: $R = \alpha B/(1 + \beta B)$ (Beverton and Holt, 1957); or $R = \alpha B - \beta^B$ (Ricker, 1954). The maximum productivity (in the absence of density-dependent effects) is represented by the parameter α, whilst β determines the strength of density-dependent limitations. In some cases environmental variation may be so high that recruitment is represented in the model as being independent of the biomass. These are alternative structural assumptions of the model that will influence how it behaves.

The Beverton and Holt (1957) and Ricker (1954) functions are both special cases of the more general $R = \alpha B(1 - \beta\gamma B)^{1/\gamma}$ (Schnute, 1985), obtained by setting $\gamma = -1$ or $\gamma \to 0$ respectively (Quinn and Deriso, 1999b). Despite the fact that the Schnute (1985) model provides a more flexible form, γ can rarely, if ever, be estimated from fisheries data (i.e. statistical uncertainty is very high), and by convention it is fixed at either of these two values. Similarly for the Pella and Tomlinson (1969) production function: $g(B) - (r/p)B(1-(B/K)^p)$; which has a shape parameter p that determines the relative biomass at which productivity is maximised. Again, because it is difficult to estimate m from fisheries data it is often fixed by the modeller to yield different structural assumptions. Most often it is assumed that $p = 1$, giving the logistic Schaefer (1954) model. The boundary between statistical uncertainty and structural uncertainty can therefore be somewhat arbitrary, depending on whether a parameter is fixed or estimated.

Choices on what processes to include and how are typically made with reference to the intended use of the model and whether or not data are available to estimate the component parameters. For instance, stochastic temporal variation in both recruitment and natural mortality may contribute to process uncertainty. Stock assessment models routinely estimate the former but not the latter, due to limitations on what information can be extracted from fisheries data. As a consequence of these choices, different models are quite capable of producing different results when applied to the same data (as demonstrated by Deroba et al., 2014), largely due to their different underlying assumptions about what and how biological processes should be represented.

Structural uncertainty is extremely intractable, often requiring the use of multiple models and implying that there may be many influential processes in the system that are not even known to be relevant. It is therefore practically difficult to deal with, and it is not clear to what extent it is reducible, since the degree of effort required to fully resolve this type of uncertainty is unknown (Charles, 1998). Because of this, any model should only ever be considered a working hypothesis or statement of what we think we need to know (scientifically) for the current purposes of management, the structure of which should be updated as our knowledge of the system improves.

Risk and precaution in fisheries management

Uncertainty should be distinguished from the concept of *risk* (Shotton, 1993, Francis and Shotton, 1997). Even though risk depends on uncertainty, it includes a measurement of consequence (the *cost* or *loss*). Technically, risk is defined as the product of probability and the cost associated with a particular event, summed over all possible events (i.e. the "expected loss"; Rosenberg and Restrepo, 1994). However this cost can rarely be made explicit in fisheries, and the more common working definition is "the probability of an undesirable event" (Francis and Shotton, 1997). For example, the "risk of stock depletion" may be equated with the probability that the spawning stock biomass will drop below a level at which recruitment is impaired. However a meaningful definition of risk has still not been universally accepted. An alternative definition has been adopted, for example, by Kell et al. (Chapter 17, this volume). Managers are often interested in risk (broadly defined here but made specific for that particular management situation), and much of the literature on uncertainty has arisen as a preliminary step towards risk-based management (e.g. Rosenberg and Restrepo, 1993, Shotton, 1993, Rosenberg and Restrepo, 1994, Francis and Shotton, 1997). This is a theme taken up in more detail by Kell et al. (Chapter 17, this volume).

Within the context of environmental management the *precautionary approach* (PA) is a politically negotiated framework designed to deal with the sustainability risk associated with decision-making. Following its inclusion in the Rio Declaration of the United Nations (UN) Conference on Environment and Development (1992) − which states in Principle 15 that "lack of full scientific certainty shall not be used as a reason for postponing cost-effective [management] measures" − it has been adopted by the FAO (Food and Agricultural Organization of the UN) as a guiding principle for the sustainable management of fisheries (FAO, 1995a, 1995b). The FAO guidelines stipulate that management should exercise "prudent foresight to avoid unacceptable or undesirable situations, taking into account that changes in fisheries systems are . . . not well understood" (FAO, 1995b). The PA would therefore require a manager to make decisions despite uncertainty in the system, but nevertheless ensure it is managed sustainably for the long term. Short of extremely conservative measures (such as closing the fishery), management action can only be taken according to the PA if the risk associated with that action is understood and can therefore be minimised.

Translating system uncertainty into the risk associated with a decision (or range of decision options) is one of the key objectives of management science. If process uncertainty makes feedback control the most appropriate form of management for fisheries, how can the risk associated with a particular HCR be measured? More generally, management science should be able to measure performance of a management strategy against a range of metrics, including but not limited to the risk of overexploitation or depletion (Smith, 1993). How should we choose a control rule that allows for exploitation of the resource but without an unacceptable risk to sustainability?

Building feedback control strategies

The design of feedback control strategies, what could be loosely called *harvesting theory*, has a rich theoretical literature that has coalesced with a more pragmatic approach to produce the current state of the art. Formal analytical methods have used the approaches of optimal control theory, a branch of the engineering sciences, and a technique known as dynamic programming (Bellman, 1957) has arguably proved the most useful, due to its ability to accommodate the nonlinear and nonequilibrium characteristics of ecological systems. This theoretical approach has produced some guiding principles but is restricted in terms of the complexity it can consider, and requires a precise definition of optimality, which is not always possible given the trade-offs inherent in managing for multiple objectives. In fisheries at least, it has been superseded by a heuristic approach that now represents the majority of applications.

Theoretical approaches

Dynamic programming is based on the *principle of optimality* (Bellman, 1957), which states that management is only optimal if at each sequential time step the optimal decision is chosen, irrespective of the trajectory taken to get there. In a deterministic system, for any given initial state, a solution to the problem of optimal control is found by selecting a temporal sequence of decisions that will lead to the best possible outcome over the time horizon considered. The result is a control rule that will return an optimum decision given any state and time. In a stochastic setting, which is much more relevant to biological systems, an optimal decision at any given time can still lead to a range of outcomes. As a consequence of this uncertainty, stochastic dynamic programming, when applied over an infinite time horizon, will yield a state-dependent but time-independent decision rule (Walters, 1986b, Williams et al., 2002b). This corresponds to our current definition of feedback control. Moreover, it is guaranteed to be the optimum solution given the assumptions made during the analysis. Although complicated to apply, the technique is therefore extremely powerful, and is widely applied to decision-making problems in the broader field of natural resource management (e.g. Mangel, 2000, McCarthy et al., 2001, Martin et al., 2009, Marescot et al., 2013).

Within the context of harvesting theory, the most well-developed example is the control of recreational hunting effort for waterfowl in the United States (Williams et al., 2002a, Conroy and Peterson, 2013). In this instance, stochastic dynamic programming is used to derive an optimal relationship between population survey results (both a census population size and a measure of the habitat available for breeding that year) and duration of the hunting season, which in turn determines the level of exploitation. Application of these methods requires that the concept of optimality is explicitly defined by a *utility function*, which is a mathematical representation of preference for any given combination of management outcomes. In this instance, the objectives of management are clearly stated as an intent to maximise the harvest whilst maintaining the population at or above a target size. For American black ducks (*Anas rubripes*), for example, the utility of a harvest for the current year is calculated using the model-predicted population size for the following year: if the population is predicted to be above the target population size, then utility is equal to the harvest, otherwise it declines to equal zero at population extinction (Conroy and Peterson, 2013). This simple definition allows control theory to be applied, and stochastic dynamic programming has been used to select an optimum decision rule.

Walters (1975) was the first to apply stochastic dynamic programming to a fisheries system, using the example of a Pacific sockeye salmon (*Oncorhynchus nerka*) fishery to define different optimal relationships between the harvest rate and estimated stock size. These control strategies depended on the optimality criteria (i.e. the definition of utility). If the utility function equalled the average catch, then the optimum control rule was one that ensured *constant escapement*. In other words, stop harvesting when the population drops below some threshold level, but catch any surplus fish above this threshold (Figure 2.1). However if the objective is to minimise the variance in catches around some target level, the optimal harvest rate increases rapidly from a much lower threshold and then plateaus, so that the harvest rate becomes independent of the stock size. At this point it becomes known as a *constant harvest rate* strategy (Figure 2.1).

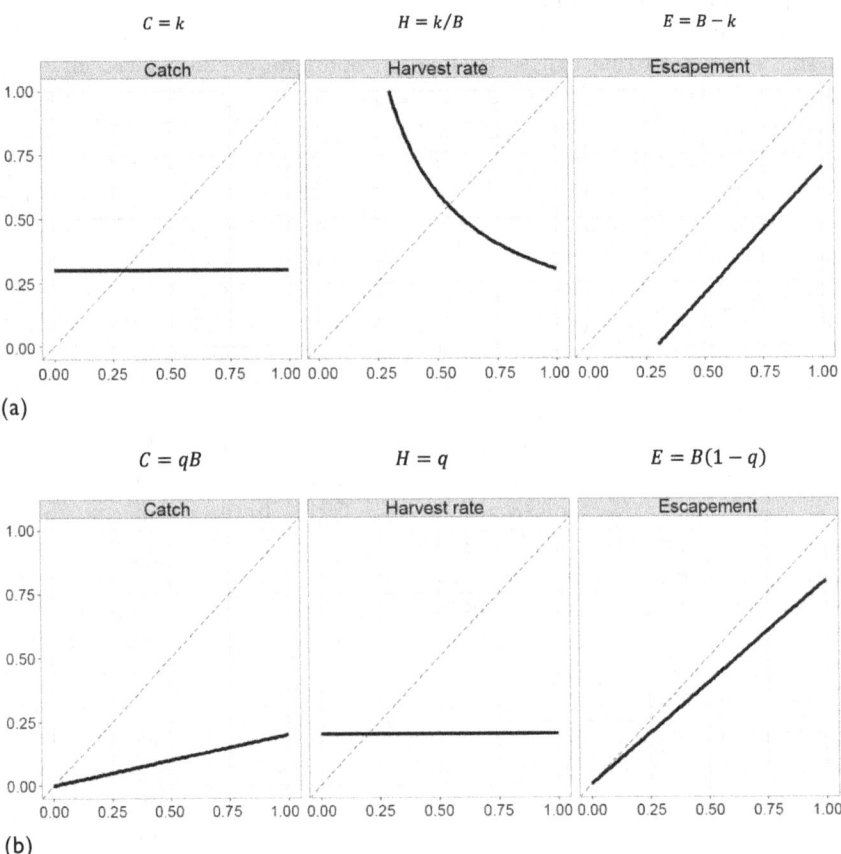

(a)

(b)

Figure 2.1 Harvest control rules proposed in the theoretical fisheries management literature: (a) constant catch; (b) constant harvest rate; (c) threshold harvest rate; (d) constant escapement; and (e) proportional threshold. The output parameters C, H and E refer to the catch, harvest rate (defined as catch over exploitable biomass) and escapement (defined as the biomass that remains unharvested at the end of the fishing season), and are shown on the vertical axis of each plot. The control rule inputs are the biomass B (on the horizontal axis) and constants q and k which define how the control rule behaves. For all examples, fixed values of q = 0.2 and k = 0.3 are assumed. Biomass, catch and escapement are all represented on the same scale between 0 and 1.

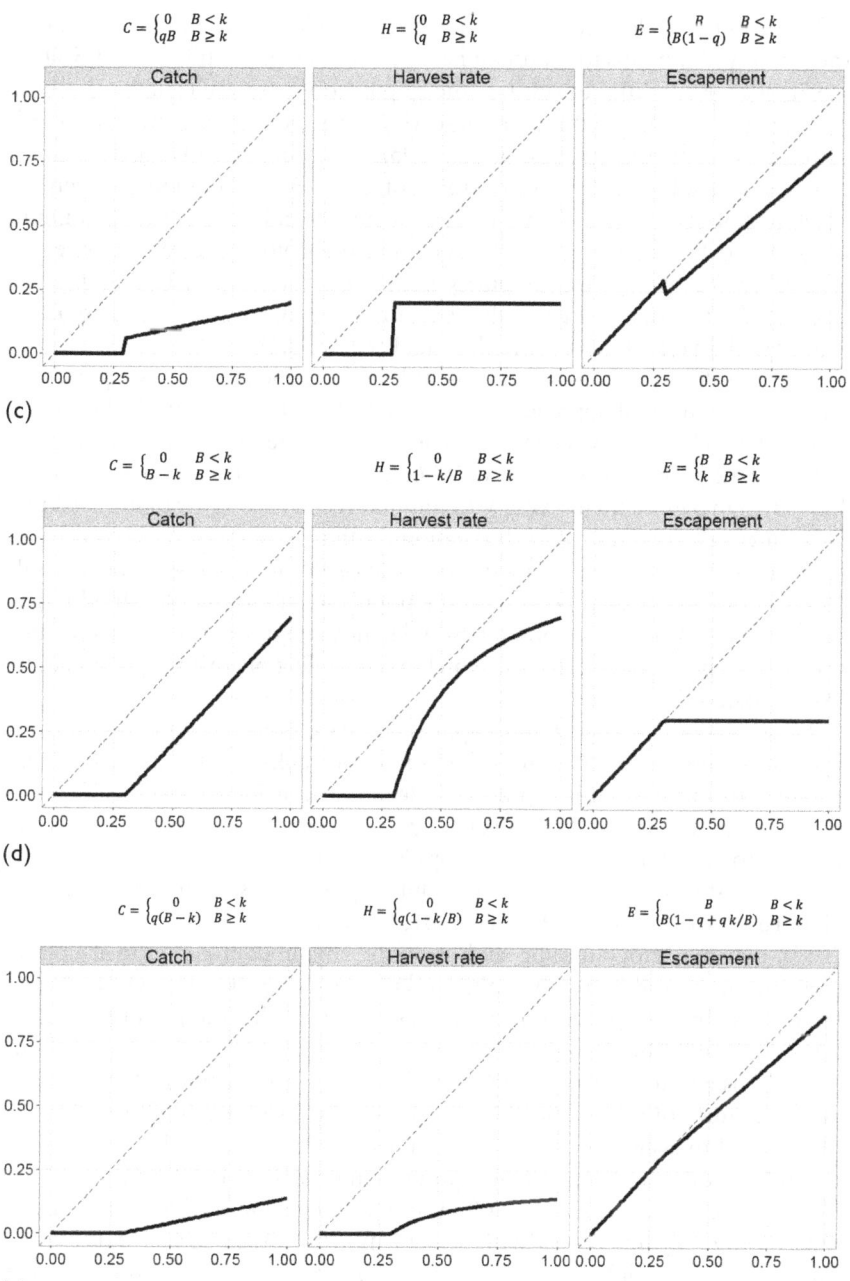

Figure 2.1 (Continued)

Maintaining stable catches and therefore a viable fishing industry is one of the primary arguments for constant harvest rate strategies (Hilborn and Walters, 1992b, Walters, 1975). However as the population gets smaller it can become increasingly volatile (Beddington and May, 1977), creating a theoretical risk of resource extinction (Lande et al., 1995), and a strategy that strictly advocates a constant harvest rate is therefore not viable. A compromise between the objectives of maintaining a minimum spawning stock biomass (i.e. a minimum escapement) but reasonably stable catches at high biomass levels was used to justify the *threshold harvest rate* policy of Quinn et al. (1990). This policy was developed intuitively and prescribes no fishing below a biomass threshold, and a constant harvest rate above the threshold (Figure 2.1).

The need for a minimum escapement level has generally been supported by a range of theoretical approaches (e.g. Reed, 1979, Lande et al., 1995, Sæther et al., 1996). However when stock size is uncertain the optimum escapement is not constant (Clark and Kirkwood, 1986, Engen et al., 1997, Sethi et al., 2005). Instead, to maximise the harvest with minimum risk to long-term sustainability, escapement should increase with the biomass above a minimum threshold. This type of strategy has been termed a *proportional threshold strategy* (Engen et al., 1997), because only a proportion of the population above the threshold is harvested. An advantage over constant escapement strategies is that they allow for a smaller threshold population size, which has the effect of creating more stability in the catches (Lande et al., 1997).

These insights from theoretical approaches have provided some useful guidance concerning management strategy design (Trenkel et al., Chapter 18, this volume). In particular, it is common for feedback control rules to specify a rectilinear shape whereby fishing mortality is constant at higher biomass levels, but declines to zero when the biomass drops (see Figure 2.3), which is similar to the proposal first made by Walters (1975). But despite some sporadic examples (e.g. Hilborn, 1976, Smith and Walters, 1981, Kope, 1992, Hawkshaw and Walters, 2015), dynamic programming and associated methods from optimal control theory have proved to be of limited value in fisheries due to two important limitations. First, applications have typically assumed that the population status can be measured by a single state variable, namely the biomass or total numbers. This is because the approach quickly becomes computationally intractable as the number of state variables increases (Williams et al., 2002b) and forces the method to assume a very simple population dynamics model with limited control options. Second, dynamic programming and optimal control theory in general requires specification of a utility function. Given the multiplicity of objectives across different stakeholders, formulation of an agreed utility function is usually very difficult if not impossible in a fisheries context. It is of course universally agreed that an objective of management is long-term sustainability of both the exploited population and the fishery, however different stakeholders may place different emphasis on potentially conflicting objectives, such as the quantity and quality of catch, the rate at which it can be caught or the impact of fishing on the wider ecosystem. For example, it is not possible to maximise both the average catch and catch rate: higher catches will be associated with a lower catch rate because the average stock biomass will be lower, and vice versa. In principle, this trade-off could be represented by a utility function that combines both catch rate and the total catch, with appropriate weights given to each. But in practice, it is typically very difficult to reach agreement between different stakeholders on how this weighting should be achieved, increasingly so as more performance criteria are added.

Because of the limitations of control theory, fisheries have adopted an alternative approach. When faced with a complex problem in which the nature of an optimal solution is only broadly defined across multiple objectives, searching for an appropriate decision rule requires a heuristic process of exploration (from the Greek *heurískō* meaning "I find, discover"): an approximate relationship between management action and the system state is proposed and further refined by simulation testing with a model that can reproduce fishery dynamics over time. This procedure decouples the form of the control rule from complexity of the population dynamics model upon which explorations are based, allowing a much greater degree of system detail to be considered (Williams et al., 2002b). Furthermore it does not require a utility function to be specified, in which case the concept of optimality is not properly defined. The benefits of a simulation-based exploratory approach are recognised in natural resource management science (Williams et al., 2002b) but have been most developed and applied within a fisheries context. On this point fisheries management science has largely diverged from other resource management problems, which have retained a stronger emphasis on optimal control theory.

Heuristic approaches

Under an heuristic paradigm, fisheries management science has focused on quantifying the performance of different strategies across a range of potentially conflicting performance measures, but without necessarily advocating that any particular strategy is optimal (Smith, 1993). The choice of control rule with its associated performance trade-offs is left to the wider stakeholder group, and the evaluation process can therefore be thought of as an informed search intended to identify the preference of decision-makers for a particular combination of performance outcomes. Although simulation-based evaluations had already appeared in the fisheries literature (e.g. Southward, 1968, Hilborn, 1979, Hightower and Grossman, 1985, Hall et al., 1988, Quinn et al., 1990), Smith (1993) coined the now popular term management strategy evaluation (MSE) to describe an overall framework that included simulation-based experiments as a central component. In this respect it converged with contemporaneous, independent developments by scientists working for the International Whaling Commission (IWC; Cooke, 1994).

The IWC introduced the concept of a management procedure (MP), which includes an assessment method, the data on which it is based, and a decision rule (HCR) that translates the assessment output into a management action

(Rademeyer et al., 2007). This concept largely overlaps with our current definition of a management strategy, but for purposes of clarity it is worth emphasising that an MP is more strictly defined, always including a HCR and a precise definition of how the control rule inputs should be calculated that is maintained throughout the period of implementation. In contrast a management strategy may simply state that an assessment should be conducted every year, without specifying how it should be conducted.

The first MP was the New Management Procedure (NMP) adopted by the IWC in 1974, in which catch was calculated as a function of the surplus production and estimated population size (Punt and Donovan, 2007). This rule was successful in arresting the depletion of heavily exploited whale stocks (Cooke, 1994), but soon proved unworkable for cases in which either stock structure was unknown or parameters for the production function could not be estimated − uncertainty prevented its effective application. Furthermore, there was no commercial incentive to collect additional information on these cases, since insufficient data generally equated to maintenance of the status quo. The problem of how to make decisions when confronted with uncertainty led to development of the Revised Management Procedure (RMP), which was finalised in 1994 and first encapsulated many of the principles now considered aspirational for MP design in capture fisheries (Punt and Donovan, 2007). By including a specification of how decisions could be made even with very little data, it represented a paradigm shift away from the idea that lack of information should be a barrier to the decision-making process, thereby conforming to a central tenet of the precautionary approach that uncertainty should not incapacitate management (Cooke, 1999).

The framework for MP evaluation first developed by the IWC has been summarised by Punt and Donovan (2007) in the following points (see also Punt et al., 2014):

1   Specification of management objectives with associated statistics to measure performance of an MP against these objectives;
2   Specification of candidate MPs;
3   Specification of population dynamic *operating models* with which to represent status and behaviour of the resource under alternative MPs;
4   Simulation of candidate MP performance into the future, including at each time step: (a) generation of observational data; (b) application of the MP; (c) operating model projection assuming MP outputs;
5   Collation of performance statistics and selection of the MP most likely to reach management objectives.

At the core of this approach, a computer simulation is used to determine the likely performance of a given MP by attempting to replicate how it will be applied (Figure 2.2). By repeating this exercise across a range of candidate MPs and operating models, a heuristic search process is used to locate the MP most likely to reach prespecified management objectives (a practical example of this

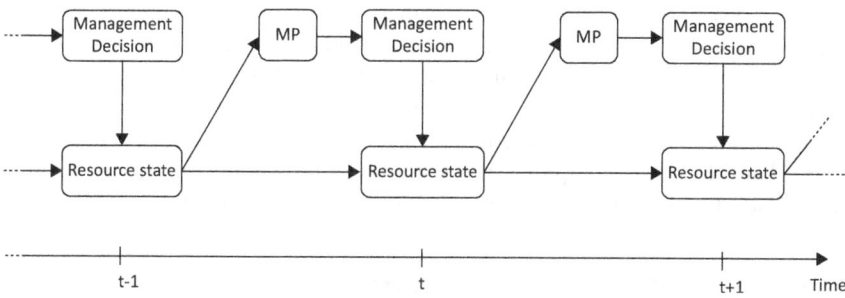

Figure 2.2 Schematic representation of the simulation-based evaluation of an MP. Application of the MP is iterated forward over time for a defined period, following which performance criteria are collected for comparison of the MP with other candidates. Based on a similar figure given in Williams (2011).

methodology is given by Hamon and Poos, Chapter 3, this volume). Because the simulations can explicitly incorporate uncertainty in the resource dynamics (process and structural uncertainties within the operating model or models), observation and assessment of the resource status (statistical uncertainty within the MP) and implementation of management actions (implementation uncertainty); to the extent that these uncertainties can be quantified, the risk associated with any given MP can be measured. This allows MPs to be designed in a manner that is fully consistent with the precautionary approach. This MP evaluation framework is very similar to MSE, with the exception that MSE was conceived as a way to evaluate a wider range of management options (i.e. not just feedback policies; Smith et al., 1999). Management strategy evaluation can also include qualitative evaluations (i.e. not simulation based; Trenkel et al., Chapter 18, this volume) and the term is more often used in the context of ecosystem-based fisheries management (Sainsbury et al., 2000, Plagányi, Chapter 15, this volume).

Because the RMP included a stock assessment model to estimate population size, it would now be referred to as a *model-based* MP (Rademeyer et al., 2007). Simplified, *empirical* MPs (Rademeyer et al., 2007) have subsequently been proposed that do not include a process-based model of the resource, but are instead based on summarised representations of the data, such as an abundance index trend. The relative merits of these two approaches is discussed by Hillary (Chapter 13, this volume).

If simulation is able to show that an MP can achieve management objectives despite uncertainty, particularly process and structural uncertainties that will influence how the system behaves in the future, it is considered *robust* (Charles, 1998). However whether an MP performs well in reality cannot be taken for granted. In particular, the value of simulation studies will be defined by the modeller's ability to conceive and quantify the various sources of uncertainty assumed during the simulation experiment (Rochet and Rice, 2009, Trenkel et al., Chapter 18, this volume), which may not be complete. However, over time we would expect that the collection of additional data would improve understanding of both the resource and the fishery. Improved understanding, combined with a retrospective evaluation of how an MP has performed over a period of actual implementation, is what allows for adaptive management (Dankel and Edwards, Chapter 1, this volume), whereby our capacity to design robust and effective harvest strategies is continuously advanced.

Adaptive management in context

The most difficult source of uncertainty for the scientist to deal with is structural uncertainty, meaning that it is not known whether the operating model includes all the system components necessary for prediction, or if they are included, whether they are represented in an appropriate manner. The operating model describes our best understanding of the system being managed and how it will respond to the MP being evaluated. It includes a model of the biological resource dynamics that can predict how the population will respond to fishing, the collection of monitoring data for the MP via an observation model, and an implementation model that reflects how management regulations (the MP output) are applied in practice (Punt et al., 2014). Of these three components, the resource and observation models often receive the most attention during development of the operating model. The biological model is usually based on a stock assessment model that has been fitted to available data, and the residual errors (i.e. the difference between fitted model values and empirical observations) can then be used to predict the observation error during projections of the system into the future. This allows the collection of monitoring data to be simulated (Hamon and Poos, Chapter 3, this volume). The implementation model may not be dealt with explicitly within the operating model, being represented simply, if at all, by an implementation error. However in more complicated applications, particularly to a multispecies system (e.g. Dichmont et al., Chapter 10, and de Moor and Butterworth, Chapter 11, this volume), the implementation model can form a significant part of the operating model.

When constructing an operating model, a *reference case* is typically chosen as the most plausible representation of the system (Rademeyer et al., 2007). Parameterisation of the operating model, particularly the biological component, involves fitting the model to data, as already mentioned. This will yield an uncertainty distribution around the model parameters – statistical uncertainty – which should be incorporated in the simulations (e.g. by sampling from the respective parameter distributions). However the model will also incorporate a range of structural assumptions, and to fully capture this uncertainty it is advocated that multiple operating models should be used (Punt et al., 2014) – a *reference set* of alternative models that are considered equally likely (Rademeyer et al., 2007). For example, it is not always possible to estimate all the parameters within the operating model. In such situations, unestimated parameters could be fixed at different values considered equally likely. In the southern bluefin tuna case study, for example, the reference set includes alternative assumptions regarding the natural mortality and stock-recruitment relationship, since the data do not provide information that would allow these parameters to be estimated (Hillary et al., Chapter 8, this volume).

The reference set provides the primary basis for selecting an MP. Further evaluations can make use of a *robustness set* (Rademeyer et al., 2007), which represents less likely but potentially drastic scenarios (Punt et al., 2014). For example, the robustness set used in the New Zealand rock lobster case study includes alternative assumptions regarding the noncommercial catches, since these are largely unknown, but could be important for the predicted system dynamics (Breen et al., Chapter 6, this volume).

The performance of any given MP against each management objective should be specified quantitatively as a *performance statistic* (e.g. the average catch per year) that can be derived from the operating model outputs (Punt et al., 2014). However, when multiple models are being used it is not always clear how performance of a particular MP should be summarised. One approach is to require the MP to perform well against all models in the reference set using the full suite of performance statistics, but to use only measures of the risk to sustainability when evaluating against the robustness set. For example, the stakeholder group may decide that the catch, catch rate, catch variability and risk of the resource dropping to a level at which recruitment might be impaired (depletion) should all be considered when selecting an MP. In this example, if an MP leads to a high risk of resource depletion when tested against both the reference and robustness sets, it would be immediately dropped. However if this risk is small (i.e. acceptable), its performance would then be assessed using the various statistics related to catch, but using only the reference set of operating models.

The reference case, or reference set of operating models, represents our current state of knowledge concerning the system being managed – more specifically how the resource (and fishery) dynamics will respond to different management interventions. It is the predictive capacity of operating models that allows them to be used for MP design and naturally we would hope, but not necessarily expect (Hilborn and Walters, 1992a), this capacity to be improved as more scientific data are collected. Periodic updates of the operating model(s) is the process of learning by doing that is known as *adaptive management* (Walters and Holling, 1990).

The relationship between learning and decision-making is a defining feature of adaptive management. Learning contributes to management by informing the decision-making process (in this case selection of an MP out of the range of alternatives), and the decision should facilitate learning (Williams, 2011). At the very least the management strategy should be intended to ensure sustainable harvest, and thus learning is usually achieved through a passive process of long-term data collection. In some cases a more active approach, involving deliberate management manipulations to facilitate learning, may also be considered (e.g. Smith and Walters, 1981; but see the review by Walters, 2007).

In fisheries for which a management strategy has been selected using MSE, adaptive management is usually achieved through an *implementation review* (Punt and Donovan, 2007, Punt et al., 2014). An implementation review would typically take place every 5 to 10 years (Punt et al., 2014), and at this point the operating model or models can be recalibrated or even recast. Although adaptive management traditionally refers to this updated understanding of resource dynamics (Walters, 1986b), the review could also include a retrospective appraisal of how the MP has performed against management objectives, or how these targets could be better defined (Irwin and Conroy, 2013). This information might suggest alternative strategies not included in the previous evaluation, better parameterisation of the control rule, or improvements to the MP selection process itself. In fisheries therefore, the definition of objectives, the design of an MP using simulation, subsequent implementation, and then periodic review broadly represents the adaptive management cycle (Dankel and Edwards, Chapter 1, this volume).

Current state of the art

Starting from a broader perspective of natural resource management and control theory, fisheries management science now places a strong emphasis on the design of management strategies using MSE, which usually includes computer-based simulation as a central component. Punt et al. (2014) give a useful introduction to a best practice approach and Deroba and Bence (2008) summarise a range of HCRs currently in use. Model-based MPs often use one of two generic rectilinear control rules, which relate either fishing mortality or catch to the spawning stock biomass (Figure 2.3). Good examples are given by Hicks et al. (Chapter 4, this volume), Cox and Kronlund (Chapter 5, this volume) and Needle (Chapter 12, this volume). Generic rules of this type are considered to have intrinsically desirable properties (Froese et al., 2011), to the extent that they are sometimes not even evaluated formally prior to implementation (Hillary, Chapter 13, this volume). The selection of control parameters that describe the shape of the rule is often informed by reference points used to define management targets and limits. In particular, maximum sustainable yield (MSY) is a frequently cited management objective, and in the United States and New Zealand for example, the maximum (target) fishing mortality rate (corresponding to parameter q_2) is specified as F_{MSY}, which is the fishing mortality associated with MSY (Restrepo et al., 1998, MFISH, 2011). If the fishing mortality rate increases above this level ($F > F_{MSY}$), it is known as *overfishing*. The combination of F_{MSY} and B_{MSY} represent the basis for most reference points in fisheries management, and the control parameter k_2 is often set close to, but below, B_{MSY} (e.g. Cox et al., 2013). If the stock declines below k_2, the intensity of exploitation is reduced to allow it to recover more quickly. Two additional reference points, known as limit reference points, are B_{LIM} and F_{LIM}, which can be used to inform appropriate values for control parameters k_1 and q_1, respectively.

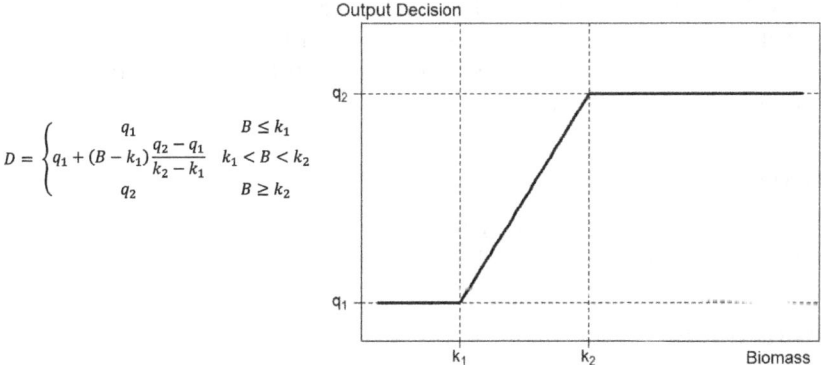

$$D = \begin{cases} q_1 & B \leq k_1 \\ q_1 + (B - k_1)\dfrac{q_2 - q_1}{k_2 - k_1} & k_1 < B < k_2 \\ q_2 & B \geq k_2 \end{cases}$$

Figure 2.3 Generic rectilinear harvest control rule type that is popular in fisheries management and relate an estimated spawning biomass B to a decision D. The decision is either a catch or fishing mortality. Control parameters q_1, q_2, k_1 and k_2 define the shape of the rule and how it behaves.

The reference points B_{LIM} and F_{LIM} approximate the minimum tolerable spawning biomass, below which future recruitment will be endangered, and the fishing mortality that would eventually result in a stock of that size. If $B < B_{LIM}$ it would be considered *overfished*.

The definition of reference points is a science in itself, and often requires strong assumptions concerning productivity of the stock (Mace, 1994). How to build a model-based MP when these reference points are poorly defined is one of the many questions that can be answered using MSE. Other questions currently being addressed by fisheries management science, and MSE in particular, include: how to proceed in situations with limited data (Bentley and Stokes, 2009a, 2009b, Berkson and Thorson, 2014, Carruthers et al., 2014, Moore and Curtis, Chapter 14, this volume); how to accommodate the spatial dimension in fisheries management (Goethel et al., Chapter 16, this volume); and how to accommodate the wider ecosystem when formulating management advice (Plagányi, Chapter 15, this volume). The field has experienced major advances over a relatively recent period of time (Hilborn, 2012), and the emergence of MSE and its growing global appeal has motivated the contributions collated in this volume. It is intended to provide a useful reference for future work and an accessible entry point for an increasingly complex area of research.

Acknowledgements

I am most grateful for reviews of early drafts of this chapter by, and subsequent discussions with, Anthony D.M. Smith (CSIRO, Hobart, Australia) and R.I.C. Chris Francis (NIWA, Wellington, New Zealand), who thereby greatly improved its content. My own understanding was further helped through many discussions with Dorothy J. Dankel (Centre for the Study of the Sciences and the Humanities, University of Bergen, Norway). Finally, I would like to acknowledge NIWA (Wellington, New Zealand) for providing the time that was necessary for the work.

Note

1 "Remember that all models are wrong; the practical question is how wrong do they have to be to not be useful" (Box and Draper, 1987, p. 74).

References

Beddington, J.R. & May, R.M. 1977. Harvesting natural populations in a randomly fluctuating environment. *Science,* 197, 463–465.

Bellman, R. 1957. *Dynamic Programming,* Princeton, NJ, Princeton University Press.

Bentley, N. & Stokes, K. 2009a. Contrasting paradigms for fisheries management decision making: How well do they serve data-poor fisheries? *Marine and Coastal Fisheries: Dynamics, Management, and Ecosystem Science,* 1, 391–401.

Bentley, N. & Stokes, K. 2009b. Moving fisheries from data-poor to data-sufficient: Evaluating the costs of management versus the benefits of management. *Marine and Coastal Fisheries,* 1, 378–390.

Berkson, J. & Thorson, J.T. 2014. The determination of data-poor catch limits in the United States: Is there a better way? *ICES Journal of Marine Science,* 72, 237–242.

Beverton, R.J.H. & Holt, S.J. 1957. *On the Dynamics of Exploited Fish Populations,* London, Chapman and Hall.

Box, G.E.P. & Draper, N.R. 1987. *Empirical Model Building and Response Surfaces,* New York, John Wiley & Sons.

Carruthers, T.R., Punt, A.E., Walters, C.J., Maccall, A., Mcallister, M.K., Dick, E.J. & Cope, J. 2014. Evaluating methods for setting catch limits in data-limited fisheries. *Fisheries Research,* 153, 48–68.

Charles, A.T. 1998. Living with uncertainty in fisheries: Analytical methods, management priorities and the Canadian groundfishery experience. *Fisheries Research,* 37, 37–50.

Clark, C.W. & Kirkwood, G.P. 1986. On uncertain renewable resource stocks: Optimal harvest policies and the value of stock surveys. *Journal of Environmental Economics and Management,* 13, 235–244.

Conroy, M.J. & Peterson, J. 2013. Case studies. In: *Decision Making in Natural Resource Management: A Structured, Adaptive Approach.* Hoboken, NJ: Wiley – Blackwell. 263–293.

Cooke, J.G. 1994. The management of whaling. *Aquatic Mammals,* 20, 129–135.

Cooke, J.G. 1999. Improvement of fishery-management advice through simulation testing of harvest algorithms. *ICES Journal of Marine Science,* 56, 797–810.

Cox, S.P., Kronlund, A.R. & Benson, A.J. 2013. The roles of biological reference points and operational control points in management procedures for the sablefish (*Anoplopoma fimbria*) fishery in British Columbia, Canada. *Environmental Conservation* 40 (4): 318–328.

Deroba, J.J. & Bence, J.R. 2008. A review of harvest policies: Understanding relative performance of control rules. *Fisheries Research,* 94, 210–223.

Deroba, J.J., Butterworth, D.S., Methot, R.D., De Oliveira, J.A.A., Fernandez, C., Nielsen, A., Cadrin, S.X., Dickey-Collas, M., Legault, C.M., Ianelli, J., Valero, J.L., Needle, C.L., O'malley, J.M., Chang, Y.J., Thompson, G.G., Canales, C., Swain, D.P., Miller, D.C.M., Hintzen, N.T., Bertignac, M., Ibaibarriaga, L., Silva, A., Murta, A., Kell, L.T., De Moor, C.L., Parma, A.M., Dichmont, C.M., Restrepo, V.R., Ye, Y., Jardim, E., Spencer, P.D., Hanselman, D.H., Blaylock, J., Mood, M. & Hulson, P.J.F. 2014. Simulation testing the robustness of stock assessment models to error: Some results from the ICES strategic initiative on stock assessment methods. *ICES Journal of Marine Science*, 72, 19–30.

Engen, S., Lande, R. & Sœther, B.-E. 1997. Harvesting strategies for fluctuating populations based on uncertain population estimates. *Journal of Theoretical Biology*, 186, 201–212.

FAO 1995a. *Code of Conduct for Responsible Fisheries*, FAO, Rome.

FAO 1995b. Precauationary approach to fisheries. Part 1. Guidelines on the precautionary approach to capture fisheries and species introduction. *Fisheries technical paper.* FAO.

Fogarty, M.J., Mayo, R.K., O'Brien, L., Serchuk, F.M. & Rosenberg, A.A. 1996. Assessing uncertainty and risk in exploited marine populations. *Reliability Engineering & System Safety*, 54, 183–195.

Francis, R.I.C.C. & Shotton, R. 1997. "Risk" in fisheries management: A review. *Canadian Journal of Fisheries and Aquatic Sciences*, 54, 1699–1715.

Froese, R., Branch, T.A., Proelß, A., Quaas, M., Sainsbury, K. & Zimmermann, C. 2011. Generic harvest control rules for European fisheries. *Fish and Fisheries*, 12, 340–351.

Fulton, E.A., Smith, A.D.M., Smith, D.C. & van Putten, I.E. 2011. Human behaviour: The key source of uncertainty in fisheries management. *Fish and Fisheries*, 12, 2–17.

Hall, D.L., Hilborn, R., Stocker, M. & Walters, C.J. 1988. Alternative Harvest Strategies for Pacific Herring (*Clupea harengus pallasi*). *Canadian Journal of Fisheries and Aquatic Sciences*, 45, 888–897.

Hawkshaw, M. & Walters, C. 2015. Harvest control rules for mixed-stock fisheries coping with autocorrelated recruitment variation, conservation of weak stocks, and economic well-being. *Canadian Journal of Fisheries and Aquatic Sciences*, 72, 759–766.

Hightower, J.E. & Grossman, G.D. 1985. Comparison of Constant Effort Harvest Policies for Fish Stocks with Variable Recruitment. *Canadian Journal of Fisheries and Aquatic Sciences*, 42, 982–988.

Hilborn, R. 1976. Optimal exploitation of multiple stocks by a common fishery: A new methodology. *Journal of the Fisheries Research Board of Canada*, 33, 1–5.

Hilborn, R. 1979. Comparison of fisheries control systems that utilize catch and effort data. *Journal of the Fisheries Research Board of Canada*, 36, 1477–1489.

Hilborn, R. 1987. Living with Uncertainty in Resource Management. *North American Journal of Fisheries Management*, 7, 1–5.

Hilborn, R. 2012. The Evolution of Quantitative Marine Fisheries Management 1985–2010. *Natural Resource Modeling*, 25, 122–144.

Hilborn, R. & Mangel, M. 1997. *The Ecological Detective: Confronting Models with Data*, Princeton, NJ, Princeton University Press.

Hilborn, R. & Walters, C. 1992a. Designing adaptive management policies. In *Quantitative Fisheries Stock Assessment: Choice, Dynamics and Uncertainty.* New York: Chapman and Hall.

Hilborn, R. & Walters, C. 1992b. Harvest strategies and tactics. In *Quantitative Fisheries Stock Assessment: Choice, Dynamics and Uncertainty.* New York: Chapman and Hall.

Hilborn, R. & Walters, C. 1992c. *Quantitative Fisheries Stock Assessment: Choice, Dynamics and Uncertainty,* New York, Chapman and Hall.

Holling, C.S. 1973. Resilience and Stability of Ecological Systems. *Annual Review of Ecology and Systematics,* 4, 1–23.

Holling, C.S. 1978. *Adaptive Environmental Assessment and Management,* Caldwell, NJ: Blackburn Press.

Irwin, B.J. & Conroy, M.J. 2013. Consideration of reference points for the management of renewable resources under an adaptive management paradigm. *Environmental Conservation,* 40, 302–309.

Kope, R.G. 1992. Optimal Harvest Rates for Mixed Stocks of Natural and Hatchery Fish. *Canadian Journal of Fisheries and Aquatic Sciences,* 49, 931–938.

Lande, R., Engen, S. & Sœther, B.-E. 1995. Optimal harvesting of fluctuating populations with a risk of extinction. *American Naturalist,* 728–745.

Lande, R., Sœther, B.-E. & Engen, S. 1997. Threshold harvesting for sustainability of fluctuating resources. *Ecology,* 78, 1341–1350.

Mace, P.M. 1994. Relationships between common biological reference points used as thresholds and targets of fisheries management strategies. *Canadian Journal of Fisheries and Aquatic Sciences,* 51, 110–122.

Mangel, M. 2000. On the fraction of habitat allocated to marine reserves. *Ecology Letters,* 3, 15–22.

Marescot, L., Chapron, G., Chadès, I., Fackler, P.L., Duchamp, C., Marboutin, E., Gimenez, O. & Freckleton, R. 2013. Complex decisions made simple: A primer on stochastic dynamic programming. *Methods in Ecology and Evolution,* 4, 872–884.

Martin, J., Runge, M.C., Nichols, J.D., Lubow, B.C. & Kendall, W.L. 2009. Structured decision making as a conceptual framework to identify thresholds for conservation and management. *Ecological Applications,* 19, 1079–1090.

McCarthy, M.A., Possingham, H.P. & Gill, A.M. 2001. Using stochastic dynamic programming to determine optimal fire management for *Banksia ornata. Journal of Applied Ecology,* 38, 585–592.

MFISH 2011. *Operational Guidelines for New Zealand's Harvest Strategy Standard,* Ministry of Fisheries New Zealand.

Newman, K.B. & Lindley, S.T. 2006. Accounting for demographic and environmental stochasticity, observation error, and parameter uncertainty in fish population dynamics models. *North American Journal of Fisheries Management,* 26, 685–701.

Pella, J.J. & Tomlinson, P.K. 1969. A generalized stock production model. *IATTC Bulletin,* 13, 421–458.

Polacheck, T., Hilborn, R. & Punt, A.E. 1993. Fitting surplus production models: Comparing methods and measuring uncertainty. *Canadian Journal of Fisheries and Aquatic Sciences,* 50, 2597–2607.

Punt, A.E. 2003. Extending production models to include process error in the population dynamics. *Canadian Journal of Fisheries and Aquatic Sciences,* 60, 1217–1228.

Punt, A.E., Butterworth, D.S., de Moor, C.L., De Oliveira, J.A.A. & Haddon, M. 2014. Management strategy evaluation: Best practices. Fish and Fisheries DOI: 10.1111/faf.12104.

Punt, A.E. & Donovan, G.P. 2007. Developing management procedures that are robust to uncertainty: Lessons from the International Whaling Commission. *ICES Journal of Marine Science,* 64, 603–612.

Quinn, T.J., II & Deriso, R.B. 1999a. *Quantitative fish dynamics,* USA, Oxford University Press.

Quinn, T.J., II & Deriso, R.B. 1999b. Stock and recruitment. *Quantitative fish dynamics,* USA, Oxford University Press.

Quinn, T.J., II, Fagen, R. & Zheng, J. 1990. Threshold management policies for exploited populations. *Canadian Journal of Fisheries and Aquatic Sciences,* 47, 2016–2029.

Rademeyer, R.A., Plagányi, E.É. & Butterworth, D.S. 2007. Tips and tricks in designing management procedures. *ICES Journal of Marine Science,* 64, 618–625.

Reed, W.J. 1979. Optimal escapement levels in stochastic and deterministic harvesting models. *Journal of Environmental Economics and Management,* 6, 350–363.

Regan, H.M., Colyvan, M. & Burgman, M.A. 2002. A taxonomy and treatment of uncertainty for ecology and conservation biology. *Ecological Applications,* 12, 618–628.

Restrepo, V.R., Thompson, G.G., Mace, P.M., Gabriel, W.L., Low, L.L., Maccall, A.D., Methot, R.D., Powers, J.E., Taylor, B.L., Wade, P.R. & Witzig, J.F. 1998. *Technical guidance on the use of precautionary approaches to implementing national standard 1 of the Magnuson-Stevens Fishery Conservation and Management act,* National Marine Fisheries Service National Oceanic and Atmospheric Administration of the US Department of Commerce.

Ricker, W.E. 1954. Stock and recruitment. *Journal of the Fisheries Research Board of Canada,* 11, 559–623.

Rochet, M.-J. & Rice, J.C. 2009. Simulation-based management strategy evaluation: Ignorance disguised as mathematics? *ICES Journal of Marine Science,* 66, 754–762.

Rosenberg, A.A. & Brault, S. 1993. Choosing a management strategy for stock rebuilding when control is uncertain. *In:* Smith, S.J., Hunt, J.J. & Rivard, D. (eds.) *Risk Evaluation and Biological Reference Points for Fisheries Management.* Ottowa, Canada: NRC Research Press.

Rosenberg, A.A. & Restrepo, V. 1993. The eloquent shrug: Expressing uncertainty and risk in stock assessments. *ICES CM,* D:12, 1–15.

Rosenberg, A.A. & Restrepo, V.R. 1994. Uncertainty and risk evaluation in stock assessment advice for U.S. Marine fisheries. *Canadian Journal of Fisheries and Aquatic Sciences,* 51, 2715–2720.

Sæther, B.-E., Engen, S. & Lande, R. 1996. Density dependence and optimal harvesting of fluctuating populations. *Oikos,* 76, 40–46.

Sainsbury, K.J., Punt, A.E. & Smith, A.D.M. 2000. Design of operational management strategies for achieving fishery ecosystem objectives. *ICES Journal of Marine Science,* 57, 731.

Schaefer, M.B. 1954. Some aspects of the dynamics of populations important to the management of commercial marine fisheries. *Bulletin of the Inter-American Tropical Tuna Commission,* 1, 26–56.

Schnute, J. 1985. A general theory for analysis of catch and effort data. *Canadian Journal of Fisheries and Aquatic Sciences,* 42, 414–429.

Sethi, G., Costello, C., Fisher, A., Hanemann, M. & Karp, L. 2005. Fishery management under multiple uncertainty. *Journal of Environmental Economics and Management,* 50, 300–318.

Shotton, R. 1993. Risk, uncertainty and utility: A review of the use of these concepts in fisheries management. *ICES CM,* D:71, 1–14.

Smith, A. & Walters, C.J. 1981. Adaptive management of stock-recruitment systems. *Canadian Journal of Fisheries and Aquatic Sciences,* 38, 690–703.

Smith, A.D.M. 1993. Risk assessment or management strategy evaluation: What do managers need and want? *ICES CM,* D:18, 1–6.

Smith, A.D.M., Sainsbury, K.J. & Stevens, R.A. 1999. Implementing effective fisheries-management systems – management strategy evaluation and the Australian partnership approach. *ICES Journal of Marine Science,* 56, 967–979.

Southward, G.M. 1968. A simulation of management strategies in the Pacific Halibut fishery. *International Pacific Halibut Commission Report,* 47, 1–70.

Walters, C.J. 1975. Optimal harvest strategies for salmon in relation to environmental variability and uncertain production parameters. *Journal of the Fisheries Research Board of Canada,* 32, 1777–1784.

Walters, C.J. 1986a. *Adaptive Management of Renewable Resources,* New York, NY, McMillan Publisher.

Walters, C.J. 1986b. Feedback policy design. *Adaptive Management of Renewable Resources,* New York, NY, McMillan Publisher.

Walters, C.J. 2007. Is adaptive management helping to solve fisheries problems? *AMBIO: A Journal of the Human Environment,* 36, 304–307.

Walters, C.J. & Hilborn, R. 1978. Ecological optimization and adaptive management. *Annual Review of Ecology and Systematics,* 9, 157–188.

Walters, C.J. & Holling, C.S. 1990. Large-scale management experiments and learning by doing. *Ecology,* 71, 2060–2068.

Williams, B.K. 2011. Adaptive management of natural resources – framework and issues. *Journal of Environmental Management,* 92, 1346–1353.

Williams, B.K., Nichols, J.D. & Conroy, M.J. 2002a. Case study: Management of the sport harvest of North American waterfowl. *Analysis and Management of Animal Populations: Modeling, Estimation and Decision Making,* United States of America, Academic Press.

Williams, B.K., Nichols, J.D. & Conroy, M.J. 2002b. Modern approaches to decision analysis. *Analysis and Management of Animal Populations: Modeling, Estimation and Decision Making,* United States of America, Academic Press.

## Chapter 3

The practical evaluation of feedback control strategies

Katell G. Hamon and Jan-Jaap Poos

Introduction

Management procedures (MPs) represent a defined framework for making immediate decisions regarding the level of exploitation in a given fishery. An MP will always include a harvest control rule (HCR), plus specification of an assessment method and the requisite data for providing the control rule input. The output of an MP is usually a recommended catch level (the total allowable catch, TAC) or an effort prescription (the total allowable effort, TAE). The design of MPs using a simulation-based approach was introduced by Dankel and Edwards (Chapter 1, this volume) and Edwards (Chapter 2, this volume), and the current chapter will further develop this idea by providing an example simulation exercise.

Simulation testing is central to the MP evaluation process. However evaluation cannot be considered a purely technical exercise, since important components of the full evaluation cycle include the setting of management objectives, the definition of uncertainties, and a decision on which MP should be selected for implementation. These all require input from the wider stakeholder group, most importantly including the fishers and management personnel. This broader scientific engagement process, which is so critical for a credible evaluation, is well illustrated by the chapters in Part 2 of this volume. For the current chapter we will use the context provided by an example ICES (International Council for the Exploration of the Sea) case study and focus only on the technical aspects of the simulation. Noting that MPs are a type of management strategy, the former having a stricter definition (Edwards, Chapter 2, this volume), this chapter therefore provides a simple introduction to the simulation-based, scientific component of management strategy evaluation (MSE). Code is presented in example boxes, written in R (R Core Team, 2014), and can be used by the reader to run the evaluation described in this chapter (noting that the code in each box should be run sequentially).

Simulation testing of the MP requires a number of model components, representing the MP and the operating model (OM). These are illustrated in Figure 3.1. The OM includes the population dynamics and the exploitation

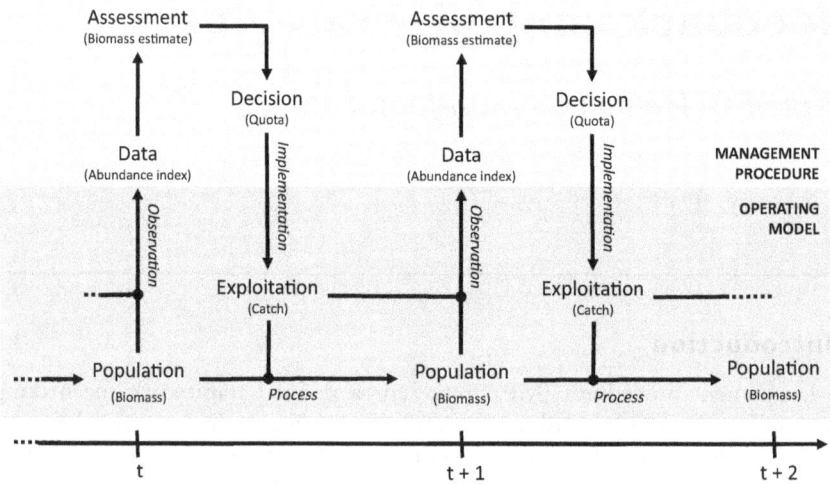

Figure 3.1 Illustration of the primary system components that can be included in the simulation evaluation of a management procedure (MP), given a particular operating model (OM). The arrows represent processes that need to be modelled and the labels in italics represent some of the different components of uncertainty that can be included in the evaluation process. Statistical uncertainty is not shown, but would represent the difference between the estimated biomass in the MP and the underlying operating model biomass. Statistical uncertainty may also be present in parameterisation of the OM. Structural uncertainty could be included by using multiple OMs.

of the stock, but is also required to emulate the observation process, whereby monitoring data are simulated for input into the MP, and to implement the management recommendation output from the MP (Punt et al., 2014). The OM will usually translate the output from an MP into an effect on the population. For example, a TAC biomass output from an MP may need to be converted into the numbers at age removed from the population. However, in more advanced settings a more elaborate implementation module is required, which may include for example a partitioning of the catch between species in a multispecies fishery (e.g. Dichmont et al., Chapter 10, and de Moor and Butterworth, Chapter 11, this volume).

The correct representation of uncertainty is a key component of simulation-based evaluation of an MP. These are discussed here based on the definitions given by Edwards (Chapter 2, this volume). Arguably the most important form of uncertainty is the structural uncertainty associated with the OM. In the current chapter we assume a simple production model (Box 3.1 on page 43) and do not examine alternative OM formulations. But to better account for structural uncertainty, parallel simulations could be conducted using a range of production functions. For simplicity of presentation we have also made no attempt to account for process uncertainty in our projections (Figure 3.1), and have instead assumed that the change in biomass over time is deterministic.

Parameters within the OM are obtained by fitting the model to currently available data (Box 3.1 on page 43), and the statistical uncertainty around these parameter values is included in the OM projections (Box 3.2 on page 50). To illustrate the concept of feedback control, we begin with a simple HCR and assume perfect knowledge of the system (i.e. the stock biomass predicted by the OM is used directly as a control rule input; Box 3.3 on page 55). Next we introduce the statistical uncertainty associated with assessment of the resource during implementation of the MP. Since our MP is model based (i.e. it includes a stock assessment model for estimation of the biomass), this type of statistical uncertainty represents the difference between the biomass estimated at each simulated time step and the actual biomass represented by the OM. This could be included via an observation component (with error) and a simulated assessment procedure (Figure 3.1). However, because this is technically more challenging, we have instead represented statistical uncertainty within the MP as an error distribution around the underlying OM value (Box 3.4 on page 58). Finally we have included implementation error as a deterministic multiplier of the MP output (Box 3.4). The effect of different implementation error assumptions is examined using multiple runs (Box 3.5 on page 60).

The R code given in Boxes 3.1–3.5 outlines the basic steps involved in the simulation-based evaluation of an MP, beginning with the development of an OM, including population dynamics (Box 3.1) and exploitation (Box 3.2), followed by initial investigations of the MP (Box 3.3), and finally a comparison of two MPs assuming different structural assumptions in the OM (in this case the implementation error; Box 3.5). These steps are described in more detail later in this chapter.

Building the operating model

Population dynamics

The first component of the evaluation framework consists of the population dynamics of the fish stock being fished. Because complete, direct observations are impossible, models are used to infer the abundance and mortality of fish stocks (Fournier and Archibald, 1982, Hilborn, 2003, Cotter et al., 2004). Population dynamic models vary in complexity regarding the processes that are included. Since the processes governing the population dynamics need to be estimated, the level of complexity should depend on the availability of data to inform this estimation.

In situations where information on total catches and relative biomass indices are available, stock dynamics can be described using biomass dynamic models (Schaefer, 1954, Pella and Tomlinson, 1969, Hilborn and Walters, 1992, chap. 8). These models assume that the productivity of the stock depends on the size of the stock (an example productivity curve is given in Figure 3.2). Productivity is measured in terms of the new biomass added to the population at each time step as a result of the recruitment of new individuals to the population and the somatic growth of individuals already present. Productivity first increases with the biomass as the reproductive power of the stock increases, and then declines as competition increases until the carrying capacity of the stock is reached (Figure 3.2). At this point, natural death is exactly replaced by reproduction and somatic growth. In an unexploited system there are only two equilibrium points: when the stock is at carrying capacity or when the stock is extinct. In an exploited system, catches decrease the biomass every year moving the biomass away from carrying capacity. Catches are sustainable as long as they are not above the productivity of the stock. The maximum sustainable yield (MSY) is reached if exploitation can maintain the stock at its maximum productivity. In our example, we estimate the stock dynamics of *Sebastes mentella* in the Irminger Sea with such a model (Box 3.1).

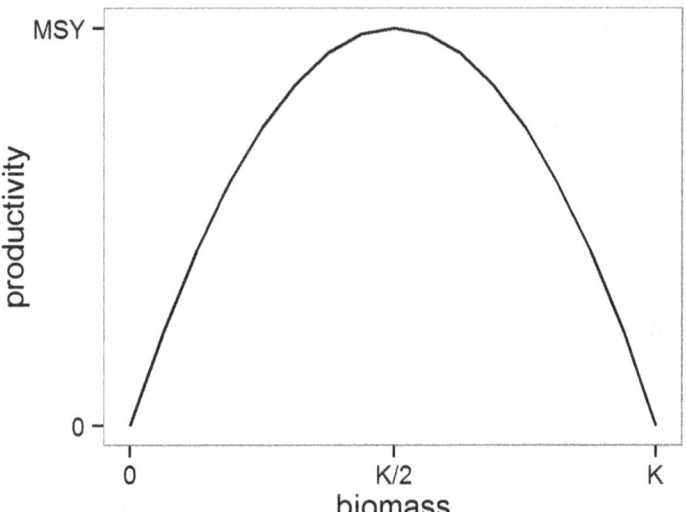

Figure 3.2 Schaefer (1954) production model: productivity of the fish stock is dependent on the biomass of the stock. The parameter K refers to the carrying capacity of the population. For this model, maximum productivity occurs at half of K. The maximum sustainable yield (MSY) occurs at the point of maximum productivity.

Box 3.1 Estimating population dynamics using a production model

The example provided in the subsequent boxes is based on an MSE done for Sebastes mentella in the Irminger Sea and adjacent waters by ICES WKREDMP (2014). The MSE was done in response to a request for the evaluation of a proposed Harvest Control Rule submitted to ICES by the North East Atlantic Fisheries Commission (NEAFC), the responsible management body for the stock. The fishery for this stock only began in the early 1990s.

The available data on the stock includes catch data (harvest) since 1991 and biomass survey indices (index) since 1999. Catch data suggests that annual catches rose quickly from 59 t in 1991 to nearly 140,000 t in 1996, stabilising at 85,000–105,000 t during the period 1997–2004, when some countries ceased fishing. From 2005 onwards, annual landings have declined, being in the range 30,000 and 68,000 t. In the biennial surveys, a trawl was used to calculate abundance of deep-pelagic beaked redfish. The method is based on a combination of standardised survey catches and the hydroacoustic data. The surveys result in an estimate of total stock biomass.

```
# specify input data and associated years
data.years <- 1991:2013
harvest <- c(0.1,3,15,52,76,139,95,93,84,93,86,103,104,
             92,46,67,59,30,54,59,47,33,44)
index <- c(NA,NA,NA,NA,NA,NA,NA,NA,935,NA,1057,NA,678,NA,
           420,NA,554,NA,458,NA,474,NA,280)
```

The dynamics of the stock are simulated using a Schaefer (1954) model for which the carrying capacity K and the intrinsic growth rate r of the stock must be estimated. To estimate these parameters the catch data and survey indices can be used. In order to do the stock assessment, we cut the algorithm in several functional pieces, each becoming a function in R. The first function (schaefer) returns the biomass that results from the Schaefer equation. The second function (dynamics) is the actual "engine" of the assessment, repeating the Schaefer function over time, taking the r and K parameters in combination with a time series of catches. The assess function tries to find those r, K and sigma parameters that minimise the negative log-likelihood of the observed survey biomass indices, and predicted biomass indices based on the dynamics function. The assess function uses the optimiser in R, applying to the nll function that actually computes the negative log-likelihood. Results from the stock assessment are plotted below (Figure 3.3) illustrating the catches and estimated stock biomass over time.

```
# library definitions #
library(mvtnorm)
```

```
# function definitions #

# logistic production function
schaefer <- function(B,C,K,r)
{
  res <- B + B * r * (1 - B/K) - C
  return (max(0.001,res))
}

# biomass projection function
dynamics <- function(pars,C,yrs)
{
  K <- exp(pars[1])
  r <- pars[2]

  nyr <- length(C) + 1

  if(missing(yrs)) yrs <- 1:nyr

  B <- numeric(nyr)

  B[1] <- K
  for (y in 2:nyr)
  {
    B[y] <- schaefer(B[y-1],C[y-1],K,r)
  }
  return(B[yrs])
}

# function to calculate the negative log-likelihood
nll <- function(pars,C,U)
{
  sigma  <- pars[3]
  B      <- dynamics(pars,C)
  Uhat   <- B
  output <- -sum(dnorm(log(U),log(Uhat),sigma,log=TRUE),
            na.rm=TRUE)
  return(output)
}

# function to perform an assessment
# (i.e. to fit the logistic model to abundance data)
assess <- function(catch,index,calc.vcov=FALSE,pars.init)
{
  # fit model
  res <- optim(pars.init,nll,C=catch,U=index,hessian=TRUE)

  # output
  output <- list()
  output[['pars']]          <- res$par
```

```
    output[['biomass']]       <- dynamics(res$par,catch)
    output[['convergence']]   <- res$convergence
    output[['nll']]           <- res$value
    if(calc.vcov)
      output[['vcov']]        <- solve(res$hessian)

    return(output)
}


# perform assessment #

# initial parameter vector for: log(K), r, sigma
ini.parms    <- c(log(1200), 0.1, 0.3)

# fit logistic model to data
redfish       <- assess(harvest,index,T,ini.parms)

# extract maximum likelihood biomass and parameter
# estimates
biomass.mle <- redfish$biomass
pars.mle    <- redfish$pars

# obtain statistical uncertainty by sampling 500
# iterations from
# a multivariate normal distribution with the
# variance covariance matrix estimated during the model fit
niter                    <- 500
pars.iter                <- matrix(NA, nrow = niter,
                            ncol = 3)
colnames(pars.iter)   <- c("logK",'r','sigma')

for (i in 1:niter)
{
#random multivariate normal distribution in package
# mvtnorm
  pars.iter[i,] <- rmvnorm(1, mean = redfish$pars,
                     sigma = redfish$vcov)
}


# generate replicate model outputs that represent
# the statistical uncertainty associated with the model fit
biomass.iter <- data.frame()

for (i in 1:niter)
{
  biomass.iter <- rbind(biomass.iter, data.frame
                     (year = seq (min(data.years), max
                     (data.years)+1), biomass = dynamics
                     (pars.iter[i,], harvest), iter=i))
}
```

```
# plot the estimated biomass time series with ggplot2
library(ggplot2)

Fig3.3 <- ggplot(data=biomass.iter,aes(x = year,
        y = biomass))
Fig3.3 + stat_summary(fun.data = "median_hilow",
        geom = "smooth", fun.ymin = function(x)0,
        conf.int=0.95,col="black")+ geom_line(aes(y =
        harvest, x = year), data = data.frame(harvest =
        harvest, year = data.years), lty=2) +
        geom_point(aes(y = index, x = year),data =
        data.frame (index = index, year = data.years)) +
        ylab("Estimated B and C (million tonnes)") +
        theme_bw()
```

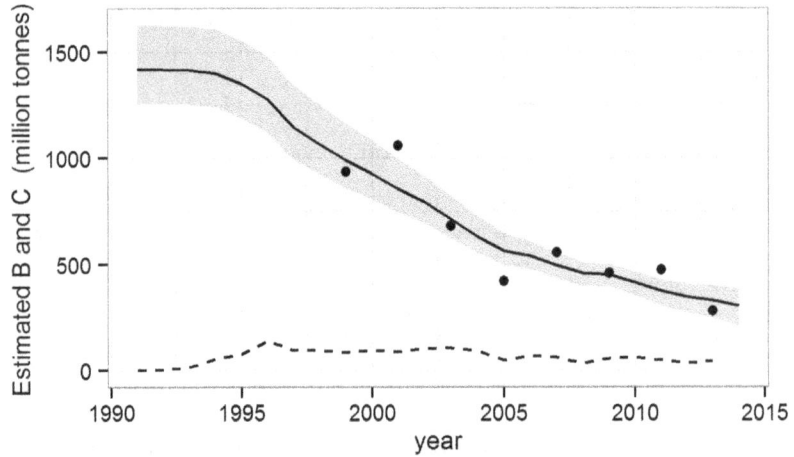

Figure 3.3 Estimated biomass from assessment black full line with 95% confidence interval (grey area), historical catch (dashed line) and survey index (dots).

Sometimes, the generic qualities of the biomass dynamic model are not appropriate: the life cycle of individuals in fish stocks can be described in much more detail – for example, from the amount of fish in the larval phase and juvenile stage, to their recruitment to the fishery, together with growth, sexual maturation, reproduction and finally the death of the fish. The individuals grow in length and weight as they age (which can be represented, for example, using the von Bertalanffy growth model; e.g. Cox and Kronlund, Chapter 5, this volume). As they grow older the fish become mature and able to reproduce. The mature part of fish stocks are referred to as spawning stock biomass (SSB), and the size of the SSB represents the reproductive power of the stock. In some models it is assumed that there is a relationship between SSB and future recruitment (e.g. Hicks et al., Chapter 4, this volume). However the natural variation may be so large that any such relationship, if it exists at all, cannot be estimated (e.g. Needle, Chapter 12, this volume). Aleatoric uncertainty of this magnitude makes it more sensible to estimate a mean or median recruitment value that is independent of the SSB. In both instances, deviations around the expected value can be estimated within the model (often referred to as process error; see Edwards, Chapter 2, this volume).

The final step of the life cycle of fish is mortality. Fish die of many causes that can be natural or anthropogenic. The predation of fish by other animals or the competition for food and habitat are the main natural causes of mortality (Sparholt, 1990, Gislason et al., 2010), and are an important component of ecosystem models (see Plagányi, Chapter 15, this volume). In some cases, fishing is the main anthropogenic source of mortality, but other human activities influencing the environment and the habitat can also contribute. While it is relatively easy to obtain data on the fishing mortality from fish landed and sold in harbours through available records, other sources of mortality are difficult to assess because they are generally not directly observable (Sparholt, 1990).

Whichever model is used, uncertainty around the life cycle of fish stocks remains. Some of it lies in the parameter estimation (statistical uncertainty) or the extent to which the underlying dynamics of the stock is captured by the model (structural uncertainty). These uncertainties can sometimes be reduced with additional data, but process uncertainty around recruitment or natural mortality, which are partly due to natural environmental variation, is hard to reduce (see Fogarty et al., 1996).

Exploitation

The exploitation of fish stocks happens through the general categories of commercial, recreational and subsistence fishing. The fishers (hereinafter referred to as fishing fleets) and the fish stocks mutually influence each other: the fleets influence the fish stock by catching fish, and the fish stock influences the location and amount of fishing, which is a short-term effect on each of the fleets, and longer-term entry and exit of fishers from the fishery.

The effect of catches on the fish stock is measured by the fishing mortality. Simple models assume that fishing mortality is directly proportional to fishing effort, but it has long been recognised that this relationship can be influenced by how fleets react and adapt to changes in the fishery (Hilborn, 1985). The link between the fishing effort (measured using units of time, area or fishing gear) and the resulting fishing mortality is known as the *catchability*. In reality, catchability is determined by factors that include the distribution of fishing effort in time and space and the composition of the fleets in terms of gear and vessel size. A growing number of fleet dynamic models take these into account when estimating the impact on fish stocks (van Putten et al., 2012). While the added value of incorporating fishers' behaviour in management models has long been demonstrated (Wilen et al., 2002, Fulton et al., 2011), in practice only a few examples are available (Venables et al., 2009, Andersen et al., 2010, Little et al., 2011).

Depending on the underlying conceptual model being assumed, catchability can be defined in a number of ways (Arreguín-Sánchez, 1996), such as the probability of catching an individual or unit of biomass per unit of effort, and is affected by a number of biological and technical factors: availability, vulnerability and selectivity (Marchal et al., 2003). Availability represents the overlap in time and space of the stock and the fishery. Vulnerability encompasses the behavioural factors affecting the catchability, such as hiding during moult periods for lobsters, or the tendency of fish to form aggregations that can be easily caught. Selectivity is the susceptibility of a fish to be retained by a certain gear type. Selectivity is usually size specific, and selectivity designs (typically the size of the fishing net's mesh or hook) are used to avoid unwanted catch. The catch is then divided into two categories: the retained catch that will be landed in the harbour and sold (*landings*) and the unwanted catch that is thrown back because of market conditions or management regulations (*discards*).

Both landings and discards contribute to the fishing mortality of a stock. Estimating the mortality caused by discarding is generally more difficult for two reasons: first, sample sizes for discard estimates are often low because of the need to observe the estimates on-board fishing vessels. Second, the survival rate of discarded fish is often unknown and difficult to estimate. Even if a fish is alive when thrown back in to the sea, there may be predation by birds in the moments after release (Garthe et al., 1996), or injuries during fishing and handling (van Beek et al., 1990, Bartholomew and Bohnsack, 2005) may increase mortality after release. For some stocks, the major part of the catch is discarded, and ignoring discards in scientific advice is risky (Rijnsdorp et al., 2007).

Implementation

Implementation refers to the steps between a management recommendation and actual exploitation of the fishery. It can include a variety of processes, including deviation of the actual management regulation from the recommendation, or how the regulation affects the behaviour of fishers. The effectiveness of management measures strongly depends on their implementation.

A range of environmental, economic and institutional factors affects the behaviour of fishers and fleet decisions. Introducing new management rules can lead to unforeseen change in behaviour. For example, fishers tend to target more valuable parts of the stock when individual transferrable quotas (ITQs) are introduced (Hamon et al., 2009). This change in behaviour can have unexpected negative impacts such as an increase of discards. In most cases the choice to discard is based on economic and regulatory reasons. If the ex-vessel price (the average price for an individual species, harvested by a specific gear, in a specific area) of a fish is lower than the handling and landing costs, it is more profitable for fishers to discard it. In some cases, when the quota for a species is limited, fishers can choose to discard certain size classes to reserve their quota for more valuable fish sizes (so-called high-grading). Discarding can also be a direct effect of management measures, for example the discarding of fish smaller than the minimum landing size, or if there is a mismatch between quotas of species caught together in mixed fisheries (Poos et al., 2010). This is because quotas often limit landings in harbours, rather than total catches on-board vessels (Ulrich et al., 2011). Landing obligation regulations (meaning that discards are banned) have been implemented around the world to reduce this practice, but compliance with fishing regulations is often based on weighing the expected benefits of infraction against the costs and risk of getting caught (Hatcher et al., 2000).

It is argued that some management measures based on incentives lead to better compliance than others. Those are generally based on property rights, either allocated to individuals (individual transferable quotas are believed to increase stewardship; Garrity, 2011) or to groups (Basurto and Ostrom, 2009). Compliance with regulation is difficult to assess. For existing rules, noncompliance can be empirically estimated through interviews and surveys but predicting the compliance rate for new rules can be challenging, as the response of fishers depends on social factors (Hatcher et al., 2000) and individual risk profiles (Wildavsky and Dake, 1990).

In the example code of this chapter, no complex fleet dynamics are modelled. Rather the harvest rate is assumed either to be constant (Box 3.2), to perfectly follow advice (Box 3.3), or implementation error is included as a percentage of TAC overshoot (Box 3.4 and Box 3.5).

The management procedure

Data and assessment

The current state of the stock cannot be directly observed, and we must assume that this is also true during future implementations of the MP. The MSE, which simulates this future implementation, must take this into account. Just like when we started setting up the operating model, the state of stocks are estimated using fisheries-dependent data (catch and effort data collected from the fishery) and independent data collected regularly through research surveys. Most models assess the state of the stocks based on the change in catch per unit of effort in surveys or the fishery (i.e. if the catch per unit of effort increases, the underlying biomass of the stock is increasing). While fisheries-dependent data are easier to collect in most places, the fleet behaviour can affect the quality of the data (e.g. in case of nonreported discards or noncompliance); even when the data is correctly reported, the changes occurring in the fishery in terms of fishing grounds, fishing gears or vessel power will affect the average catchability (through structural uncertainty) and the outcome of the assessment model. To correct for these, data from scientific surveys using the same fishing technique year after year and fishing in the same areas at the same season are used as indices. Of course, the scientific surveys are costly to implement and are not available for every fishery.

Depending on the future management strategy, the target harvest rates can be set based on biomass estimates coming from a stock assessment, or based on a proxy for biomass such as catch per unit of effort observations coming from the fishery. In the example shown in Box 3.4, the management plan was based only on the biennial biomass survey. We can model the statistical uncertainty in this survey from the stock assessment that was used to set up the operating model.

The decision rule

The decision rule should be developed according to the management targets and objectives determined as part of the initial stages of the MSE, and must be appropriate for the data being collected from the fishery. Environmental and economic objectives can be translated into numerical targets like minimum biomass levels and harvest rates in turn can be transformed into effort limits or TACs. In this example, the MP is based on the biennial biomass survey. As a simplified example in Box 3.2, we set the management target as a fixed harvest rate independent from the status of the stock.

Box 3.2 Projection of the operating model at a constant harvest rate and assuming perfect knowledge of the resource biomass

We now build on our knowledge of the stock from Box 3.1 and make a future stock prediction based on a set of assumptions. To project the stock into the future, we should account for exploitation. Here we assume that the exploitation rate in the future will remain at 0.1 (control), that is the TAC will be fixed at 10% of the biomass assuming perfect observation of the available biomass (observe) and perfect implementation of the TAC, that is the total catch matches the TAC (implement). This level of exploitation is close to the current harvest rate observed in the fishery.

Statistical uncertainty around the stock dynamics is taken into account through sampling of the parameter values for K and r. The results of the projection are given in Figure 3.4.

```
# define the years for which projection is done
proj.years <- 2014:2034

# perfect biomass estimate
observe <- function(biomass, ...)
{
  biomass
}
```

```
# fixed target exploitation rate
control.pars <- list()
control.pars[['Htarg']] <- 0.1
control <- function(estimated.biomass, control.pars)
{
  control.pars[['Htarg']]
}

# implementation
implement <- function(TAC)
{
  TAC
}

# evaluation function that projects the operating model
# forward
# and implements the management procedure at each
# time-step
evaluate <- function(pars.iter, biomass.iter,
control.pars, data.years, proj.years, iterations, ...)
{
  iyr <- length(data.years)+1
  pyr <- length(proj.years)
  yrs <- c(data.years, proj.years, max(proj.years)+1)

  res <- data.frame()

  for(i in 1:iterations)
  {
    K.i   <- exp(pars.iter[i,1])
    r.i   <- pars.iter[i,2]
    sig.i <- pars.iter[i,3]

    biomass.i <- c(subset(biomass.iter, iter == i)
                     $biomass, numeric(pyr))
    index.i   <- c(index,numeric(pyr))
    catch.i   <- c(harvest,numeric(pyr))
    TAC.i     <- numeric(pyr)

    for(y in iyr : (iyr + pyr - 1))
    {
      index.i[y]     <- observe(biomass.i[y], sig.i)
      TAC.i[y]       <- control(index.i[y],
                         control.pars) * index.i[y]
      catch.i[y]     <- implement(TAC.i[y], ...)

      biomass.i[y+1] <- schaefer(biomass.i[y],catch.i[y],
                         K.i,r.i)
    }

    res <- rbind(res, data.frame(year = yrs[-length(yrs)],
            value = index.i, type="index", iter=i),
```

```
                    data.frame(year = yrs[-length(yrs)],
                    value = catch.i, type = "catch", iter=i),
                    data.frame(year = yrs, value = biomass.i,
                    type = "biomass", iter = i))
    }
    return(res)
}

# project with fixed exploitation rate for all iterations
# and 20 years
project.fixed <- evaluate(pars.iter,
biomass.iter, control.pars, data.years, proj.years, niter)

# plot the catch and biomass time series
Fig3.4 <- ggplot(data = subset(project.fixed, type
        != "index"), aes(x = year, y = value))
Fig3.4 + stat_summary(fun.data = "median_hilow",
        geom="smooth", col = "black", lty = 2,
        conf.int = 0.95) + stat_summary(fun.y = median,
        fun.ymin = function(x)0, geom="line",
        data = subset(project.fixed, type != "index" &
        year %in% data.years)) + ylab("million tonnes") +
        facet_wrap(~type, scale="free_y")+theme_bw()
```

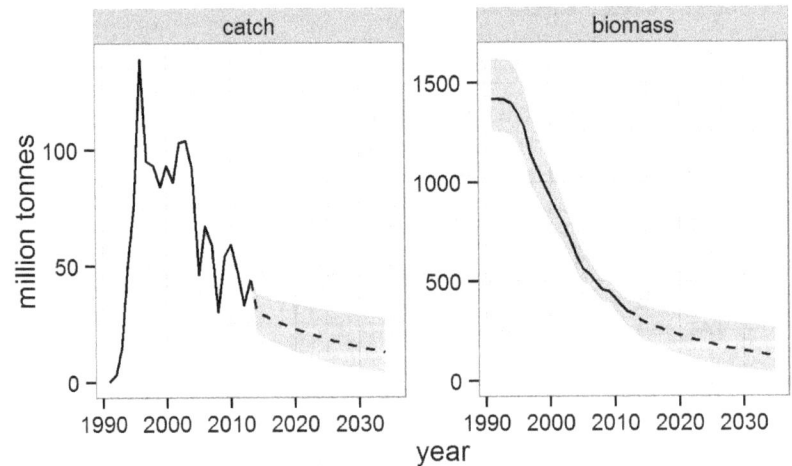

Figure 3.4 Twenty-year projection of *Sebastes mentella* stock, with constant harvest rate of 10% and assuming perfect knowledge of *r* and *K*: catches (left panel) and stock biomass (right panel). Dashed lines correspond to projections, grey areas to the 95% interval and full lines to historical data.

There are many instruments used to control the harvest rate exerted by fisheries. These can broadly be distinguished according to two categories: input and output controls. Input controls are restrictions put on the intensity of use of gear that fishers use to catch fish (Pope, 2002). Generally these refer to restrictions on the fleet size (fishing capacity), the amount of time fishing vessels are allowed to fish or restrictions on the gear size. In contrast, output controls are direct limits on the amount of fish coming out of a fishery (Pope, 2002). Generally output controls are limits placed on the amount of fish that may be caught within a given time frame, like a TAC. Often, however, the limits are placed on the landings, so that the output controls do not encompass potential discarding.

As clearly stated in the case study chapters in this book (Part 2), the definition of objectives is a key step during which all stakeholders (or as many as possible) should be involved for the management process to be successful. Clearly defined objectives can then be used to design harvest control rules, which state a set of rules, formulated as an algorithm, to set the exploitation level (using an input or output control). The HCR is defined to reach a target (optimum/safe level of biomass or exploitation) and defines the path to achieve it. To achieve the ecological objectives, a basic HCR can be defined as a segmented rule linking the targeted short-term exploitation rate for a given level of estimated biomass (see Figure 3.5 and Edwards, Chapter 2, this volume for the mathematical representation). In model-based MP, an estimate of the biomass is converted into a harvest rate then defining the appropriate level of exploitation. Such an HCR is applied to our example (Box 3.3). In addition to this simple rule, additional constraints can be used to comply

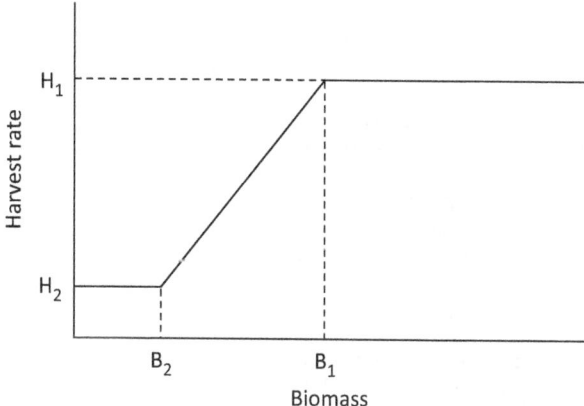

Figure 3.5 Example of a harvest control rule (HCR) with two biomass trigger points B_1 and B_2. If the biomass is larger than B_1, then the target harvest rate is set to H_1. If the biomass falls between B_1 and B_2 then the target harvest rate is an increasing function of biomass. If the biomass is lower than B_2, then the target harvest rate is equal to H_2.

with economic and social objectives, for example limiting the interannual variation of the TAC. This type of TAC constraint aids the preservation of the fishing fleet and the associated employment through stable catches and creates a minimum disturbance on the market.

Box 3.3 Introducing management by a harvest control rule

We now continue to develop our example by introducing a harvest control rule (HCR). This HCR was formulated by NEAFC when its parties agreed to implement a long-term management plan, and requested ICES to evaluate the rule (ICES WKREDMP, 2014). In the absence of well-defined biological reference points, the NEAFC parties agreed to use proxies based on the trawl survey indices. These proxies were used to define the shape of the HCR. The HCR here sets two target harvest rates expressed as fractions of the estimated biomass. Associated with these target harvest rates are two biomass triggers using the biomass survey as proxy. The upper biomass trigger point (B1) is located at 50% of the highest observed index value (B_{max}). Above this trigger point the harvest rate is 0.05 (H1). The lower biomass trigger point (B2) is located at 20% of the highest observed index value. If the biomass falls below this trigger point then the target harvest rate is set at 0.01 (H2). In between the two trigger points the harvest rate is an increasing function of the estimated biomass, as shown in Figure 3.5. Due to the low biomass value in 2014, applying the HCR results in a sudden drop in catches allowing the stock biomass to rebuild (Figure 3.6).

```
# define HCR that converts an estimated biomass
# into a harvest rate using a functional form determined
# by the
# values in 'control.pars'
control <- function(estimated.biomass, control.pars)
{
  H1 <- control.pars[['H1']]
  H2 <- control.pars[['H2']]
  Bmax <- control.pars[['Bmax']]
  B2 <- control.pars[['B2']]
  B1 <- control.pars[['B1']]

  harv <- ifelse(estimated.biomass >= B1, H1,
          ifelse(estimated.biomass < B2, H2, (H1 - H2)/
          (B1 - B2) * (estimated.biomass - B2) + H2))

  return(harv)
}
```

```
# Define control parameters for HCR using reference points
control.pars <- list()
control.pars[['H1']] <- 0.05
control.pars[['H2']] <- 0.01
control.pars[['Bmax']] <- max(index, na.rm=T)
control.pars[['B2']] <- 0.2 * control.pars[['Bmax']]
control.pars[['B1']] <- 0.5 * control.pars[['Bmax']]

# perform evaluation by projecting system forward in time
project.hcr <- evaluate(pars.iter, biomass.iter,
control.pars, data.years, proj.years, niter)

# plot catch and biomass times series
Fig3.6 <- ggplot(data = subset(project.hcr,
        type != "index"), aes(x = year, y = value))
Fig3.6 + stat_summary(fun.data = "median_hilow",
        geom = "smooth", col = "black", lty=2,
        conf.int = 0.95)+ stat_summary(fun.y = median,
        fun.ymin = function(x)0, geom = "line",
        data = subset(project.hcr, type != "index" &
        year %in% data.years)) + ylab("million tonnes") +
        facet_wrap(~type, scale = "free_y") +
        theme_bw()
```

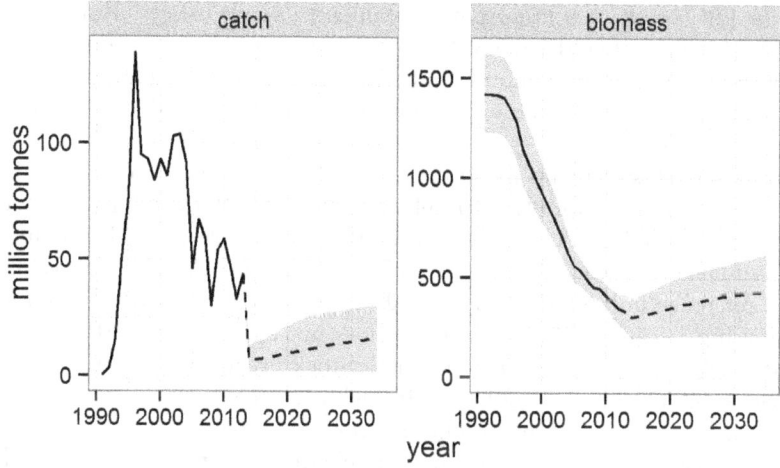

Figure 3.6 Projection of stock with perfect knowledge and fishing mortality as determined by the HCR: catches (left panel) and stock
biomass (right panel). Dashed lines correspond to projections,
grey areas to the 95% interval and full lines to historical data.

Management strategy evaluation

Management objectives and performance indicators

Management of fisheries is about managing the fishing fleets to sustainably exploit its resources while achieving a set of ecological, economic and social objectives. In most cases the ecological objectives are clearly stated with reference points and target levels of exploitation, whereas economic and social objectives are not always clearly stated. There are a few exceptions such as Australia, where the management target of some fisheries is based on the maximum economic yield (MEY, see Dichmont et al., Chapter 10, this volume).

Ecological objectives usually include healthy fish stocks measured mainly through SSB rather than total biomass, as a minimum reproductive capacity is needed to maintain the stock at sustainable levels. If the SSB level falls too low then there is a risk of recruitment failure; this is called *recruitment overfishing*. While the concept of ecosystem management is increasingly used, ecological objectives are still largely assessed at the species level.

Because they often conflict with ecological objectives, economic and social objectives are usually less transparent than ecological objectives and are dealt with by decision-makers through trade-offs (Voss et al., 2014). For example, to prevent a fishery from going bankrupt during transition periods, the ecological objectives can temporally be set slightly lower than the optimal levels. It is recognised that stability in the management measures such as TAC or effort is important to maintain a viable fishery (Butterworth, 2007). The level of landings and the variability in TAC or allowed effort are therefore seen as socioeconomic indicators. Additionally, multiannual management can be used to allow fisheries to make long-term decisions such as investment decisions.

International agreements under the United Nations Convention on the Law of the Sea (Johannesburg Plan of Implementation of 2002) state that fish stocks should be restored or maintained at levels that are capable of producing maximum sustainable yield (MSY; see Figure 3.2). Unfortunately estimates of MSY-based reference points such as B_{MSY} and F_{MSY} depend on a stock–recruitment relationship, because if the recruitment does not decline with SSB then it is theoretically possible to harvest the whole stock every year. For many stocks therefore, MSY cannot be estimated. In such cases reference points are simply assumed, usually based on precedent or theoretical work.

The management objective for the example stock was to develop a plan that would be in accordance with international agreements on sustainable harvest. The workshop from which the data is taken was set to evaluate different HCRs. In this chapter, the performance of each HCR is evaluated in terms of catches and stock biomass in the projected years. In the workshop, a wider range of indicators was used, such as the probability of biomass increase and interannual variability in catches (ICES WKREDMP, 2014).

Evaluation of the management procedure

Simulation-based evaluation is not isolated to individual examples such as the ones described so far. Rather it constitutes multiple simulations that take into account a range of structural uncertainties and alternative control rule parameterisations. One important structural deficiency in many model-based fishery representations is that the fleet dynamics and implementation of regulations by the fleet are poorly understood. Implementation uncertainty can be difficult to detect and thereby quantify. However, the management strategy evaluation allows testing the robustness of the foreseen management to implementation error. In Box 3.4, the MSE is run assuming that the catches exceed the TAC by 10% annually. The result of having this excess catch can be compared to a situation without overshoot, to see how sensitive the management strategy is to implementation error. We also model the statistical uncertainty in the survey that is used to inform the harvest control rule about the status of the stock. This uncertainty is parameterised using the stock assessment that was used to set up the operating model.

Box 3.4 Simulate observational data and the assessment process

In this box we try to capture some of the uncertainty in the system. First we assume that the biomass survey doesn't perfectly capture the state of the stock. We use the variability estimated in the stock assessment (Box 3.1) to add statistical uncertainty around the biomass estimates using the standard deviation (sigma) calculated in the stock assessment. In addition, we added implementation uncertainty about the level of catch. We assume that the catch exceeds the TAC set using the HCR by 10% (overshoot). The resulting catch has higher variability than with perfect knowledge and implementation while the biomass is slightly lower (Figure 3.7).

```
observe <- function(biomass, sigma)
{
    biomass * rlnorm(1, -0.5*sigma^2, sigma)
}

implement <- function(TAC, overshoot)
{
    TAC * (1 + overshoot)
}

project.error <- evaluate(pars.iter, biomass.iter,
                 control.pars, data.years, proj.years,
                 niter, overshoot = 0.1)
```

```
Fig3.7 <- ggplot(data = subset(project.error,
       type != "index"),aes(x = year, y = value))
Fig3.7 + stat_summary(fun.data = "median_hilow",
       geom = "smooth", col = "black", lty = 2,
       conf.int = 0.95)+ stat_summary(fun.y = median,
       fun.ymin = function(x) 0,geom="line",
       data = subset(project.error, type != "index"
       & year %in% data.years)) + ylab("million
       tonnes") + facet_wrap(~type, scale="free_y") +
       theme_bw()
```

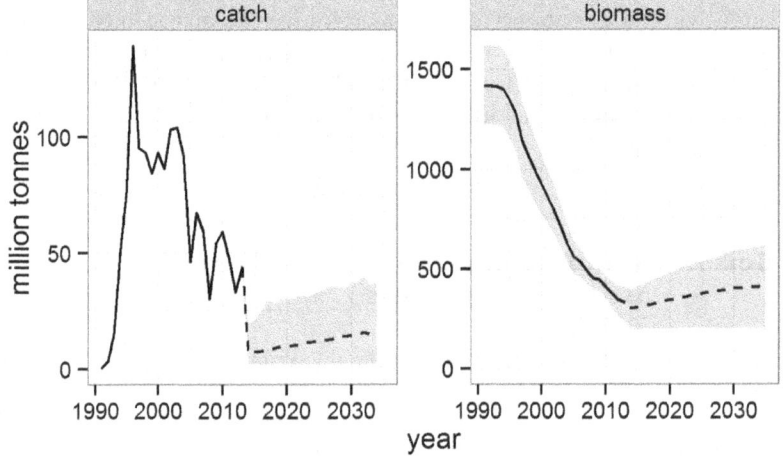

Figure 3.7 Projection of stock with TAC overshoot and imperfect knowledge of stock status. Fishing mortality is determined by the
HCR: catches (left panel) and stock biomass (right panel).
Dashed lines correspond to projections, grey areas to the 95%
interval and full lines to historical data.

In this final section we compare the performance of alternative control
rules under different implementation error assumptions. An example of the
code used is in Box 3.5. It can be clearly seen that a large assumed implementation error requires a more conservative control rule. This is the type
of information that MSE is designed to provide to decision-makers, allowing them to make informed choices on what an appropriate control rule
might be. Clearly, there is room for a great deal of judgement, which can
only be informed by continual dialogue between scientific modellers and
decision-makers.

Box 3.5 Comparing different HCRs and accounting for the possible TAC overshooting

In this box we evaluate a combination of two HCR rules, and two assumptions of a possibility of having a consistent overshoot of the TACs. The first HCR is equal to the HCR introduced in Box 3.3. The alternative HCR sets a higher target harvest rate (H1). The two options for the TAC overshoot are either to have catches that follow the TAC exactly (overshoot=0), or assuming an annual TAC overshoot of +20%. The results are summarised in Figure 3.8, which shows the biomass status of the stock and the catches over the projected period. The results suggest that HCR1 (with H1 = 0.05) results in larger biomasses than HCR2 (with H1 = 0.15). If applied perfectly, HCR2 on the other hand leads to larger median catches, but the variability of the catches is also larger. In case catch exceeds the TAC set by strictly applying the HCR (20% overshoot), the biomass is consistently lower than if the rule is perfectly applied.

```
# Use HCR with reference points, as in Box 3
control.pars <- list()
control.pars[['H1']] <- 0.05
control.pars[['H2']] <- 0.01
control.pars[['Bmax']] <- max(index, na.rm=T)
control.pars[['B2']] <- 0.2 * control.pars[['Bmax']]
control.pars[['B1']] <- 0.5 * control.pars[['Bmax']]

proj.hcr1.noerror <- evaluate(pars.iter, biomass.iter,
                      control.pars, data.years, proj.years,
                      niter, overshoot= 0)
proj.hcr1.error <- evaluate(pars.iter, biomass.iter,
                      control.pars, data.years, proj.years,
                      niter, overshoot=0.2)

#Increase the target Harvest ratio in the HCR
control.pars[['H1']] <- 0.15

proj.hcr2.noerror <- evaluate(pars.iter, biomass.iter,
                      control.pars, data.years,
                      proj.years, niter, overshoot = 0)
proj.hcr2.error <- evaluate(pars.iter, biomass.iter,
                      control.pars, data.years, proj.years,
                      niter, overshoot=0.2)

MSE <- rbind(cbind(proj.hcr1.noerror, HCR="hcr1",
                  implement="no overshoot"),
             cbind(proj.hcr1.error, HCR="hcr1",
                  implement="20% overshoot"),
             cbind(proj.hcr2.noerror, HCR="hcr2",
                  implement="no overshoot"),
```

```
                    cbind(proj.hcr2.error, HCR="hcr2",
                        implement="20% overshoot"))

Fig3.8 <- ggplot(data=subset(MSE, type != "index" &
            year %in% proj.years), aes(x = HCR, y = value,
            ymin = 0))
Fig3.8 + geom_hline(aes(yintercept = c(B2,B1)),
            data = data.frame(type = c("biomass",
            "catch"), B2 = c(control.pars[['B2']], NA),
            B1 = c(control.pars[['B1']], NA)))+
            geom_boxplot(aes(fill = implement), width = 1) +
            facet_wrap(~type, scale = "free_y") +
            ylab("Million tonnes") +
            scale_fill_grey(start = 0.5) + theme_bw()
```

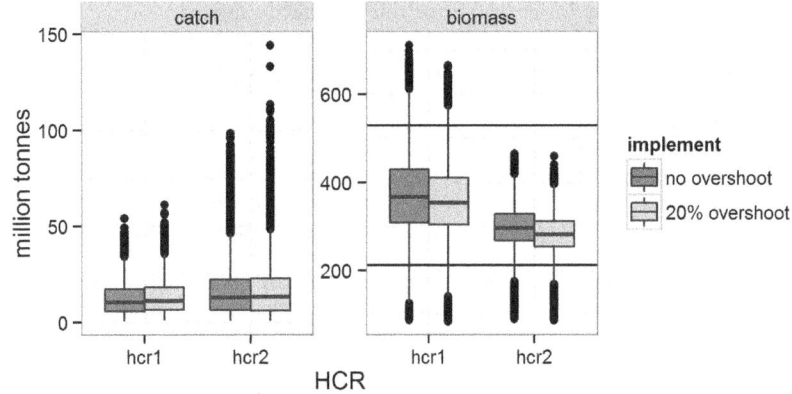

Figure 3.8 Comparison of the two harvest control rules combined with the possibility of having a consistent TAC overshoot in terms of catches (left panel) and stock biomass (right panel) in the projected years. Horizontal lines within the boxes indicate medians. The upper and lower hinges correspond to the first and third quartiles. The whiskers extend from the hinge to the highest value that is within 1.5 times the interquartile range. Horizontal lines in the stock biomass panel indicate U_{min} and U_{max}.

Acknowledgements

The management strategy evaluation used in this chapter was developed at the ICES Workshop on Redfish Management Plan Evaluation. The actual evaluation was simplified to allow for a better demonstration of the principles of simulations in MSE. We thank the participants of the workshop for their invaluable input in the development of the model. In addition, we thank Charles T.T. Edwards for his extensive contribution to the R code and both editors for their valuable comments to the text.

References

Andersen, B.S., Vermard, Y., Ulrich, C., Hutton, T. & Poos, J.-J. 2010. Challenges in integrating short-term behaviour in a mixed-fishery Management Strategies Evaluation frame: A case study of the North Sea flatfish fishery. *Fisheries Research*, 102, 26–40.

Arreguín-Sánchez, F. 1996. Catchability: A key parameter for fish stock assessment. *Reviews in fish biology and fisheries*, 6, 221–242.

Bartholomew, A. & Bohnsack, J.A. 2005. A review of catch-and-release angling mortality with implications for no-take reserves. *Reviews in Fish Biology and Fisheries*, 15, 129–154.

Basurto, X. & Ostrom, E. 2009. Beyond the Tragedy of the Commons. *Economia delle fonti di energia e dell'ambiente*, 52(1), 35–60.

Butterworth, D.S. 2007. Why a management procedure approach? Some positives and negatives. *ICES Journal of Marine Science: Journal du Conseil*, 64, 613–617.

Cotter, A.J.R., Burt, L., Paxton, C.G.M., Fernandez, C., Buckland, S.T. & Pax, J.X. 2004. Are stock assessment methods too complicated? *Fish and Fisheries*, 5, 235–254.

Fogarty, M., Mayo, R., O'brien, L., Serchuk, F. & Rosenberg, A. 1996. Assessing uncertainty and risk in exploited marine populations. *Reliability Engineering & System Safety*, 54, 183–195.

Fournier, D. & Archibald, C.P. 1982. A general theory for analyzing catch at age data. *Canadian Journal of Fisheries and Aquatic Sciences*, 39, 1195–1207.

Fulton, E.A., Smith, A.D.M., Smith, D.C. & van Putten, I.E. 2011. Human behaviour: The key source of uncertainty in fisheries management. *Fish and Fisheries*, 12, 2–17.

Garrity, E.J. 2011. System dynamics modeling of individual transferable quota fisheries and suggestions for rebuilding stocks. *Sustainability*, 3, 184–215.

Garthe, S., Camphuysen, K. & Furness, R. 1996. Amounts of discards by commercial fisheries and their significance as food for seabirds in the North Sea. *Marine Ecology Progress Series. Oldendorf*, 136, 1–11.

Gislason, H., Daan, N., Rice, J.C. & Pope, J.G. 2010. Size, growth, temperature and the natural mortality of marine fish. *Fish and Fisheries*, 11, 149–158.

Hamon, K.G., Thébaud, O., Frusher, S. & Little, L.R. 2009. A retrospective analysis of the effects of adopting individual transferable quotas in the Tasmanian red rock lobster, Jasus edwardsii, fishery. *Aquatic Living Resources*, 22, 549–558.

Hatcher, A., Jaffry, S., Thebaud, O. & Bennett, E. 2000. Normative and social influences affecting compliance with fishery regulations. *Land Economics*, 76, 448–461.

Hilborn, R. 1985. Fleet dynamics and individual variation – why some people catch more fish than others. *Canadian Journal of Fisheries and Aquatic Sciences*, 42, 2–13.

Hilborn, R. 2003. The state of the art in stock assessment: Where we are and where we are going. *Scientia Marina*, 67, 15–20.

Hilborn, R. & Walters, C.J. 1992. *Quantitative Fisheries Stock Assessment: Choice, Dynamics and Uncertainty*, Dortrecht, Springer.

ICES WKREDMP 2014. *Report of the Workshop on Redfish Management Plan Evaluation (WKREDMP)*, Copenhagen, Denmark.

Little, L.R., Grafton, R.Q., Kompas, T., Smith, A.D.M., Punt, A.E. & Mapstone, B.D. 2011. Complementarity of no-take Marine reserves and individual transferable catch quotas for managing the line fishery of the Great Barrier Reef. *Conservation Biology*, 25, 333–340.

Marchal, P., Ulrich, C., Korsbrekke, K., Pastoors, M. & Rackham, B. 2003. Annual trends in catchability and fish stock assessments. *Scientia Marina*, 67, 63–73.

Pella, J.J. & Tomlinson, P.K. 1969. *A Generalized Stock Production Model*, 13, 419–496, Inter-American Tropical Tuna Commission.

Poos, J.J., Bogaards, J.A., Quirijns, F.J., Gillis, D.M. & Rijnsdorp, A.D. 2010. Individual quotas, fishing effort allocation, and over-quota discarding in mixed fisheries. *ICES Journal of Marine Science*, 67, 323–333.

Pope, J.G. 2002. Input and output controls: The practice of fishing effort and catch management in responsible fisheries. *In:* Cochrane, K.L. (ed.) *A Fishery Manager's Guidebook: Management Measures and Their Application*. Food and Agriculture Organization of the United Nations.

Punt, A.E., Butterworth, D.S., De Moor, C.L., De Oliveira, J.A.A. & Haddon, M. 2014. Management strategy evaluation: best practices. *Fish and Fisheries*, n/a–n/a.

R CORE TEAM 2014. *R: A Language and Environment for Statistical Computing*, Vienna, Austria, R Foundation for Statistical Computing.

Rijnsdorp, A.D., Daan, N., Dekker, W., Poos, J.J. & van Densen, W.L.T. 2007. Sustainable use of flatfish resources: Addressing the credibility crisis in mixed fisheries management. *Journal of Sea Research*, 57, 114–125.

Schaefer, M.B. 1954. Some aspects of the dynamics of populations important to the management of the commercial marine fisheries. *Inter-American Tropical Tuna Commission Bulletin*, 1, 23–56.

Sparholt, H. 1990. Improved estimates of the natural mortality rates of nine commercially important fish species included in the North Sea multispecies VPA model. *Journal du Conseil: ICES Journal of Marine Science*, 46, 211–223.

Ulrich, C., Reeves, S.A., Vermard, Y., Holmes, S.J. & Vanhee, W. 2011. Reconciling single-species TACs in the North Sea demersal fisheries using the Fcube mixed-fisheries advice framework. *ICES Journal of Marine Science: Journal du Conseil*, 68, 1535–1547.

van Beek, F.A., van Leeuwen, P.I. & Rijnsdorp, A.D. 1990. On the survival of plaice and sole discards in the otter-trawl and beam-trawl fisheries in the North Sea. *Netherlands Journal of Sea Research*, 26, 151–160.

van Putten, I.E., Kulmala, S., Thebaud, O., Dowling, N., Hamon, K.G., Hutton, T. & Pascoe, S. 2012. Theories and behavioural drivers underlying fleet dynamics models. *Fish and Fisheries*, 13, 216–235.

Venables, W.N., Ellis, N., Punt, A.E., Dichmont, C.M. & Deng, R.A. 2009. A simulation strategy for fleet dynamics in Australia's northern prawn fishery: Effort allocation at two scales. *Ices Journal of Marine Science*, 66, 631–645.

Voss, R., Quaas, M.F., Schmidt, J.O., Tahvonen, O., Lindegren, M. & Moellmann, C. 2014. Assessing Social–Ecological Trade-Offs to Advance Ecosystem-Based Fisheries Management. *PLoS ONE*, 9(9), e107811.

Wildavsky, A. & Dake, K. 1990. Theories of risk perception – who fears what and why. *Daedalus,* 119, 41–60.

Wilen, J.E., Smith, M.D., Lockwood, D. & Botsford, L.W. 2002. Avoiding surprises: Incorporating fisherman behavior into management models. *Bulletin of Marine Science,* 70, 553–575.

Part II

Evaluating the feedback control of exploitation

Fisheries management science is a practical exercise, designed to meet the needs of resource managers who are in turn accountable to the social, economic and political bodies that they represent. Part 2 of this book describes the application of management science to nine case studies, selected to include a wide array of globally representative approaches. The precise techniques used are as diverse as the jurisdictions in which they are developed, reflecting institutional dogma, the complexities of the resource in question and scientific opinion on how the problem should be tackled. However at their core they retain a central focus on simulation-based evaluation.

Chapters 4, 5 and 6 provide a good introduction to how management science is applied. Chapter 4 concerns the Pacific hake resource (*Merluccius productus*), giving an example of how to parameterise and implement a generic model-based control rule, which provides an output fishing mortality given an estimated stock biomass. Chapter 5 details a similar approach for sablefish (*Anoplopoma fimbria*) in British Columbia, Canada. The evaluation of model-based control rules is complicated by the need to simulate the assessment process at each implementation of the management procedure (MP), and the authors of both chapters demonstrate how this can be done. In Chapter 6, an empirical rule is developed for management of red rock lobster (*Jasus edwardsii*) in New Zealand. Management procedures are now used to manage seven of the nine New Zealand lobster stocks, and the authors describe their history and effect, with a focus on one particular implementation. These MPs use catch per unit of effort as an input, assumed to reflect stock abundance, and output an allowable catch limit.

The calculation of sustainable catches in fisheries management requires an expectation of future recruitment. Spawning biomass can be used calculate this, but proxies such as long-term average recruitment are often used for stocks for which there is no observable stock-recruit relationship. This can present problems when managing a system in which environmental drivers display regime-like behaviour, and in Chapter 7 the authors test a regime-based harvest control rule (HCR) for snow crab (*Chionoecetes opilio*) in the eastern Bering Sea. This modifies expectations for future recruitment based on the current environmental regime and generates an appropriate fishing mortality, which can then be used to set the quota.

Chapter 8 provides an entry point into the considerable scientific debate surrounding the current state and future of tuna stocks worldwide, and the capacity of regional fisheries management organizations to manage the associated fisheries effectively. The authors focus on adoption by the Commission for the Conservation of Southern Bluefin Tuna (CCSBT) of a fully evaluated MP for the southern bluefin tuna (*Thunnus maccoyii*) stock. The MP uses abundance indices for both the juvenile and subadult populations in a unique hybrid structure with both model-based and empirical elements. This is the first time that a comprehensively evaluated MP has been adopted for an internationally managed tuna stock.

Moving from an oceanic scale to a regional inshore fishery, Chapter 9 details the development of an MP for walleye (*Sander vitreus*) on one of the Great Lakes that lie on the border between the United States and Canada. This case study is unusual in that it deals with a freshwater ecosystem, and the authors are able to describe in detail the unusually well-structured co-management process that allowed development and acceptance of an MP by commercial and recreational fishers in both countries.

Chapter 10 describes an example of effort-based management, unique to the case studies presented here, concerning the Northern Prawn Fishery in Australia. This is a multispecies and spatially complex fishery, managed according to the principle of maximum economic yield. Prawn fisheries are often criticised for the damage they can cause to the wider ecosystem, through the high incidental catch of nontarget species. The authors have been able to address this concern by modelling the impact of fishing on benthic species, and demonstrate how the fishery can be managed to be economically viable with minimal damage to the ecology of the region.

Multispecies systems are particularly hard to manage and represent a formidable challenge for the scientists charged with providing management advice. In the South African pelagic fishery, vessels target primarily adult sardine (*Sardinops sagax*) and juvenile anchovy (*Engraulis encrasicolus*). But juvenile sardine shoal together with anchovy, at least for the first part of their southward winter migration along the west coast of South Africa, and become an unavoidable bycatch in the anchovy-directed fishery. This bycatch impacts negatively on possible future adult sardine yields, requiring a trade-off decision for the pelagic fishery as directed anchovy and sardine catches cannot be simultaneously and independently maximised. In Chapter 11 the authors describe how these fisheries are managed jointly using a complex set of control rules, with the simulation-testing framework taking explicit account of this bycatch and the changing amounts of juvenile sardine shoaling with anchovy throughout the year.

Finally, in Chapter 12 the author describes one of the first applications of management strategy evaluation to a European fishery. North Sea haddock

(*Melanogrammus aeglefinus*) are a key component of the demersal fisheries of several northern European nations, and this chapter details evaluation of a model-based MP, first conducted during 2008, which contributed to widespread acceptance and implementation of the procedure. In this chapter the author was also able to consider a post hoc (and more descriptive) retrospective evaluation of the MP, which concluded that all of its main objectives had been achieved – although not necessarily due to the MP itself.

Chapter 4

Conservation and yield performance of harvest control rules for the transboundary Pacific hake fishery in US and Canadian waters

Allan C. Hicks, Sean P. Cox, Nathan Taylor, Ian G. Taylor, Chris Grandin and James N. Ianelli

Introduction

Pacific hake (*Merluccius productus*) are the most abundant groundfish in the California Current ecosystem off the west coast of the United States and Canada. Four large hake cohorts (1980, 1984, 1999, and 2010) have supported approximately 41% of the cumulative catch taken by US and Canadian hake fisheries since 1981. These cohorts created wide fluctuations in abundance as they passed through the fishable stock, yet the total catch of hake in US and Canadian fisheries has remained mostly between 250,000 and 360,000 mt per year since 1991 when the domestic vessels dominated the fisheries (Figure 4.1).

The Pacific hake fishery is currently managed as a single stock via an international agreement between the United States and Canada.[1] The agreement defines a default harvest control rule, percentages of the total allowable catch (TAC) allocated to each country (73.88% US and 26.12% Canada), and four committees that collectively guide the stock assessment and harvest decision processes. The harvest control rule consists of an $F_{40\%}$ target fishing mortality rate combined with a 40:10 linear adjustment on catch (Equation H4.1; see Table 4.2) that reduces the target fishing mortality rate from $F_{40\%}$ when spawning stock biomass is above 40% of its equilibrium unfished level to zero when the spawning stock biomass is below 10% of its unfished level. $F_{40\%}$ is the fishing mortality rate that would result in a 40% spawning potential ratio (SPR) or spawning biomass per recruit that is 40% of the unfished level. This is determined by calculating equilibrium spawning biomass-per-recruit at different fishing mortality rates given the parameters and assumptions of the model (Methot and Wetzel 2013). This SPR approach is used because the stock-recruitment relationship for Pacific hake, along with F_{MSY} and B_{MSY} reference points, cannot be estimated reliably from historical data. To simplify the notation in the rest of this chapter, we refer to the $F_{40\%}$ with a 40:10 adjustment harvest control rule as the $F_{40\%}$ rule.

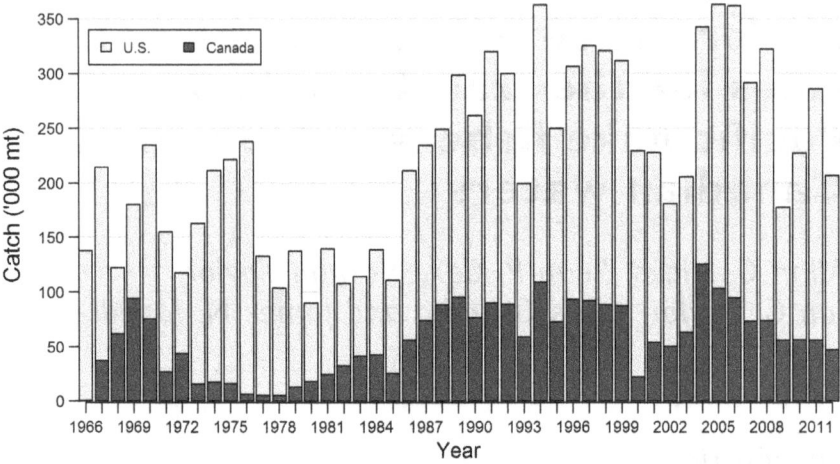

Figure 4.1 Annual catch (thousands of mt) taken from US waters (light grey) and Canadian waters (dark grey).

Although the default $F_{40\%}$ rule is consistent with both US and Canadian fisheries policy, as well as fishery eco-certification standards, the range of realized TACs over the last decade is substantially constrained compared to TACs output from the harvest control rules applied (Figure 4.2). Specifically, it appears that large TACs suggested by harvest control rules have been consistently adjusted downward to possibly reflect more conservative stakeholder, management, and business objectives. Although hake TACs have historically been more conservative than required under the $F_{40\%}$ rule, it is difficult to determine at this time whether these decisions have sacrificed potential yield or have protected the hake stock against survey and stock assessment errors. Furthermore, it is unclear how TACs in the observed 180,000–375,000 mt range would affect conservation and yield performance in the future. For instance, if annual assessment model biomass estimates are positively biased, then the 180,000–375,000 mt range may not be as conservative as it appears, despite an apparently successful fishery up to this point. Finally, we are uncertain how management will behave when the $F_{40\%}$ rule would suggest TACs below 200,000 mt, since we have no empirical experience with this range of values (Figure 4.2).

Since 2012, hake fishery scientists (via the Joint Technical Committee), the hake fishing industry (via the Joint Management Committee and Advisory Panel), and fishery managers from the US and Canada have collaborated on the design and implementation of a management strategy evaluation (MSE) process aimed at defining objectives of the fishery and performance metrics to evaluate those objectives, as well as developing a better understanding of the short- and long-term implications of harvest control rules for Pacific hake. In particular,

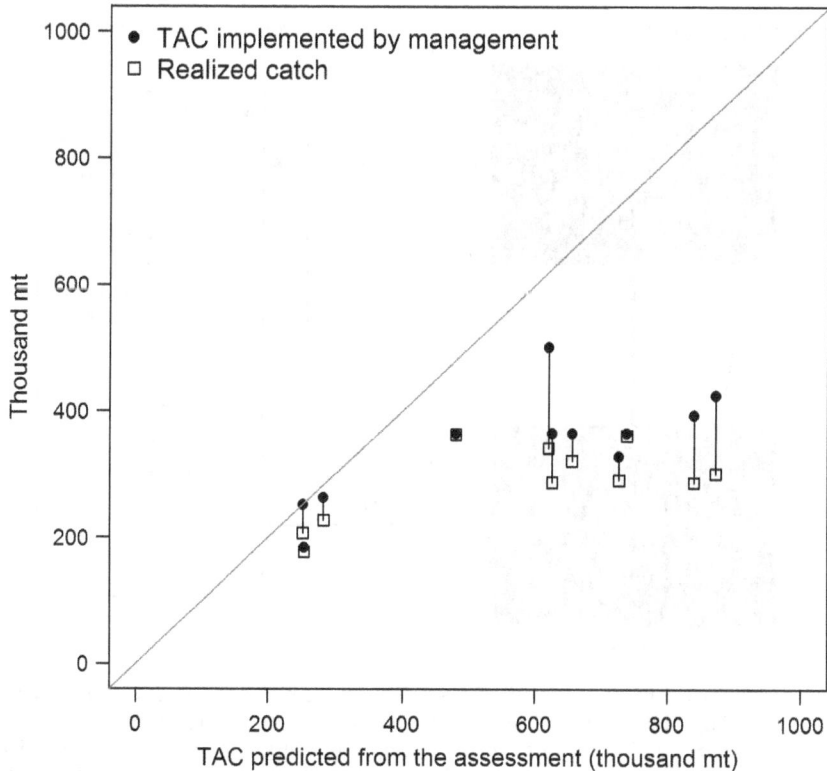

Figure 4.2 Relationship between the actual TACs set by management in specific years (closed circles) or the realized catch (open squares) and the total allowable catch (TAC) outputs from historical management procedures (2004–2014).

evaluating the expected conservation performance of the harvest strategy is required as per condition of the Pacific hake fishery eco-certification under the Marine Stewardship Council (Marine Stewardship Council 2015). Large-scale acoustic/midwater-trawl surveys for Pacific hake are also expensive, consuming substantial portions of funding and human resources available for fisheries surveys in both the US and Canada. Quantifying the effects of cost reductions via less frequent surveys requires completing the inference chain from surveys to harvest advice; harvest control rules are a key link in this chain (Figure 4.3). Finally, the Canadian hake industry is particularly concerned that a shift in hake age structure toward younger fish limits the biomass of older hake making their way into Canadian waters (Bailey, et al. 1982). These initial MSE results are the beginning of a better understanding of the overall fishery management system

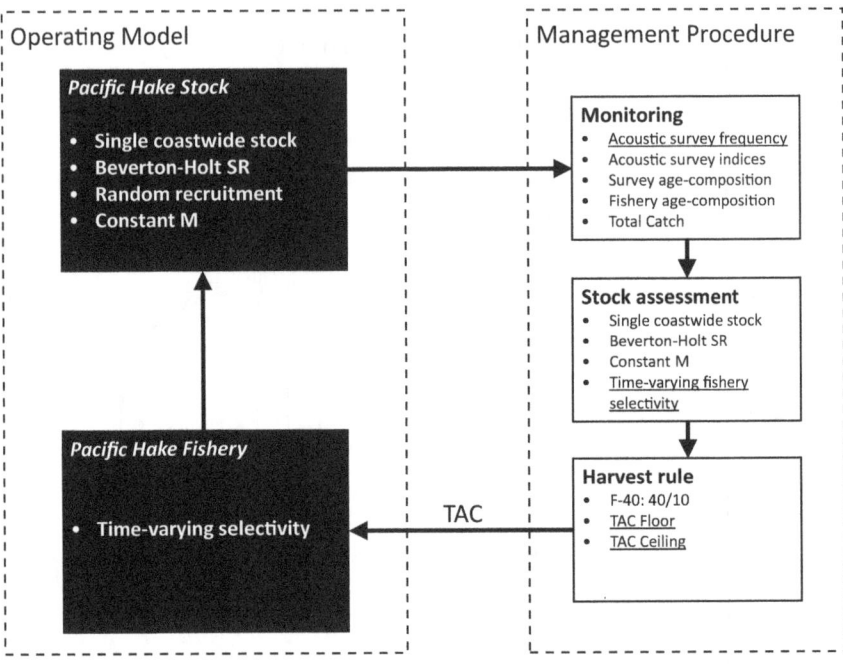

Figure 4.3 Structure of the Pacific hake fishery system simulation test. The Operating Model (left) represents the biological functioning of the Pacific hake stock and the process driving temporal changes in fishery selectivity. The Management Procedure (right) specifies the flow of information from raw data collection through the Stock Assessment and Harvest Rule to determine the total allowable catch (TAC) by the fishery. Population dynamics models of Pacific hake occur in both the Operating Model and in the Stock Assessment.

that will inform managers if further exploration of data collection programs and harvest control rules could improve catch advice and allow each country the best possible opportunity to fully access their portion of the TAC.

The decision rules

In this chapter, we evaluate how interactions between hake population dynamics and stock assessment model biases affect conservation and yield performance of four harvest control rules: (1) the $F_{40\%}$ harvest control rule (Equation H4.1); (2–3) partially constrained $F_{40\%}$ rules in which TACs ceilings are limited to either (2) 375,000 mt or (3) 500,000 mt; and (4) a fully constrained $F_{40\%}$ rule with a TAC floor of 180,000 mt and ceiling of 375,000 mt, similar to the realized range of TACs based on historical decisions. Ishimura et al. (2005)

evaluated the Pacific hake "40–10 harvest strategy," but they did not explicitly incorporate future data or assessment model biases into their management procedure evaluations. Here, we use Stock Synthesis software (SS version 3.24s; Methot and Wetzel 2013) as both a stochastic operating model and an age-structured stock assessment model for annual biomass estimation. The operating model is able to include processes not typically found in an assessment model (e.g. time-varying selectivity). This allows for the evaluation of inaccuracies in stock assessment model biomass estimates which propagate to TACs via the harvest control rules. We also include a rather unrealistic "Perfect" $F_{40\%}$ rule in which the spawning and exploitable biomasses used in the rule are known exactly (i.e. no assessment model). Performance under this rule provides a base line from which to evaluate the effects of stock assessment model errors in the presence of observation errors (in surveys and age-composition) and large recruitment variability.

Constant recruitment and time-invariant fishery age-selectivity assumptions involved in estimation of the $F_{40\%}$ fishing mortality rate are both violated for the Pacific hake fishery, which means that consistently applying $F_{40\%}$ is not guaranteed to reduce spawning stock biomass to 40% of the unfished level. Therefore, we also conducted a series of simulations to examine the relationship between spawning stock biomass relative to the unfished level (i.e. relative spawning biomass) and harvest control rules that consistently apply $F_{X\%}$, where $X = 30\%$, 32%, . . ., 50% (i.e. from $F_{30\%}$ to $F_{50\%}$ by 2% increments).

Evaluation methods

We used a closed-loop simulation approach to evaluate hake population and fishery responses to alternative harvest control rules (Figure 4.3). Hake population dynamics were represented in an operating model (OM) that was similar in structure to the stock assessment model used in the simulated hake management procedure. The primary difference between the two is that the OM generates fishery age-composition data based on a time-varying selectivity model, whereas the assessment assumes that fishery selectivity is constant over time. OM simulations resulted in variability in stock trajectories that were similar to historical observations. The closed-loop simulation proceeded as follows.

1   Draw a set of OM parameters from the posterior distribution of an SS model using Pacific hake data up to and including 2012 and estimating time-varying fishery selectivity for all past and future years. Simulations begin in 2013 with a catch of 365,112 mt, based on the 2013 TAC decision.

2   Simulate biennial acoustic survey biomass and age composition, and fishery age-composition data that were generally comparable to the real data collection system ("Monitoring" in Figure 4.3).

3   Fit the SS assessment model to acoustic survey, age-composition, and catch data ("Stock assessment" in Figure 4.3).

4   Determine a TAC by applying the default $F_{40\%}$ harvest strategy to the maximum posterior density estimate of the spawning biomass from the stock assessment ("Harvest rule" in Figure 4.3).
5   Apply floor and ceiling constraints on TACs, if applicable.
6   Input TAC to the OM, assuming that the entire TAC was taken by the fishery.
7   Project Steps 1–6 forward for 30 years.
8   Repeat Steps 1–7 999 times.
9   Repeat this process using the same parameter vectors from the operating model posterior distribution for each decision rule that was tested.

Operating model

Stock Synthesis (Methot and Wetzel 2013) was used as the operating model to simulate the hake population dynamics and was similar to the 2013 stock assessment model for Pacific hake (Hicks et al. 2013), with the addition of time-varying selectivity in the fishery for all past and future years (Tables 4.1 and 4.2). The model equations are described in Methot and Wetzel (2013), but notable equations are repeated here in Table 4.2. A Bayesian posterior distribution was reestimated in SS using the same data as the 2013 assessment (1966–2012), where time-varying selectivity was modelled with age- and year-specific selectivity parameters (Equations O4.10–O4.12). For simulated selectivity-at-age in future years, parameters were randomly generated from a multivariate normal distribution with the covariance matrix derived from historical (1966–2012) estimates. OM simulations produced variability in stock trajectories that were greater than those predicted by the formal 2013 assessment.

Table 4.1 Notation used to describe the management strategy evaluation for Pacific hake. Values are given for parameters that were fixed. Parameters that were drawn from a posterior distribution conditioned to data are noted by a range of values or by "Distribution" when a range would not be informative.

Symbol	Value	Description
a		Age subscript
t		Time (i.e. year) subscript
l		Fleet subscript for fishery or survey
A	20	Plus-group age
$F_{X\%}$		The fishing mortality rate that would reduce the spawning potential ratio (SPR) to X% of the unfished level
C_t		Total catch in year t
$N_{t,a}$		Number-at-age (a) in year t

Symbol	Value	Description
R_0	(1.25–7.09) billion	Unfished equilibrium recruitment
R_t		Recruitment in year t
$\tilde{R}_t$	$\sim N\left(0, \sigma_R^2\right)$	Recruitment deviation in year t
M	(0.167–0.310)	Natural mortality
h	(0.415–0.995)	Steepness in the stock-recruit relationship
$V_{t,l}$		Biomass vulnerable to fleet l (survey or fishery) at the beginning of year t
B_0		Unfished equilibrium spawning biomass
B_t		Spawning biomass
F_t		Fishing mortality (harvest rate) in year t
$Z_{t,a}$		Total mortality (fishing and natural) at age in year t
$w_{t,a}$		Weight-at-age (kg) in year t
$\bar{w}_{t,a}$		Mean weight-at-age (kg), averaged over the years 1975–2012
c		Sex ratio of age-0 recruits
$S_{t,l,a}$		Selectivity-at-age for fleet l (fishery or survey)
$S'_{t,l,a}$		Untransformed selectivity at age for each fleet in year t
$\rho_{l,a}$	Distribution	Selectivity parameters
$p_{t,l,a}$		Time-varying selectivity parameter
A_{min}	1 or 2	Minimum age where selectivity > 0 (age 1 for fishery, age 2 for survey)
$\Phi_{fem,a}$		Maturity-at-age for females
f_a	See Hicks et al. (2013)	Fecundity-at-age (mean weight multiplied by maturity for females only)
Q	(0.5–1.5)	Catchability coefficient for the survey biomass index
σ_I	0.42	Standard error of the survey index
σ_R	1.4	Recruitment variability (standard deviation)
φ^2	0.2	Variance of Gaussian penalty on selectivity parameters
Ω	See Hicks et al. (2013)	Aging error without cohort effects
I		Middle of the year biomass index from the survey
$n_{t,l,a}$		Numbers-at-age for fleet l

Harvest Control Rule

H4.1

$$C_t = 0 \qquad\qquad \frac{B_t}{B_0} < 0.10$$

$$C_t = F_{X\%}V_{t,\,fishery}\left[\left(B_t - 0.1B_0\right)\left(\frac{0.4B_0\,/\,B_t}{0.4B_0 - 0.1B_0}\right)\right] \qquad\qquad 0.10 \le \frac{B_t}{B_0} \le 0.40$$

$$C_t = F_{X\%}V_{t,\,fishery} \qquad\qquad \frac{B_t}{B_0} > 0.40$$

State dynamics

O4.1

$$N_{t=0,a} = cR_0 e^{-aM} \qquad a = [0,1,\ldots,3A-1]$$

O4.2

$$N_{t=0,A} = \sum_{a=A}^{3A-1} N_{0,a} + \frac{N_{0,3A-1}e^{-M}}{1 - e^{-M}}$$

O4.3

$$N_{t+1,a} = \begin{cases} cR_{t+1,0} & a = 0 \\ N_{t,a-1}e^{-Z_{t,a}} & 1 \le a \le A-1 \\ N_{t,A-1}e^{-Z_{t,A-1}} + N_{t,A}e^{-Z_{t,A}} & a = A \end{cases}$$

O4.4

$$R_t = \frac{4hR_0B_t}{B_0(1-h) + B_t(5h-1)} e^{-0.5\sigma_R^2 + \tilde{R}_t} \qquad \tilde{R}_t \sim N\left(0, \sigma_R^2\right)$$

Biomass

O4.5

$$V_{I,t} = \sum_{a=0}^{A} N_{t,a}S_{t,I,a}w_{t,a}$$

O4.6

$$B_t = 0.5\sum_{a=0}^{A} N_{t,a}f_a$$

O4.7

$$f_a = \overline{w}_a \Phi_{fem,a}$$

Mortality and selectivity

O4.8

$$F_t = C_t \Big/ V_{t,\,fishery}$$

O4.9

$$Z_{t,a} = M + S_{t,\,fishery,a}F_t$$

O4.10

$$S_{t,I,a} = e^{S'_{t,I,a} - S'_{t,I,max}}$$

O4.11

$$S'_{t,I,a} = \sum_{i=A_{min}}^{a} p_{t,I,i} \qquad a \le 6$$

O4.12

$$p_{t,I,a} = p_{I,a} + \varepsilon_{t,a} \qquad \varepsilon_{t,a} \sim N\left(0, \varphi^2\right)$$

Data generation

D4.1

$$I_t = qV_{survey,t} e^{\varepsilon_t - 0.5\left(\sigma_I^2 + M\right)} \qquad \varepsilon_t \sim N\left(0, \sigma_I^2\right)$$

D4.2

$$n_{t,I,a} = N_{t,a}S_{t,I,a}\Omega$$

For each of 1,000 closed-loop simulation replicates in the evaluation, we sampled a vector of parameters from the Bayes posterior distribution consisting of fishery and acoustic survey selectivity-at-age ($S_{l,a}$ where l indexes either the fishery or survey), survey catchability (q), natural mortality (M), steepness (h), unfished equilibrium recruitment (R_0), and annual recruitment deviations. This sampling approach accounted for the covariance structure among these parameters.

Data generation

The Pacific hake acoustic survey provides the only fishery-independent index of abundance and age composition. Survey abundance index and age-composition data for projection years 2014–2042 were generated biennially (on even years), which is similar to the data collection frequency for Pacific hake stock assessments. The acoustic survey index of abundance, I_y, for year y was assumed to follow a log-normal distribution as in Equation D4.1. The natural mortality adjustment (0.5) is used because the survey happens approximately midway through the calendar year. The survey biomass is calculated using Equation 4.5, where $w_{l,a}$ for future years is the average hake weight-at-age between 1975 and 2012 (Hicks et al. 2013).

Proportion-at-age data for the fishery (annual) and survey (biennial) were simulated from a multinomial distribution with sampling probabilities proportional to Equation D4.2, where l indexes either the survey or fishery and Ω is the estimated aging error matrix. Sample sizes for the fishery and survey were set equal to the effective sample sizes used in the 2013 Pacific hake assessment (Hicks et al. 2013). The aging error matrix contains the probabilities of assigned ages for each true age, where the probabilities are determined from a normal distribution centered on the true age with standard deviation increasing with true age (Hicks et al. 2013).

The plus-group age, A is 15 years for data generation. Weight-at-age and maturity-at-age were identical in the operating and assessment models.

Management procedures

A management procedure is defined as a combination of data (acoustic survey, fishery age-composition, and survey age-composition), stock assessment model, target fishing mortality rate ($F_{x\%}$), and harvest control rule, which all combine to determine annual TACs. The harvest control rules we investigated, including the "Perfect" case, used the 40:10 adjustment (Equation H4.1) as defined in the hake agreement.

Assessment model

The assessment model, also configured in SS, was used to estimate spawning stock and exploitable biomass by fitting historical and simulated index and age-composition data. The model mimicked the 2013 hake assessment (Hicks et al. 2013), and differed from the OM by assuming time-invariant selectivity. Estimates from the simulated data sets were obtained by maximizing the joint Bayes posterior density.

Simulated stock assessments were performed in each future projection year. To help model convergence, we initialized the assessment model parameters at values estimated in the previous year. In some simulations, gradient-based assessment convergence criteria were still unacceptable (i.e. the global minimum may not have been found), so we made small random perturbations to the initial parameters and repeated the assessment. If convergence was not achieved within three attempts the assessment was adopted, despite lack of clear convergence.

Performance measures

Ten conservation and yield performance metrics were developed based on Joint Technical Committee (Hicks et al.) consultations with industry and government, as well as from criteria implied by the default harvest strategy defined in the agreement. For a given metric, we compute central tendency measures as median average values – median (over 1,000 simulation trials) of the average (over projection years t_1 to t_2 within a simulation trial) – and risk measures as probabilities of particular events occurring over the period t_1 to t_2. Probabilities (reported as percentages) are derived from the proportion of times an event occurs, or a condition is met (e.g. the fishery is closed), over all simulations within a projection period.

Conservation metrics include (*i*) median average depletion and probabilities of spawning biomass being (*ii*) below 10% of the unfished equilibrium level (B_0), (*iii*) between 10% and 40% of B_0, and (*iv*) above 40% of B_0. Thresholds of 10% and 40% were chosen because they are the default endpoints of the 40:10 adjustment defined by the agreement.

Yield metrics are the median average catch, the average annual variability in catch (AAV; Cox and Kronlund 2008), the probability that the fishery is closed (i.e. catch = 0), and the probability that catch is above or below thresholds of 180,000 and 375,000 mt. The AAV metric, which provides a measure of the average year-to-year change in TACs, is computed via

$$AAV = \frac{\sum_{i=t_1}^{i=t_2} |C_{i+1} - C_i|}{\sum_{i=t_1}^{i=t_2} |C_i|},$$

where t_1, t_2 are the first and last years defining the period over which performance is measured and C_i is the TAC output for the harvest control rule applied in year i.

We measured performance of each harvest control rule over short- ($t_1 = 2014, t_2 = 2023$) and long-term ($t_1 = 2033, t_2 = 2042$) periods. Short-term performance is highly dependent on the starting conditions in 2014. Long-term performance statistics provide insight into the equilibrium performance related to conservation and sustainability objectives, as long as the operating model is representative of future events (e.g. recruitment variability).

Outcomes and discussion

Perfect knowledge simulations

Perfect $F_{40\%}$ rule

The initial conditions of the operating model, which included a large 2010 cohort estimated in the 2013 stock assessment, led to short-term simulation outcomes that were characterized by high spawning biomasses, high average catch outcomes, and probabilities less than 5% that the stock may decline lower than 10% of the unfished equilibrium level (Table 4.3). This result was consistent across all management procedures, but is best illustrated by the Perfect $F_{40\%}$ rule, which applied the $F_{40\%}$ control rule with the 40:10 adjustment to the true exploitable biomass, and shows how the initial condition effects influence outcomes in the absence of stock assessment errors or TAC constraints (Figure 4.4a and 4.4f). Under this management procedure, there was approximately a 1% chance that the fishery would be closed or that spawning biomass would be reduced below 10% of the unfished level. Variability from year to year in the catch was high, with a median AAV = 32% measured over the short term or over the long term. The AAV from realized past catches (1991–2012), given management constraints and less than 100% utilization, is 21%.

Median relative spawning biomass under the Perfect $F_{40\%}$ rule was approximately 26% (Table 4.3) measured over the long term with a median long-term average catch of 242,000 mt. The simplifying assumptions used to calculate $F_{40\%}$ for this highly variable population leads to a fishing mortality rate which is expected to deplete Pacific hake to a level that is almost overfished by past Pacific Fishery Management Council Standards for Pacific hake, on average.

Perfect $F_{X\%}$ rules

Substituting alternative SPR targets for 40% in the Perfect $F_{X\%}$ harvest control rule had greater effects on spawning biomass than catch (Figure 4.5). Median relative spawning biomass declined, as expected, as the target fishing mortality rate increase from $F_{50\%}$ to $F_{30\%}$ (lower $X\%$ implies higher F). In general, realized mean relative spawning biomass was considerably lower than implied by $X\%$. Median long-term average catch was relatively insensitive to alternative SPR values, ranging from 242,000 mt at $F_{SPR=30\%}$ to 230,000 mt at $F_{SPR=50\%}$, with a slight peak at 243,000 mt using $F_{SPR=38\%}$. On the other hand, interannual variability in catch measured by AAV was highly sensitive to changes in $F_{X\%}$ mainly because maintaining spawning

Table 4.3 Conservation and yield performance for the $F_{40\%}$ with 40:10 adjustment harvest control rule combined with alternative floor and ceiling TAC constraints. The "Perfect" rule computes annual TACs from the $F_{40\%}$ with 40:10 adjustment combined with the true operating model biomass. The remaining harvest control rules use annual stock assessment model biomass estimates in computing TACs. The base $F_{40\%}$ rule sets annual TACs to the exact output from the $F_{40\%}$ rule with 40:10 adjustment. The alternatives further constrain TACs to lie within the indicated floor and ceiling bounds. Performance rankings are given in bold for selected indicators. The overall average rank (last row) is calculated by first averaging the three yield performance rankings (metrics used in ranking are given in italic font) and then averaging the result with the Conservation rankings. For ranking conservation, all probabilities less than or equal to 5% were ranked 1.0, while values greater than 5% were ranked in order beginning with rank 2. All catch units are thousands of metric tons.

	Short term (2014–2023)					Long term (2033–2042)				
	Perfect	F_{40}	$F_{40}{:}0{-}500$	$F_{40}{:}0{-}375$	$F_{40}{:}180{-}375$	Perfect	F_{40}	$F_{40}{:}0{-}500$	$F_{40}{:}0{-}375$	$F_{40}{:}180{-}375$
Conservation										
Median average depletion	37%	51%	56%	62%	61%	26%	39%	42%	45%	35%
$Pr(B < B_{10\%})$	1%	5%	3%	2%	4%	2%	6%	5%	5%	19%
$Pr(B_{10\%} \le B \le B_{40\%})$	59%	39%	32%	28%	27%	77%	48%	47%	44%	41%
$Pr(B > B_{40\%})$	40%	57%	64%	70%	69%	21%	45%	49%	51%	41%
Rank		**1.00**	**1.00**	**1.00**	**1.00**		**2.00**	**1.00**	**1.00**	**3.00**
Yield										
Median average catch	413	388	368	335	344	242	199	203	216	233
Median AAV	32%	53%	28%	15%	9%	32%	52%	41%	34%	19%
Pr(catch = 0)	<1%	10%	8%	6%	0%	1%	13%	12%	10%	0%
Pr(catch < 180)	24%	32%	23%	16%	5%	44%	52%	50%	44%	21%
Pr(180 ≤ catch ≤ 375)	30%	19%	16%	84%	95%	31%	27%	25%	56%	79%
Pr(catch > 375)	46%	49%	61%	0%	0%	25%	21%	26%	0%	0%
Rank		**2.67**	**3.00**	**2.67**	**1.67**		**3.67**	**3.33**	**2.00**	**1.00**
Average rank		**1.83**	**2.00**	**1.83**	**1.33**		**2.83**	**2.17**	**1.50**	**2.00**

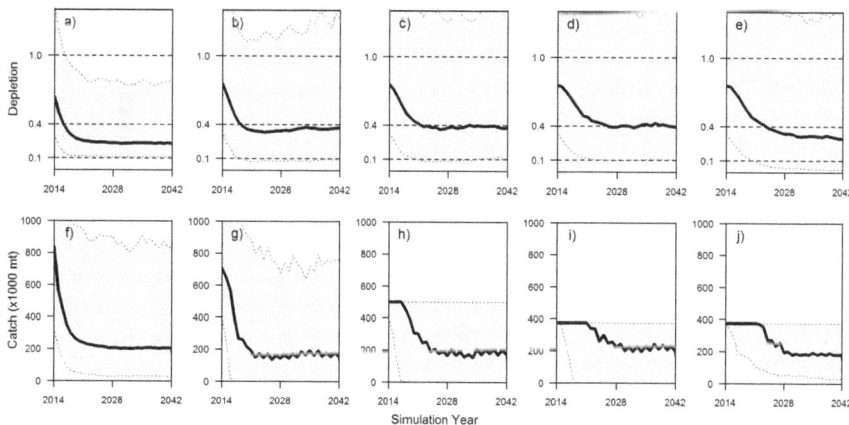

Figure 4.4 Time series of median depletion (top row) and catch (bottom row)
with 90% intervals (shaded regions) for the cases shown in Table 4.3
(columns from left to right: Perfect, F40, F40:0–500, F40:0–375,
F40:180–375).

biomass well above 40% of B_0 reduced the frequency of downward adjustments in
fishing mortality required when biomass was less than 40% of B_0.

Assessment simulations

Unconstrained $F_{40\%}$ rule with a 40:10 adjustment

The range of simulated spawning biomass and catch trajectories was considerably wider when the $F_{40\%}$ rule was computed from simulated stock assessment
estimates. Therefore, conservation and yield performance were both substantially different compared to those of the Perfect $F_{40\%}$ rule (Table 4.3; short
term). Although the median short-term spawning biomass was higher for $F_{40\%}$
(51%) compared to the Perfect $F_{40\%}$ rule (37%), the chance of spawning biomass
declining below 10% of B_0 increased from 1% to 5% and the chance of a fishery closure increased from less than 1% to 10%. Uncertainty in the application
of the $F_{40\%}$ rule also cost the fishery 25,000 mt in short-term median average
annual yield compared to the Perfect $F_{40\%}$ rule. Long-term metrics showed
similar effects (43,000 mt) of stock assessment errors on conservation and yield
performance of unconstrained $F_{40\%}$ rules (Table 4.3; long term).

In contrast to the Perfect $F_{40\%}$ rule, which produced long-term median
spawning biomass depletion of approximately 26%, the $F_{40\%}$ rule computed
from stock assessments produced median average depletion of 39% (Table 4.3),
which is surprisingly close to the target value and indicates that the assessment
model often underestimates biomass, resulting in lower catches. These stock
assessment errors also led to more frequent adjustments to the catch determined

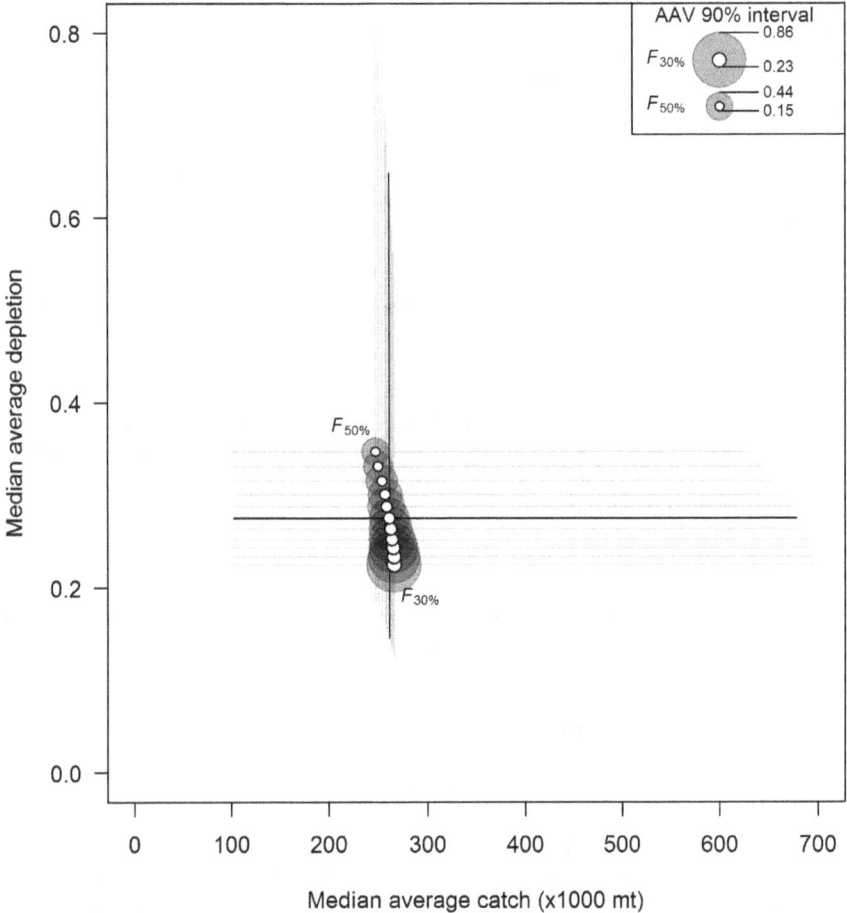

Figure 4.5 Trade-off between depletion, catch, and AAV for alternative SPR targets ($F_{30\%}$ to $F_{50\%}$) when managing with perfect information. The 90% intervals for depletion and catch are shown as lines. The 90% interval for AAV is represented by the shaded area of the circle, where the larger circles indicate more annual variability in catches. The dark lines show the current $F_{40\%}$ harvest strategy.

from the target fishing mortality rate (both upward and downward) as indicated by the greater interannual variability in yield (long-term AAV = 52%).

Modified $F_{40\%}$ harvest control rules: catch ceilings

Constraining the upper limit to annual TACs via ceilings increased both the median and range of simulated hake spawning biomass over the short and long term (Figure 4.4c, d, and e), decreased expected short-term yield, and increased the long-term expected yield (Figure 4.4h, i, and j). The 500,000 mt ceiling used in the $F_{40\%}$:0–500 rule generally improved all conservation and yield performance metrics over the $F_{40\%}$ rule, although most of the differences were relatively minor (Table 4.3). The 375,000 mt ceiling, which is closer to the apparent ceiling implied by historical decisions and represented in the $F_{40\%}$:0–375 rule, had more substantial improvements over $F_{40\%}$ alone. For instance, median average long-term relative spawning biomass and median average yield increased 15% (39% to 45%) and 8.5% (199,000 mt to 216,000 mt), respectively, compared to unconstrained $F_{40\%}$ (Table 4.3). Interannual variability in yield (AAV) was also substantially lower in both the short and long term. Most notably, long-term AAV for $F_{40\%}$:0–375 (34%) was nearly equal to that obtained from the Perfect $F_{40\%}$ rule (Table 4.3), essentially offsetting the effects of stock assessment errors on annual variability in yield.

Modified $F_{40\%}$ harvest control rules: catch ceiling and floor

Constraining hake TACs to the 180,000 to 375,000 mt range ($F_{40\%}$: 180–375) improved upon all the yield performance metrics of $F_{40\%}$:0–375, while decreasing performance on all the conservation metrics (Table 4.3). Perhaps the most striking difference is the increase from 5% to 19% in probability of the stock falling below 10% of unfished equilibrium biomass, which is a result of allowing at least 180,000 mt of catch regardless of whether the stock assessment suggests that it should be lower.

Discussion

Within an ideal fishery management strategy, harvest control rules provide a repeatable way to link information about fish stock abundance to the catch limits available to commercial fisheries. But in real fisheries, harvest control rules are not always implemented so literally. For the Pacific hake fishery off the west coast of North America, the $F_{40\%}$ default harvest control rule defined under the Pacific hake treaty has rarely been applied in practice. Instead, most TACs historically adopted by decision-makers have actually been lower than those computed via strict application of the rule, implying that decision-making has been influenced by other motives than the Pacific hake treaty. Whether this was justified is unclear, and perhaps several years from now we will have a more complete understanding of past recruitment dynamics that will allow us to examine the implied optimality of historical decisions, as Martell, et al. (2008) did for highly variable Pacific salmon fisheries. Instead, our investigation involved a prospective evaluation of this and other related management procedures to inform continued management of the resource.

In our simulations, limiting TACs to levels below those suggested by the $F_{40\%}$ rule (sometimes substantially) benefitted both long-term fishery yield and spawning biomass conservation. When taken literally, the $F_{40\%}$ rule (with 40:10 adjustment) led to the highest interannual variability in yield, as well as the highest probability that biomass would be below biological reference points such as $B_{40\%}$ and $B_{10\%}$. Although these greater risks led to 20,000–50,000 mt more short-term yield, they were also associated with long-term yields that were 4,000–34,000 mt lower than any other harvest control rule we tested. In contrast, management procedures that included catch ceilings, more closely mimicking the realized pattern of hake TACs, had higher long-term median average catch, maintained higher spawning biomasses with negligible risks of being below $B_{10\%}$, and generated lower annual variability in catch. A catch ceiling of 375,000 mt combined with a catch floor set to the industry-preferred minimum TAC of 180,000 mt (i.e. the $F_{40\%}$:180–375 rule) had the best long-term yield performance, but generated a substantially greater long-term risk (19%) of biomass depletion being below $B_{10\%}$. The same ceiling with a floor equal to zero (i.e. $F_{40\%}$:0–375) sacrificed only 8% in median long-term average yield with only 5% risk of depleting spawning biomass below $B_{10\%}$. These results indicate that a ceiling for the TAC may be prudent, but imposing a lower limit might threaten conservation objectives.

A crude ranking of performance based on the probability of spawning biomass depletion below $B_{10\%}$ combined with and average rank of three yield performance metrics (Table 4.3) ranked the $F_{40\%}$:0–375 rule second (tied with $F_{40\%}$) in overall short-term performance and first in overall long-term performance. This is interesting because it suggests that the current decision-making procedures leading to realized hake TACs are probably better in the face of large uncertainties in estimated hake abundance and recruitment than the default $F_{40\%}$ rule defined in the Pacific hake treaty. Of course, our conclusion that $F_{40\%}$:0–375 performed best in the simulations is predicated on using this rule consistently over all levels of estimated hake abundance. For making real TAC decisions for the hake fishery, this would mean applying the $F_{40\%}$ rule literally whenever the TAC calculation is below 375,000 mt, which is an event that occurred few times in the last decade (Figure 4.2). Even though the evidence from each of these events is reasonably consistent with $F_{40\%}$:0–375 behavior, other reasons, such as recent trends in spawning biomass, could result in managers deviating from this rule.

It is unclear how hake decision-makers modify stock assessment advice to arrive at the realized TACs. Each year there are probably many biological, economic, and social factors to consider. For instance, uncertainty in spawning and exploitable biomass estimates, the magnitude of recent estimates of recruitment, and the history of retrospective stock assessment errors typically cast some doubt on the TACs output from the $F_{40\%}$ rule. The industry Advisory Panel and Joint Management Committees must also consider the economic implications of TACs and how they will be apportioned among shore-based trawl fleets, offshore factory trawl fleets, and indigenous peoples that all operate within both countries. Each of these fleets experiences different bycatch constraints, hake availability, market prices, and fleet capacity. Finally, setting annual hake TACs requires consensus among a diverse mix of stakeholders and managers from two countries, so it is possible that different objectives may also play a role. For instance, some Canadian industry members are concerned that a hake age-distribution skewed towards younger fish may limit their ability to extract the Canadian portion of the TAC because hake tend to move north as they get older. Setting TACs well below those output from the $F_{40\%}$ rule in recent years could be the result of a compromise aimed at maintaining goodwill between USA and Canadian stakeholders.

We found Stock Synthesis to be a useful tool for evaluating Pacific hake management procedures because it allowed us to investigate the risks and benefits of alternative harvest control rules while taking into account key stock assessment uncertainties and feedbacks within the management system. Furthermore, by making some relatively simple modifications to the general Stock Synthesis package, we were able to spend less time setting up closed-loop simulations and more time working with stakeholders and managers. In particular, we are able to continue spending the necessary effort communicating the benefits of an MSE approach to fisheries research for the Pacific hake fishery. Our examination of alternative harvest control rules in this chapter provides the foundation for future simulation research as we peel away layers of uncertainty surrounding interactions between, for instance, Pacific hake biology and movement, stock assessment errors, fishery dynamics, and climate. Incorporating these processes into Stock Synthesis models provides an objective way to examine how these processes might affect the future robustness of Pacific hake harvest strategies.

Note

1 US Department of State, *Fisheries – Agreement Between the United States of America and Canada* (Seattle, WA: 2003). www.state.gov/documents/organization/187769.pdf

References

Bailey, K.M., Francis, R.C. & Stevens, P.R. 1982. The life history and fishery of Pacific whiting, *Merluccius productus*. *CalCOFI Reports*, XXIII: 81–98.

Cox, S.P. & Kronlund, A.R. 2008. Practical stakeholder-driven harvest policies for groundfish fisheries in British Columbia, Canada. *Fisheries Research*, 94(3): 224–237.

Hicks, A. C., Taylor, N., Grandin, C. J., Taylor, I. G., Cox, S. (2013). Status of the Pacific hake (whiting) stock in U.S. and Canadian Waters in 2013. Prepared by the Joint Technical Committee of the U.S. and Canada Pacific Hake/ Whiting Agreement; National Marine Fishery Service; Canada Department of Fisheries and Oceans 190 p.

Marine Stewardship Council, (MSC). 2015. Pacific hake mid-water trawl. Available at www.msc.org/track-a-fishery/fisheries-in-the-program/certified/pacific/pacific-hake-mid-water-trawl (last accessed November 26, 2015).

Martell, S.J.D., Walters, C.J. & Hilborn, R. 2008. Retrospective analysis of harvest management performance for Bristol Bay and Fraser River sockeye salmon (Oncorhynchus nerka). *Canadian Journal of Fisheries and Aquatic Sciences*, 65(3): 409–424.

Methot, R.D., Jr. & C.R. Wetzel 2013. Stock synthesis: A biological and statistical framework for fish stock assessment and fishery management. *Fisheries Research*, 142: 86–99.

Chapter 5

Model-based management procedures for the sablefish fishery in British Columbia, Canada

Sean P. Cox and A. Robert Kronlund

Introduction

The sablefish (*Anoplopoma fimbria*) fishery in British Columbia (BC), Canada, is managed via a collaborative process involving fishing industry stakeholders and Fisheries and Oceans Canada (DFO). Sablefish inhabit shelf and slope waters throughout the North Pacific Ocean from the Aleutian Islands and Bering Sea south to Baja California. In BC waters, the sablefish total allowable catch (TAC) is set annually in proportion to estimated total available biomass using a model-based management procedure (Cox and Kronlund, 2013). Industry and government have shared science and assessment responsibilities in support of TAC decisions since 1988, creating a perception through the 1990s of a healthy stock and sustainable fishery. However, declines of 50% in BC sablefish survey and fishery catch-per-unit effort indices between 1990 and the early 2000s generated concern about stock status, the future of the fishery, and ultimately growing contention between stakeholders and government. These distractions lead to misdirected arguments over the "correctness" of the stock assessment models used to estimate sablefish stock size rather than focusing on long-term fishery performance relative to a predetermined suite of objectives.

In 2006, BC sablefish fishery stakeholders and DFO initiated a collaborative management strategy evaluation (MSE; la Mare, 1998, Sainsbury and Punt, 2000) process in which scientific data, assessments, and decision rules were coordinated to achieve specific conservation and fishery catch objectives (described later). Stakeholders expressed that their primary goal was to stop the decline of the BC sablefish stock as soon as possible. Furthermore, they wanted to achieve that goal using decision rules that linked monitoring data to TAC decisions in as few transparent steps as possible. Based on preliminary discussions with the science team (Sean P. Cox and A. Robert Kronlund), it was agreed that computer simulation would be used to test performance of candidate management procedures (MPs) that could meet these design criteria while also being consistent with conservation objectives implied in Canadian fishery policy (DFO, 2009). Simulation modelling is a technical topic foreign to most fisheries stakeholders, and therefore achieving acceptance of the simulation approach was needed before it could be used as a basis for decision-making. Therefore, much of the early MSE process (i.e. from 2006 to 2009) focused on developing a common understanding among scientists, stakeholders, and fishery managers of concepts ranging from survey design to fish population dynamics and closed-loop computer simulations of whole fishery management systems. The evolution of stakeholder understanding and contribution to the MSE process for BC sablefish is reflected in the transition from empirical MPs that used only survey and landings data in TAC computations (Cox and Kronlund, 2008) to more recent model-based MPs that integrate multiple abundance indices, landings, and harvest control rule reference points derived from a stock assessment model (Cox and Kronlund, 2013). The operating models used to test MP robustness also evolved in complexity from basic production models to age/fleet-structured models to, more recently, age/fleet/growth-structured models that account for size-based discarding at sea.

This chapter describes the current BC sablefish management strategy and, in particular, the role of computer-based simulation in revising the harvest control rule component to better meet stakeholder concerns about the economic consequences of low TACs. We describe specific harvest strategy elements including fishery objectives that guide MP choices, the existing MP that has been used to set TACs since 2011, and the operating models used in computer simulation tests. Then, we describe how simulations and performance diagnostics were used in 2013 to evaluate a specific question about revising the existing MP to include a TAC floor.

Current management and fishery objectives

The commercial sablefish TAC is shared between the combined long-line trap and hook (91.25%) and trawl (8.75%) sectors, which both utilize an individual transferable quota system to allocate TAC among individual harvesters (DFO, 2013). Sablefish captured in nondirected fisheries targeting Pacific halibut (*Hippoglossus stenolepis*), rockfishes (*Sebastes* sp.), and lingcod (*Ophiodon elongatus*) must also be accounted for within the TAC. Individual quota accounting is monitored via 100% at-sea video or observer coverage with independent auditing and 100% fishery-independent dockside landings validation.

Quantitative fishery objectives guide the BC sablefish fishery management system and annual TAC decision process. Objectives have been developed via consultations between fishery managers, scientists, and industry stakeholders, and are specifically chosen to be consistent with Canada's Fisheries Decision-making Framework Incorporating the Precautionary Approach (DMF, www. dfo-mpo.gc.ca/fm-gp/peches-fisheries/fish-ren-peche/sff-cpd/precaution-eng.htm). They have been revised in several phases of the MSE process (Cox et al., 2010, Cox et al., 2011, Cox and Kronlund, 2013) to reflect both industry concerns and updates to Canadian fishery policy. Upon completing the scientific analyses and review in 2010, an MP was chosen and implemented that met the following objectives in computer simulation tests:

1   Maintain spawning stock biomass (SSB) above the limit reference point, LRP = $0.4B_{MSY}$, in 95% of years measured over two sablefish generations (36 years).
2   When SSB is between $0.4B_{MSY}$ and $0.8B_{MSY}$, limit the probability of decline over the next 10 years from very low (5%) at the LRP to moderate (50%) at B_{MSY}. At intermediate stock status levels, define the tolerance for decline by linearly interpolating between these probabilities.
3   Maintain SSB above (a) B_{MSY}, or (b) $0.8B_{MSY}$ when rebuilding from below $0.8B_{MSY}$, in 50% of the years measured over two sablefish generations.
4   Maximize the average annual catch over 10 years.

Objectives 1–4 establish the overall intent of the fishery management strategy, but they are not involved in routine decision-making about regulations and annual TACs. Instead, these objectives exert influence by establishing the criteria that candidate MPs must meet in computer simulation tests. Each objective is therefore associated with a quantitative performance indicator computable from simulation output. With the exception of Objective 4, all performance indicators are computed from the operating models used to represent the true stock dynamics.

Sablefish management procedure in British Columbia

The MP meeting fishery objectives applies a Schaefer surplus production stock assessment model (Punt, 2003) to total sablefish landings from all BC fisheries and catch per unit effort (CPUE) indices from the commercial trap fishery (1979–2009), standardized trap survey (1991–2009), and stratified random trap survey (2003–present). The assessment model estimates a time series of historical exploitable biomass along with a set of harvest control rule parameters (Equation H5.1, Table 5.1) consisting of the exploitation rate and biomass at maximum sustainable yield ($\hat{U}_{MSY}$ and $\hat{B}_{MSY}$, respectively), and the exploitable biomass forecast for the coming year ($\hat{B}_{T+1}$). These estimates are used in a piecewise linear harvest control rule (henceforth U60–40) that reduces the target exploitation rate from a maximum $\hat{U}_{MSY}$ when the exploitable biomass forecast is at or above the upper control point $0.6\,\hat{B}_{MSY}$ to 0 at the lower control point $0.4\,\hat{B}_{MSY}$ (Equation H5.2, Figure 5.1). This form of harvest control rule was chosen based on DFO fishery policy that requires a reduction in the harvest rate as stock status approaches low levels so as to ensure that sufficient spawning biomass is maintained.

Model-based control rules use an assessment to estimate stock status at each annual implementation, but reference points such as B_{MSY} and U_{MSY} are typically based on external analyses (e.g. spawning potential ratio, yield-per-recruit, ref.

Table 5.1 British Columbia sablefish harvest control rule for computing a precautionary TAC (Q_{T+1}) based on inputs (H5.1) from surplus production stock assessment model and proposed TAC floor (Q_{Floor}). The assessment estimates are the optimal harvest rate $\hat{U}_{MSY}$, biomass producing maximum yield $\hat{B}_{MSY}$, and exploitable biomass forecast, $\hat{B}_{T+1}$. Multipliers (0.4, 0.6) were selected based on previous simulation testing (Cox et al., 2013).

H5.1
$$\Psi = \left(\hat{U}_{MSY}, \hat{B}_{MSY}, \hat{B}_{T+1}, Q_{Floor}\right)$$

H5.2
$$U_{T+1} = \begin{cases} 0 & \hat{B}_{T+1} < 0.4\hat{B}_{MSY} \\ \hat{U}_{MSY}\left(\dfrac{\hat{B}_{T+1} - 0.4\hat{B}_{MSY}}{0.6\hat{B}_{MSY} - 0.4\hat{B}_{MSY}}\right) & 0.4\hat{B}_{MSY} \leq \hat{B}_{T+1} < 0.6\hat{B}_{MSY} \\ \hat{U}_{MSY} & \hat{B}_{T+1} \geq 0.6\hat{B}_{MSY} \end{cases}$$

H5.3
$$Q_{T+1} = \max\left(Q_{Floor}, U_{T+1}\hat{B}_{T+1}\right)$$

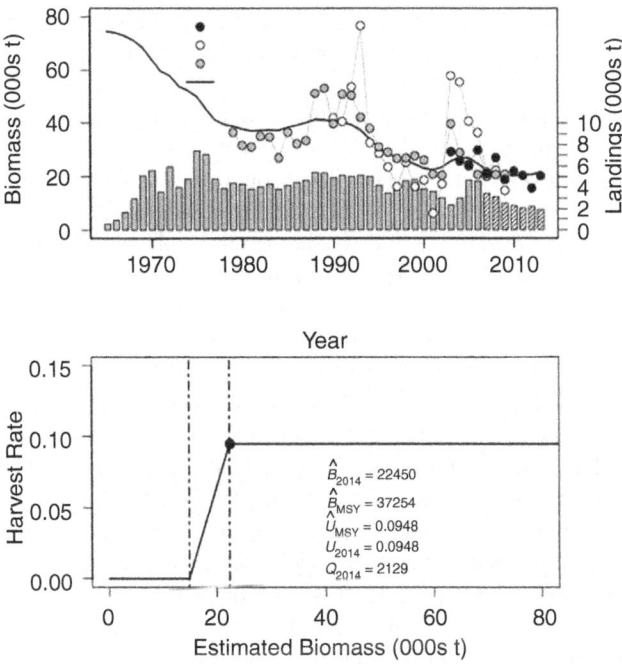

Figure 5.1 The BC sablefish management procedure implemented to determine the 2014–2015 TAC. Top panel: Annual landings (bars) and surplus production model fit (line) to biomass indices from the commercial trap fishery (CPUE; 1979–2009), standardized trap survey (StdS; 1991–2009), and stratified random trap survey (StRS; 2003–2013). Biomass and landings are given in units of 1,000 metric tons. The hatched bars indicate catch levels set as part of the MSE process since 2006. Bottom panel: The harvest control rule mapping estimated exploitable biomass to a target harvest rate. The legend lists the quantities used in the harvest control rule calculation and 2014 implementation.

haddock case study). The control rule described here is different because the reference points defining its structure are also estimated by the assessment. These are distinct from the reference points assumed by the operating model, and are sometimes referred to as "control points" (Cox et al., 2013). An advantage of estimating control points is that the MP becomes self-contained and independent of strong prior assumptions regarding stock productivity. This could make the MP more robust to structural uncertainty in the operating model. However, it also means that small changes in estimates of $\hat{U}_{MSY}$ or $\hat{B}_{MSY}$ could interact with the harvest control rule to cause relatively large annual changes in catch. Here, we developed a compromise via the use of informative Bayesian priors in the estimation procedure, so that variation in $\hat{U}_{MSY}$ and $\hat{B}_{MSY}$ was reduced to acceptable levels. It was also necessary to account for estimation uncertainty in $\hat{B}_{T+1}$, which in particular could lead to catches being set too high. To ensure adequate precaution, control points 0.6 $\hat{B}_{MSY}$ and 0.4 $\hat{B}_{MSY}$ were chosen based on fishery performance in computer simulation tests that accounted for stock dynamics and stock assessment model estimation errors (Cox and Kronlund, 2013).

In a 2013 review of U60–40 performance and fishery objectives, industry stakeholders expressed concern that further application of the procedure would reduce catch below thresholds needed for economic viability (DFO, 2014). Stakeholders contemplated a temporary hiatus on U60–40 MP and, in the interim, suggested fixing the TAC at the 2013 level. However, fixing the TAC could have two undesirable effects: (*i*) it could cost harvesters catch if the U60–40 MP would have recommended a TAC increase or (*ii*) it could compromise fishery conservation objectives if a further TAC reduction was warranted. In the spirit of the MSE process, a computer simulation experiment was proposed to determine whether a revised U60–40 MP would continue to meet Objectives 1–4 as listed earlier. The revised procedure included a TAC floor that limited further catch reductions, while still allowing for TAC increases if warranted by the new 2013 data and stock assessment. This so-called U60–40+Floor harvest control rule would therefore set the sablefish TAC to the maximum of either Q_{Floor} = 1,992 metric tons (i.e. the 2013 TAC) or the usual product of the target harvest rate and exploitable biomass forecast (Equation H5.3). Later we describe our simulation approach and results used to determine whether the U60–40+Floor procedure is consistent with the fishery strategy as defined by Objectives 1–4. We also included a Perfect U60–40, which computes the TACs using the true operating model biomass and optimal harvest rate. This MP provides a way to separate the role of stock assessment model errors and harvest control rules on conservation and catch performance.

Evaluation process

Operating model

A closed-loop feedback simulation approach was used to evaluate U60–40+Floor performance against a sablefish operating model that included alternative hypotheses about population dynamics, at-sea release mortality rates, individual growth rate, and recruitment autocorrelation. Model notation is given in Table 5.2. The operating model represents a sablefish population that is structured by age, size, and growth group, where the latter dimension is used to model the effects of size-based discarding at sea. Equations describing unfished equilibrium quantities and reference points are given in Table 5.3 where they are labeled with the prefix "E" to associate them with the equilibrium condition. Equations specifying the operating model dynamics and computational algorithm are given in Table 5.4 where they are given the prefix "O."

Table 5.2 Notation for the BC sablefish operating model.

Symbol	Value	Description
T_1	47	Year in which the management procedure begins
T_2	83	Total number of years to simulate
A	35	Number of operating model age-classes
t	$1, 2, \ldots, T$	Time step ($T = T_1$ for conditioning, $T = T_2$ for simulating)
a	$1, 2, \ldots, A$	Age-class
l	$1, 2, \ldots, n_l$	Growth-group index
n_l	12	Number of growth groups
g	$1, 2, \ldots, G$	Fishery/gear index
B_0		Unfished spawning biomass (1,000s of metric tonnes)
h		Stock-recruitment function steepness
q_g		Catchability coefficient for gear g
σ_R	0.60	Standard error of log-recruitment
γ_R	**0.0, 0.4**	Lag-1 autocorrelation in log-recruitment deviations
M	**0.06, 0.08**	Estimated and fixed natural mortality rates, respectively (per year)
L_l^∞		Asymptotic length (cm) for growth group l
σ_{L_∞}	7.8	Standard error of asymptotic length (cm)
$L_{1,l}$	30	Length-at age-1 for all growth groups (cm)
k	0.30	Von Bertalanffy growth constant
c_1, c_2	8.48e-6, 3.05	length-weight coefficients
$\tilde{A}_{50}, \tilde{A}_{95}$	5, 8	Age-at-50% and 95% maturity
$\tilde{L}_{lim}$	55	Minimum size limit (cm)

(Continued)

Table 5.2 (Continued)

Symbol	Value	Description
$\tilde{L}^C_{50,g,1}$, $\tilde{L}^C_{95,g,1}$		Length-at-50% and length-at -95% selectivity: ascending limb
$\tilde{L}^C_{95,g,2}$, $\tilde{L}^C_{50,g,2}$		Length-at-95% and length-at -50% selectivity: descending limb
$\tilde{L}^D_{95,g}$, $\tilde{L}^D_{50,g}$		Length-at-50% and length-at -95% discard probability
$\pi_{1,l}$	$1/n_1$	Proportion of age-1 recruits assigned to growth group l
d_g	$(0.15, 0.30, 0.8, 0, 0)$	Discard mortality rates by gear type (per year)
$L_{a,l}$		Length-class of growth group l at age a (cm)
$w_{a,l}$		Weight-at-age for fish in growth group l
m_a		Proportion mature-at-age
$P_{a,l,g}$		Proportion of age a, growth group l discarded
$S_{a,l,g}$		Selectivity for age a, growth group l by gear g
R_0		Unfished equilibrium recruitment
$\phi^{SSB}_{\tilde{F}}$		Spawning biomass per recruit given vector $\tilde{F}$
ϕ^L_g		Landed yield per recruit for gear g
ϕ^D_g		Discarded yield per recruit for gear g
$N_{a,l,t}$		Number of age a fish in growth group l in year t
$\omega_{R,t}$		Autocorrelated log-normal recruitment process deviation
B_t		Spawning biomass in year t
$C_{a,t,g}$		Landed catch-at-age in fishery g
$D_{a,t,g}$		At-sea releases-at-age fishery g
$F_{t,g}$		Fishing mortality rate for gear g in year t
$Z_{a,l,t}$		Total mortality rate for age a, growth group l in year t
$I_{t,g}$		Model biomass index for gear g
$\hat{I}_{t,g}$		Simulated biomass index for gear g
$u^C_{a,t,g}$		Proportion of age a in year t landed catch
$u^D_{a,t,g}$		Proportion of age a in year t dead discarded catch
δ_t	$N(0, 1)$	Standard normal error in log-recruitment
$\varepsilon_{t,g}$	$N(0, 1)$	Standard normal error for biomass index g

Table 5.3 Equilibrium functions of an assumed fishing mortality rate vector $\tilde{\mathbf{F}} = \left(\tilde{F}_1, \tilde{F}_2, \ldots, \tilde{F}_G\right)$ and estimated parameters Θ (Equation O5.1).

Equation	Formula	Description
E5.1	$\Omega = \left(\tilde{\mathbf{F}}, \Theta\right)$	Parameters
	$Z_{a,l} = M + \sum_{g=1}^{G} s_{a,l,g} \tilde{F}_g \left(d_g P_{a,l,g} - P_{a,l,g} + 1\right)$	Total mortality-at-age/ growth group
E5.2		Survivorship to age a
	$\ell_{a,l} = \begin{cases} 1 & a = 1 \\ \ell_{a,l} e^{-\tilde{Z}_{a-1,l}} & 2 \le a < A \\ \ell_{a-1,l} e^{-\tilde{Z}_{a-1,l}} / \left(1 - e^{-\tilde{Z}_{a-1,l}}\right) & a = A \end{cases}$	
E5.3	$\phi_g^L = \sum_{a=1}^{A} \sum_{l=1}^{n} \ell_{a,l} w_{a,l} S_{a,l,g} \tilde{F}_g \left(1 - P_{a,l,g}\right)\left(1 - e^{-\tilde{Z}_{a,l}}\right)/Z_{a,l}$	Landed yield per recruit for gear g
	$\phi_g^D = \sum_{a=1}^{A} \sum_{l=1}^{n} \ell_{a,l} w_{a,l} S_{a,l,g} \tilde{F}_g P_{a,l,g} \left(1 - e^{-\tilde{Z}_{a,l}}\right)/Z_{a,l}$	Discarded yield per recruit for gear g
E5.4	$\phi_F^{SSB} = \sum_{a=1}^{A} \sum_{l=1}^{n} \ell_{a,l} m_a w_{a,l}$	Spawning stock biomass per recruit
E5.5	$R = \left(4 h R_0 \phi_F^{SSB} - B_0 (1-h)\right) / \left((5h-1)\phi_F^{SSB}\right)$	Age-1 recruitment
E5.6	$B = R \phi_F^{SSB}$	Spawning stock biomass
E5.7	$C_g^L = R \sum_{g=1}^{G} \phi_g^L$	Total landed yield for gear g
	$C_g^D = R \sum_{g=1}^{G} \phi_g^D$	Total discarded yield for gear g

Table 5.4 BC sablefish operating model. Equations sequentially define the population dynamics and simulated observations for a set of input parameters as defined by scenarios in Table 5.5. The subset of parameters given by Θ were estimated during operating model conditioning.

Parameters

O5.1	$\Theta = \left\{ B_0, h, M, \{\delta_t\}_{t=2}^{T-3}, \left\{\tilde{L}_{50,g,l}^C\right\}_{g=1}^{5}, \left\{\tilde{L}_{95,g,l}^C\right\}_{g=1}^{5}, \tilde{L}_{50,g=3,2}^C, \tilde{L}_{95,g=3,2}^C \right\}$

Growth, maturity, and selectivity

O5.2	$L_{a,l} = L_l^\infty + \left(L_{1,l} - L_l^\infty\right) e^{(-k(a-1))}$
O5.3	$w_{a,l} = c_1 L_{a,l}^{c_2}$
O5.4	$m_a = \left(1 + \exp\left[-\log(19)\left(a - \tilde{A}_{50}\right)/\left(\tilde{A}_{95} - \tilde{A}_{50}\right)\right]\right)^{-1}$

(Continued)

Table 5.4 (Continued)

Growth, maturity, and selectivity

O5.5
$$S_{a,l,g} \propto \left(1+\exp\left[-\log(19)\left(L_{a,l}-\tilde{L}^C_{50,g,1}\right)/\left(\tilde{L}^C_{95,g,1}-\tilde{L}^C_{50,g,1}\right)\right]\right)^{-1}$$
$$\times\left(1+\exp\left[-\log(19)\left(L_{a,l}-\tilde{L}^C_{50,g,2}\right)/\left(\tilde{L}^C_{95,g,2}-\tilde{L}^C_{50,g,2}\right)\right]\right)^{-1}$$

O5.6
$$P_{a,l,g} = \begin{cases} 1.0 & L_{a,l} < L_{lim} \\ \left(1+\exp\left[-\log(19)\left(L_{a,l}-\tilde{L}^D_{50,g}\right)/\left(\tilde{L}^D_{95,g}-\tilde{L}^D_{50,g}\right)\right]\right)^{-1} & L_{a,l} \geq L_{lim} \end{cases}$$

State dynamics

O5.7
$$R_0 = B_0 / \phi^{SSB}_{F=0}$$

O5.8
$$N_{a,l,1} = \begin{cases} \pi_{1,l}R_0 e^{-M(a-1)} & 1 \leq a \leq A-1 \\ N_{a-1,l,1}/\left(1-e^{-M}\right) & a = A \end{cases}$$

O5.9
$$\omega_{R,t} = \begin{cases} \dfrac{\sigma_R}{\sqrt{1-\gamma_R^2}}\delta_t & t = 1 \\ \gamma_R\omega_{R,t-1}+\sigma_R\delta_t & t > 1 \end{cases}$$

O5.10
$$N_{1,l,t} = \pi_{1,l}\frac{4R_0 B_{t-1}}{B_0(1-h)+(5h-1)B_{t-1}}e^{\omega_{R,t}-0.5\sigma_R^2/(1-\gamma_R^2)}$$

O5.11
$$N_{a,l,t} = \begin{cases} N_{a-1,l,t-1}e^{-Z_{a-1,l,t-1}} & 2 \leq a \leq A-1 \\ N_{a-1,l,t-1}e^{-Z_{a-1,l,t-1}}+N_{a,l,t-1}e^{-Z_{a,l,t-1}} & a = A \end{cases}$$

O5.12
$$B_t = \sum_{a=1}^{A}m_a\sum_{l=1}^{n_l}w_{a,l}N_{a,l,t}$$

O5.13
$$C_{a,t,g} = \sum_{l=1}^{n_l}w_{a,l}N_{a,l,t}\frac{S_{a,l,g}F_{t,g}\left(1-P_{a,l,g}\right)}{Z_{a,l,t}}\left[1-e^{-Z_{a,l,t}}\right]$$

O5.14
$$D_{a,t,g} = \sum_{l=1}^{n_l}w_{a,l}N_{a,l,t}\frac{S_{a,l,g}F_{t,g}P_{a,l,g}}{Z_{a,l,t}}\left[1-e^{-Z_{a,l,t}}\right]$$

O5.15
$$Z_{a,l,t} = M+\sum_{g=1}^{G}S_{a,l,g}F_{t,g}\left(d_gP_{a,l,g}-P_{a,l,g}+1\right)$$

Observations

O5.16
$$I_{t,g} = q_g\sum_{a=1}^{A}\sum_{l=1}^{n_l}w_{a,l}S_{a,l,g}N_{a,l,t}$$

O5.17
$$\hat{I}_{t,g} = I_{t,g}\exp\left[\tau_{t,g}\varepsilon_{t,g}-\tau_{t,g}^2/2\right]$$

O5.18
$$u^C_{a,t,g} = C_{a,t,g}/\sum_{a'=1}^{A}C_{a,t,g}$$

Table 5.5 Sablefish operating model scenario parameters: stock-recruitment steepness, h; natural mortality rate, $\hat{M}$; and unfished spawning biomass, $\hat{B}_0$. Equilibrium quantities and reference points: maximum sustainable yield, $\widehat{MSY}$; optimal legal harvest rate, $\hat{U}_{MSY}$; spawning biomass and depletion at MSY, $\hat{B}_{MSY}$, $\hat{D}_{MSY}$; and depletion at the limit reference point, $0.4\hat{D}_{MSY}$. Biomass units are thousands of metric tons.

Scenario	$\hat{h}$	$\hat{M}$	$\hat{B}_0$	$\widehat{MSY}$	$\hat{U}_{MSY}$	$\hat{B}_{MSY}$	$\hat{D}_{MSY}$	$0.4\hat{D}_{MSY}$
S1: BaseMPD	0.88	0.06	114.77	3.23	0.11	27.68	0.24	0.10
S2: FixedM	0.50	0.08	147.73	3.21	0.06	53.13	0.36	0.14
S5: S1+AR	0.88	0.06	114.77	3.22	0.11	27.68	0.24	0.10
S6: S2+AR	0.50	0.08	147.73	3.21	0.06	53.13	0.36	0.14
S7: S1-Posterior Mean	0.75	0.06	120.05	3.06	0.09	33.70	0.28	0.11
S8: S1-10th	0.59	0.06	121.08	2.53	0.06	40.02	0.33	0.13

Sablefish mean length (cm) for age a and growth group l is modelled using a von Bertalanffy growth function (O5.2) in which each group l has a unique asymptotic size based on 12 equally spaced quantiles of a $N\left(\bar{L}_{\infty}, \sigma_L^2\right)$ distribution, where $\bar{L}_{\infty}$ is the mean asymptotic length across growth groups. Growth groups were used to model size-based at-sea discarding in two parts: (*i*) fish are first brought onboard fishing vessels according to gear-specific selectivity functions that depend on sablefish length (O5.5) and then (*ii*) fish smaller than the legal size limit are all released and subject to discard mortality while fish greater than the size limit are released according to a high-grading relationship for which we use a descending logistic function of length (O5.6). Discarding relationships in O5.6 were developed in consultation with the multisectoral stakeholders and DFO groundfish managers. Growth, maturity, and fishery selectivity schedules (O5.2–5.6), and stock-recruitment parameters are then used to determine how the landed yield per recruit (E5.3), spawning stock biomass per recruit (E5.4), equilibrium spawning biomass (E5.6), and average total landed yield (E5.7) vary in response to a particular combination of gear-specific fishing mortality rates, referenced by the vector $\tilde{F}$ (Beverton and Holt, 1957).

Fishing mortality rate reference points were derived by numerically solving for the vector of fishing mortality rate parameters in E5.1 that maximize equilibrium landed yield in E5.7. We assume that the vector parameter $\tilde{F}_{MSY}$ is separable into an overall scalar F_{MSY} value multiplied by a vector of sector-specific coefficients $f_1, f_2, \dots f_G$, that is $\tilde{F}_{MSY} = f_1 F_{MSY}, f_2 F_{MSY}, \dots, f_G F_{MSY}$. Biomass and yield reference points, B_{MSY} (E5.6) and MSY (E5.7), respectively, are derived once the optimal vector $\tilde{F}_{MSY}$ is known. We determined $\tilde{F}_{MSY}$ by (*i*) computing equilibrium total landed yield over a grid of 50 F values ranging from 0 to 4 times the natural mortality rate, M; (*ii*) for each F, solving for the vector of gear-specific multipliers f_g that minimize the sum-of-squared differences between the modelled catch allocation among gear types, i.e.

$$
\left. C_g^L \middle/ \sum_{j=1}^{G} C_j^L \right. ,
$$

and the predetermined target proportions (i.e. our assumed future allocation of catch among gear types); (*iii*) fitting a cubic spline between the scalar fishing mortality rate F (i.e. giving $\tilde{\mathbf{F}} = f_1 F, f_2 F, \ldots, f_G F$) and the resulting total landed yield values; and (*iv*) using the first derivative of the cubic spline to obtain the root, $\tilde{\mathbf{F}}_{MSY}$. In practice, we only complete step (*ii*) once because the optimal multipliers for this sablefish fishery are insensitive to the values of F.

All operating model scenarios assume that the BC sablefish recruitment and spawning stock were at their respective unfished, deterministic equilibria in 1965 prior to the development of directed fisheries. Equations O5.7 and O5.8 give these initial conditions. Subsequent recruitment of age-1 individuals in each length-class $N_{1,l,t}$ occurs on January 1st of the model year with interannual variability involving a deterministic Beverton-Holt relationship to spawning stock biomass and log-normally distributed, lag-1 autocorrelated random variation about the expected recruitment (O5.9–5.10). Parameters representing the unfished spawning stock biomass (B_0) and steepness (h) controlling the Beverton-Holt relationship are estimated in the operating model conditioning step along with annual recruitment deviations δ_t for the historical period 1965–2009. We do not include recruitment deviations for the 2007–2009 cohorts in the operating model fitting procedures because individual cohort sizes are only estimable for years 1966–2006. Therefore, recruitment deviations in operating model projection for 2007–2009 are generated in the same manner as future recruitment deviations in the projection period (i.e. O5.9).

For both the historical and future periods, we solved the catch equation (O5.13) for $F_{t,g}$ given the landed catch by gear ($C_{a,t,g}$ for historical period) or total landed quota (Q_T in H5.3 for future projections). Four iterations of a Newton-Raphson algorithm are used to solve the catch equations, achieving an accuracy to within 2–4% of the observed catches. Predicted values for at-sea discards $D_{a,t,g}$ in O5.14 are computed based on the resulting $F_{t,g}$ values during both historical and projection periods.

Operating model scenarios and conditioning based on historical data

For this chapter, we used six of the original eight operating model scenarios for evaluating U60–40+Floor performance relative to the existing U60–40 procedure. We retain the original numbering of operating model scenarios that appear in Cox et al. (2011) in case the reader wishes to cross-reference that paper. The six scenarios include the baseline operating model (S1: BaseMPD), which is parameterized by the maximum posterior density estimates of all parameters, including the natural mortality rate. The natural mortality rate, M, has an informative normal prior with mean $M = 0.08/\text{yr}$ and standard deviation $0.005/\text{yr}$. Scenario S2: FixedM is identical to S1 except that it uses a fixed natural mortality rate of $M = 0.08/\text{yr}$. Scenarios S5: S1+AR and S6: S2+AR are identical to S1 and S2, respectively, except that recruitment autocorrelation is set to $\gamma = 0.4$ for projection years 2011–2046 ($\gamma = 0.0$ for S1 and S2). Scenario S7: S1-Posterior Mean uses posterior means of all operating model parameters, while the worst-case S8: S1–10th uses parameters underlying the 10th percentile of the posterior distribution for MSY. The latter operating model is a robustness test reflecting a case in which current sablefish productivity is grossly overestimated.

Each operating model scenario was individually fitted to a combination of biomass indices, age-proportion data, total at-sea releases, and tag recovery-at-length data. This process creates the conditions associated with a particular suite of assumptions behind stock and fishery dynamics as well as statistical properties of simulated data. Historical biomass indices include commercial trap fishery catch per unit effort (1979–2009), a standardized trap survey (StdS, 1991–2009), and a stratified random trap survey (StRS, 2003–2009). Age-proportion data included the commercial trap fishery (1979–2009), StdS (1990–2009), and StRS (2003–2009). The operating model was not updated to include 2010–2013 data because the request to revise the U60–40 MP came too late in the year to arrange for otolith aging and data quality control.

Predicted values of biomass indices were generated from O5.16 and age-proportions were computed using O5.18. Equation O5.17 was used to generate pseudo-data for the StRS biomass index during the operating model projection period using the estimate of $\tau_{1,S,tRS}$ obtained during operating model conditioning. Total at-sea discard estimates were available for commercial trap and long-line hook gear (2006–2009) and trawl (1996–2009). Equation O5.14 was used to predict at-sea discards for these years. Recoveries from the sablefish tagging program that were captured within their first year at-liberty ($N = 43,929$ releases, $n = 2,309$ recoveries pooled over years 1996–2009) were used to provide direct estimates of length-based selectivity for each gear by length-at-release. These data create a relatively strong prior distribution on length-based selectivity parameters for each gear type. AD Model Builder (Fournier et al., 2012) was used to fit the model and generate Bayes posterior distributions for model parameters and derived equilibrium quantities.

Closed-loop simulation algorithm

The following algorithm was used to simulate MP performance against each operating model scenario (the step indicating the proposed TAC Floor appears in bold font):

1   Define a management procedure based on (a) data types, (b) assessment method, (c) harvest control rule.

2   Initialize a preconditioned operating model scenario for the period (1965–2010) based on historical data.
3   Project the operating model population and fishery one time step into the future and apply the following:

a   Generate and append a new survey biomass index (1.i) data point and landed catch to the existing stock assessment data set;
b   Estimate control rule quantities $\hat{U}_{MSY}$, $\hat{B}_{MSY}$, and $\hat{B}_{T+1}$ by applying the assessment model (1.ii) to the updated data;
c   Apply the harvest control rule (1.iii) to generate a candidate quota $U_{T+1}\hat{B}_{T+1}$;
d   Set the quota to the maximum of 3.c or the TAC Floor Q_{Floor};
e   Update the operating model population given the total mortality rate (i.e. fishing + discard + natural mortality) generated by the final catch limit and sublegal regulation (1.iv), and new recruitment;
f   Repeat Steps 3a–3e until the projection period ends.

4   Calculate quantitative performance statistics for the simulation replicate.
5   Repeat Steps 2–4 for 100 replicates, each of which applies a new sequence of random recruitment deviations from the Beverton-Holt relationship and index observation errors.

Performance diagnostics

Each fishery objective is associated with a unique, quantitative performance indicator computable from simulation output (Table 5.6). All performance indicators were computed from the operating models used to represent the true stock and fishery dynamics. The existing U60–40 MP was chosen from eight original candidate MPs by applying Objectives 1–4 in priority order such that Objectives 2–4 were not considered if a procedure failed to meet Objective 1. Where MPs were not rejected at Objectives 1 or 2, we then considered performance for stock growth towards $0.8B_{MSY}$ (Objective 3) and finally catch performance over the short-term (Objective 4). The two final performance measures (P5.6–5.7) provided information about minimum expected catch levels and interannual variability in yield, respectively, which were both important to industry stakeholders.

Sablefish generation time, which we used to define the timeframes of fishery Objectives 1 and 3, is calculated as the average age of the unfished spawning stock, that is

$$\bar{a}_0 = \left.\sum_{a=1}^{A} a l_a m_a \middle/ \sum_{a=1}^{A} l_a m_a \right.,$$

where maturity-at-age m_a is given by Equation O5.4 and the age-specific survivorship l_a is defined by Equation E5.2 for $F = 0$ (and ignoring growth groups). Generation time estimates vary between 16 and 18 years depending on the estimated natural mortality rate for each model scenario.

Table 5.6 Performance indicators derived from the operating model and used to compare closed-loop simulation outcomes to objectives for each candidate management procedure.

Eq.	Objective	Indicator	Probability or statistic	Definition
P5.1	Objective 1	Proportion of projection years where spawning biomass exceeds $0.4B_{MSY}$ (Period: $t_1 = 2011$, $t_2 = 2046$)	$P(B > 0.4B_{MSY})$	$P(B > 0.4B_{MSY}) = \dfrac{\sum_{t_1}^{t_2} I(B_t > 0.4B_{MSY})}{t_2 - t_1 + 1}$
P5.2	Objective 2	Proportion of 10-year trends that are declining (Period: $t_1 = 2011$, $t_2 = 2020$)	$P(\beta < 0) < P(\text{decline})$	$P(\beta < 0) = \dfrac{1}{100} \sum_{1}^{100} I(\beta < 0)$
P5.3	Objective 3a	Proportion of projection years where spawning biomass exceeds B_{MSY} (Period: $t_1 = 2011$, $t_2 = 2046$)	$P(B > B_{MSY})$	$P(B > B_{MSY}) = \dfrac{\sum_{t_1}^{t_2} I(B_t > B_{MSY})}{t_2 - t_1 + 1}$
P5.4	Objective 3b	Proportion of projection years where spawning biomass is in the Healthy Zone (Period: $t_1 = 2011$, $t_2 = 2046$)	$P(B > 0.8B_{MSY})$	$P(B > 0.8B_{MSY}) = \dfrac{\sum_{t_1}^{t_2} I(B_t > 0.8B_{MSY})}{t_2 - t_1 + 1}$
P5.5	Objective 4	Mean of annual landed catch (Period: $t_1 = 2011$, $t_2 = 2020$)	$\overline{C^L}$	$\overline{C^L} = \dfrac{1}{t_2 - t_1 + 1} \sum_{t_1}^{t_2} C_t^L$
P5.6	Min and Max	Minimum and maximum landed catch (Period: $t_1 = 2011$, $t_2 = 2020$)	Min C Max C	$\min(C_{2011}^L, C_{2012}^L, \ldots, C_{2021}^L)$ $\max(C_{2011}^L, C_{2012}^L, \ldots, C_{2021}^L)$
P5.7	Industry preference	Average annual absolute change in landed catch (Period: $t_1 = 2011$, $t_2 = 2020$)	AAV	$AAV = \sum_{t=t_1}^{t_2} |C_t^L - C_{t-1}^L| / \sum_{t=t_1}^{t_2} C_t^L$

Results and decision-making

Conservation performance of the revised (U60–40+Floor) and existing MPs (U60–40) was similar across operating model scenarios (Table 5.7). Both U60–40+Floor and U60–40 procedures maintained simulated sablefish biomass above the limit reference point under all operating model scenarios, including the worst case (S8), as required under Objective 1.

Both procedures only failed Objective 2 in the two worst-case scenarios (Table 5.7). When the operating model recruitment process was autocorrelated (S6) and the U60–40+Floor MP was applied, the simulated probability of decline over 10 years (0.33) was greater than the acceptable probability (0.24) given by Objective 2 (Figure 5.2). This result is only slightly worse than the U60–40 procedure probability (0.31), which also failed Objective 2 under this scenario. In the worst-case scenario (S8), U60–40+Floor and U60–40 procedures both failed Objective 2 with probabilities of future biomass decline of 0.46 and 0.45, respectively, as compared to an acceptable probability of 0.21.

Table 5.7 Performance summary for three management procedures tested against selected operating model scenarios. Objectives are satisfied by a procedure (marked •) if the performance indicator given in Table 5.6 meets the criterion given by the objective. For example, a procedure meets Objective 2 if the proportion of declining stock trajectories is less than P(decline) over the first 10 projection years. Values of performance indicators are provided where a procedure fails under Objectives 1–3.

| Scenario | Objective | | | | | | Min C | AAV |
Management procedure	1	2	3a	3b	4			
S1: BaseMPD								
Perfect U60–40	•	•	0.40	0.83	2.50		2.21	4.25
U60–40	•	•		•	0.88	2.37	1.97	7.18
U60–40+Floor	•	•		•	0.87	2.38	1.98	6.51
S5: S1+AR								
Perfect U60–40	•	•	0.38	0.73	2.48		2.19	4.91
U60–40	•	•		•	0.81	2.36	1.91	8.13
U60–40+Floor	•	•		•	0.78	2.38	1.98	6.55
S7: S1-Posterior Mean								
Perfect U60–40	•	•	0.28	0.70	2.15		1.88	5.41
U60–40	•	•	0.28	0.65	2.27		1.90	7.70
U60–40+Floor	•	•	0.27	0.62	2.28		1.98	6.30

Scenario Management procedure	Objective					Min C	AAV
	1	2	3a	3b	4		
S2: FixedM							
Perfect U60–40	●	●	0.16	0.50	2.14	1.96	4.74
U60–40	●	●	0.12	0.42	2.37	2.10	6.36
U60–40+Floor	●	●	0.12	0.40	2.37	2.10	5.70
S6: S2+AR							
Perfect U60–40	●	●	0.19	0.43	2.11	1.94	5.26
U60–40	●	0.31 > 0.24	0.17	0.40	2.36	2.07	6.74
U60–40+Floor	●	0.33 > 0.24	0.16	0.36	2.38	2.07	5.74
S8: S1–10th							
Perfect U60–40	●	●	0.17	0.52	1.65	1.48	8.17
U60–40	●	0.45 > 0.21	0.02	0.15	2.23	1.92	7.03
U60–40+Floor	●	0.46 > 0.21	0.01	0.11	2.24	1.98	5.44

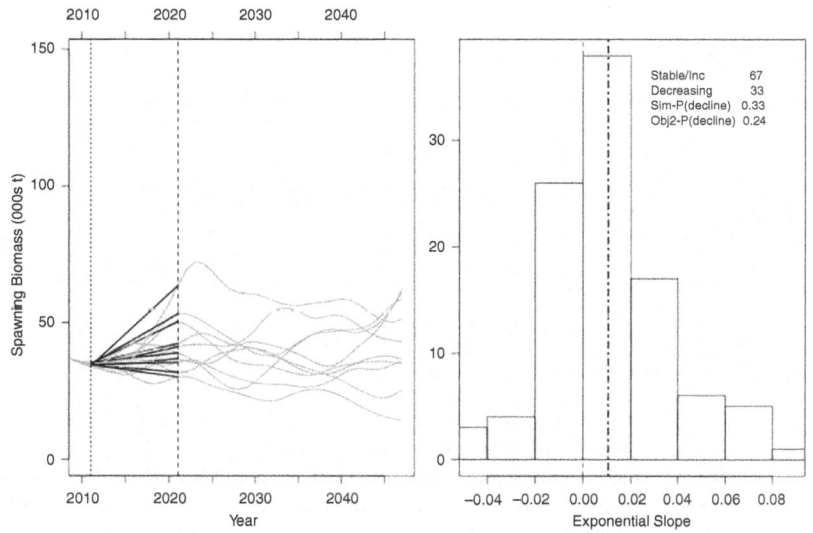

Figure 5.2 Ten-year stock trends (Objective 2) simulated under the U60–40+Floor management procedure and autocorrelated recruitment operating model scenario (S6:S2+AR). Left panel shows 10 randomly chosen biomass trajectories (grey lines) and their corresponding 10-year trend estimates (black lines). Biomass is given in units of 1,000 metric tons. Right panel shows histograms of all 100 stock trend statistics, the tolerance for decline under Objective 2 "Obj2-P(decline)," and the proportion of simulated declining trends "Sim-P(decline)." The vertical dashed line indicates the mean growth trend.

Both the revised and existing MPs failed to maintain the stock at or above a target reference point of B_{MSY} at least 50% of the time (Objective 3a) under all scenarios except S1: BaseMPD and S5: S1+AR (Table 5.7). This result reflects the combination of sablefish stock status below B_{MSY} as of 2010 and population growth limited by a relatively long generation time of approximately 18 years. Even with perfect information about the operating model exploitable biomass and optimal harvest rate (Table 5.7; Perfect U60–40), the stock would not grow fast enough to meet Objective 3a under most scenarios. The alternative $0.8B_{MSY}$ (Objective 3b) was established as an alternative rebuilding target (Cox et al., 2011). Under the revised and existing MPs, the simulated stock was able to recover to at least $0.8B_{MSY}$ with greater than 50% probability in three (S1, S5, S7) out of six scenarios and with greater than 36% probability in two others (S2, S6). Only the Perfect U60–40 procedure met Objective 3b under the worst-case scenario (S8).

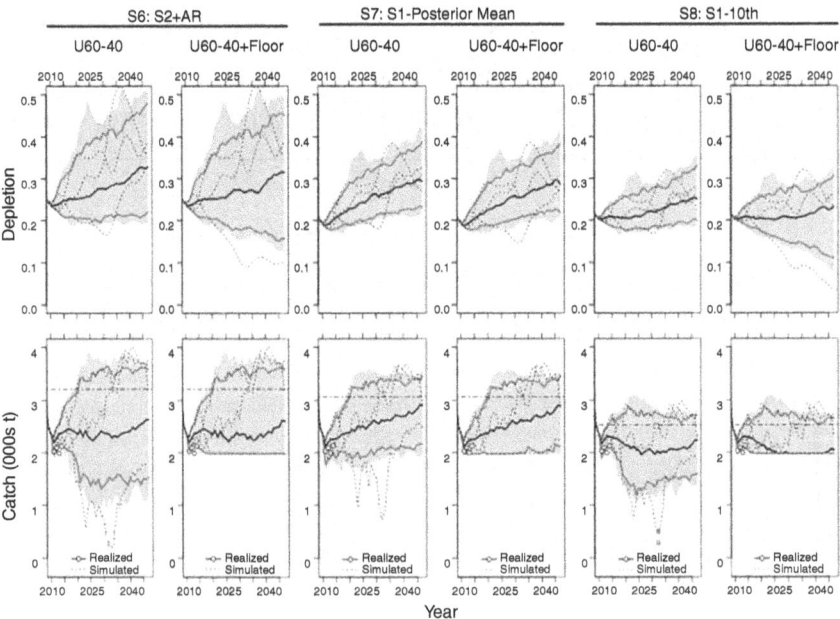

Figure 5.3 Spawning biomass depletion (upper panels) and total fishery landings (lower panels) simulated under the U60–40 and U60–40+Floor proce-dures for each of three operating model scenarios. Thick black lines indi-cate the median trajectories, outer red lines give 10th and 90th percentiles, and the shaded region covers the central 90% of trajectories. Dotted black lines show three randomly selected individual trajectories. From top to bottom, the horizontal dashed lines indicate depletion at B_{MSY}, $0.8B_{MSY}$, and $0.4B_{MSY}$, and the S7 operating model MSY. Catch is given in units of 1,000 metric tons. White circles in the catch panels indicate realized TACs for the 2011–2012 through 2014–2015 fishing years.

Realized TACs obtained via the U60–40 MP between 2011–2012 and 2013–2014, and the U60–40+Floor MP in 2014–2015 all lie within the central 90% projection interval of simulations conducted in 2010 (Figure 5.3), suggesting that the simulated catch performance used in 2010 decision-making was a reasonable representation of short-term fishery performance. This audit of MSE simulations is similar to the retrospective analyses used in stock assessment where model estimates are compared as new data are added to the assessment. Comparing predicted catch to realized catch in an MSE context provides a mechanism to help industry and managers assess simulation model performance based on concrete and directly observable metrics.

Implementation and conclusions

Based on performance in simulation tests under the S7: S1-Posterior Mean scenario, the U60–40+Floor procedure was adopted by industry and DFO. While there was some concern about failures in the worst-case robustness tests, U60–40+Floor provided nearly identical performance to the original U60–40 without violating the most important conservation objective under any operating model scenario. Implementing the U60–40+Floor MP resulted in a 2014–2015 TAC of 2,129 metric tons, which is 6.5% above the proposed TAC floor, thus avoiding both undesirable effects that could have resulted from fixing the TAC at the 2013–2014 level. Computer simulation testing provided reasonable assurance that fishery conservation objectives would be met under U60–40+Floor to a similar degree as the existing U60–40 MP, while also not forcing the fishing industry to forgo an increase in yield.

Since 2006, the BC sablefish industry has aimed to rebuild sablefish biomass to more productive levels. Although the industry was aware that a rebuilding goal would be challenged by the short-term financial cost of lower TACs needed for biomass growth, realized TACs between 2011 and 2014 raised concerns about economic risks. Rather than engaging in a debate over correctness of the original simulation models or abandoning the MP altogether, industry and government instead turned to the MSE process to design a set of feasible actions that could potentially address short-term economic challenges. The actions involved revising objectives and initiating a search for alternative MPs that could alleviate the economic risks while continuing to promote stock growth. These steps reflect industry's ownership of the MSE process (Smith et al., 1999) as well as growing understanding that MSE is a process, not a product; in other words, the MSE approach is a vehicle for linking science and decision-making in a strategic way, rather than focusing myopically on short-term stock assessment modelling projects aimed at setting TACs. Recently, the BC sablefish industry further extended the MSE approach to develop research strategies for nontarget species and sensitive benthic areas, which increasingly demand equal consideration alongside traditional single-species economic benefits. In this way, the BC sablefish industry is using MSE to prepare for anticipated future challenges and risks.

The MSE process for BC sablefish sets fishery objectives, identifies challenges and risks to meeting those objectives, devises a set of feasible actions that will help the industry overcome those challenges, and then uses computer simulation as needed to evaluate the actions against plausible uncertainties. This strategic process strives to meet current standards of acceptable scientific and management practice and, in our experience, is the best way to provide ecological and economic stability in uncertain times.

References

Beverton, R.J.H. & Holt, S.J. 1957. *On the Dynamics of Exploited Fish Populations.* Dordrecht: Springer Netherlands. doi:10.1007/978-94-011-2106-4

Cox, S.P. & Kronlund, A.R. 2008. Practical stakeholder-driven harvest policies for groundfish fisheries in British Columbia, Canada. *Fisheries Research, 94*(3), 224–237. doi:10.1016/j.fishres.2008.05.006

Cox, S.P., Kronlund, A.R. & Benson, A.J. 2013. The roles of biological reference points and operational control points in management procedures for the sablefish (Anoplopoma fimbria) fishery in British Columbia. *Environmental Conservation, 40*(4), 318–328. doi:10.1017/S0376892913000271

Cox, S.P., Kronlund, A.R. & Lacko, L. 2011. *Management Procedures for the Multi-gear Sablefish (Anoplopoma fimbria) Fishery in British Columbia, Canada* (No. 2011/063) (pp. 1–140). Canadian Science Advisory Secretariat Research Documents. Available at www.dfo-mpo.gc.ca/Csas-sccs/publications/sar-as/2011/2011_025-eng.pdf (last accessed 19 November 2015).

Cox, S.P., Kronlund, A.R. & Wyeth, M.R. 2010. *Development of Precautionary Management Strategies for the British Columbia Sablefish (Anoplopoma fimbria) Fishery* (No. 2009/043). DFO Canadian Science Advisory Secretariat Science Response (pp. vi–145).

de la Mare, W.K. 1998. Tidier fisheries management requires a new MOP (management-oriented paradigm). *Reviews in Fish Biology and Fisheries, 8*(3), 349–356.

DFO. 2013. *Pacific Region Integrated Fisheries Management Plan: Groundfish.* Effective February 21, 2013, Version 1.1. Fisheries and Oceans Canada.

DFO. 2014. *Performance of a Revised Management Procedure for Sablefish in British Columbia* (No. 2014/025). DFO Canadian Science Advisory Secretariat Science Response (pp. 1–13).

Punt, A.E. 2003. Extending production models to include process error in the population dynamics. *Canadian Journal of Fisheries and Aquatic Sciences, 60,* 1217–1228.

R Development Core Team. 2008. *R: A Language and Environment for Statistical Computing.* R Foundation for Statistical Computing, Vienna, Austria. ISBN 3–900051–07–0. Available at www.R-project.org (last accessed 19 November 2015).

Sainsbury, K.J., Punt, A.E. & Smith, A.D.M. 2000. Design of operational management strategies for achieving fishery ecosystem objectives. ICES Journal of Marine Science, 57(3), 731–741.

Smith, A.D.M., Sainsbury, K.J. & Stevens, R.A. 1999. Implementing effective fisheries management systems – management strategy evaluation and the Australian partnership approach. ICES *Journal of Marine Science,* 56, 967–979.

Chapter 6

Management procedures for New Zealand rock lobster stocks

Paul A. Breen, Nokome Bentley, Vivian Haist,
Paul J. Starr and Daryl R. Sykes

Introduction

The red rock lobster (*Jasus edwardsii*) supports the most valuable inshore commercial fishery in New Zealand, with exports worth about NZ$200 million (New Zealand Ministry for Primary Industries (MPI) unpublished data) and is also valuable to recreational and customary Maori fishers. The commercial trap or pot fisheries have been managed since 1991 with individual transferable quotas in nine arbitrary stocks (Figure 6.1; see Annala and Esterman 1986) and with minimum legal sizes (MLS; general across all stocks with three stock-specific exceptions), protection of ovigerous females and regulations for escape gaps that allow sublegal lobsters to escape from pots.

For major stocks, the total allowable commercial catch (TACC) is reviewed periodically in quantitative stock assessments (e.g. Starr et al. 2014), using the integrated Bayesian length-based model of Haist et al. (2009). For two smaller stocks a surplus production model has been used. Assessments are done by a small team of contractors to the NZ Rock Lobster Industry Council Ltd. (NZ RLIC Ltd.), which holds the contestable MPI contract for data collection and analyses. The contractors are independent from each other but work collegiately to produce the stock assessment. Management is the responsibility of a stakeholder entity: the National Rock Lobster Management Group with representatives from MPI, NZ RLIC Ltd., Maori and recreational fishers.

Each stock assessment is time-consuming. Financial and scientific resources are limited: the costs of assessment, management and compliance are borne largely by levies on the commercial fishery plus some government contribution for noncommercial fisheries. Accordingly, major stocks are assessed in rotation, about once every 5 years, and less important stocks irregularly. Historically, changes to the catch limit were associated with a stock assessment, but a stock could experience large changes in abundance between assessments without management action being taken.

The stakeholder management group mitigates this problem by adopting management procedures (MPs), which allow a change in the TACC every year if necessary, even with no stock assessment. MPs (see Butterworth and

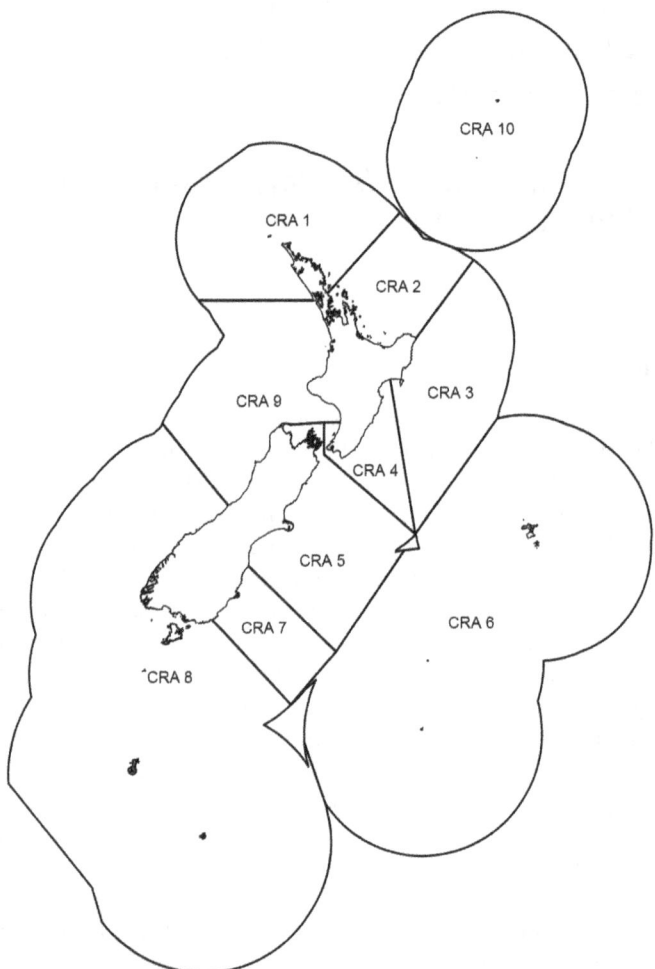

Figure 6.1 New Zealand red rock lobster (*Jasus edwardsii*) stocks or Quota Management Areas CRA 1 through CRA 9; the prefix "CRA" in the stock
name comes from the incorrect common name "crayfish."

Punt 1999) are decision rules that have been extensively simulation tested; they
are used worldwide; for example Johnston and Butterworth (2005) and Johnson et al. (2014) describe MPs for rock lobsters in South Africa. MPs specify
what data will be used as input, what the output will be and the mathematical relation between the two. Rock lobster MPs in New Zealand use standardised catch per unit effort (CPUE) as input and they output a catch limit:
either a TACC or Total Allowable Catch (TAC), comprising the TACC plus
noncommercial allowances. Usually the MP evaluation process takes place in

conjunction with a stock assessment and the fitted assessment model is used as an operating model.

The first New Zealand lobster MP was adopted in 1996 to rebuild the depleted CRA 8 stock and to manage the volatile CRA 7 stock (Starr et al. 1997). For both stocks, this MP specified changes in TAC from CRA 8 CPUE, a procedure justified by apparent migration of lobsters from CRA 7 to CRA 8. This MP compared CPUE with a target value, compared CPUE trend with a target rate of increase and combined these comparisons. Explorations by Breen et al. (2003) showed that long-term behaviour of this MP was likely to be unstable. Subsequent MPs have used CPUE in a much simpler way, described in more detail below. In 2002, separate MPs were developed for CRA 7 and CRA 8 and were reviewed and revised in 2007 and 2012 (Haist et al. 2013).

The input to all New Zealand lobster MPs evaluated so far has been standardised CPUE (see Starr 2014), which is assumed to be an index of vulnerable biomass. Some work has investigated the use of MPs with additional inputs (e.g. settlement indices, Bentley et al. 2005, Bentley unpublished), but so far other inputs have not been used. Before 2007, the input CPUE was from the preceding fishing year, which runs from April through March and is named by the first year, thus "2011–2012" is called "2011." This approach created a 1-year lag between observed CPUE and resulting catch limit: the fishing year ended on the 31st of March and any new catch limit from the MP was applied in April of the next year. To shorten the lag to 6 months, "offset-year" CPUE is calculated using the October through September year.

The need to rebuild depleted stocks has been a strong motivation for developing MPs, which appear to have been successful; results in CRA 7 and CRA 8 impressed stakeholders and led to MP development in other stocks. In the CRA 4 fishery described later, industry adopted an MP, before any formal adoption by MPI, to reduce their catches voluntarily (quota "shelving"; Breen et al. 2009a). In 2007 the CRA 5 industry also commissioned an MP to be operated as a voluntary quota shelving tool; they were concerned about a decline in their CPUE and worried that their stock might decline sharply as seen in other stocks near that time. MP evaluations were made with a surplus-production operating model and CRA 5 adopted a quota-shelving MP for the 2008 fishing year onwards (Breen 2009). This was reviewed in association with a stock assessment in 2010 (Haist et al. 2011), and after more evaluations in 2011 an MP was agreed by the stakeholder management group and adopted by the Minister.

The CRA 3 MP was developed in 2009 to rebuild the depleted stock (Breen et al. 2009b), agreed by the stakeholder management group and adopted by the Minister for the 2010 fishing year; it was used to set catch limits for the next 4 years, then was reviewed in 2014 and is likely to be replaced with a new MP. The CRA 2 MP was developed in 2013 in association with a stock assessment (Starr et al. 2014) that indicated the stock was above the biomass associated with maximum sustainable yield (B_{MSY}) but below the level desired by stakeholders and the stakeholder management group. The MP chosen by stakeholders and the stakeholder management group is intended to rebuild the stock and was adopted by the Minister for the 2014 fishing year.

For CRA 9, MPs were evaluated in 2013 with a surplus-production based operating model outside the formal stock assessment schedule (Breen 2014). An MP was agreed by the stakeholder management group and adopted by the Minister for the 2014 fishing year. Fits of a surplus-production model suggested this stock was well above B_{MSY} with a low exploitation rate. The adopted MP allows increased harvest with minimised risk and gave an immediate TACC increase of about 30%.

Fishery objectives and current management

The New Zealand fisheries management goal is defined formally but loosely in legislation: the Minister for Primary Industries must set catch limits that move the stock towards a level at or above B_{MSY} (sometimes B_{MSY} is used as a target reference point). MPI has published guidelines for additional reference points, such as a limit of 20% of the unexploited spawning stock biomass, below which the stock should not fall (Ministry of Fisheries 2011). For lobsters, B_{MSY} is taken to be vulnerable biomass: that which is available to be retained legally by the fishery. Because B_{MSY} is difficult to define operationally, an empirical proxy is sometimes used, based on a past vulnerable biomass level where the stock appeared to be safe and productive (B_{REF}). The lowest observed vulnerable biomass level of the stock, B_{MIN}, is often used as a limit reference point.

There are no operationally defined social goals for fisheries management in New Zealand. MPI has an unstated goal to see fisheries managed to produce something like maximum economic yield (MEY), but stock assessments are not asked to address this. Many New Zealand stocks (all species) are thought to be at or above their nominal management target (Mace 2014), reflecting an appreciation that MEY comes from higher stock levels than B_{MSY}.

In discussions with stakeholders, high yield does not feature as a management goal. Most commercial fishers understand the trade-off between yield and abundance and they indicate a preference for an average catch substantially less than MSY, taken from a stock substantially larger than B_{MSY}. A larger stock size allows higher catch rates and reduced costs of fishing. Commercial fishers therefore appear to favour an approach that delivers good economic yield rather than high yield. Safety (a low risk of stock depletion) is also often promoted in stakeholder discussions. Fishers have seen depleted stocks and in some fisheries have struggled through the rebuilding; they therefore want healthy stocks to stay healthy. Commercial lobster fishers value stability of catch limits, high abundance and CPUE, and having a wide size range of lobsters so they can respond to changes in market demand. They have stated their willingness to trade off potential catch for stability and abundance goals.

The next section focuses specifically on the development of an MP for CRA 4, which has now been managed in this way for nearly a decade.

Management procedures for CRA 4

In 2005, before adoption of the first MP, CRA 4 commercial catch was 504 t while the TACC was 577 t (Figure 6.2). In late 2006, it was obvious that 2006 commercial catch would be even lower – it turned out to be 445 t. A series of industry-organised meetings discussed remedial options, including using a MP to specify how much TACC should be voluntarily shelved each year. After a CRA 4 stock assessment in 2005 (Breen et al. 2006), a large set of MPs had been evaluated with an operating model based on the stock assessment model (Breen and Kim 2006); these authors evaluated simple MPs that used only CPUE and more complex MPs that used CPUE, its gradient over several years, and a target value.

A CPUE-only harvest control rule was chosen (National Rock Lobster Management Group 2009) to set the voluntary catch limit (SCC) for 2007 after considering yield, abundance and stability indicators (Figure 6.3; Breen et al. 2006):

$$SCC_{y+1} = 500\left(\frac{I_y}{0.9}\right)^{1.4} \tag{6.1}$$

where I_y is the standardised CPUE from April through September in the preceding year. The MP originally evaluated by Breen and Kim (2006) had an

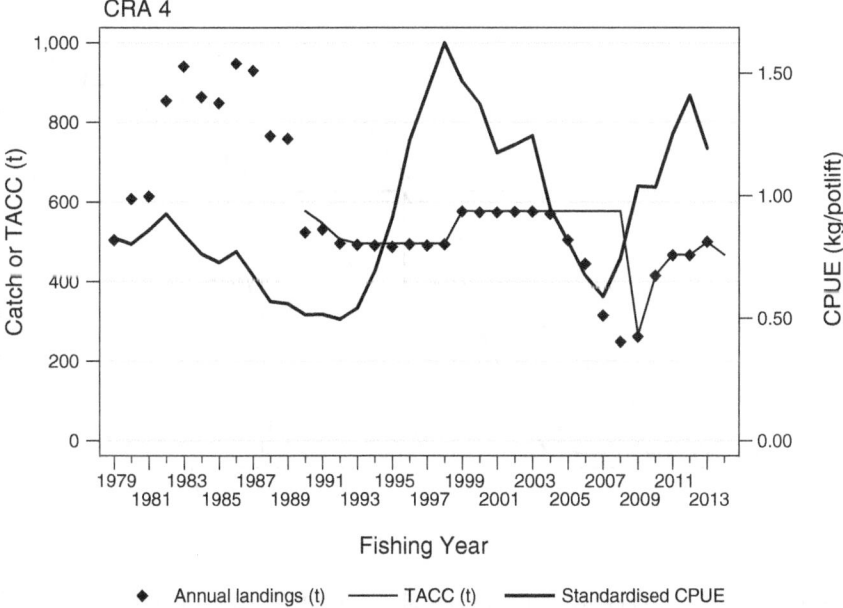

Figure 6.2 CRA 4 catch (diamonds), TACC (thin solid line) and standardised annual CPUE (thick solid line, using right-hand y-axis).

asymmetric latent year (SCC could decrease but not increase if a change had been made in the previous year). The latent year was dropped, at the request of NZ RLIC Ltd., after examination of the performance of the MP without it. The maximum allowable change was 75% and the minimum change was 5%. Because of the urgency of the situation and the absence of a CRA 4 industry body, NZ RLIC Ltd. made the final choice of MP on behalf of CRA 4 stakeholders.

In this period, harvest control rules were represented by a simple relation such as that described earlier. The line could be straight or curved, could be in two sections and could go through the origin or not. As CPUE increased the catch limit increased and vice versa, subject to the latent year and change thresholds. Such MPs were very responsive to apparent changes in stock size (i.e. changes in CPUE), but they changed the catch limit often.

When operated voluntarily in late 2006, the CRA 4 MP specified a catch limit of 321 t. Most, but not quite all, quota owners shelved the required proportion (44%) of their individual quota, resulting in an operational limit of 339 t, a 41% reduction from the TACC set by MPI. CPUE decreased in 2007, and in late 2007 the MP specified a catch limit of 228.9 t, a reduction of 60% of the

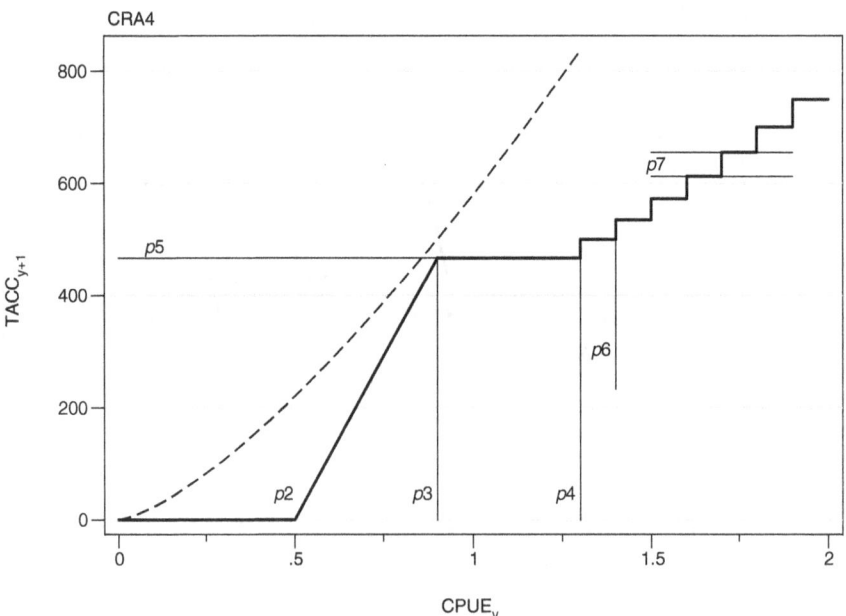

Figure 6.3 CRA 4 harvest control rules showing TACC (t) as functions of CPUE (kg/pot). Dashed line: the original rule; solid line: the newer rule developed in 2011. The new rule is an example of a generalised plateau step rule described in the text.

TACC set by MPI. Again, most but not all quota owners agreed to shelve, and the operational limit was 245 t, a 57% reduction from the TACC. In this period, some recreational fishers voluntarily reduced their bag limit.

Voluntary shelving of commercial quota on this scale was a major achievement of NZ RLIC Ltd. and a successful outcome of the property-rights management regime in New Zealand, which was designed to encourage stewardship by the resource users (Hilborn 2007). For the first shelving year, most commercial quota owners agreed that the stock was declining quickly and that something needed to be done; most accepted the proposed MP on that basis, knowing that the catch limit would increase as the stock increased. For the second shelving year, many owners were persuaded by the idea that either they could accept the MP's result for voluntary reduction or MPI would cut the TACC anyway and they would have no mechanism to recover the cut when the stock increased. However, some quota owners thought that the specified reduction was too great. Most of those owners who refused to shelve (owning about 5% of the quota) considered that the MP was unnecessary despite declining CPUE. Some argued that their catch rates were much better than those of the average fishers, whose rates dragged the overall CPUE down; they were not persuaded by plots that showed little difference in CPUE trends between the top 50% and the rest of the fishers. Industry agreed to the voluntary shelving only if at least 95% of quota were shelved, so a 5% nonparticipation rate was tolerated.

In late 2008, CPUE had increased as the stock responded to the decreased commercial catch and the rule specified a catch limit of 265.9 t. The Minister formally adopted the CRA 4 MP to guide statutory TAC setting, beginning with the 2009 fishing year. The TAC was calculated by adding the noncommercial allowances to the TACC, which for 2010 was 266 t, a 54% reduction from the original 577 t TACC. In late 2009, the harvest control rule specified a TACC of 477.59 t, but the maximum change threshold resulted in a TACC of 465.50 t for 2010.

A new stock assessment was made in 2011 (Breen et al. 2012) and new MP evaluations were made with a variety of harvest control rules. After their experience with the original MP, CRA 4 fishers requested a more stable MP that was more cautious at high CPUE than the original had been. At the suggestion of the CRA 8 industry in 2007, the "plateau rule" had been developed (Figure 6.3). This rule form tries to maintain a catch limit over a range of CPUE values and to respond safely at low CPUE; it has a mechanism for catch limit increase at high CPUE. The plateau can be eliminated, giving a simpler rule with two sections. This rule family has nine parameters, defining the rule type ($p1$), the intercept ($p2$), the CPUE at both ends of the plateau ($p3, p4$), the plateau height ($p5$), the width of CPUE steps above the plateau ($p6$), the height of steps above the plateau as a proportion of the preceding TACC ($p7$) and thresholds for minimum and maximum change ($p8, p9$).

The provisional TACC (before operation of thresholds) is given by:

$$TACC_{y+1} = 0 \qquad \text{for } I_y \leq p2 \tag{6.2}$$

$$TACC_{y+1} = p5\left(\frac{I_y - p2}{p3 - p2}\right) \qquad \text{for } p2 < I_y \leq p3 \tag{6.3}$$

$$TACC_{y+1} = p5 \qquad \text{for } p3 < I_y \leq p4 \tag{6.4}$$

$$TACC_{y+1} = p5\left((1 + p7)^{floor\left(\left(I_y - p4\right)/p6\right)+1}\right) \qquad \text{for } I_y > p4 \tag{6.5}$$

where I_y is the standardised CPUE from October through September calculated in year y.

Management procedure evaluations

Operating model

The 2011 stock assessment of CRA 4 (Breen et al. 2012) was made with the length-based Bayesian model described by Haist et al. (2009), using a 6-month time step. It estimates annual recruitment to the model at a size well below MLS; an initial exploitation rate if the model is not fitted to the earliest catch data from 1945; natural mortality rate; four growth parameters for each sex for growth; two parameters for female maturation-at-size; four parameters for each sex for size-selectivity of the commercial fishery in two different periods; four parameters for the relative seasonal vulnerability of each sex (with mature and immature females considered separately); catchability coefficients for two CPUE series and for larval settlement; and a parameter for the shape of the relation between CPUE and abundance.

Although the model can be fit to multiple stocks, the CRA 4 stock assessment used a single stock model. Instantaneous model dynamics were chosen, with iterative estimation of the fishing mortality rate, F. The model was implemented with AD Model Builder (Fournier et al. 2012) and fit to two CPUE series, length frequency data from observers and voluntary commercial logbooks, growth increments from tag-recaptures, and larval settlement indices. The estimated CRA 4 stock status was based on posterior distributions of estimated and derived parameters, determined with Markov chain Monte Carlo simulations (McMC).

The stock assessment model was modified to make an operating model to evaluate simulated MP performance, addressing a range of uncertainties. Parameter uncertainty was addressed by sampling from the joint posterior distribution of estimated parameters and environmental uncertainty was addressed by using stochastic variation in both recruitment and CPUE observation error. The scales of stochastic uncertainties were based on the variance and autocorrelation seen in estimated recruitment deviations and CPUE residuals in each sample of the joint posterior. Estimated recruitment deviations are not distributed uniformly in time, so the scale of variation depends on the period chosen as representative. Using a shorter series would reflect a belief that average recruitment has changed from the long-term to the short-term average, while using a long series would imply belief in stationary average recruitment with annual variation. In recent work we have used the most recent 10 years of estimates, reflecting our belief that average recruitment may change over time. Although MPs are intended to be put into place for 5 years, projections are made for 20 years so that each run can experience the effects of stochastic variation in recruitment.

The operating model predicted vulnerable biomass for each future 6-month time step (autumn-winter and spring-summer) and used its catchability estimate, sampled from the posterior, to calculate offset-year CPUE for input to the MP under evaluation, which returned a TACC for the next fishing year. The model used a regression developed outside the model to predict the autumn-winter catch proportion as a function of autumn-winter CPUE. Recreational catch was assumed to be proportional to vulnerable biomass in spring-summer, when most of the recreational catch is taken. The model projected annual recreational catch from a constant exploitation rate calculated from each sample of the joint posterior. Customary Maori and illegal catches, about which little was known, were fixed at the best available values used in the stock assessment.

The base case CRA 4 stock assessment McMC, with 1,000 samples of the joint posterior, was used for the base case MP evaluations. Sets of "robustness trials" involved MP evaluations with alternative operating models to address major uncertainties in the assessment. We used arbitrarily reduced recruitment projections to compare how rules performed when conditions were poor, an alternative noncommercial catch vector and a less optimistic set of posterior samples from an alternative fit; the last two trials involved alternative McMCs.

Performance indicators

We measured MP performance against four criteria specified by the stakeholder management group: (1) yield – mean commercial and recreational catches and the minimum commercial and recreational catches from each run; (2) stock abundance – the minimum CPUE and vulnerable biomass; (3) safety – the number of years when spawning biomass was less than B_{MSY}, B_{REF} and B_{MIN}; and (4) stability – the average annual change in TACC, the number of years when CPUE was near an arbitrary reference value and the number of TACC changes during the run. Other indicators were also calculated, but these were the main ones used for rule comparisons.

For yield and abundance indicators there were posterior distributions at the end of the 1,000 runs for each rule. For example, commercial catch was averaged for each 20-year run, and for minimum commercial catch there was one value from each run. Annual variation in TACC was also averaged for each run. Posterior distributions of indicators were then summarised by their median and

5th and 95th quantiles. Safety indicators, which are the proportion of years in which biomass is below some reference level, and the proportion of years with TACC change, were calculated simply as the proportion of total years in a set of runs.

The indicators in these four groups have strong trade-offs. Most importantly, higher average catch is associated with lower average abundance and tends to be associated with lower safety and lower stability. Safety and abundance are positively related; safety and stability are negatively correlated when average abundance is low.

Evaluating MPs

With many candidate MPs, many performance indicators and several alternative operating models, the MP evaluations generated large quantities of results. We used several approaches to choose a subset of MPs to present to the stakeholder management group (Breen et al. 2012).

In theory, utility functions could be developed to score MPs by their various performance indicators. Bentley et al. (2003a) describe a multiplicative utility function that gives zero utility to a MP when any component has zero utility. This approach requires explicit quantitative utility functions for each relevant indicator. Although it could be useful, the utility functions would be very hard to define with any objectivity and stakeholders are probably not yet equipped to make the necessary decisions about utility functions.

A second, loosely related, approach is to define minimum acceptable levels for key performance indicators. MP evaluation results are then analysed to determine the joint probability, for each tested MP, of meeting these minimum levels. This approach was used to rank MP candidates for the 2002 CRA 8 management procedure (Bentley et al. 2003b), but has not been used subsequently. A disadvantage is that all minimum levels must be coded into the operating model before any evaluations are made.

For the 2011 CRA 4 MPs, our approach was first to use "screening" to eliminate unsuitable MPs. Thresholds were defined for key indicators. Safety indicators have predefined thresholds: for instance, 5% for the proportion of years less than B_{MIN}; 50% for the proportion of years less than B_{MSY} or B_{REF}. Thresholds for other indicators were chosen after exploring aggregated results across all simulations. MPs that failed to meet a threshold in the base case results were discarded. This initial screening ensures that any MP eventually chosen would be consistent with the safety thresholds required by MPI.

Next, we relied on the "choice frontier" described by Bentley et al. (2003a). When MP performances were plotted against each other (e.g. Figure 6.4), trade-offs became obvious. In this example, the average annual variation in TACC showed wide variation at a given level of average commercial catch. MPs on the bottom edge of the figure were more desirable with respect to this trade-off than those above, so the bottom edge was a choice frontier.

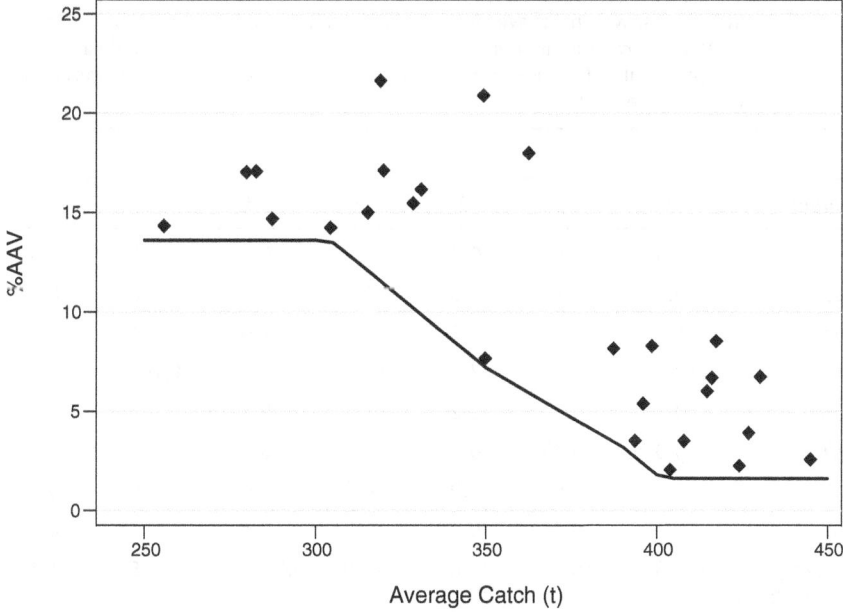

Figure 6.4 CRA 4 MP evaluations showing the trade-off between average commercial catch and average annual variation in TACC (AAV) among a set of 27 MPs considered after screening. The solid line shows the "choice frontier": MPs along this frontier have high stability for a given average commercial catch. Average commercial catch is the median of the posterior distribution of the mean catch in each run.

In MP development, the scientists' role is to meet with all stakeholder groups, discover which indicators are most important and which trade-offs are most critical, and then produce MPs that deliver the stakeholders' goals as well as MPI's safety objectives. After identifying a range of final CRA 4 candidate MPs, the scientists' job was nearly over. When the "unsafe" MPs were eliminated, all remaining MPs were likely to meet safety thresholds. Choosing a MP then involved choosing the compromise between yield and abundance, yield and stability, etc. The yield/abundance trade-off was an economic choice that commercial stakeholders were best placed to evaluate from their expert knowledge of costs and markets.

Evaluation results

The scientific team presented a set of six final candidate MPs to the stakeholder management group, listed in Table 6.1 (Breen et al. 2012). Commercial stakeholders asked to see results from a minor variation of one of these candidates;

Table 6.1 Upper part: parameters for the seven final rules from evaluations for CRA 4 (see text for parameter definitions); rule 1a was a variant requested by industry; lower part: average rule performance against six key indicators; AAV is average annual variation in TACC; the bottom two indicators are the probability (proportion of years) in which CPUE was less or more than the indicated value.

						rule	
quantity	1	1a	2	3	4	5	6
$p2$	0.5	0.5	0.3	0.3	0.3	0.3	0.3
$p3$	0.8	0.9	1.0	0.8	0.8	1.0	0.9
$p4$	1.3	1.3	1.2	1.0	1.0	1.1	1.1
$p5$	467	467	467	445	400	400	400
$p6$	0.1	0.1	0.2	0.2	0.25	0.1	0.1
$p7$	0.07	0.07	0.1	0.05	0.1675	0.05	0.05
$p8$	0.0	0.0	0.0	0.0	0.0	0.0	0.0
$p9$	0.25	0.25	0.25	0.25	0.25	0.25	0.25
average catch (t)	457.3	438.7	426.6	446.9	425.5	391.5	401.1
average CPUE (kg/pot)	0.904	0.939	0.965	0.925	0.971	1.040	1.019
average rec. catch (t)	44.3	45.7	46.8	45.2	47.1	50.0	49.1
average AAV (%)	3.6	7.6	8.3	2.6	5.0	6.6	3.8
P(CPUE < 0.8)	0.118	0.048	0.025	0.095	0.036	0.003	0.011
P(CPUE > 1.2)	0.024	0.030	0.032	0.027	0.036	0.106	0.092

Editor's note: Rules 1, 1a, 2, 3, 4, 5 and 6 correspond to rules 28, 28a, 29, 30, 31, 32 and 33 in Breen et al. (2012).

it was evaluated at the industry meeting. The seven final MPs spanned a range of aggressiveness and associated indicators. The stakeholder management group chose one MP from this set (the industry-requested variant, Table 6.1), went to public consultation and then recommended that the Minister for Primary Industries adopt it for the 2012–2013 season. The Minister did so and it has been used each year since then.

Fishery implementation

Seven New Zealand lobster stocks now have MPs and an MP has been proposed for the eighth stock (CRA 1). All are either the plateau step rule described earlier for CRA 4 or a variant with a slope above the plateau (plateau slope rule) (Figure 6.5). When CPUE is to the right of the plateau, plateau slope rules are defined by:

$$TACC_{y+1} = p5\left(1 + \frac{0.5\left(I_y - p4\right)}{p6 - p4}\right) \qquad \text{for } I_y > p4 \qquad (6.6)$$

where $p6$ determines the slope and $p6$ is the CPUE at which the TACC is 1.5 times the plateau height.

The MPs in use are summarised in Table 6.2. Two new MPs were introduced in 2014 but five MPs have been in place long enough that associated changes can be compared (Figure 6.6).

The overall number of lobster vessels has declined by a factor of two since 1990, but vessel decline in the past decade has been greater in the five stocks with MPs. The number of potlifts has been flat in stocks without MPs but has declined strongly since 2000 in stocks with MPs. Catch has been flat in stocks

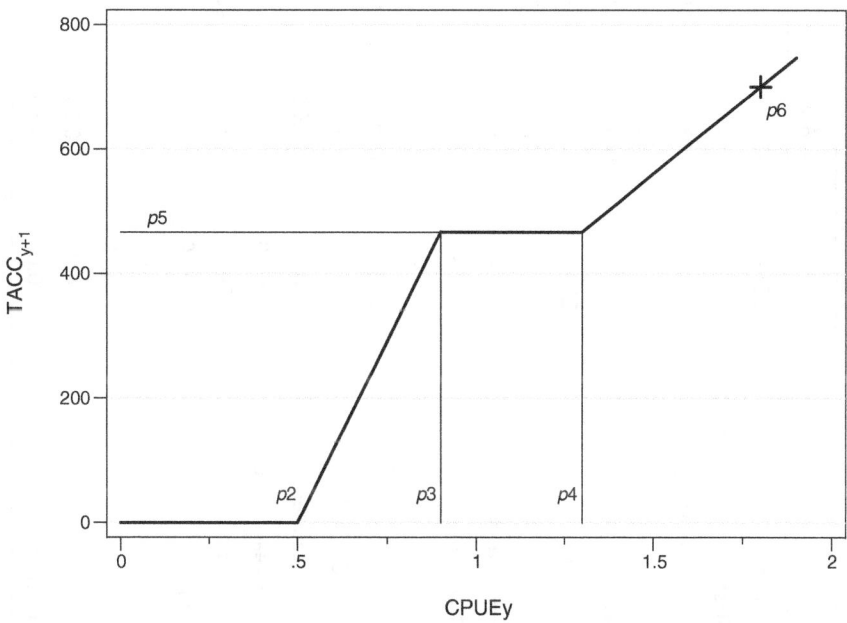

Figure 6.5 A generalised plateau slope rule; see text for parameter definitions.

Table 6.2 Summary of MPs implemented for New Zealand rock lobster stocks during 2014. Fishery information is from Ministry for Primary Industries (2014) and Starr (2014). The bottom part of the table shows parameters for the generalised plateau harvest control rules.

	CRA 2	CRA 3	CRA 4	CRA 5	CRA 7	CRA 8	CRA 9
first year of MP	2014	2010	2010	2009	1996	1996	2014
potlifts in first year (thousands)	611	149	410	228	249	1,031	–
potlifts 2013 (thousands)	–	113	359	239	29	291	13
% change	–	–25	–12	5	–89	–72	–
catch in first year (t)	–	164	415	350	63	862	–
catch in 2012 (t)	236	224	499	350	44	964	47
% change	–	37	20	0	–30	12	–
CPUE in first year (kg/pot)	–	1.10	1.01	1.54	0.25	0.84	–
CPUE in 2012 (kg/pot)	0.39	1.99	1.39	1.47	1.54	3.31	3.69
% change	–	82	37	–5	510	295	–
p1 rule type	step	slope	step	step	slope	slope	step
p2 CPUE at TACC = 0	0.00	–3.00	0.50	0.30	0.17	0.45	0.50
p3 CPUE at plateau left	0.30	1.00	0.90	1.40	1.00	1.90	1.00
p4 CPUE at plateau right	0.50	1.00	1.30	2.00	1.75	3.70	1.40
p5 plateau height	200	275	467	350	80	962	40
p6 step width	0.10	–	0.10	0.20	–	–	0.75
p6 slope	–	1.6	–	–	3	8.624	–
p7 step height	0.10	–	0.07	0.05	–	–	0.15
p8 minimum change	0.05	0.05	0.00	0.00	0.10	0.05	0.05
p9 maximum change	0.00	0.10	0.25*	0.00	0.50	0.00	0.15**

* Effective only when CPUE is left of the plateau.
** Threshold for maximum TACC increases only.

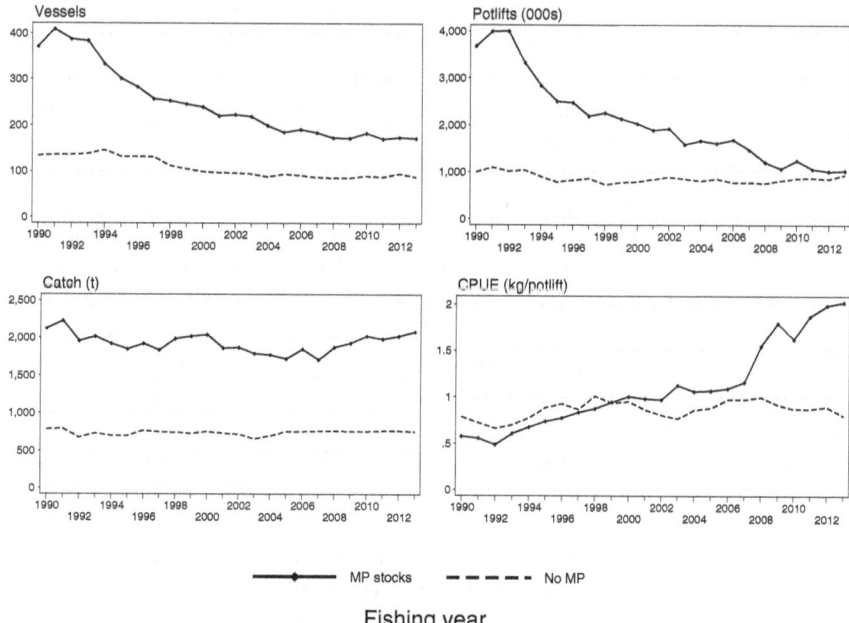

Figure 6.6 Comparing the aggregate fisheries in five *Jasus edwardsii* stocks with MPs (CRA 3, CRA 4, CRA 5, CRA 7 and CRA 8) with the four stocks that did not have MPs before 2014 (CRA 1, CRA 2, CRA 6 and CRA 9). CPUE in the bottom right is arithmetic CPUE: catch divided by the number of potlifts.

Source: Catch data from Ministry of Primary Industries (2014) and CPUE data from Starr (2014).

without MPs but has increased since 2008 in stocks with MPs. CPUE has also been flat since 1997 in stocks without MPs but has increased strongly in stocks with MPs. In the five stocks that have had MPs for some time (i.e. CRA 3, CRA 4, CRA 5, CRA 7 and CRA 8), effort has reduced in four and stayed the same in CRA 5; catch has increased in three; CPUE has increased in four and decreased slightly in CRA 5; in CRA 7 and CRA 8 the CPUE increase has been two- or threefold (Table 6.2).

Credit for these changes is not all due to MPs: some stocks have performed well without much intervention; in CRA 3 industry did voluntary TACC shelving before MPs were adopted. But much of the credit is due to MPs that were adopted successfully to rebuild depleted stocks.

Success of these MPs lies not just in rebuilding depleted stocks and maintaining healthy stocks. MP development has encouraged stakeholders to think about and discuss their goals for management, consider and choose among trade-offs and work to agree on safe and effective harvest strategies. In the rock lobster fishery, there has been a shift in thinking towards the strategic (Bentley and Stokes 2009). Commercial industry, MPI and the customary Maori stakeholders are now very supportive of MPs. Recreational stakeholders were supportive in the past but are currently dissatisfied with fisheries management generally for several reasons. It is a positive result that a series of Ministers, in whose hands the TACC and allowance decisions always lie, have also been persuaded of the benefits of MPs and have chosen to accept MP results throughout their short New Zealand history.

The darkest cloud on the horizon involves noncommercial fishing. Noncommercial fishing is not strongly regulated: illegal fishing is partly suppressed by enforcement, but lobsters have very high value and are easily taken, hidden and sold. Recreational fishing is limited only by the MLS and bag limits. If noncommercial catch responds adaptively to increasing biomass, then TACC-setting MPs will tend to pump catch away from the commercial sector (by reducing TACCs when the stock decreases) into the noncommercial sector, who can increase their share as the stock tries to recover. The current situation is manageable only because noncommercial catches are relatively small in many stocks.

Our single attempt to incorporate a noncommercial allowance into an MP (Haist et al. 2011) was rejected by the Minister for Primary Industries: this would have involved changing customary Maori, illegal catch and recreational allowances as well as the TACC component of the TAC. These allowances, unlike the TACC, do not constrain the catch but are nonetheless contentious, so an MP that made frequent changes to noncommercial allowances would have been controversial with no practical gain. Noncommercial fishing will be an increasing problem as the noncommercial catches increase, until MPI is forced to develop an effective allocation strategy.

Acknowledgements

Thanks to our esteemed colleagues with whom we worked on lobster MPs: Andrew Branson, Charles T.T. Edwards, Terese Kendrick, Susan Kim, Marine Pomarede, Adam Smith, Kevin Stokes and D'Arcy Webber; colleagues at MPI, especially Mark Edwards, Alicia McKinnon and Kevin Sullivan; members of the National Rock Lobster Management Group; and many individual fishermen and quota owners in the commercial rock lobster fishery.

References

Annala, J.H. & D.B. Esterman. 1986. Yield estimates for the New Zealand rock lobster fishery. *In:* Jamieson, G.S. & Bourne, N. (eds.) *North Pacific Workshop on Stock Assessment and Management of Invertebrates.* Canadian Special Publication of Fisheries and Aquatic Sciences 92, pp. 347–358. Government of Canada.

Bentley, N., Breen, P.A., Kim, S.W. & Starr, P.J. 2005. Can additional abundance indices improve harvest control rules for New Zealand rock lobster fisheries (*Jasus edwardsii*)? *New Zealand Journal of Marine and Freshwater Research, 39*(3): 629–644.

Bentley, N., Breen, P.A. & Starr, P.J. 2003b. Design and evaluation of a revised management decision rule for red rock lobster fisheries (*Jasus edwardsii*) in CRA 7 and CRA 8. *New Zealand Fisheries Assessment Report 2003/30*.

Bentley, N., Breen, P.A., Starr, P.J. & Sykes, D.R. 2003a. Development and evaluation of decision rules for management of New Zealand rock lobster stocks. *New Zealand Fisheries Assessment Report 2003/29*.

Bentley, N. & Stokes, K. 2009. Contrasting paradigms for fisheries management decision making: how well do they serve data-poor fisheries? *Marine and Coastal Fisheries: Dynamics, Management, and Ecosystem Science*, 1(1): 391–401.

Breen, P.A. 2009. A voluntary harvest control rule for a New Zealand rock lobster (*Jasus edwardsii*) stock. *New Zealand Journal of Marine and Freshwater Research*, 43(3): 941–951.

Breen, P.A. 2014. CRA 9 Management procedure evaluations. *New Zealand Fisheries Assessment Report 2014/20*.

Breen, P.A., Haist, V., Starr, P.J. & Kendrick, T.H. 2009b. Development of a management procedure for the CRA 3 stock of rock lobsters (*Jasus edwardsii*). Unpublished Final Research Report to the Ministry of Fisheries for CRA200601, Objective 4. NZ RLIC Ltd., 30 November 2009.

Breen, P.A., Haist, V., Starr, P.J. & Pomarede, M. 2012. The 2011 stock assessment and management procedure development for red rock lobsters (*Jasus edwardsii*) in CRA 4. *New Zealand Fisheries Assessment Report 2012/09*.

Breen, P.A. & Kim, S.W. 2006. Development of an operational management procedure (decision rule) for CRA 4. *New Zealand Fisheries Assessment Report 2006/53*.

Breen, P.A., Kim, S.W., Bentley, N. & Starr. 2003. Preliminary evaluation of maintenance management procedures for New Zealand rock lobster (*Jasus edwardsii*) fisheries. *New Zealand Fisheries Assessment Report 2003/20*.

Breen, P.A., Kim, S.W., Haist, V. & Starr, P.J. 2006. The 2005 stock assessment of red rock lobsters (*Jasus edwardsii*) in CRA 4. *New Zealand Fisheries Assessment Report 2006/17*.

Breen, P.A., Sykes, D., Starr, P.J., Haist, V. & Kim, S. 2009a. A voluntary reduction in the commercial catch of rock lobster (*Jasus edwardsii*) in a New Zealand fishery. *New Zealand Journal of Marine and Freshwater Research*, 43(1), 511–523.

Butterworth, D.S. & Punt, A.E. 1999. Experiences in the evaluation and implementation of management procedures. *ICES Journal of Marine Science*, 56, 985–998.

Fournier, D.A., Skaug, H.J., Ancheta, J., Ianelli, J., Magnusson, A., Maunder, M.N., Nielsen, A. & Sibert, J. 2012. AD Model Builder: Using automatic differentiation for statistical inference of highly parameterized complex nonlinear models. *Optimization Methods Software*, 27, 233–249.

Haist, V., Breen, P.A. & Starr, P.J. 2009. A new multi-stock length-based assessment model for New Zealand rock lobsters (*Jasus edwardsii*). *New Zealand Journal of Marine and Freshwater Research*, 43(1), 355–371.

Haist, V., Breen, P.A., Starr, P.J. & Kendrick, T.H. 2011. The 2010 stock assessment of red rock lobsters (*Jasus edwardsii*) in CRA 5, and development of an operational management procedure. *New Zealand Fisheries Assessment Report 2011/12*.

Haist, V., Starr, P.J. & Breen, P.A. 2013. The 2012 stock assessment of red rock lobsters (*Jasus edwardsii*) in CRA 7 and CRA 8, and review of management procedures. *New Zealand Fisheries Assessment Report 2013/60*.

Hilborn, R. 2007. Moving to sustainability by learning from successful fisheries. *AMBIO: A Journal of the Human Environment*, 36, 296–303.

Johnston, S.J. & Butterworth, D.S. 2005. Evolution of operational management procedures for the South African west coast rock lobster (*Jasus lalandii*) fishery. *New Zealand Journal of Marine and Freshwater Research, 39,* 687–702.

Johnston, S.J., Butterworth, D.S. & Glazer, J.P. 2014. South coast rock lobster OMP 2014: initial specifications. Unpublished Report to the South African Department of Fisheries. *Fisheries/2014/SEP/SWG_SCRL/07.* Available at www.mth.uct.ac.za/maram/pub/2014/FISHERIES_2014_SEP_SWG-SCRL_07.pdf (last accessed 19 November 2015).

Mace, P. 2014. Interim update of the status of New Zealand's fish stocks. pp. 53–59 *In* P. Mace & M. Vignaux (eds.) *Fisheries Assessment Plenary May 2014 – Supplement. A celebration of 30+ years of fisheries science.* Unpublished report, Ministry for Primary Industries, Wellington, New Zealand.

Ministry of Fisheries. 2011. *Operational guidelines for New Zealand's Harvest Strategy Standard.* Unpublished report, Ministry of Fisheries, Wellington, June 2011. Available at http://fs.fish.govt.nz/Page.aspx?pk=113&dk=22847 (last accessed 19 November 2015).

National Rock Lobster Management Group. 2009. *Annual Report to The Minister of Fisheries, Hon. Phil Heatley.* Unpublished report held by the Ministry for Primary Industries, Wellington, December 2009.

Starr, P.J. 2014. Rock lobster catch and effort data: summaries and CPUE standardisations, 1979–80 to 2012–13. *New Zealand Fisheries Assessment Report 2014/28.*

Starr, P.J., Breen, P.A., Hilborn, R. & Kendrick, T.H. 1997. Evaluation of a management decision rule for a New Zealand rock lobster substock. *Marine and Freshwater Research, 48*(8), 1093–1101.

Starr, P.J., Haist, V., Breen, P.A. & Edwards, C.T.T. 2014. The 2013 stock assessment of red rock lobsters (*Jasus edwardsii*) in CRA 2 and development of management procedures. *New Zealand Fisheries Assessment Report 2014/19.*

Chapter 7

Fisheries management for regime-based recruitment

Lessons from a management strategy evaluation for the fishery for snow crab in the eastern Bering Sea

Cody S. Szuwalski and André E. Punt

Introduction

Forming expectations of the future productivity of a stock is one of the central problems in fisheries management. Catches deemed 'sustainable' today depend on expectations of tomorrow's production. This productivity assumption can be incorporated into assessments of stock status via reference points that indicate, for example, the spawning stock biomass at maximum sustainable yield (B_{MSY}). In this case, estimates of B_{MSY} include implicit considerations of productivity, making the assessment of status (i.e. biomass relative to the reference point) strongly dependent on this assumption. Expectations for the future productivity of a stock are often based on an assumed relationship between estimated spawning biomass and recruits (e.g. Ricker, 1954, Beverton and Holt, 1957). However, there is no obvious stock-recruit relationship for many stocks (Szuwalski et al., 2014); in these cases estimation of MSY-based reference points are difficult to justify, and proxies are needed for management (e.g. Clark, 1991). Proxies based on equilibrium spawning-biomass-per-recruit (SBPR) analyses are used to manage stocks in the North Pacific by the National Marine Fisheries Service/North Pacific Fisheries Management Council (NPFMC, 2007) of the United States. Applying the fishing mortality ($F_{35\%}$) that decreases the SBPR of a stock to 35% of virgin levels (SBPR$_{35\%}$) has been shown to produce near-maximum sustainable yields for a large range of life histories (Clark, 1991). The associated spawning biomass, $B_{35\%}$, is calculated by multiplying SBPR$_{35\%}$ by the expected recruitment. This expectation can be calculated as an average over historical recruitment estimates, which implicitly represent a manager's expectations of future productivity. Determining which estimates of recruitment are likely to reflect future recruitment can be a difficult task if environmental conditions change over time, and the outcomes can influence the perceived status of a stock and sustainability of the fishery.

Snow crab (*Chionoecetes opilio*) in the eastern Bering Sea (EBS) are distributed over much of the eastern Bering Sea shelf to depths of 200 meters (Figure 7.1), in a highly variable physical environment that has undergone 'regime shifts' which influenced the productivity of stocks in the area (e.g. Adkison et al., 1996, Hare and Mantua, 2000, Bailey, 2000, Conners et al., 2002). Reference points and measures of productivity for snow crab are based on mature male biomass (MMB) because the fishery only takes males. Calculating an equivalent reference point for both sexes combined would therefore require an assumption regarding how male fishing mortality affects female reproductive success, and there is currently no way to compute this relationship. The estimated MMB has fluctuated over time and supports a lucrative fishery (Figure 7.2b). However, it was declared overfished in 1999 when MMB was estimated to be below the minimum stock size threshold set by the NPFMC (Turnock and Rugulo,

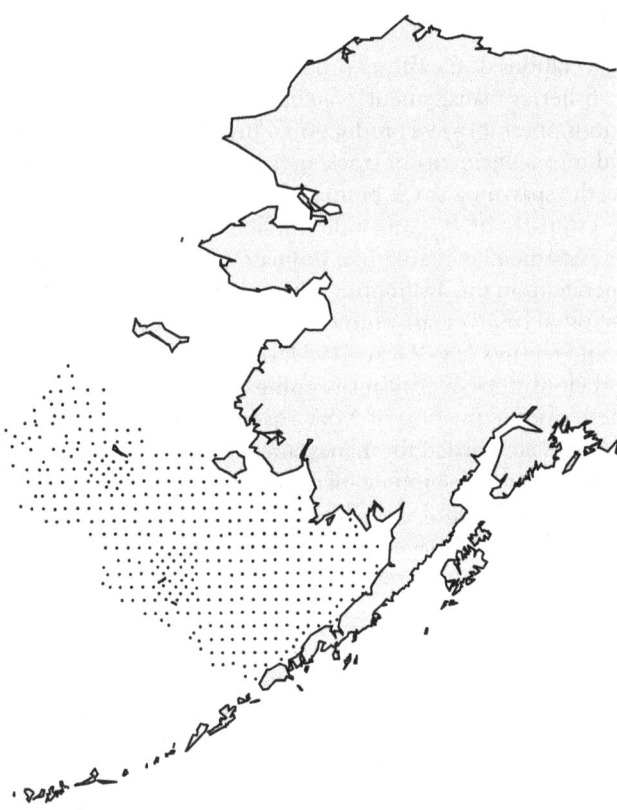

Figure 7.1 Survey location (dots) used to calculate an index of abundance for snow crab in the EBS.

2013). The overfished declaration initiated a rebuilding plan which required the stock to be rebuilt to the target biomass within 10 years (NPFMC, 2000). However, the stock did not rebuild in the allotted 10-year time period, due in large part to less than average recruitment over the rebuilding period.

It is important to have some understanding of recruitment dynamics in order to inform projected rebuilding times for overfished stocks, but no relationship between recruitment and mature male biomass is apparent for snow crab (Figure 7.2a). Consequently, proxies that depend on average recruitment are used for reference points in management. However, average recruitment for snow crab changes over time (Figure 7.2a), and this change may be related to shifts in the Pacific Decadal Oscillation (PDO; Szuwalski and Punt, 2013a). Estimates of recruitment from the past 'climate regimes' may not be relevant to expectations about future recruitment given the current state of the system if changes in average recruitment over time are a product of a shifting environment. Shifts to a lower productivity regime are particularly concerning for snow crab because using estimates of recruitment from a period of higher productivity to set management targets may have resulted in a stock that appears to be overfished.

A change in average recruitment preceded the overfished declaration for snow crab by roughly 10 years, which is approximately the amount of time it takes a fertilized egg to reach the mature male population and be selected by the fishery. Coupled with the coincidence of a change in the PDO and average recruitment, this observation suggests that the overfished declaration for snow crab could have been a result of shifting environmental conditions, rather than overfishing, because expected recruitment (and therefore the target $B_{35\%}$ reference point) would have been overestimated. Given this potential influence of the environment on the productivity of snow crab, we conducted a management strategy evaluation to compare the status quo harvest control rule to one

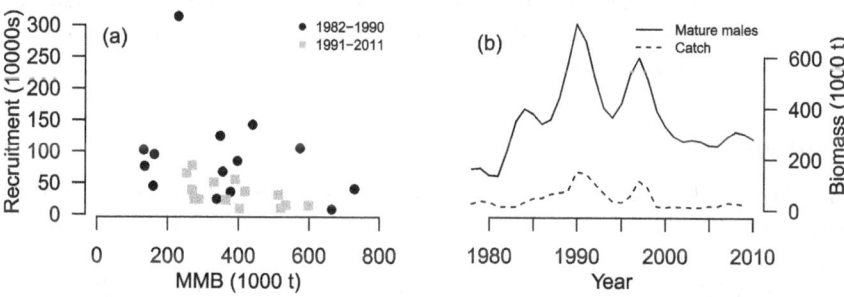

Figure 7.2 Estimated mature male biomass and observed catch (a), and estimated recruitment plotted against MMB with a period of relatively warm sea surface temperatures indicated by the circles and a relatively cold period indicated by the squares (b).

that updates expectations for future recruitment based on perceived changes in the underlying recruitment regime.

Current management and fishery objectives

The North Pacific Fishery Management Council (NPFMC) and the Alaska Department of Fish and Game (ADFG) jointly manage EBS snow crab. The national standards for fishery conservation and management in the reauthorized version of the Magnuson-Stevens Act (DOC, 2007) guide the management decisions of the NPFMC. The national standards of the Magnuson-Stevens Act state that fisheries management plans should prevent overfishing while pursuing 'optimum yield' (National Standard 1) and that management should be based on the 'best available science' (National Standard 2). The NPFMC primarily sets the overfishing level (OFL; the level of catch that corresponds to the application of $F_{35\%}$ to the mature male population), using the estimates of MMB from the NMFS assessment and a harvest control rule (HCR; Equation 7.1) (NPMFC, 2007). It then sets the Acceptable Biological Catch (ABC, less than the OFL to account for scientific uncertainty), and the minimum stock size threshold (MSST; the biomass at which the stock is declared overfished, corresponding to half of $MMB_{35\%}$). Finally, the ADFG sets the total allowable catch (TAC), which must be lower than the ABC.

Harvest control rules

We evaluated two HCRs, which differ in how average recruitment is calculated during estimation of the target biomass reference point. Both HCRs use estimates of current MMB from an assessment method that tracks crab by length, sex and maturity, and closely mimics that used for formal assessments of stock status (Turnock and Rugolo, 2013; see appendices for a description). The assessment model is fit to data from the directed fishery, bycatch from the nonpelagic trawl fishery for groundfish and data from the National Marine Fisheries Service (NMFS) summer survey.

Status quo HCR

Since 2008, federal scientists have used the current harvest control rule (i.e. the status quo HCR; Equation 7.1) to calculate the OFL. ADFG applies an additional HCR based on the total mature biomass at the time of the survey and the applied TAC is the smaller of the federally produced OFL and the state-produced catch recommendation (see Szuwalski and Punt, 2013b, for details on ADFG HCR). Management targets used in the HCR are based on proxies for the MMB at which maximum sustainable yield occurs (MMB_{MSY}) and the fishing mortality that produces that biomass at equilibrium (F_{MSY}) using spawning-biomass-per-recruit methods (e.g. Clark, 1991, NPFMC, 2007). $F_{35\%}$ is used as a target fishing mortality for the snow crab fishery based on the outcomes of a minimax analysis similar to that of Clark (1991) (NPFMC, 2007). $MMB_{35\%}$ is calculated each year as the $SBPR_{35\%}$ multiplied by an average recruitment calculated from the entire time series of estimated recruitment. Values of $F_{35\%}$ and $MMB_{35\%}$ are used in conjunction with a control rule to adjust the proportion of $F_{35\%}$ that is applied to the population based on perceived status of the population relative to $MMB_{35\%}$. The applied fishing mortality (F_{OFL}) decreases as estimated MMB decreases below $MMB_{35\%}$ and, if MMB declines to the MSST, the stock is declared overfished and a rebuilding plan needs to be implemented. This type of control rule stipulates that a constant proportion of the selected biomass will be harvested when $MMB > MMB_{35\%}$, but that this proportion is reduced if MMB declines below this threshold value, to preserve a minimum spawning stock biomass.

$$
F_{OFL} \begin{cases} \text{Bycatch only} & \text{if } \dfrac{MMB_{cur}}{MMB_{35\%}} \le \beta \\[2em] \dfrac{F_{35\%}\left(\dfrac{MMB_{cur}}{MMB_{35\%}} - \alpha\right)}{1-\alpha} & \text{if } \beta < \dfrac{MMB_{cur}}{MMB_{35\%}} < 1 \\[2em] F_{35\%} & \text{if } MMB_{cur} > MMB_{35\%} \end{cases}
$$

$$(7.1)$$

Where
MMB_{cur} = current estimated mature male biomass
$MMB_{35\%}$ = mature male biomass at the time of mating resulting from fishing at $F_{35\%}$
$F_{35\%}$ = Fishing mortality that reduce the mature male biomass per recruit to 35% of the unfished level
α = Determines the slope of the descending limb of the control rule (0.25)
β = Fraction of $MMB_{35\%}$ below which directed fishing mortality is zero (0.5)

Regime-based HCR

The control rule in Equation 7.1 relies on an estimate of average historical recruitment to calculate $MMB_{35\%}$. However, for systems in which productivity changes over time, it may be desirable to exclude some estimates of recruitment when calculating the expected recruitment because the management target should be based on productivity under the current environmental conditions

(DOC, 2007). Sliding windows (e.g. use only observations of recruitment from the most recent 15 years; A'mar et al., 2009) might be used to accomplish this, but if shifts in productivity are sudden, the average will still include estimates of recruitment not relevant to the current environmental conditions. Sudden shifts in the physical environment of the eastern Bering Sea appear to occur on a roughly decadal time scale (Hare and Mantua, 2000, Overland et al., 2008). Average recruitment from the most recent recruitment 'regime' may better represent the current productivity of the system, and therefore provide a better input for an HCR structured in the aforementioned manner. An algorithm that can identify changes in average recruitment over time is required to achieve this. Rodionov's sequential t-test analysis for regime shifts, called STARS (Rodionov, 2004), is used for this purpose here. STARS defines a 'reference regime' based on an assumed minimum regime duration and the available data. Next, the deviations of each new year's datum from the previous regime's average are compared with a critical value obtained from Student's t-distribution, assuming some significance threshold (p-value). A new regime is suspected when the deviation for the new year exceeds this threshold. Each subsequent year's normalized deviations from the reference regime average are summed after the first significant observation. A shift in regimes is 'confirmed' when the algorithm has progressed a number of years into the 'new' regime equal to the minimum assumed regime duration and the running sum does not change sign. A minimum regime length of 8 years and a threshold of alpha = 0.1 were used in this analysis.

Evaluation

The management strategy evaluation presented here considers four scenarios: two operating models (OM) and two harvest control rules (HCR). The operating models simulate virtual populations from which the data required for assessment purposes can be sampled. Aspects of the fishery (e.g. selectivity and effort) and the fished population (e.g. growth and natural mortality) are modelled in a statistical framework (see Appendix 7A for equations). Numbers at length are tracked by year, so data such as a survey index of abundance and associated length frequencies are easily drawn from the operating model. For each simulated year in the future, the assessment method is applied to the data required for assessment to estimate quantities important in management (e.g. MMB, fishing mortality), and the HCRs are applied to determine a catch limit to be applied to the population. In this analysis, the underlying population dynamics in the operating model are identical to the population dynamics in the assessment method, save the recruitment dynamics.

Each scenario is projected 50 years into the future (starting from 2011) for 65 parameter vectors defining the operating model. The parameter vector used to determine the population dynamics for each simulated population was obtained by fitting the operating model to the data currently available for assessment (spanning years 1978–2011) and then sampling the joint posterior distribution using a Markov chain Monte Carlo (MCMC) algorithm. The parameters defining a population (i.e. the draws from the joint posterior) did not change over the period of projection within a given simulation. Parameter vectors were selected by implementing a 10% burn-in on one million cycles of the MCMC algorithm, and selecting a thinning ratio that returned the desired number of samples. Evidence of nonconvergence of the MCMC algorithm was checked for using several diagnostic statistics (e.g. lack of autocorrelation and the Geweke statistic; Gelman et al., 2004).

Operating models

The population dynamics in the operating models are the same as those on which the assessment is based and determine the true state of the simulated population (see Appendix 7A). The way in which recruitment is projected is the only difference between the operating models. Two recruitment models are used to project the population forward: status quo and regime based.

Status quo recruitment

The harvest control rule currently used by management to recommend harvest limits uses $F_{35\%}$ as a target fishing mortality, which implies that $F_{35\%}$ is equal to F_{MSY}. This assumption is somewhat corroborated for snow crab, but may not be the case for other crab in the Bering Sea (Punt et al., 2014). Consequently, the status quo recruitment scenario projects recruitment using a Beverton–Holt spawner–recruit relationship in which steepness is selected such that $F_{35\%}$ is equal to F_{MSY}. Projected recruitments are subject to a multiplicative, bias-corrected log-normal error (i.e. with an expected value of 1), with a standard deviation similar to that estimated from recruitment values obtained by fitting the operating model to the actual data for EBS snow crab ($\sigma = 0.75$). The performance of the status quo control rule under status quo recruitment serves as a reference for other scenarios because all assumptions of the status quo control rule are met (i.e. there is a stock-recruit relationship and steepness is such that $F_{35\%}$ is equal to F_{MSY}) (Figure 7.3a).

Regime-based recruitment

A change in average recruitment split the assessed history of the stock into two periods: a period of high recruitment that occurred in the early part of the fishery until the late 1980s, after which recruitment declined. The regime-based operating model mimics this shift and shifts from high to low recruitment states every 10 years (Figure 7.3b). The averages during the high and low states are equal to the observed averages, meaning that there was no assumed relationship with spawning stock biomass, and the high average is roughly double the low average. Process error around the averages was generated from a bias-corrected log-normal distribution with $\sigma = 0.75$.

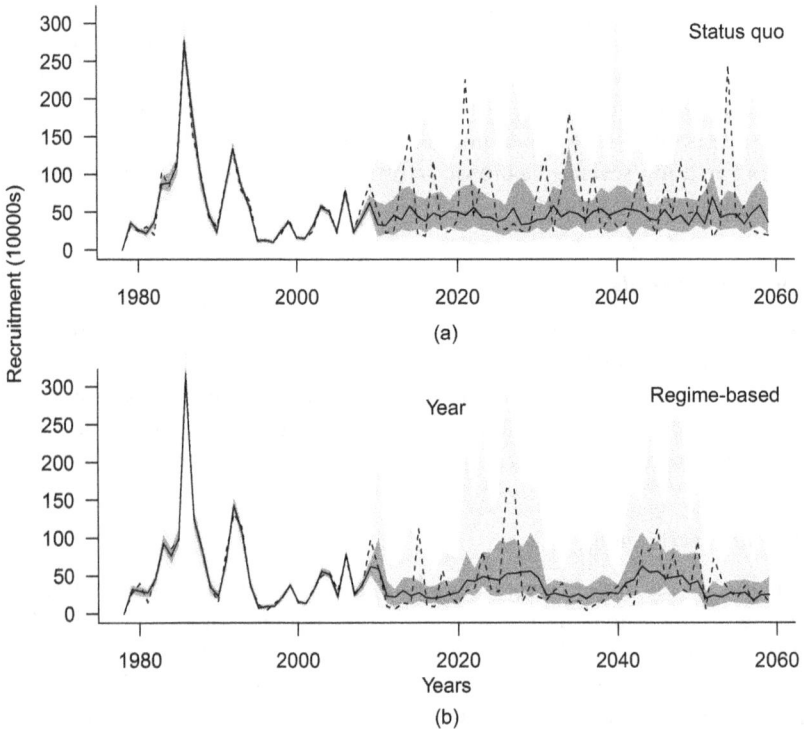

Figure 7.3 Projected recruitment for each operating model. The status quo operating model (a) represents a system in which there is a single relationship
between MMB and recruitment over the projection period. In the regime-
based operating model (b), recruitment is not related to spawning biomass
and alternates between 'high' and 'low' recruitment regimes. Light grey
represents the 5th and 95th intersimulation interval; dark grey is the 25th
and 75th. The black line is the median, and the dashed line is one random
realization of future recruitment.

Performance metrics

The NPFMC seeks to achieve maximum sustainable yield while avoid overfishing and overfished stocks (NPFMC, 2007). Consequently, long-term yield, the
probability of a stock becoming overfished ($MMB_{cur} <$ MSST) and the probability of overfishing occurring ($F_{cur} > F_{35\%}$) are the most important metrics for
evaluating potential HCRs. For the regime-based OM, $MMB_{35\%}$ was calculated using recruitment from the current regime only. Performance metrics are
calculated over the last 40 years of the 50-year projection period to allow for
two full cycles of recruitment in the regime-based operating model. Biomass
and yield metrics were calculated as a mean (over years) of the median (over

simulations). Probabilities were calculated relative to the 'true' OM dynamics as a proportion across all years and iterations.

Performance of the status quo HCR

The application of the status quo HCR to the status quo OM returns the mean median MMB to 108% of $MMB_{35\%}$ during the years 2020–2059 (Figure 7.4a). The true population was never overfished because the ADFG is conservative when compared to the federal OFL. The median estimated $MMB_{35\%}$ (from a given year over all simulations in a scenario) was essentially unbiased over the projected years and overfishing occurred in only 4% of the simulations. The

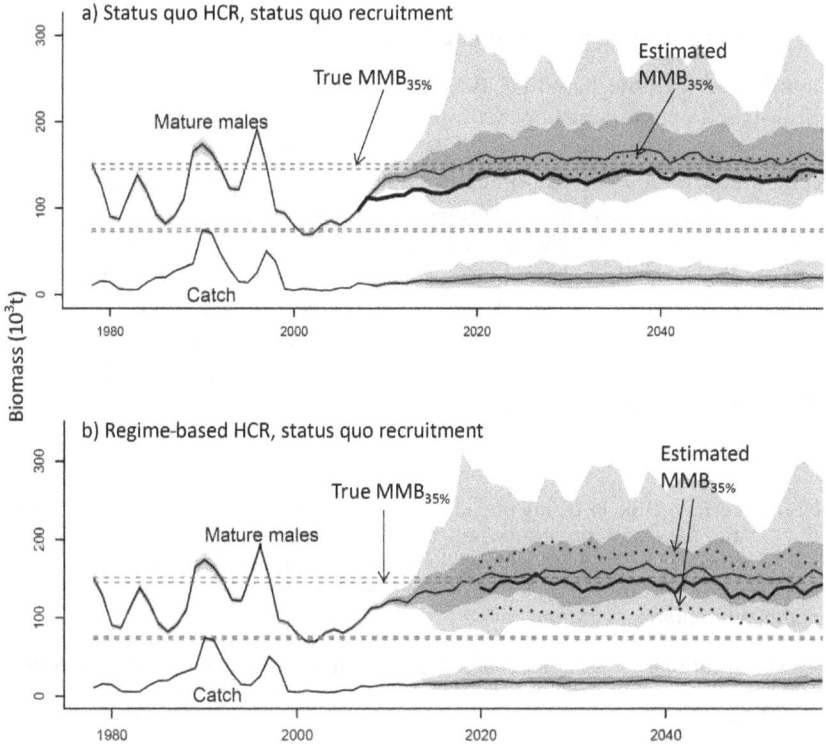

Figure 7.4 Trajectories of catch and mature male biomass under the regime-based HCR and the status quo HCR when recruitment dynamics are status quo. MMB and catch are in thousands of tonnes. Dotted black lines are the range of estimated $MMB_{35\%}$. Pairs of thin, horizontal dashed lines are ranges for the true $MMB_{35\%}$ (upper) and the true MSST (lower). Light grey outlines 5th and 95th quantiles; dark grey outlines 25th and 75th quantiles. Thin solid lines are the medians.

mean median catch over the projection period was 40,600 t, with a standard deviation of 2,200 t (Table 7.1).

The mean median MMB varied widely around the $MMB_{35\%}$ when applying the status quo HCR to data generated from the regime-based operating model (Figure 7.5a). The stock was well above the 'true' $MMB_{35\%}$ in 'high' recruitment regimes and below the 'true' $MMB_{35\%}$ in 'low' recruitment regimes. This is intuitive because the status quo HCR uses all available recruitment estimates to calculate average recruitment, so the expected future recruitment is effectively a 'medium' recruitment regime. Consequently, the bias in estimated $MMB_{35\%}$ alternated between positive (when recruitment was low), and negative (when recruitment was high). Overfishing (relative to $F_{35\%}$) never occurred within the high regimes, but occurred during 19% of the simulated years in low regimes (Figure 7.6).

Performance of the regime-based HCR

The value of a regime-based HCR is most easily seen when it is applied to the regime-based operating model. Compared to the status quo HCR, estimates of $MMB_{35\%}$ from the regime-based HCR were nearly unbiased and much more precise under regime-based population dynamics (Figure 7.5b). Consequently, the mean median catch over the projection period for the regime-based HCR was higher than that for the status quo HCR (28,500 t vs. 27,600 t, respectively; Table 7.1) and the variability in catch was lower for the regime-based HCR than the status quo HCR (standard deviations of 10,100 t vs. 12,800 t, respectively). However, overfishing occurred in 19% of simulated years on average (more often than the status quo rule), and occurred more often in the low regimes than the high regimes (Figure 7.6).

Table 7.1 Performance in terms of yield and probability of overfishing for the status quo harvest control rule and regime-based harvest control rule under status quo recruitment and regime-based recruitment scenarios.

Recruitment scenario	Status quo		Regime-based	
Harvest control rule	Status quo	Regime-based	Status quo	Regime-based
Yield Mean median (1,000 t)	40.6	39.7	27.6	28.5
St. dev. median (1,000 t)	2.2	2.1	12.8	10.1
Probability of overfishing	4%	23%	11%	19%

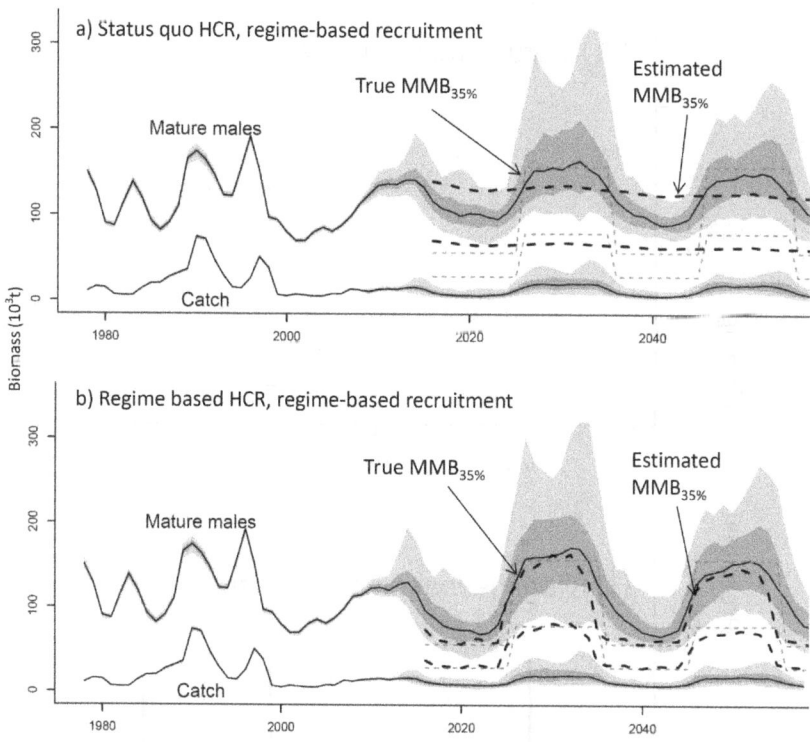

Figure 7.5 Trajectories of catch and mature male biomass under the regime-based HCR and the status quo HCR when recruitment dynamics are regime based. MMB and catch are in thousands of tonnes. Thick dashed lines are the median estimated MMB$_{35\%}$ (upper) and estimated MSST (lower). Thin, horizontal dashed lines are the median true MMB$_{35\%}$ (upper) and true MSST (lower). Light grey outlines 5th and 95th quantiles; dark grey outlines 25th and 75th quantiles. Thin solid lines are the medians.

Applying a regime-based HCR to a population with dynamics that are not truly regime based (e.g. the status quo operating model) produced poorer results than applying the status quo HCR (Figure 7.4b). Specifically, mean median yield produced by the regime-based HCR is less than the yield produced by the status quo rule (39,700 t vs. 40,600 t) with similar variability (2,200 t vs. 2,100 t). However, the probability of overfishing under a regime-based HCR was five times that of fishing under the status quo HCR (23% vs. 4%).

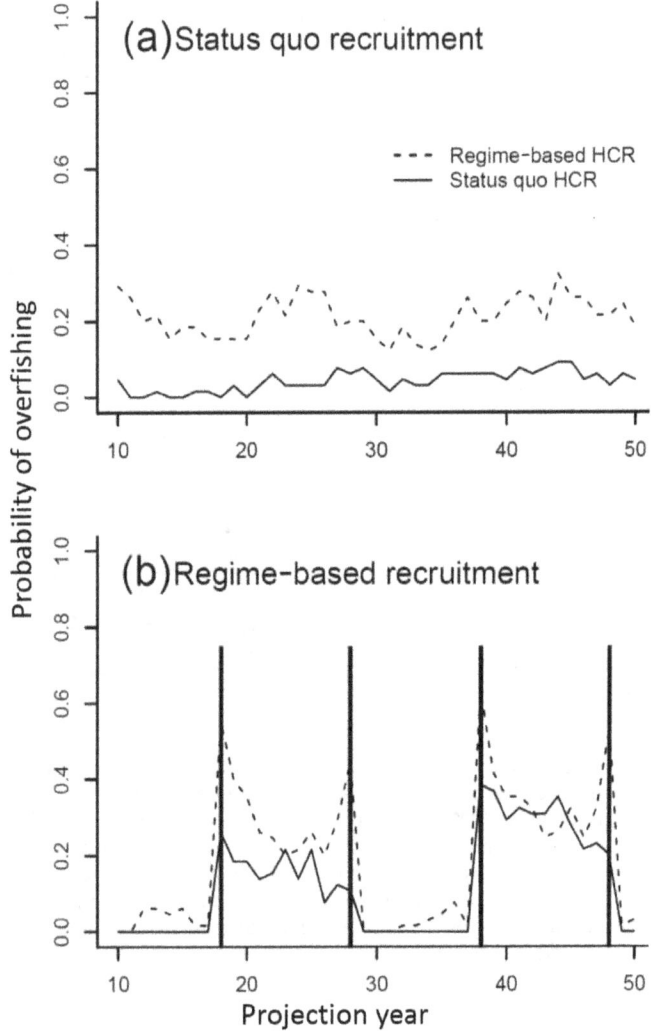

Figure 7.6 Probability of overfishing as the percentage of simulations in a given year in which the removed catch exceeded the true overfishing limit (OFL) for a regime-based HCR and the status quo HCR under future recruitment models. Vertical lines in (b) indicate the years in which the regime switched.

Implementation and conclusions

Incorporating changes in inferred productivity in management strategies can reduce the bias in calculated target biomasses, increase catches and reduce variability in regime-based systems. However, when such HCRs are applied to nonregime-based systems, misidentification of changing recruitment patterns leads to increased imprecision in estimated target biomasses. As a result, the probability of overfishing increases and yield is lost. Consequently, regime-based HCRs are most useful when the underlying dynamics of a system are truly regime-like and shifting from a high regime to a low regime. A stock may be falsely declared overfished (a type I error, though this did not occur in our simulations) under these circumstances if HCRs such as the status quo HCR presented here are used. So, the key piece of information for determining whether or not to implement a regime-based rule is identifying whether or not a population operates under regime-like dynamics or not.

Ultimately, the regime-based HCR was not considered for implementation by the NPFMC because the 'true' population dynamics of snow crab were uncertain. The coincidence of the shift in the PDO and the decrease in recruitment was compelling, but the estimated fishing mortality experienced by the population during the same time period was very high. A disruption of recruitment dynamics by fishing was a possibility that could not be ruled out, so the Precautionary Approach (FAO, 1996) guided managers to assume that fishing is the cause of changes in the population when evidence to the contrary was not definitive.

This MSE for snow crab met many of the criteria for best practices put forth by Punt et al. (in press), such as incorporating managers and stakeholders in the selection of objectives and performance metrics and simulating the application of the management strategy (i.e. incorporating feedback from the application of the assessment method in each year). However, this MSE could have been improved by including more sources of uncertainty. Variation in expected recruitment due to changes in environmental regime was the primary source of uncertainty here, but changes in other biological processes for snow crab may also be linked to changes in temperature (e.g. growth; Ernst et al., 2005). Ontogenetic migrations also introduce spatial structure in the population, and it is unclear how this may influence the ability of management to achieve management goals. Szuwalski and Punt (2015) show that a spatially aggregated assessment method can capture the dynamics of the population in spite of spatial structure given a set of assumptions consistent with the data available on snow crab movement. However, assessing the dynamics of a population accurately is only the first step in implementing a management strategy that will reach the goals of management.

The analyses presented in this chapter are drawn from Szuwalski and Punt (2013b) and one operating model considered in that paper was excluded for the sake of clarity. The excluded operating model linked a model of snow crab recruitment dynamics (Szuwalski and Punt, 2013a) to projections of the Pacific Decadal Oscillation (PDO) from Intergovernmental Panel on Climate Change class models (all under the A1B emission scenario) chosen for their ability to capture large-scale aspects of the climate in the EBS (e.g. sea ice area; Overland and Wang, 2007). The performance of the regime-based HCR under this scenario mirrored the performance under the regime-based operating model;

compared to the status quo yields were higher, variability in yield was lower, and the probability of overfishing increased.

Two key lessons were learned from the climate-linked operating model. First, the between-model variability in projections of the PDO was so large that making any prediction about the future state of the population was difficult. Second, upon reanalyzing the relationship between the PDO and recruitment with additional years of recruitment estimates and a revised assessment method, the relationship was not as strong as before. This is perhaps unsurprising given Myers's (1999) observation that only 28 of 75 of recruitment-environment relationships hold up when analyzed with additional years of data. However, the reason for the weakened relationship was particularly intriguing: it stemmed from only a change from log space to normal space of the likelihoods of one of the data components in the stock assessment. This seemingly small change in the objective function resulted in different patterns of estimated recruitment (Figure 7.7). In spite of the altered pattern in recruitment, the shift in average recruitment near the time of the shift in the PDO persisted, lending some credibility to the hypothesis that the PDO plays some role in recruitment dynamics, even if the specific environmental variables responsible are more difficult to identify.

Punt et al. (2014) found that many studies concluded incorporating changes in climate into management strategies did not improve the ability to achieve management goals. Studies that did report improved management outcomes had very good knowledge about the dynamics of the system under management. However, many of these studies looked at stationary (but highly autocorrelated) environmental forcing, whereas climate change will likely introduce nonstationary forcing that exceeds the historically observed decadal variability in the next few decades (Solomon et al., 2011). An understanding of how and when fish stock productivity changes, in response to environmental or human-induced shifts, will be increasingly important for management science.

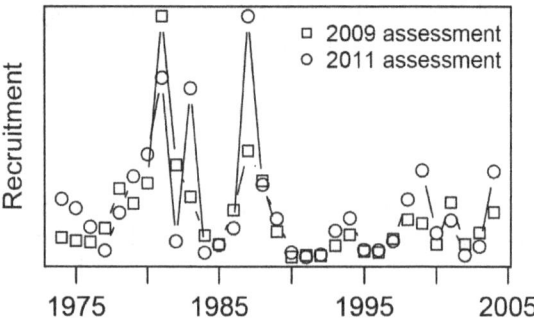

Figure 7.7 Comparison of estimated recruitment from the 2009 (circles) and 2011 (squares) assessments. The y-axis is not shown because the relative scales of the estimates differed between assesment years.

References

Adkison, M.D., Peterman, R.M., Lapointe, M.F., Gillis, D.M. & Korman, J. 1996. Alternative models of climatic effects of Sockeye salmon, *Oncorhynchus nerka*, productivity in Bristol Bay, Alaska, and the Fraser River, British Columbia. *Fiseries Oceanography*, 5, 137–152.

A'mar, Z.T., Punt, A.E., & Dorn, M.W. 2009. The impact of regime shifts on the performance of management strategies for the Gulf of Alaska walleye pollock (*Theragra chalcogramma*) fishery. *Canadian Journal of Fisheries and Aquatic Sciences*, 66, 2222–2242.

Bailey, K.M. 2000. Shifting control of recruitment of walleye pollock after a major climate and ecosystem change. *Marine Ecology Progress Series*, 198, 215–224.

Beverton, R.J.H. & Holt, S.J. 1957. On the dynamics of exploited fish. Fisheries Investigations Series II Marine Fisheries of Great Britain Ministry of Agriculture. *Fisheries and Food*, 19, 1–533.

Clark, W.G. 1991. Groundfish exploitation rates based on life history parameters. *Canadian Journal of Fisheries and Aquatic Sciences*, 48, 734–750.

Conners, M.E., Hollowed, A.B. & Brown, E. 2002. Retrospective analysis of Bering Sea bottom trawl surveys: Regime shift and ecosystem reorganization. *Progress in Oceanography*, 55, 209–222.

Department of Commerce (DOC). 2007. Magnuson-Stevens Fishery Conservation and Management Act. U.S.A. Public Law 94–265.

Ernst, B., Orensanz, J.M. & Armstrong, D.A. 2005. Spatial dynamics of female snow crab (*Chionoecetes opilio*) in the eastern Bering Sea. *Canadian Journal of Fisheries and Aquatic Sciences*, 62, 250–268.

Food and Agricultural Organization (FAO) 1996. *Precautionary Approach to Capture Fisheries and Species Introductions.* 2, Rome, FAO.

Gelman, A., Carlin, J.B., Stern, H.S., & Rubin, D.B. 2004. *Bayesian Data Analysis.* 2nd edition. Chapman and Hall, London.

Hare, S.R., & Mantua, J.M. 2000. Empirical evidence for North pacific regime shifts in 1977 and 1989. *Progress in Oceanography*, 47, 103–145.

Myers, R.A. 1998. When do environment-recruitment correlations work? *Reviews in Fish Biology and Fisheries*, 8, 285–305.

NPFMC (North Pacific Fishery Management Council) 2000. *Bering Sea Snow Crab Rebuilding Plan.* Amendment 14. Bering Sea Crab Plan Team, NorthPacific Fishery Management Council, Anchorage, AK, USA.

NPFMC (North Pacific Fishery Management Council)/NMFS (National Marine Fisheries Service) 2007. Environmental Assessment for Amendment 24. Overfishing definitions for Bering Sea and Aleutian Islands King and Tanner crab stocks. North Pacific Fishery Management Council, Anchorage, AK, USA. www.dakr.noaa.gov.

Overland, J., Rodionov, S., Minobe, S., & Bond, N. 2008. North Pacific regime shifts: definitions, issues, and recent transitions. *Progress in Oceanography*, 77, 92–102.

Overland, J.E., & Wang, M. 2007. Future climate of the North Pacific Ocean. *EOS Transactions of the American Geophysical Union*, 88, 178–182.

Punt, A.E., A'mar, T., Bond, N.A., Butterworth, D.S., de Moor, C.L., De Oliveira, J.A.A., Haltuch, M.A., Hollowed, A.B., & Szuwalski, C.S. 2014. Fisheries management under climate and environmental uncertainty: Control rules and performance simulation. *ICES Journal of Marine Science*, 71(8), 2208–2220.

Punt, A.E., Butterworth, D.S. & Haddon, M. (in press) Management strategy evaluation: Best practices. *Fish & Fisheries.* doi:10.1111/faf.12104

Punt, A.E., Szuwalski, C.S., & Stockhausen, W. 2014. An evaluation of stock-recruitment proxies and environmental change points for implementing the US Sustainable Fisheries Act. *Fisheries Research,* 157, 28–40.

Ricker, W.E. 1954. Stock and recruitment. *Journal of Fisheries Research Board of Canada,* 11, 559–623.

Rodionov, S. 2004. A sequential algorithm for testing climate regime shifts. *Geophysical Research Letters,* 31, L09204.

Solomon, A. et al. 2011. Distinguishing the roles of natural and anthropogenically forced decadal climate variability. *Bulletin of the American Meteorological Society,* 92, 141–156.

Somerton, D., Wenberg, K. & Goodman, S. 2010. Review of the research to estimate snow crab selectivity by the NMFS trawl survey. Report to NPFMC. January 2010.

Szuwalski, C.S. & Punt, A.E. (2015) Can an aggregate assessment method capture the dynamics of a spatially structured population? Snow crab in the eastern Bering Sea as a case study. *Fisheries Research,* 164, 135–142.

Szuwalski, C.S., VertPre, K, Punt, A., Hilborn, R., & Branch, T.A.(in press) Examining common assumptions about recruitment: a meta-analysis of worldwide recruitment dynamics. *Fish & Fisheries.* doi:10.1111/faf.12083

Szuwalski, C.S. & Punt, A.E. 2013a. Regime shifts and recruitment dynamics of snow crab, *Chionoecetesopilio,* in the eastern Bering Sea. *Fisheries Oceanography,* 22(5), 345–354.

Szuwalski, C.S. & Punt, A.E. 2013b. Fisheries management for regime-based ecosystems: a management strategy evaluation for the snow crab fishery in the eastern Bering Sea. *ICES Journal of Marine Science,* 70, 955–967.

Szuwalski, C.S. & Punt, A.E. 2012. Identifying research priorities for management under uncertainty: The estimation ability of the stock assessment method used for eastern Bering Sea snow crab (*Chionoecetesopilio*). *Fisheries Research,* 134–136, 82–94.

Turnock, B.J. & Rugolo, L.J. 2013. Stock assessment of eastern Bering Sea snow crab. *In: Stock Assessment and Fishery Evaluation Report for the King and Tanner Crab Fisheries of the Bering Sea and Aleutian Islands Regions.* Anchorage, AK, USA: North Pacific Fishery Management Council, pp. 39–167.

The size-structured population dynamics model

7A.1 Basic dynamics

The model considers the dynamics of the population, grouping animals by sex s, shell condition (new/old) v, maturity state m, and length-class l, $N_{s,v,m,y,l}$:

$$
N_{s,v,m,y+1,l} = \begin{cases}
\phi_{s,l} \sum_{l'} \kappa_{s,\text{imat},l'} Q_{s,\text{imat},y,l'} X_{s,l',l} & \text{if } v = \text{new}; m = \text{mat} \\[2mm]
(1-\phi_{s,l}) \sum_{l'} \kappa_{s,\text{imat},l'} Q_{s,\text{imat},y,l'} X_{s,l',l} + \bar{R} e^{\varepsilon_y} \, \text{Pr}_l & \text{if } v = \text{new}; m = \text{imat} \\[2mm]
Q_{s,\text{mat},y,l'} & \text{if } v = \text{old}; m = \text{mat} \\[2mm]
(1-\kappa_{s,\text{imat},l'}) Q_{s,\text{imat},y,l'} & \text{if } v = \text{old}; m = \text{imat}
\end{cases}
$$

$$(7A.1)$$

where $k_{s,m,l}$ is the annual probability of an animal of sex s in maturity state m and length-class l moulting (the model includes a terminal moult to maturity so $k_{s,\text{mat},l} = 1$), $Q_{s,m,y,l}$ is the number of animals of sex s in maturity state m and length-class l that survive fishing and natural mortality during year y:

$$
Q_{s,m,y,l} = \sum_v N_{s,v,m,y,l} e^{-Z_{s,v,m,y,l}}
$$

$\Phi_{s,l}$ is the proportion of animals of sex s in length-class l that are mature, $X_{s,l',l}$ is the proportion of animals of sex s in length-class l' that moult into length-class l given that they moult (i.e. the size-transition matrix), $Z_{s,v,m,y,l}$ is the rate of total mortality on animals of sex s, shell condition v, maturity state m, and length-class l during year y:

$$
Z_{s,v,m,y,l} = M_{s,m} + \sum_f F_{s,f,y,l}
$$

$M_{s,m}$ is the instantaneous rate of natural mortality for animals of sex s and maturity state m, $F_{s,f,y,l}$ is the instantaneous rate of fishing mortality by 'fleet' f (directed pot fishery and the groundfish trawl fishery) on animals of sex s in length-class l during year y, $\bar{R}$ is the median recruitment, ε_y is the recruitment deviation for year y, and Pr_l is the proportion of recruitment that occurs to length-class l.

7A.2 Fishing mortality and selectivity

Fishing mortality is caused by directed fishing (for males), discard in the directed fishery (males and females) and bycatch in the groundfish trawl fisheries (the impact of the groundfish trawl fishery by length is assumed to be independent of sex). Fishing mortality is assumed to be independent of shell type and whether an animal is mature or not. Generically, fishing mortality due to each of these sources is given by:

$$F_{s,f,y,l} = S_{s,f,l} \tilde{F}_{s,f,y} = S_{s,f,l} \bar{F}_{s,f} e^{\eta_{s,f,y}} \qquad (7A.2)$$

Where $S_{s,f,l}$ is the selectivity of fleet f on animals of sex s in length-class l, $\bar{F}_{s,f}$ is the reference level of fully selected fishing mortality for fleet f and sex s, and $\eta_{s,f,y}$ is the deviation during year y from the reference level of fully selected fishing mortality for fleet f and sex s.

Selectivity is assumed to be a logistic function of size and to be time-invariant. The selectivity patterns for the catch of males in directed fishery, the catch of females in the directed fishery, and the catch of males and females in the trawl fishery are given by:

$$S_{\text{mal,dir},l} = (1 + \exp[-S_{\text{slop,mal,dir}} (\bar{L}_l - S_{50,\text{mal,dir}})])^{-1}$$
$$S_{\text{fem,dir},l} = (1 + \exp[-S_{\text{slop,fem,dir}} (\bar{L}_l - S_{50,\text{fem,dir}})])^{-1} \qquad (7A.3)$$
$$S_{\text{trawl},l} = (1 + \exp[-S_{\text{slop,trawl}} (\bar{L}_l - S_{50,\text{trawl}})])^{-1}$$

where $\bar{L}_l$ is the midpoint of length-class l, S_{slop}, is the slope of the selectivity curve and S_{50} is the length at which 50% of the individuals encountered are selected.

The probability of a male in length-class l being retained given that it was caught in the directed fishery, $\tilde{R}_l$, is given by:

$$\tilde{R}_l = (1 + \exp[-\tilde{S}_{\text{slop}} (\bar{L}_l - \tilde{S}_{50})])^{-1} \qquad (7A.4)$$

7A.3 Catches

Tehe model predictions of the catch by fleet are given by:

$$\hat{C}_{\text{mal,dir},y} = \sum_l \sum_v \sum_m w_{\text{mal},l} \frac{\tilde{R}_l F_{\text{mal,dir},y,l}}{F_{\text{mal,dir},y,l} + F_{\text{trawl},y,l}} N_{\text{mal},v,m,y,l} e^{-\delta_y M_{s,m}} (1 - e^{-(F_{\text{mal,dir},y,l} + F_{\text{trawl},y,l})})$$

$$\hat{C}_{\text{mal,tot},y} = \sum_l \sum_v \sum_m w_{\text{mal},l} \frac{F_{\text{mal,dir},y,l}}{F_{\text{mal,dir},y,l} + F_{\text{trawl},y,l}} N_{\text{mal},v,m,y,l} e^{-\delta_y M_{s,m}} (1 - e^{-(F_{\text{mal,dir},y,l} + F_{\text{trawl},y,l})})$$

$$\hat{C}_{\text{fem,dir},y} = \sum_l \sum_v \sum_m w_{\text{fem},l} \frac{F_{\text{fem,dir},y,l}}{F_{\text{fem,dir},y,l} + F_{\text{trawl},y,l}} N_{\text{fem},v,m,y,l} e^{-\delta_y M_{s,m}} (1 - e^{-(F_{\text{fem,dir},y,l} + F_{\text{trawl},y,l})})$$

$$\hat{C}_{\text{trawl},y} = \sum_s \sum_l \sum_v \sum_m w_{s,l} N_{s,v,m,y,l} (1 - e^{-F_{\text{trawl},y,l}})$$

$$(7A.5)$$

where $\hat{C}_{mal,dir,y}$ is the model-estimate of the retained catch (in mass) during year y of males by the directed fishery, $\hat{C}_{mal,tot,y}$ is the model-estimate of the total (retained and discarded) catch (in mass) during year y of males by the directed fishery, $\hat{C}_{fem,dir,y}$ is the model-estimate of the catch (in mass) during year y of females by the directed fishery, $\hat{C}_{trawl,y}$ is the model-estimate of the catch (in mass) during year y of animals of both sexes by the trawl fishery, $w_{s,l}$ is the weight of an animal of sex s in length-class l, and δ_y is the midpoint of the fishery (in years).

7A.4 Growth

The probability of moulting as a function of maturity state for immature animals is a declining logistic function of length (mature animals are assumed not to moult given the assumption of a terminal moult at maturity):

$$\kappa_{s,m,l} = 1 - (1 + \exp[-\theta^1_{s,m}(\overline{L}_l - \theta^2_{s,m})])^{-1} \tag{7A.6}$$

The growth increment for animals that do moult is based on the gamma function, i.e.:

$$X_{s,i,j} = Y_{s,i,j} / \sum_k Y_{s,i,k} \tag{7A.7a}$$

$$Y_{s,i,j} = (\Delta_{i,j})^{(\hat{L}_{s,i} - (\overline{L}_i - 2.5))/\beta_s} e^{-\Delta_{i,j}/\beta_s} \tag{7A.7b}$$

where $\hat{L}_{s,l}$ is the expected length for an animal of sex s in length-class l given that it moults:

$$\hat{L}_{s,l} = \gamma^1_s + \gamma^2_s \overline{L}_l \tag{7A.8a}$$

γ^1_s and γ^2_s are the parameters of the relationship between length and growth increment, $\Delta_{i,j}$ is the difference in length between midpoints of length-classes i and j:

$$\Delta_{i,j} = \overline{L}_j + 2.5 \quad \overline{L}_i \tag{7A.8b}$$

β_s is the parameter which defines the variability in growth increment.

7A.5 Recruitment

The fraction of the annual recruitment which recruits to length-class l is based on a gamma distribution, that is:

$$Pr_l = (\Delta_{1,l})^{\upsilon^1/\upsilon^2} e^{-\Delta_{1,l}/\upsilon^2} / \sum_j (\Delta_{1,j})^{\upsilon^1/\upsilon^2} e^{-\Delta_{1,j}/\upsilon^2} \tag{7A.9}$$

where υ^1 and υ^2 are the parameters that define the recruitment fractions.

The annual recruitments are treated as estimable parameters.

7A.6 Initial conditions

The numbers by length-class at the start of the first year considered in the model are treated as estimable parameters, i.e.:

$$
N_{s,v,m,l,y_1} = \begin{cases}
\phi_{s,l}\lambda_{s,v,l} & \text{if } v = \text{new}; m = \text{mat} \\
(1-\phi_{s,l})\lambda_{s,v,l} & \text{if } v = \text{new}; m = \text{imat} \\
\lambda_{s,v,l} & \text{if } v = \text{old}; m = \text{mat} \\
0 & \text{if } v = \text{old}; m = \text{imat}
\end{cases}
\tag{7A.10}
$$

where $\lambda_{s,v,l}$ is the proportion of animals of sex s in length-class l which are of shell condition v.

The objective function for the size-structured model

7B.1 Likelihood components

The model is fitted to the length-frequency of the retained catch of males, the length-frequency of the total catch of males, the length-frequency of females in the directed fishery, and the length-frequency of the catch by the trawl fishery:

$$L_{1,a} = \lambda_{1,a} \sum_{y} N_{1,y}^{\mathrm{eff}} \sum_{l} p_{1,y,l}^{\mathrm{obs}} \ell n(\hat{p}_{1,y,l} / p_{1,y,l}^{\mathrm{obs}}) \tag{7B.1a}$$

$$L_{1,b} = \lambda_{1,b} \sum_{y} N_{2,y}^{\mathrm{eff}} \sum_{l} p_{2,y,l}^{\mathrm{obs}} \ell n(\hat{p}_{2,y,l} / p_{2,y,l}^{\mathrm{obs}}) \tag{7B.1b}$$

$$L_{1,c} = \lambda_{1,c} \sum_{y} N_{3,y}^{\mathrm{eff}} \sum_{l} p_{3,y,l}^{\mathrm{obs}} \ell n(\hat{p}_{3,y,l} / p_{3,y,l}^{\mathrm{obs}}) \tag{7B.1c}$$

$$L_{1,d} = \lambda_{1,d} \sum_{s} \sum_{y} N_{4,s,y}^{\mathrm{eff}} \sum_{l} p_{4,y,s,l}^{\mathrm{obs}} \ell n(\hat{p}_{4,y,s,l} / p_{4,y,s,l}^{\mathrm{obs}}) \tag{7B.1d}$$

where $\lambda_{1,a-d}$ are weighting factors (see Table 7A.1), $N_{i,y}^{\mathrm{eff}}$ is the effective sample size for year y and data-type i (i = 1: male retained catch; i = 2: male total catch; i = 3: female discards in the directed fishery; i = 4: catches in the trawl fishery), $p_{i,y,l}^{\mathrm{obs}}$ is the observed proportion of the catch during year y of data-type i that is in size-class l, $\hat{p}_{1,y,l}$ is the model-estimate of the proportion of the retained catch of males in the directed fishery during year y that in length-class l:

$$\hat{p}_{1,y,l} = \tilde{R}_l \, S_{\mathrm{mal,dir},l} \sum_{v} \sum_{m} N_{\mathrm{mal},v,m,y,l} e^{-\delta_y M_{\mathrm{mal},m}} \Big/ \sum_{v'} \sum_{m'} \sum_{l'} \tilde{R}_{l'} \, S_{\mathrm{mal,dir},l'}$$
$$N_{\mathrm{mal},v',m',y,l'} e^{-\delta_y M_{\mathrm{mal},m'}} \tag{7B.2a}$$

$\hat{p}_{2,y,l}$ is the model-estimate of the proportion of the total catch of males in the directed fishery during year y that is in length-class l:

$$\hat{p}_{2,y,l} = S_{\mathrm{mal,dir},l} \sum_{v} \sum_{m} N_{\mathrm{mal},v,m,y,l} e^{-\delta_y M_{\mathrm{mal},m}} \Big/ \sum_{v'} \sum_{m'} \sum_{l'} S_{\mathrm{mal,dir},l'}$$
$$N_{\mathrm{mal},v',m',y,l'} e^{-\delta_y M_{\mathrm{mal},m'}} \tag{7B.2b}$$

$\hat{p}_{3,y,l}$ is the model-estimate of the proportion of the catch of females in the directed fishery during year y that is in length-class l:

$$\hat{p}_{3,y,l} = S_{\text{fem,dir},l} \sum_{v} \sum_{m} N_{\text{fem},v,m,y,l} e^{-\delta_y M_{\text{fem},m}} / \sum_{v'} \sum_{m'} \sum_{l'} S_{\text{fem,dir},l'}$$
$$N_{\text{fem},v',m',y,l'} e^{-\delta_y M_{\text{fem},m'}}$$

(7B.2c)

$\hat{p}_{4,y,l}$ is the model-estimate of the proportion of the trawl catch during year y that is sex s and is in length-class l:

$$\hat{p}_{4,y,s,l} = S_{\text{trawl},l} \sum_{v} \sum_{m} N_{s,v,m,y,l} e^{-\delta_y M_{s,m}} / \sum_{s'} \sum_{v'} \sum_{m'} \sum_{l'} S_{\text{trawl},l'}$$
$$N_{s',v',m',y,l'} e e^{-\delta_y M_{s',m'}}$$

(7B.2d)

The model is fit to the survey length-frequency data by sex, shell condition, and maturity state, that is:

$$L_2 = \lambda_2 \sum_{s} \sum_{v} \sum_{m} \sum_{y} N^{\text{eff}}_{5,s,v,m,y} \sum_{l} p^{\text{obs}}_{5,s,v,m,y,l} \ell n(\hat{p}_{5,s,v,m,y,l} / p^{\text{obs}}_{5,s,v,m,y,l})$$

(7B.3)

where $N^{\text{eff}}_{s,v,m,y}$ is the effective sample size for the survey length-frequency data for animals of sex s in shell condition v, and maturity state m during year y; $p^{\text{obs}}_{5,s,v,m,y,l}$ is the observed proportion of the survey catch for animals of sex s in shell condition v and maturity state m that is in length-class l; and $\hat{p}_{5,s,v,m,y,l}$ is the model-estimate corresponding to $p^{\text{obs}}_{5,s,v,m,y,l}$:

$$\hat{p}_{5,s,v,m,y,l} = S^l_{s,y\star} N_{s,v,m,y,l} / \sum_{s'} \sum_{v'} \sum_{m'} \sum_{l'} S^l_{s',y\star} N_{s',v',m',y,l'}$$

(7B.4)

where $S^l_{l,y\star}$ is the selectivity of the survey gear for animals in length-class l during year $y\star$ (years are grouped into three epochs: before 1982, 1982–1988, and 1989+):

$$S^l_{y\star,l} = q^l_{y\star}(1 + \exp[-S^l_{\text{slop},y\star}(\overline{L}_l - S^l_{50,y\star})])^{-1}$$

(7B.5)

$q^l_{y\star}$ is the survey catchability coefficient for the years represented in the set $y\star$, $S^l_{50,y\star}$ is the length-at-50%-survey selectivity for the years represented in the set $y\star$, and $S^l_{\text{slop},y\star}$ is the slope of the survey selectivity ogive for the years represented in the set $y\star$.

The model is fit to the survey indices separately by sex, that is:

$$L_3 = 0.5\lambda_3 \sum_{s} \sum_{y} (\ell n I_{s,y} - \ell n \hat{I}_{s,y})^2 / \sigma^2_{s,y}$$

(7B.6)

where $I_{s,y}$ is the survey index of abundance for sex s and year y, $\hat{I}_{s,y}$ is the model–estimate corresponding to $I_{s,y}$:

$$\hat{I}_{s,y} = \sum_l \sum_v S^I_{y*,l}\, w_{s,l}\, N_{s,v,\text{mat},y,l}$$

$\sigma_{s,y}$ is the standard error of $I_{s,y}$.

The contribution of the 'large' males (102mm+) to the likelihood function is given by:

$$L_4 = \lambda_4 \sum_y \left(\ell n J_y - \ell n \hat{J}_y \right)^2 \tag{7B.7}$$

where J_y is the number of 'large' males in the survey during year y, and $\hat{J}_y$ is the model–estimate corresponding to J_y:

$$\hat{J}_y = \sum_v \sum_m \left(0.5 S^I_{\text{mal},16} N_{s,v,m,y,16} + \sum_{l>16} S^I_{\text{mal},l} N_{s,v,m,y,l} \right) \tag{7B.8}$$

The contribution of the catch data to the likelihood function is given by:

$$L_{5,a} = \lambda_{5,a} \sum_y \left(\ell n C^{obs}_{\text{mal,dir},y} - \ell n \hat{C}_{\text{mal,dir},y} \right)^2 \tag{7B.9a}$$

$$L_{5,b} = \lambda_{5,b} \sum_y \left(\ell n C^{obs}_{\text{mal,tot},y} - \ell n \hat{C}_{\text{mal,tot},y} \right)^2 \tag{7B.9b}$$

$$L_{5,c} = \lambda_{5,c} \sum_y \left(\ell n C^{obs}_{\text{fem,dir},y} - \ell n \hat{C}_{\text{fem,dir},y} \right)^2 \tag{7B.9c}$$

$$L_{5,d} = \lambda_{5,d} \sum_y \left(\ell n C^{obs}_{\text{trawl},y} - \ell n \hat{C}_{\text{trawl},y} \right)^2 \tag{7B.9d}$$

where $C^{obs}_{f,y}$ is the observed catch (in mass) by fleet f during year y.

7B.2 Penalty components

There are several penalty components in the objective function.

1   A penalty is placed on the deviations in recruitment from average recruitment:

$$P_1 = \lambda_6 \sum_{y=1979}^{2008} \varepsilon_y^2 \tag{7B.10}$$

2   A penalty is placed on the between-length-class variation in the initial size structure:

$$P_2 = \lambda_7 \sum_s \sum_v \sum_{l=1}^{n_{L,s}-1} (ln\lambda_{s,v,l} - ln\lambda_{s,v,l+1})^2 \qquad (7B.11)$$

where $n_{L,s}$ is the number of size classes in the first year for sex s (12 for females and 22 for males).

3   Penalties are placed on the extent of interannual variation in deviations in fishing mortality:

$$P_{3,a} = \lambda_8 \sum_y \tilde{n}_{mal,dir,y}^2 \qquad (7B.12)$$

Table 7A.1 The weighting factors applied to the penalties and likelihood components.

Description	Value
Likelihood components	
Catch proportions-at-length	
Direct male catch, $\lambda_{1,a}$	1
Total male catch, $\lambda_{1,b}$	1
Direct female catch, $\lambda_{1,c}$	0.2
Trawl catch, $\lambda_{1,d}$	0.25
Survey proportions-at-length, λ_2	1
Survey biomass data, λ_3	0.25
Survey proportions-at-length (large males), λ_4	0.001
Catches	
Direct male catch, $\lambda_{5,a}$	1,000
Total male catch, $\lambda_{5,b}$	1,000
Direct female catch, $\lambda_{1,c}$	10
Trawl catch, $\lambda_{1,d}$	100
Penalty components	
Recruitment deviations, λ_6	2
Initial size structure, λ_7	1
Deviations in fishing mortality, λ_8	0.1

Chapter 8

Managing international tuna stocks via the management procedure approach

Southern bluefin tuna example

Richard M. Hillary, Ann L. Preece, Campbell R. Davies, Hiroyuki Kurota, Osamu Sakai, Tomoyuki Itoh, Ana M. Parma, Douglas S. Butterworth, James N. Ianelli and Trevor A. Branch

Introduction

There is considerable international concern, controversy and scientific debate on the current state and future prospects for tuna stocks worldwide, and what constitute effective policy and management measures for rebuilding stocks which are heavily depleted (Polacheck 2002, Myers and Worm 2003, Sibert et al. 2006, Worm and Tittensor 2011). In the case of the highly prized bluefin tunas, Atlantic and southern bluefin tuna (SBT) stocks have been described by some as "collapsed" with little chance of rebuilding (Collette et al. 2011), and the Pacific bluefin tuna has been estimated to have declined rapidly recently. Proposed alternatives to current fisheries management measures include commercial catch moratoria and large-scale marine protected areas (Cullis-Suzuki and Pauly 2010, Collette et al. 2011). For the most part, however, these recommendations have not been supported by rigorous scientific analysis of their likely performance, or realistic assessment of the governance, operational management and scientific monitoring issues that would need to be addressed for their effective implementation.

Southern bluefin tuna (SBT) are a long-lived (to near 40 years), late-maturing (from about age 8 to 20) and highly migratory tuna found throughout the southern temperate oceans except for the more easterly regions of the South Pacific (Farley et al. 2007). Commercial surface and long-line fisheries for SBT began in the 1950s, with annual catches reaching a maximum of 81,750 t in 1961 and remaining relatively high until catch restrictions were first introduced in 1989 (see Figure 8.1). The impetus for these early management arrangements was evidence of high fishing mortality rates, low estimates of recruitment, and the demise of the purse seine surface fishery off the east coast of Australia in the mid-1980s (Caton 1990). Since the turn of the century, both spawning biomass and median recruitment have remained at historically low levels (see Figure 8.2, Anonymous (2011b) and Hillary et al. (2015)).

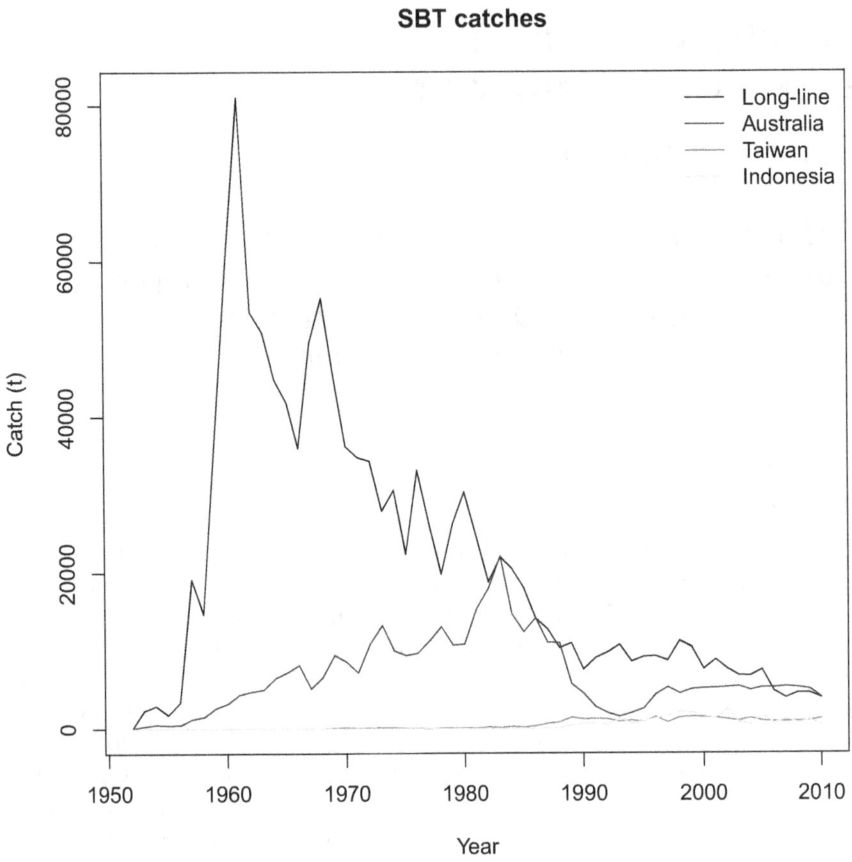

SBT catches

Figure 8.1 SBT catch history by gear or national fleet.

Southern bluefin tuna were first managed since the early 1980s through informal tripartite agreements between the fishing nations most concerned, namely Australia, Japan and New Zealand, but became the responsibility of the Commission for the Conservation of Southern Bluefin Tuna (CCSBT) in 1993 (Anonymous 1994). The CCSBT is one of five global regional fisheries management organisations (RFMOs) for tuna. These RFMOs are convened by the Food and Agriculture Organization (FAO), and consist of both scientific and political representatives from nation–states involved in the fishery. Typically, the science is conducted by working groups acting under a Scientific Committee, which then formulates recommendations for the Commission. The Commission is a political body responsible for making decisions, with each Commissioner acting on behalf of the jurisdiction they represent. These decisions are made by consensus only, and subsequently adopted by the member states.

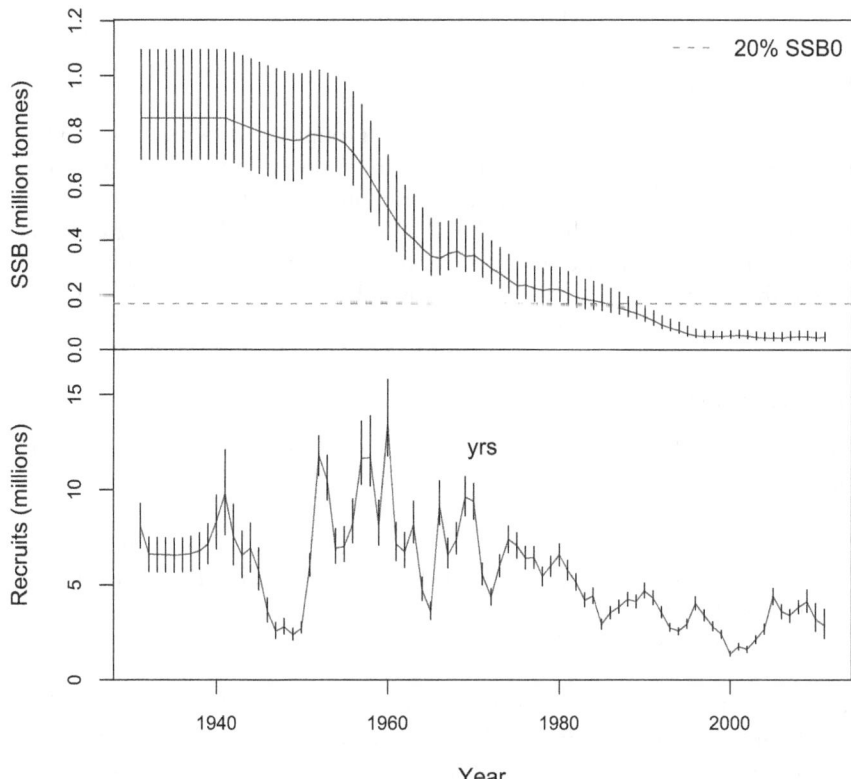

Figure 8.2 Median and 80%PI for SBT spawning stock biomass (top) and recruitment (bottom) in the 2011 operating model (OM).

Despite the introduction of catch restrictions, management prior to 2011 was, at most, only able to halt the decline of the stock at a low level, but not reverse the trend. At this time spawning stock biomass was estimated to be as low as 3–7% of unfished levels (Anonymous 2011b), which resulted in agreement that strong management intervention/corrective action was required.

A difficulty in implementing timely corrective action, faced by other RFMOs in addition to the CCSBT, has been that the setting of annual catch limits, often informed by time-consuming stock assessments, requires a negotiated consensus between all members based on stock status advice. This tends to maintain the status quo, as a no-change decision is often the only one able to achieve consensus. In contrast a management procedure (MP) for setting the catch limit represents a default recommendation that is automatically adjusted in response to monitored indicators of stock status. Provided the MP can be agreed upon by all members, it then facilitates responsive management in subsequent years.

This chapter details development of a MP that was adopted by the Commission in 2011 following a 2-year process of development and testing.

The management procedure

Scientific work carried out by the workings groups in any RFMO is guided by the Scientific Committee, and in turn, the Commission. For the CCSBT, this guidance has stipulated which data series are agreed to be valid for use by the MP, and directed the choice of HCR. We discuss only the final agreed data inputs and HCR structure in this section but do try to include the reasons behind those final choices, which were driven by the preceding HCR incarnations and available data sources.

The input data

Initially, the following data sources were permitted to be used by any candidate MP:

* Japanese long-line catch per unit effort (CPUE) (relative abundance index);
* Aerial surveys of juvenile SBT in the Great Australian Bight in summer (relative biomass index);
* Catch composition of the long-line fleet (age structure).

At the first iteration of the evaluation process it was agreed that any candidate MP should use *both* the CPUE and survey indices (Anonymous 2010b), which eventually led to the agreement that these two indices should be the only data inputs to the MPs (Anonymous 2010a).

Targets and constraints

A number of settings and constraints, as well as alternative interim target rebuilding criteria, were defined for any candidate MP:

* Frequency of change: every 1, 2 and 3 years were initially permitted;
* Maximum and minimum TAC change: 3,000 and 5,000 t for the maximum change, and 100 t for the minimum, were allowed;
* Interim rebuilding target: recover to 20% of unfished spawning biomass B_0 with probabilities 0.6, 0.7 or 0.9 by 2030, 2035 or 2040.

During the evaluation process some decisions were made on reducing the number of possible MP setting options. It was fairly clear that there was little performance difference between MPs changing the TAC every 2 or every 3 years, so 3 years was taken as the default (Anonymous 2010b). Certain combinations of target year and probability also aliased for each other quite strongly; accordingly, for the final evaluation cycle the 70% probability rebuilding level of risk was agreed upon (Anonymous 2011b). This left the maximum change in TAC (3,000 or 5,000 t) and the target year (2035 or 2040) as axes remaining to be explored.

The harvest control rule

Initially, a wide range of MP types were explored:

- Empirical: using both indices as well as catch composition, only the survey, and fuzzy-logic decision structures;
- Model-based: using modified production models, sometimes utilising both indices, as well as a random-effect two-stage filter approach.

Coincident with the decisions made about data inputs to the MPs, the first evaluation round selected two candidate MPs to be taken forwards in the next phase of evaluation (Anonymous 2010b). The first (called MP1) was model-based, using the following integrated population dynamics model to filter (smooth) the aerial survey and CPUE indices:

$$B_{y+1} = J_y + g_y B_y, \tag{8.1}$$

$$J_y = \exp\left(\mu_J + \varepsilon_y^J\right), \tag{8.2}$$

$$g_y = \exp\left(\mu_g + \varepsilon_y^g\right), \tag{8.3}$$

$$\varepsilon_y^{J,g} \sim N\left(0, \sigma_{J,g}^2\right). \tag{8.4}$$

The term B_y represents the adult relative biomass (covering a mix of low, medium and fully mature animals and assumed proportional to the long-line CPUE); the term J_y represents the juvenile relative biomass (covering fully immature ages 2 to 4 assumed proportional to the previous year's aerial survey index); the g_y are basically a combination of total mortality and surplus production effects within the adult biomass range. This is a variant of a model first proposed in Trenkel (2008). To avoid identifiability issues, the B_y and CPUE were assumed to scale exactly (i.e. the catchability was fixed at one, making the B_y in some sense the "true" CPUE). The ratio between the long-line CPUE and the survey catchability was scaled using life-history information to ensure the mean contribution from incoming juveniles to the adult biomass was plausible. Estimated parameters are the mean and year-specific random effects for the J_y and g_y terms – the $\mu_{J,g}$ and $\varepsilon_y^{J,g}$, respectively – with the variance terms $\sigma_{J,g}^2$ fixed given the available life-history and observation error information. More specifically, the observation error term σ_J^2 was approximated from variability in the abundance index, whereas choice of the process error variance σ_g^2 was informed by short-term productivity assumptions that are dependent on the life-history strategy of bluefin tuna. The HCR for MP1 had a target-based structure for the B_y (effectively specifying that the CPUE level seen in the 1980s should be reached around the interim rebuilding target), with limit-type terms for the juvenile biomass and adult biomass growth terms based on observed minima in the historic data.

The TAC for MP1 was defined using the TAC from the previous year and an adjusted target catch as follows:

$$TAC_{y+1}^{[MP1]} = 0.5 \times \left(TAC_y + C_y^{\text{targ}} \Delta_y^J\right) \tag{8.5}$$

$$C_y^{\text{targ}} = \begin{cases} \delta \left[\dfrac{B_y}{B^*}\right]^{1-\varepsilon_b} & \text{for } B_y \geq B^* \\[2em] \delta \left[\dfrac{B_y}{B^*}\right]^{1+\varepsilon_b} & \text{for } B_y < B^* \end{cases} \tag{8.6}$$

and $\varepsilon_b \in [0,1]$ represents the degree to which the response to a biomass level above or below the target level B^* is asymmetric. The target catch level δ is the tuning parameter of the HCR. The recruitment adjustment Δ_y^J is defined as follows:

$$\Delta_y^J = \begin{cases} \left[\dfrac{\bar{J}}{\mathcal{I}}\right]^{1-\varepsilon_J} & \text{for } \bar{J} \geq \mathcal{I} \\[2em] \left[\dfrac{\bar{J}}{\mathcal{I}}\right]^{1+\varepsilon_J} & \text{for } \bar{J} < \mathcal{I} \end{cases} \tag{8.7}$$

and $\varepsilon_J \in [0,1]$ is the level of asymmetry in response to the current moving (arithmetic) average – and this has been changed to include up to year y – recruitment levels, $\bar{J}$:

$$\bar{J} = \frac{1}{T_J} \sum_{i=y-T_J+1}^{y} J_i \tag{8.8}$$

of length T_J relative to the average, J, calculated over the years for which the estimates are based on observed data.

The second candidate MP (called MP2) was an empirical MP with a slope-in-the-index structure for the CPUE part of the HCR:

$$TAC_{y+1}^{[MP2]} = TAC_y \times \begin{cases} 1 - k_1 |\lambda|^{\gamma} & \text{for } \lambda < 0 \\ 1 + k_2 \lambda & \text{for } \lambda \geq 0 \end{cases} \tag{8.9}$$

where λ is the slope in the regression of $\ln B_y$ against year (from years $y - \tau_{B+1}$ to year y).

Table 8.1 Fixed values and tuning parameters for MP3 and their respective values for the two original MPs, MP1 and MP2.

Parameter	MP3	MP1/MP2
δ	Tuned	Tuned (MP1)
k_1	1.5	1.5 (MP2)
k_2	3	5 (MP2)
Γ	1	1 (MP2)
τ_R	7	7 (MP2)
$B*$	1.2	1.2 (MP1)
ε_b	0.25	0.5 (MP1)
ε_J	0.75	0.75 (MP1)
τ_r	5	5 (MP1)

Part way though the evaluation process, as described in more detail later, the decision was made that given certain features of MP1 and MP2 performed better in certain circumstances, issues around being able to tune to some of the Commission's candidate rebuilding targets, and the advantage of giving the Commissioners fewer factors upon which they would need to decide, the key parts of MP1 and MP2 were combined (Anonymous 2011b). This new combined MP, called MP3, retained the model-based part of MP1 and used the "filtered" variables B_y and J_y in the new HCR:

$$TAC_{y+1}^{[MP3]} = 0.5 \times \left(TAC_{y+1}^{[MP1]} + TAC_{y+1}^{[MP2]} \right)$$

(8.10)

All the settings, for both the original MPs and the combined MP, can be found in Table 8.1. For time frame parameters $(T_{J,B})$ these were set based on simple statistical arguments for the associated likely precision in the trends they relate to. For the reactivity asymmetry parameters $(\varepsilon_{J,B})$ these were both chosen to favour stronger responses for negative recruitment and biomass trends than for positive ones, given the rebuilding nature of the problem at hand.

The evaluation process

The operating model and reference set

The SBT operating model (OM) is based on the previous stock assessment model. It is a complicated integrated age–based model that fits to the following observations using (penalised) maximum likelihood:

- Catch biomass for each fleet (six in total)
- Length composition of the four main long-line fleets

- Age composition of the surface school and spawning ground fisheries
- Japanese long-line CPUE abundance index
- Aerial survey of estimates of juvenile tuna biomass in the mid- to late summer
- Multicohort mark-recapture data from the 1990s.

The model is a statistical catch-at-age model, with a two-season time step. The age range is 0 to 30+ and with a time-dependent distribution for length-at-age, given strong observed changes in growth over time. Natural mortality is strongly age-structured, with an asymmetrical U-shape functional form to deal with both higher M at early ages and senescence. Directly estimated parameters include B_0, annual recruitment deviates, selectivity for each fleet (which can change over time) and natural mortality at ages 4 and 30. In addition to those parameters there is what is termed the "Grid": influential parameters for which reliable information is lacking (i.e. they are poorly estimated), with the consequences of this uncertainty needing to be taken into account during MP evaluation. These include the steepness of the stock-recruit relationship (h), natural mortality at ages 0 and 10 (M_0 and M_{10}, respectively), alternative CPUE standardisations (w.5 and w.8), the age range to which CPUE applies (ages 4–18 or 8–12), nonlinearity of the CPUE vs. abundance relationship (described by the elasticity term ω), and the effective sample sizes of the composition data.

From this array of grid permutations (320 in this case), 2,000 samples are drawn for use in projections and MP evaluations, representing the "reference set" of OM parameterisations. Table 8.2 defines the grid settings (both values and resampling specifications) for the reference set of OMs against which all MPs must initially be tuned – that is to be considered as a candidate, an MP

Table 8.2 Table of settings for the SBT OM grid. CumulN refers to the cumulative number of levels in the grid and OF signifies objective function–based resampling of that given grid element. The SQRT label denotes the case where the effective sample sizes of the catch composition data were the square root of the original values (i.e. downweighted).

	Levels	CumulN	Values	Prior	Weighting
h	5	5	0.55, 0.64, 0.93, 0.82, 0.9	uniform	OF
M_0	4	20	0.3, 0.35, 0.4, 0.45	uniform	OF
M_{10}	3	60	0.07, 0.1, 0.14	uniform	OF
ω	1	60	1	NA	NA
CPUE	2	120	w.5, w.8	uniform	prior
q age range	2	240	4–18, 8–12	0.67, 0.33	prior
sample size	1	240	SQRT	NA	NA

must be capable of meeting the rebuilding targets across the full reference set. The resampling probability for each permutation could be based on a prior specification or a likelihood-based weighting obtained from alternative fits of the OM to the data.

Robustness tests

Robustness tests can be used to test candidate MPs against a range of hypothetical future scenarios. The original suite of robustness tests was extensive (Anonymous 2010b, 2011b). There were structural tests that considered the lack of complete mixing in the mark-recapture data, alternative age selectivity for the aerial survey, historical recruitment regime shifts, a nonlinear relationship between CPUE and abundance (where CPUE changes are stronger than in abundance), future possible recruitment failure and changes in historic and future long-line catchability. There were data uncertainty tests, given historical uncertainty in both catch mass and composition for various fleets as well as alternative CPUE series and the inclusion of a trolling survey of 1-year-old SBT off Western Australia. There were also observation error tests looking at increases in the sampling variance for both the CPUE and the aerial survey. The suite of robustness tests included both more optimistic and more pessimistic stock status outlooks, relative to the reference OM. However, given the primary goal of any MP was to rebuild the stock from its depleted state while maintaining a viable fishery, the pessimistic trials eventually took precedence in the final selection process. The base-case is the "reference set" as defined in Table 8.2. The key robustness tests were:

- *lowR*: future recruitment failure (50% of the mean for 4 years);
- *upq*: positive bias (35%, estimated within the OM for 2008–2001 period) in the most recent and future CPUE catchability, resulting in a more depleted stock historically and biased indices in the future;
- *STWin*: alternative spatiotemporal weighting scheme for the CPUE calculation resulting in a more depleted stock;
- *omega75*: a nonlinear (with power 0.75) relationship between CPUE and abundance, resulting in a more depleted stock historically and bias in the indices in the future.

Performance statistics and communication

In terms of performance criteria there was, as in almost all fisheries, a natural dichotomy between biomass rebuilding and catch-based performance measures, which often trade off with each other. In terms of rebuilding performance the focus was placed firmly on the spawning stock biomass (SSB). Factors such as probability of future declines, the probability that the SSB has reached the target level halfway towards the interim target year, and the probability that the future SSB drops below the minimum estimated level were also evaluated. Fishery performance was assessed through various catch-based statistics such as average interannual variation (AAV, time-averaged percentage change in catch from year to year), probability that the TAC will go down after going up over the first 2 and first 4 TAC decisions, and average catches for various periods. Figure 8.3 shows an example of the primary summary plots used to assess performance across multiple robustness trials and for a range of performance statistics. This particular plot is for the adopted MP (MP3) for the base-case, recruitment failure (*lowR*) and catchability change (*upq*) robustness trials, for a range of maximum TAC change and initial TAC increase scenarios.

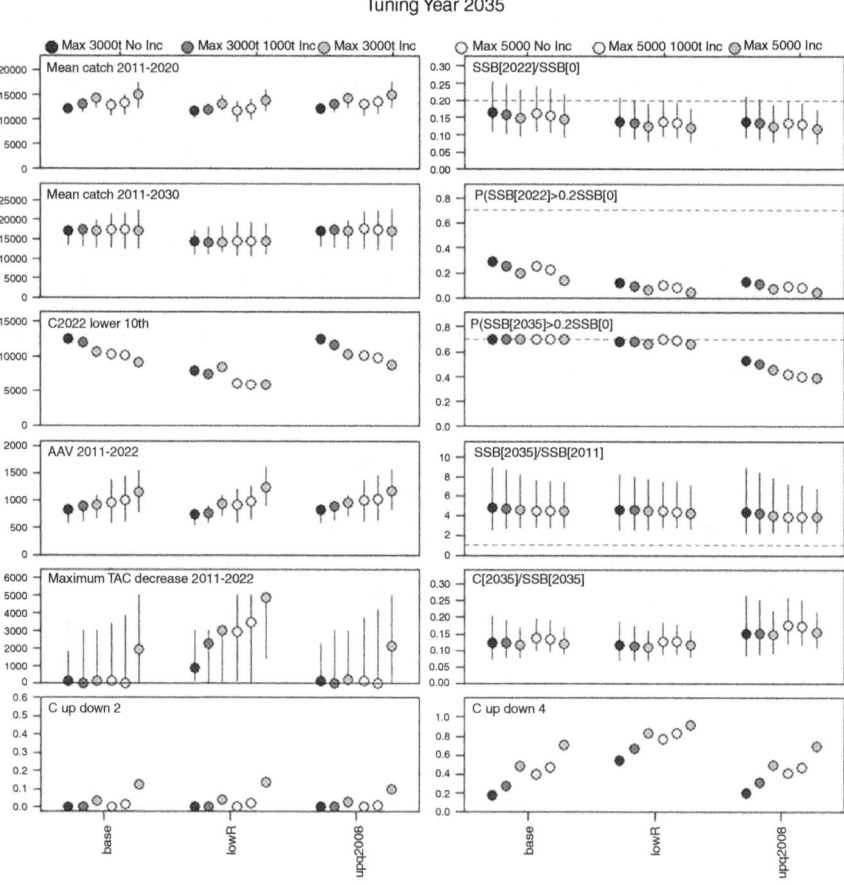

Figure 8.3 Performance statistics summary for the combined MP across the reference OM and two key robustness trials and for a range of maximum TAC and initial TAC increase options.

Plots such as these, while displaying a vast amount of information, are vital for the discussions (both inside and outside of the meeting process) that occur in the evaluation of an MP in an RFMO setting, with often greatly differing degrees of process understanding within the RFMO Scientific Committee and Commission. Summary tables do not have the same communicative effectiveness with such large amounts of performance information, and are better suited for higher level summaries (e.g. for one MP and a handful of robustness trials and settings). Another very important reality to communicate to all representatives – and especially to Commissioners – is that, whatever MP is chosen, the actual outcomes will *never* look like the means or medians they see in projections and statistical summaries. Figure 8.4 shows an example of a "worm" plot that is frequently used to make this point clear: future SSB and catch (median and 80% probability intervals) are shown alongside 10 random draws from the 2,000 projections.

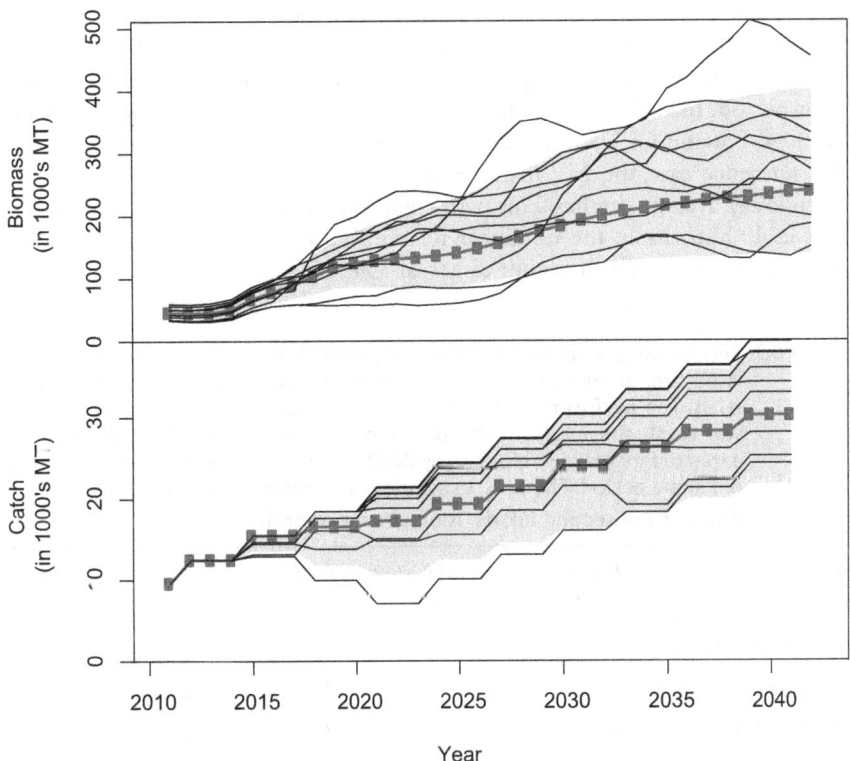

Figure 8.4 Projected median and 80%PI and 10 randomly selected "worms" for SSB (top) and TACs (bottom) for the MP permitted a full increase in 2012 and for a maximum TAC change of 3,000 t.

The final evaluation

Given positive signals in both the MP input indices, as well as a more optimistic reconditioned OM in 2011 compared to 2009 (Anonymous 2011b), all candidate MPs predicted an increase in the first TAC decision (2012 quota). One final additional constraint was added to the remaining suite (TAC change, target rebuilding year) which limited the initial increase in TAC to a maximum of 1,000 t. A very general summary of the performance of MP3 (3,000 t maximum TAC change, 2035 target year, 1,000 t maximum initial TAC increase) can be found in Table 8.3 which reports on the key robustness tests listed earlier.

For the recruitment failure test (*lowR*), although the probability of approaching the rebuilding target by 2022 is low, the MP almost manages to make the rebuilding target by 2035 by virtue of lower catches after the initial increase when the recruitment failure signal appears in the input indices. For the catchability change scenario the MP fails to make the rebuilding target, driven by a slightly less optimistic starting point in 2011 and above-average catches taken as a result of the bias in the CPUE index. For the alternative spatiotemporal CPUE trial, the starting point is noticeably less optimistic than for the reference case and, even with appreciably lower average catches over the projection period, the MP can attain the rebuilding target only with a probability of 0.34. For the CPUE vs. abundance test, while the starting point is below the reference case, the MP manages to attain a rebuilding probability of 0.48 with lower average catches. This performance is enhanced by the nature of the test itself: historically the nonlinear relationship means the decline in CPUE is even stronger when it comes to actual abundance; in the projections when

Table 8.3 Performance summary, in terms of probabilities and expectations (calculated over all iterations), for the adopted MP (MP3) for the key robustness trials. The performance statistics show the probabilities that the biomass will reach the interim rebuilding target of $0.2B_0$ by 2022 and 2035, the expected increase in biomass by 2022, and the expected annual catch over the same period. The $p(C_{\uparrow\downarrow})$ statistic is the probability that the TAC will go down at the second MP decision after it went up at the first.

Scenario	$P(SSB_{2022} > 0.2B_0)$	$P(SSB_{2035} > 0.2B_0)$	$E(SSB_{2022}/SSB_{2011})$	$E(C_{2012-2022})$	$p(C_{\uparrow\downarrow})$
Reference, 3,000 t	0.19	0.7	2.76	15,200	0.49
Reference, 5,000 t	0.14	0.7	2.65	15,600	0.71
lowR, 3,000 t	0.06	0.66	2.32	13,200	0.83
upq, 3,000 t	0.08	0.45	2.58	15,300	0.50
STWin, 3,000 t	0.01	0.34	2.39	12,872	0.81
omega75, 3,000 t	0.06	0.48	2.74	13,304	0.74

the abundance increases the CPUE does so but at a slower rate, so that the MP reacts slower to the positive signals (in terms of increasing the TAC) thereby attaining a better level of rebuilding than if future CPUE was directly proportional to abundance. In all cases, the MP manages to more than double the SSB relative to 2011 by 2022 (range of 2.3–2.8).

In 2011, the Scientific Committee (SC) of the CCSBT recommended that the combined MP (MP3) be adopted by the Commission, which would also have to decide on appropriate settings and constraints (Anonymous 2011b). A special meeting of the CCBST Commission met immediately after the SC and a number of additional runs for different settings were requested prior to the meeting of the Commission later that year. Member scientists began the additional simulations requested and collaborated across the various member countries to produce a final MP performance summary for the Commission, given the requests from the special meeting. The CCSBT formally adopted the combined MP (known now as the Bali Procedure given the location where the SC meeting it arose from was held), with a target year of 2035 and a maximum change of 3,000 t (Anonymous 2011a).

Fishery implementation of feedback control

For the first MP decision in 2011, which concerned the TAC from 2012 to 2014, the MP indicated that the global TAC could be raised to 12,449 t, given increasing trends in both monitoring indices (Fig. 8.2). The CCSBT decided to phase in this increase (Anonymous 2011a), given wide concern at the time about the low status of the SBT spawning stock (Anonymous 2011b, Hillary et al. 2015). The next TAC calculation took place in 2013 (Anonymous 2013). Given continued increasing index trends, the MP returned an annual TAC of 14,647 t for the 2015–2017 period.

The adoption of an MP has thus aided the CCSBT considerably in breaking previous deadlocks which had led to failure to reach consensus on TACs historically, leaving catches unchanged. Under an "ideal" MP process, the MP output is always implemented. In 2011, with the stock heavily depleted and the first application of a new process for setting TACs, the phasing of the TAC increase was understandable, and served to improve on the planned recovery rate for the population. However, the real proof of the efficacy of the MP process in this case will stand or fall on what action the CCSBT takes if and when the MP indicates that a decrease in the TAC is required to meet the rebuilding target. Tempering decreases in the TAC would clearly endanger the rebuilding program, and would not be consistent with an adopted MP framework where the objectives are clear and agreed.

Moreover, the value of the MP framework has been further demonstrated within the CCSBT with two challenges that, one can easily imagine, would have caused serious problems for the previous paradigm of annual assessment and negotiation. The first was the inclusion of results from a novel genetic technique that permits estimation of the absolute abundance of the spawning population, termed close-kin genetics (Bravington et al. 2012). The second was above expectation variation in the aerial survey index, with a very low point in 2012 (Anonymous 2012) and a very high point in 2014 (Anonymous 2014). Given the MP framework both of these issues were quite simple to analyse, in terms of their respective impacts on the rebuilding of the stock under the action of the adopted MP. For the close-kin genetics data, their inclusion in the OM decreased current uncertainty in the SSB, increased the estimates in absolute terms, and reduced the level of biomass depletion (Hillary et al. 2014). When projecting this revised OM forward with the MP, although the more optimistic outlook for the stock meant the MP would overshoot the rebuilding target $(p(B_{2035} > 0.2B_0)) = 0.75)$ and with higher average catches, this was not by so much that the MP needed to be retuned (Anonymous 2014). It also showed that the performance of the MP on the key robustness trials considered in 2011 could be expected to be noticeably better with the inclusion of the close-kin and most recent monitoring data (Hillary et al. 2014). With the survey, while the very low and very high points observed in 2012 and 2014, respectively, fell outside the range simulated in the reference set in 2011, a number of key robustness trials did contain these points within their simulated range. Under the meta-rules process set up for the CCSBT MP – which codifies clearly the process for dealing with situations outside the range tested in the original MP evaluation – the decision was made that these points did not trigger Exceptional Circumstances and that the MP had been tested to such extrema and been adopted as such. That is not to say that the influence of such points is not important (Hillary et al. 2014), but that their effect, once they are adjudged to be acceptable for inclusion in the OM and MP, is again relatively simple to explore using the MP in projection mode.

One can very easily imagine that, in a stock assessment setting, a new set of data that nearly doubles the previous SSB estimates, or two instances in 3 years of input data outside the bounds seen previously, would cause problems. Again, this is not to say that the impacts of these developments are not important – they quite clearly were in terms of MP projections. It is more the fact that with a tuned MP with well-defined objectives, clear performance measures with a wide-enough suite of plausible robustness tests, and a clearly codified process for dealing with meta-rule type events, such issues need not reopen debates on status, targets and acceptable or consensus TAC levels. The benefits of the MP framework are also beginning to be observed in terms of restarting the Scientific Research Program (SRP) for the CCSBT (Bravington and Davies 2013, Preece et al. 2013, Stobutzki et al. 2013), as scientists have more time both intersessionally and at the meetings to construct a future cost-effective monitoring program for the stock. For the CCSBT this is the start of a long process, with another TAC decision due in 2016, a full stock assessment and comprehensive review of the MP in 2017. Only time will tell if the MP will ultimately be successful in ensuring stock recovery, but the most recent indications are positive and the capacity of the MP framework to permit efficient and science-based decision-making, in place of time-consuming consensus-based approaches that had clearly failed in the past as regards to stock recovery, has become abundantly clear within the CCSBT.

References

Anonymous 1994. First meeting of the commission for the conservation of southern bluefin tuna, Technical report, CCSBT.

Anonymous 2010a. Report of the fifteenth meeting of the scientific committee, Technical report, CCSBT.

Anonymous 2010b. Report of the third operating model and management procedure technical meeting, Technical report, CCSBT.

Anonymous 2011a. Report of the eighteenth meeting of the commission, Technical report, CCSBT.

Anonymous 2011b. Report of the sixteenth meeting of the scientific committee, Technical report, CCSBT.

Anonymous 2012. Report of the seventeenth meeting of the scientific committee, Technical report, CCSBT.

Anonymous 2013. Report of the eighteenth meeting of the scientific committee, Technical report, CCSBT.

Anonymous 2014. Report of the nineteenth meeting of the scientific committee, Technical report, CCSBT.

Bravington, M.V. & Davies, C.R. 2013. Close-kin: Where to now?, Technical Report CCSBT- ESC/1309/17, CCSBT.

Bravington, M.V., Grewe, P.G. & Davies, C.R. 2012. Report of the close-kin project: estimating the absolute spawning stock size of sbt using genetics, Technical Report CCSBT-ESC/1208/19, CCSBT.

Caton, A.E. 1990. Southern bluefin tuna, in 'World Meeting on stock assessment of Bluefin Tunas: strengths and weaknesses', IATTC.

Collette, B.B., Carpenter, K.E., Polidoro, B.A., Juan-Jorda, M. J., Boustany, A.,... Yanez, E. 2011. 'High value and long-life – double jeopardy for tunas and billfishes', Science, 333, 291–292.

Cullis-Suzuki, S. & Pauly, D. 2010. 'Marine protected area costs as beneficial fisheries subsidies: a global evaluation', Coastal Management, 38, 113–121.

Farley, J.H., Davis, T.L.O., Gunn, J.S., Clear, N.P. & Preece, A.L. 2007. 'Demographic patterns of southern bluefin tuna', Fisheries Research 83, 151–161.

Hillary, R.M., Preece, A.L. & Davies, C.R. 2014. Assessment of stock status of southern bluefin tuna in 2014 with reconditioned operating model, Technical Report CCSBT-ESC/1409/21, CCSBT.

Hillary, R.M., Preece, A., Davies, C.R., Kurota, H., Sakai, O., Itoh, T., Parma, A., Butterworth, D.S., Ianelli, J. & Branch, T.A. 2015. A scientific alternative to moratoria for rebuilding depleted international tuna stocks. Fish and Fisheries, DOI: 10.1111/faf.12121.

Myers, R.A. & Worm, B. 2003. Rapid worldwide depletion of predatory fish communities, Nature, 423, 280–282.

Polacheck, T. 2002. Experimental catches and the precautionary approach: The southern bluefin tuna dispute, Marine Policy, 26, 283–294.

Preece, A.L., Davies, C.R., Bravington, M.V., Hillary, R.M., Eveson, J.P. & Grewe, P.G. 2013. Preliminary cost and precision estimates of sampling designs for gene-tagging for sbt, Technical Report CCSBT-ESC/1309/18, CCSBT.

Sibert, J.R., Hampton, J., Kleiber, P. & Maunder, M.N. 2006. 'Biomass, size, and trophic status of top predators in the pacific ocean', *Science,* 314, 1773–1776.

Stobutzki, I., Davies, C.R. & Preece, A.L. 2013. Scientific research program for ccsbt: suggestions for structuring and points for discussion, Technical Report CCSBT-ESC/1309/20, CCSBT.

Trenkel, V. 2008. 'A two-stage biomass random effects model for stock assessment without catches: What can be estimated using only biomass survey indices?', *Canadian Journal of Fisheries and Aquatic Sciences,* 65, 1024–1035.

Worm, B. & Tittensor, D.P. 2011. Range contraction in large pelagic predators, *Proceedings of the National Academy of Sciences USA,* 108, 11942–11947.

Chapter 9

Stakeholder-centered development of a harvest control rule for Lake Erie walleye

Michael L. Jones, Matthew J. Catalano,
Lisa K. Peterson and Aaron M. Berger

Introduction

Lake Erie is the southernmost of the five Laurentian Great Lakes, and the smallest by volume (Hartman 1972). These characteristics, along with the preponderance of urban and agricultural land within the watershed, result in Lake Erie being the most productive of the Great Lakes, and consequently the location of the largest freshwater commercial fisheries in North America. Two percid species dominate the contemporary Lake Erie commercial fishery, walleye (*Sander vitreus*) and yellow perch (*Perca flavescens*; Baldwin et al. 2009), which for walleye operates exclusively in Canadian waters. Both species are also the target of valuable recreational fisheries in Lake Erie, concentrated in US waters. Since the late 1970s, conflicts between commercial and recreational percid fishers, and between fishers and fishery managers, have resulted in an ongoing lack of consensus on management procedures. In this chapter we describe a process, initiated in 2010, whose goal was to reduce conflict by engaging fishery stakeholders (including resource users and managers) in the development of a harvest policy for Lake Erie walleye using a management strategy evaluation (MSE) approach.

Lake Erie consists of three bathymetrically distinct basins (Figure 9.1). The shallow western basin (mean depth 7.4 m) and its tributaries are the location of the primary spawning habitats for Lake Erie walleye. Juvenile and adult walleye migrate into the larger central basin (mean depth 18.5 m) and deeper eastern basin (mean depth 24.5 m) during summer with larger walleye tending to move further east (Wang et al. 2007). There is some walleye production from tributaries flowing into the eastern basin, but this production is relatively small and not considered part of the population being managed using the harvest rules discussed in this chapter. Each of the lake basins are nearly equally divided between Canadian and US territorial waters. All commercial fishing for walleye occurs in Canadian (Ontario) waters, while > 95% of recreational harvest is taken in US waters of Michigan, Ohio, Pennsylvania, and New York. Percid management in Lake Erie is coordinated by the Lake Erie Committee (LEC). The LEC consists of representatives from each of the five jurisdictions,

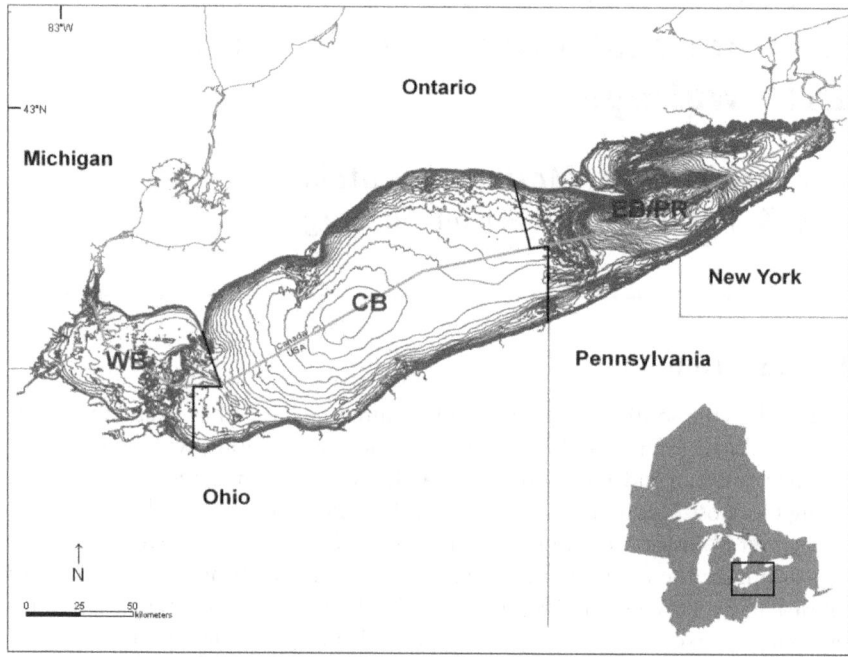

Figure 9.1 The west basin (WB), central basin (CB), and east basin/Pennsylva-
nia ridge (EB/PR) geomorphologic regions of Lake Erie have distinct
bathymetric and water quality attributes. The main walleye population
occurs in the west and central basins.

all of whom have management authority for the fisheries that operate in their
waters. These five agencies are not obligated to reach consensus on allowable
harvests each year, but have consistently endeavored to do so since the signing
of the Joint Strategic Plan for Management of Great Lakes Fisheries in 1981
(GLFC 2007).

The history of Lake Erie commercial fisheries can be divided into three periods. From the late 1800s until the early 1960s the fishery was open access, and
harvest rates rose to unsustainable levels, first for lake herring (now called cisco,
Coregonus artedii) in the 1920s, and then for walleye and a subspecies known as
blue pike (*Sander vitreusglaucus*) in the 1950s and 1960s (Figure 9.2). Cisco and
blue pike stocks have never recovered. During the 1970s the walleye fishery was
closed, primarily due to excessive mercury levels, while yellow perch harvests
rose to historic high (and likely unsustainable) levels. In the late 1970s the commercial walleye fishery was reopened, after which a period of increasingly conservative controls on commercial and recreational exploitation of walleye and
yellow perch ensued. This involved rationalization of the commercial fishery

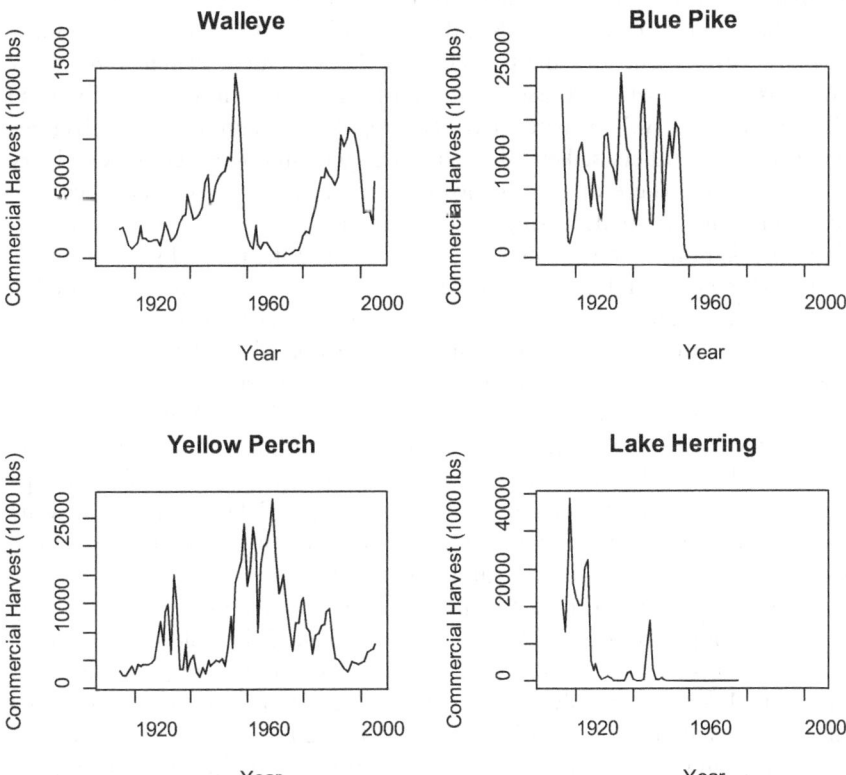

Figure 9.2 Historical commercial harvest of the four most important species in the historical Lake Erie commercial fishery. Lake herring (*Coregonus artedii*) are now called cisco. Rainbow smelt (*Osmerus mordax*; not shown) are also an important component of the contemporary commercial fishery.

and the establishment of tradable catch shares, development of stock assessment models for both species, and application of precautionary limits on commercial fishing mortality together with progressively lower bag limits in the recreational fishery. From the 1980s to the present the walleye population in Lake Erie has fluctuated considerably (Figure 9.2), with abundance strongly influenced by three exceptional year-classes, the most recent occurring in 2003. During the past decade the stock has remained above thresholds for concern, but the absence of a strong year-class since 2003 has limited the growth of the population despite low fishing mortality rates.

These changes to the management of Lake Erie percids have enabled a more sustainable fishery, and provided a basis for allocating harvest between recreational and commercial interests. Cooperation among Great Lakes fishery management agencies has been a hallmark of fishery management in the region since the establishment of the US and Canadian Great Lakes Fishery Commission in 1955 and especially the signing of the Joint Strategic Plan in 1981. Nevertheless, the annual stock assessment and harvest allocation process for Lake Erie walleye and yellow perch has been regularly encumbered by disagreements between fishery stakeholders and managers, among fishery stakeholders, and – in part because of the binational division of fishery sectors – between managers who are endeavoring to represent the interests of stakeholders in their jurisdiction. Each year the LEC uses stock assessment results to determine an overall recommended allowable harvest (RAH), which is then allocated as total allowable catches (TACs) for Canadian commercial fisheries and US recreational fisheries according to a sharing formula based on the relative surface areas of the Canadian and US regions of Lake Erie's western and central basins (43.1% goes to Canada, and 56.9% to the United States). Stakeholders have viewed the RAH/TAC determination process as seriously lacking in transparency, which has led to very little trust of decision-makers and a lack of acceptance of resulting harvest recommendations (Gaden 2007, chap. 5). In some instances this has resulted in litigation on the part of Canadian commercial fishing interests, challenging the quotas set by the LEC.

In 2001, the LEC decided to support a decision analysis exercise to help inform development of a new harvest policy for Lake Erie walleye. The exercise involved managers, agency fishery biologists, stock assessment specialists, and modelers. It did not involve fishery stakeholders. Managers were concerned that some stakeholders would be able to unduly influence the process, and to a much greater degree than others. The commercial fishing interests employ fishery biologists with expertise in stock assessment methods and population dynamics, while this is not the case for recreational fishing interests; because of this disparate technical expertise among fishery representatives, managers concluded that their participation would not benefit the process. The exercise resulted in the LEC adopting a biomass-based harvest control rule that was demonstrably effective at meeting stated management objectives (see Jones and Bence, 2009, for a brief description of this control rule); however the new policy was not viewed as a positive outcome by the fishery stakeholders, largely because of the lack of transparency in its formulation.

In 2009, the LEC decided to revisit the process of developing a harvest policy for Lake Erie percids, and this time acknowledged that the process needed to explicitly involve all stakeholders. The LEC solicited the services of the Quantitative Fisheries Center (QFC) at Michigan State University to provide both third-party facilitation of the process and analytical support. Having a third party lead the process was viewed as essential to engendering trust and to managing the uneven level of expertise among participants. Starting in November 2010, the QFC organized and led an iterative process of analysis and deliberation that involved managers, commercial and recreational fishery stakeholder representatives, and walleye fishery experts. In this chapter we briefly describe this process, including the details of the retrospective (assessment) and prospective (forecasting) models used to inform the evaluation of alternative harvest control rules. In January 2014, the LEC announced adoption of a new harvest control rule for Lake Erie walleye that was developed through this process.

Fishery objectives and management practices

Between November 2010 and January 2014, a group of fishery managers and stakeholders met 12 times to discuss management of the Lake Erie walleye fishery. This group came to be known as the Lake Erie Percid Management Advisory Group (LEPMAG). The meetings generally lasted 1 day and were facilitated by the modeling team from the QFC. Over the course of the complete set of meetings we (1) developed a set of "rules of engagement" to govern the process; (2) generated lists of management options, management objectives, and critical areas of uncertainty; (3) debated refinements to the existing walleye assessment models; (4) developed a candidate set of harvest control rules (HCRs) to consider; (5) identified a set of quantitative performance indicators that stakeholders could use to evaluate the expected performance of each candidate HCR; (6) debated alternative configurations and assumptions for the forecasting model used to simulate the candidate HCRs; and (7) gauged areas of consensus and of continued disagreement on the assessment models and the merits of alternative HCRs. Only recognized LEPMAG members were allowed to engage in discussions at the meeting, but all members were entitled to request permission from the rest of the group to invite other stakeholders as observers. Throughout the 12-meeting sequence, all observer requests were unanimously approved by LEPMAG. The process was guided by the following overarching vision statement:

> Lake Erie percid fisheries will be transparently managed using sound science and partnerships to achieve stable and sustainable harvests from shared stocks providing broad and equitable benefits for all jurisdictions.

At the fifth LEPMAG meeting in February 2012, a panel of six external stock assessment and walleye experts joined the group, heard presentations on the assessment and forecasting models from the modeling team, and then provided constructive advice to LEPMAG for refinements to the models or alternative analytical and simulation tactics to consider. LEPMAG members nominated this panel of experts. At the following meeting (June 2012) the modeling team reported back to LEPMAG on analyses and model changes that were completed in response to the panel's advice. As noted later, this process was very helpful for building trust in the analytical work, especially on the part of stakeholders lacking the technical expertise to fully critique the analyses.

Harvest control rules

Extensive discussions with LEPMAG resulted in selection of a broad class of HCRs known as biomass feedback policies. Feedback policies impose dynamic control rules, usually set as a function of some state of the fishery or population (Hilborn and Walters 1992, chap. 15; Deroba and Bence 2008). Specifically, LEPMAG preferred model-based HCRs that set target fishing mortality rates conditional on reference points related to F_{MSY} and the stock's position relative to the unfished spawning stock biomass ($S0$). LEPMAG adopted a limit reference point (LRP) for spawning stock biomass (S_y) such that fishing would be curtailed if S_y dropped below some threshold. Target fishing mortality rates were set as a percentage of F_{MSY} and were used to calculate the TAC as long as estimated biomass was above the LRP threshold.

Commercial fishery stakeholders successfully advocated for adoption of a probabilistic version of feedback HCRs (Prager et al. 2003, Shertzer et al. 2008) because they believed this would assist their efforts to achieve Marine Stewardship Council certification, which requires participating fisheries to demonstrate an explicit accounting for uncertainty. Probabilistic HCRs impose no specific functional relationship between estimated stock biomass and the fishing mortality rate, but instead involve an iterative search process for fishing mortality rates (or TACs) that ensure limit reference points are not violated with a probability that does not exceed some prespecified probability $P\star$. The probability $P\star$ represents an a priori risk tolerance of stakeholders, society, and/or management authorities. The task of computing probabilities of violating LRPs is repeated each time the TAC is set and involves estimating uncertainty in the projected stock and also the reference points themselves. LEPMAG chose to consider HCRs with a probabilistic LRP in spawning stock biomass and a nonprobabilistic target reference point in fishing mortality. Thus, three policy variables defined each candidate HCR: a target fully selected F (Ft) as a percentage of F_{MSY} ($X\%$ F_{MSY}), a LRP in spawning stock biomass as a percentage of unfished levels ($X\%$ $S0$), and a probability threshold that characterized risk tolerance relative to the LRP ($P\star$). In addition, the commercial sector sought HCRs that would reduce interannual fluctuations in the TAC, to enhance financial certainty. As such, the modeling team evaluated control rules that applied a cap (δTAC) on the interannual percent change in the TAC.

In total, the modeling team evaluated 96 different HCRs that represented combinations across a range of Ft, LRP, $P\star$, and δTAC values selected by LEPMAG. A range of eight different Ft values from 30% to 100% of F_{MSY} (in 10% increments) were evaluated. Values relating to the spawning stock biomass LRP that were considered were: 20, 30, and 40% of $S0$. The modeling team also considered a range of $P\star$ values: 0.05, 0.15, 0.30, and 0.50, which is related to the level of uncertainty incorporated within the forecasting procedures (see the later section, "Forecasting model"). Stakeholders agreed that a 20% cap on the interannual change in the TAC was acceptable, so other caps were not simulated.

Management strategy evaluation

As with many other managed commercial fisheries, Lake Erie walleye are managed using an assessment model that informs decision-makers of the current and past status of the stock, and a short-term projection to determine what harvest level is most likely to meet current management objectives. To evaluate the expected longer-term performance of candidate HCRs, the modeling team combined the assessment model into a forecasting model that both forecasted the future state of the walleye population conditional on the policy in place and simulated future assessment data to inform harvest decisions in future years. In this section, we first describe the assessment model used to reconstruct historical walleye population dynamics, and then the model that forecasted the expected outcome of alternative management strategies.

Assessment model

Since about 1990, Lake Erie walleye stocks have been assessed using a statistical catch-at-age (SCA) model (Fournier and Archibald 1982, Berger et al. 2012). During the early stages of the LEPMAG process the modeling team reviewed and discussed modifications to the existing assessment models with the LEPMAG participants, in an effort to ensure the assessment model being used in the future had stakeholder support. Our general approach was to develop and compare alternative assessment models, present the comparisons to LEPMAG, and then invite the group to choose the model that they judged to be most appropriate. Because many of the stakeholders did not feel they had the technical knowledge to make this judgment, we also invited a panel of walleye and stock assessment experts, selected by LEPMAG, to review and comment on the assessment models. The end result was that LEPMAG unanimously supported a small set of changes to the assessment model (described later) that the modeling team recommended. We believe this review process was critical to building overall stakeholder trust in the MSE process.

The assessment model consisted of population and observation submodels (Tables 9.1 and 9.2). Walleye population dynamics were based on annual time steps beginning in 1978 and included six age-classes (2, 3, 4, 5, 6, and 7+) comprising a single, spatially aggregated Lake Erie population. Recruitments to age 2 were estimated as parameters of the model and abundance-at-age was predicted by applying fishing and natural mortality rates (Equations 9.2.8–10 in Table 9.2). Based on previous tagging studies (unpublished data; but see Locke et al. 2005), the natural mortality rate was assumed to be constant and known without error ($M = 0.32$ yr^{-1}). Biomass was calculated as the product of estimated abundance-at-age and observed mean weight-at-age, where mean weights were obtained from fishery independent surveys (Equation 9.2.11). The observation model linked the population model with the observed data via estimated catchability and selectivity parameters while treating fishing effort as known. The model was fitted to (1) catch and effort data from Ohio recreational,

Table 9.1 Description of symbols in Table 9.2 describing the retrospective and prospective assessment models.

Symbol	Description

Subscript indicators

Symbol	Description
y	year (1978–2011)
a	age (2–7+)
f	fishery (commercial = 1; recreational (OH) = 2; recreational (MI) = 3)
i	survey (Ontario = 1, Ohio/Michigan = 2)
k	recreational vulnerability time block (1978–2003; 2004–2011)

Assumed values

Symbol	Description
M	instantaneous rate of natural mortality (0.32 yr^{-1})
m_a	proportion of walleye mature at age a

Observed data

Symbol	Description
$C_{y,f}$	total numbers of walleye caught by fishery
$I_{y,i}$	survey abundance index
$P_{y,a,f}$	proportions of catch at age by fishery
$P_{y,a,i}$	proportions at age from survey abundance index
N	sample size (number of years data)
$E_{y,f}$	fishery effort
$w_{y,a}$	mean weight

Estimated parameters

Symbol	Description
$s_{a,f,k}$	selectivity at age for each fishery and time block
$s_{a,i}$	selectivity at age for each survey
r_f	coefficient of recreational fishing effort-abundance power model
$r1_f$	exponent of recreational fishing effort-abundance power model
$r2_f$	time trend coefficient of recreational fishing effort-abundance power model
σ_E	standard deviation of recreational fishing effort-abundance power model
c_f	coefficient of recreational catchability-abundance power model
$c1_f$	exponent of recreational catchability-abundance power model
σ_q	standard deviation of recreational catchability-abundance power model
α	stock-recruitment initial slope
β	stock-recruitment density dependence parameter
P	stock-recruitment post Dreissenid time block effect on productivity
Φ	autoregressive coefficient
σ_R	standard deviation of recruitment process

Derived parameters

R_y	recruitment for each year
ω_y	autoregressive term for recruitment dynamics
$q_{y,f}$	fishery catchability in year y
$q_{y,i}$	survey catchability in year y
$\hat{E}_{y,f}$	recreational fishing effort
$F_{y,a,f}$	instantaneous fishing mortality rate
$Z_{y,a}$	instantaneous total mortality rate
S_y	spawning stock biomass
$N_{y,a}$	abundance at age in year y
B_y	total biomass
$V_{y,f}$	vulnerable abundance for fishery f
$\hat{C}_{y,f}$	model predicted total catch
$\hat{C}_{y,a,f}$	model predicted catch at age
$\hat{P}_{y,a,f}$	model predicted proportions of catch at age
$\hat{I}_{y,i}$	model predicted survey abundance index (catch per unit effort)
$\hat{I}_{y,a,i}$	model predicted survey abundance index at age (catch per unit effort)
$\hat{P}_{y,a,i}$	model predicted proportions at age from survey abundance index

HCR parameters

F_{MSY}	instantaneous fishing mortality rate that results in maximum sustained yield
F_{τ}	target reference point in fishing mortality rate set at X% of FMSY
F_{HCR}	fishing mortality rate set by the HCR and used to set the TAC
S_0	unfished spawning stock biomass
LRP	limit reference point in spawning stock biomass set at X% of S0
P*	probability threshold for probabilistic HCR
ΔTAC	limit on the interannual percentage change in the TAC

Table 9.2 Equations used in the statistical catch-at-age model (SCA), stock-recruitment model (SR), and/or the operating model (OM).

Recruitment	Model	Equation #
$N_{y,a=2} = R_y$	SCA, SR, OM	9.2.1
$R_y = \begin{cases} \alpha S_{y-2} e^{-\beta S_{y-2} + \varphi \omega_{y-1} + \varepsilon_y}, & y < 1993 \\ \alpha S_{y-2} e^{-\beta S_{y-2} + p + \varphi \omega_{y-1} + \varepsilon_y}, & y \geq 1993 \end{cases}$	SR, OM	9.2.2

(Continued)

Table 9.2 (Continued)

Recruitment	Model	Equation #
$\omega_y = \begin{cases} \ln\left(R_y\right) - \ln\left(\alpha S_{y-2} e^{-\beta S_{y-2}}\right), & y < 1993 \\ \ln\left(R_y\right) - \ln\left(\alpha S_{y-2} e^{-\beta S_{y-2}+\rho}\right), & y \geq 1993 \end{cases}$	SR, OM	9.2.3

Catchability		
$q_{y+1,f} = q_{y,f} e^{\varepsilon_{y,f}}$	SCA	9.2.4
$q_{y+1,i} = q_{y,i} e^{\varepsilon_{y,i}}$	SCA, OM	9.2.5
$q_{y,f} = c_f V_{y,f}^{c_{2f}} e^{\varepsilon_{y,f}}$	OM	9.2.6

Recreational fishing effort		
$\hat{E}_{y,f} = r_f V_{y-1,f}^{r_{1f}} e^{r_{2f}y+\varepsilon_y}$	OM	9.2.7

Mortality rates		
$F_{y,a,f} = \begin{cases} q_{y,f} s_{a,f} E_{y,f}, & f = 1 \\ q_{y,f} s_{a,f,k} E_{y,f}, & f > 1 \end{cases}$	SCA, OM	9.2.8
$Z_{y,a} = M + \sum_f F_{y,a,f}$	SCA, OM	9.2.9

Population model		
$N_{y,a} = \begin{cases} N_{y-1,a-1} e^{Z_{y-1,a-1}}, & a < 7 \\ N_{y-1,a-1} e^{Z_{y-1,a-1}} + N_{y-1,a} e^{Z_{y-1,a}}, & a = 7 \end{cases}$	SCA, OM	9.2.10
$B_y = \sum_a N_{y,a} w_{y,a}$	SCA, OM	9.2.11
$S_y = \sum_a N_{y,a} w_{y,a} m_a$	SCA, OM	9.2.12
$V_{y,f} = \sum_a N_{y,a} s_{a,f}$	SCA	9.2.13

Calculated quantities for the observation model		
$\hat{C}_{y,a,f} = \dfrac{F_{y,a,f}}{Z_{y,a}}\left(1 - e^{-Z_{y,a}}\right) N_{y,a}$	SCA, OM	9.2.14
$\hat{C}_{y,f} = \sum_a \hat{C}_{y,a,f}$	SCA, OM	9.2.15
$\hat{P}_{y,a,f} = \dfrac{\hat{C}_{y,a,f}}{\hat{C}_{y,f}}$	SCA, OM	9.2.16
$\hat{I}_{y,a,i} = q_{y,i} s_{a,i} N_{y,a}$	SCA, OM	9.2.17
$\hat{I}_{y,i} = \sum_a \hat{I}_{y,a,i}$	SCA, OM	9.2.18
$\hat{P}_{y,a,i} = \dfrac{\hat{I}_{y,a,i}}{\hat{I}_{y,i}}$	SCA, OM	9.2.19

Michigan recreational, and Ontario commercial fisheries; (2) catch-per-effort (CPE) data from US and Ontario fishery-independent gillnet surveys; (3) bottom trawl CPE data targeting age-0 walleye to inform recruitment estimates; and (4) age-composition observations from all fisheries and gillnet surveys. The recreational and commercial harvest at-age and survey CPE data were recast in the form of annual totals and proportions by age, so that totals for a fishery or survey could be modeled as log-normal with proportions at age treated as arising from sampling a multinomial distribution (Equations 9.2.14–19), as suggested by Fournier and Archibald (1982).

During our review of the assessment model, alternative representations of two parameters, catchability and age-specific selectivity, were determined to have a substantial influence on the estimated quantities (e.g. biomass, fishing mortality) that are used to set harvest levels. Group discussions guided the modeling team to evaluate models that estimate selectivity-at-age freely (no functional form assumed). These models fit the data better than alternative formulations, but led to estimated patterns of selectivity-at-age that were at odds with stakeholder expectations (e.g. recreational fishery selectivity increasing across a wide range of ages despite anglers belief that walleye are "fully vulnerable" to their gear by age 3). After extensive discussion and closer examination of the data that informed the assessment model, we concluded that the estimated patterns of selectivity reflected the interaction of gear-specific selectivity with spatial variations in age-specific availability of fish to the gear. The various fisheries and surveys that contribute data to the assessment model are not uniformly distributed throughout Lake Erie, so spatial variation in walleye age composition (specifically the tendency for older walleye to migrate further east during the summer) leads to apparent patterns in gear-specific selectivity (Berger et al. 2012). At the end of these discussions, LEPMAG agreed that age-specific selectivity should be modeled as a time-invariant free parameter across a single unit stock; the only exception was to model a change in selectivity for age-2 fish caught by recreational fishermen as a result of a change in size limit regulations in 2004.

Due to temporal changes in fishing efficiency and environmental conditions (e.g. water clarity), there was general agreement that catchability needed to be allowed to vary over time, but there was considerable discussion on how best to incorporate this into the assessment model. Prior to LEPMAG, the historical period was divided into "time blocks": periods during which catchability was constant, but between which it could vary. These time blocks represented expert judgment about points in time when either management changes or environmental changes (e.g. Dreissenid mussel invasion) might have been expected to substantially influence catchability. LEPMAG members raised concerns about the confidence with which these time blocks could be identified, independent of the data. As an alternative, the modeling team explored modeling catchability using a random-walk process, based on experience with this approach in other Great Lakes fishery assessments (e.g. Wilberg and Bence 2006). In general, the patterns of catchability trends over time that emerged matched the predetermined time blocks reasonably closely. Nevertheless, our discussions of this issue with LEPMAG participants resulted in unanimous support for adoption of random-walk catchability for the MSE analysis and for future assessments (Equations 9.2.4–6).

LEPMAG was also concerned with the methods for predicting recruitment in the upcoming TAC year. Prior to LEPMAG, projected age-2 abundance was estimated external to the SCA by regressing age-2 abundance estimates against age-0 trawl CPE lagged 2 years. LEPMAG agreed that a more integrated approach would be preferable. The modeling team addressed this issue by incorporating the age-0 trawl CPE data set into the SCA in a similar fashion to all of the other surveys used in the model. The trawl survey data informed recruitment estimates for all of the years of the model plus an additional projected 2 years. This approach led to an increase in precision and reduction of age-2 projection errors. LEPMAG agreed this was a reasonable approach and recommended integration of the age-0 trawl data into the SCA.

The assessment also included a stock-recruitment analysis to inform the prospective model forecasts of future recruitment. We conducted the stock-recruitment analysis on the outputs from the SCA (estimated age-2 recruitment and spawning stock biomass time series), rather than integrating it into the SCA because we have found this results in more reliable estimates of the SR model parameters and their uncertainty (Haeseker et al. 2003, Tsehaye et al. 2014). Estimates of age-2 recruitment and spawning stock biomass from the SCA were modeled using a Ricker function with lognormal process errors. Spawning stock biomass was taken as the sum over ages of the product of age-specific abundance, average weight (kg), and maturity (Wang et al. 2009; Equation 9.2.12). An additive time block effect on productivity was incorporated to account for differences in prerecruit survival due to full colonization of the lake by nonnative Dreissenid mussels and round goby (*Neogobius melanostomus*) after 1993. In addition, recruitment residuals were modeled as a first-order autoregressive process to account for apparent alternating strong and weak year-classes (Equations 9.2.1–3). The time block effect and autoregressive process had strong empirical support as indicated by Akaike's Information Criterion.

Forecasting model

Closed-loop simulations of the entire management process were used to compare how different harvest control rules performed across a range of plausible states of nature. Walleye population dynamics were simulated using a stochastic age-structured operating model (OM) that followed the general structure of the assessment model with the exception that the OM incorporated density-dependent recreational fishing effort and catchability. We parameterized the OM using estimates from the most recent assessment model and best available knowledge, and, as with the assessment model, implemented it using AD Model Builder software (Fournier 2011). We conducted 250 individual 25-year simulations of the OM, from which performance metrics were computed for each candidate HCR. Exploratory analyses indicated that 250 simulations were sufficient to characterize the central tendency and variance of projected outcomes of the closed-loop simulation. The 25-year time horizon was selected to strike a balance between long-term equilibrium HCR performance and shorter-term performance conditional on the current state of the stock.

The OM tracked a single population of age-2 through 7+ walleye over time. Recruitment to the fishery was assumed to occur at the beginning of the year for age-2 fish and was generated as a Ricker function of spawning stock biomass 2 years prior with log-normal process error, first-order autoregressive residuals, and post-Dreissenid productivity, as previously described. An upper bound of 33% above the largest historically estimated recruitment was imposed to prevent unrealistically large recruitment events associated with the tail of the log-normal distribution.

Individual cohorts declined through time via year, age and fishery-specific fishing mortalities and natural mortality. Historical recreational fishing effort observations (1978–2011) were positively associated with retrospective SCA model estimates of vulnerable abundance, but have also declined gradually over time, irrespective of abundance. Therefore, recreational fishing effort (Ohio and Michigan fisheries) in the OM was modeled as a power function (Equation 9.2.7) of vulnerable abundance (Equation 9.2.13) and year with lognormal process error. Recreational catchability was also density-dependent (power function, lognormal process error) in the OM because SCA random-walk catchability and vulnerable abundance estimates were negatively related. Thus, recreational fishing mortality was density-dependent, self-regulating, and not explicitly set by the HCR. This assumption is consistent with the observation that recreational harvest has not exceeded the TAC in recent years despite the lack of hard controls on total harvest (i.e. recreational fishing is regulated with bag, size, and season limits, but not constraints on total effort). The instantaneous commercial fishing mortality rate was obtained by iteratively solving the catch equation (Equation 9.2.14) for F to satisfy the commercial TAC (see later), conditional on natural mortality and recreational fishing mortality rates.

The commercial TAC was set by applying one of the candidate HCRs. At each time step, the OM stochastically generated observational data (harvest, effort, catch rates, age composition) conditioned by the simulated population structure. The data were passed to an assessment model that was identical to the retrospective SCA model used to parameterize the OM. Recruitment and spawning stock biomass estimates from the SCA model were used to estimate the Ricker stock-recruitment model parameters. These were in turn used to estimate F_{MSY} and $S0$ using an equilibrium solution to the Ricker stock-recruitment function (Walters and Martell 2004, chap. 3, box 3.1).

The fishing mortality rate set by the HCR (F_{HCR}) was capped at $F\tau = X\%$ F_{MSY} but was adjusted downward if the projected spawning stock biomass after harvest (S_{y+1}) was less than the LRP with a probability that exceeded

$P\star$ (Figure 9.3). Specifically, the value of F_{HCR} was varied iteratively until the condition $\Pr(S_{y+1} \leq \text{LRP}) \leq P\star$ was satisfied (Figure 9.3). At each iteration, $\Pr(S_{y+1} \leq \text{LRP})$ was obtained from a cumulative normal distribution with

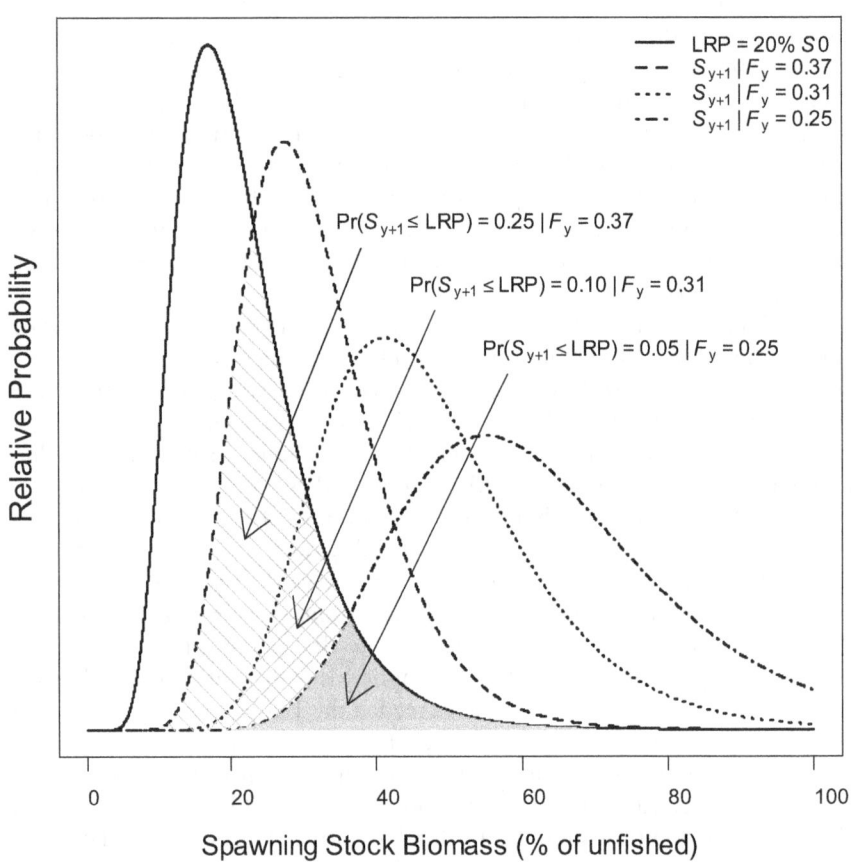

Figure 9.3 Example demonstration of the iterative F-setting procedure of the probabilistic HCR. The solid line depicts the estimated probability distribution of the LRP, which was set at 20% of the unfished spawning stock biomass (S_0) for this example. The thick dashed line represents the probability distribution for the projected spawning stock biomass (S_{y+1}) for a target fishing mortality rate (F_r) of 60% of F_{MSY}, which was $F = 0.37$ yr^{-1} in this particular example. The probability that S_{y+1} was less than the LRP ($\Pr(S_{y+1} \leq \text{LRP})$; shown as the striped area) given that $F = 0.37$ yr^{-1} was 0.25, which exceeded the a priori $P\star$ value of 0.05. Thus, F was reduced iteratively from 0.37 to 0.25 yr^{-1} in increments of 0.06 yr^{-1} until $\Pr(S_{y+1} \leq \text{LRP})$ no longer exceeded $P\star$. The probability distributions for S_{y+1} given $F = 0.31$ and 0.25 yr^{-1} are shown as the fine dashed and stippled lines respectively, and their corresponding $\Pr(S_{y+1} \leq \text{LRP})$ as the cross-hatched and grey shaded areas. Thus, in this example, the TAC would be set using $F = 0.25$ yr^{-1} to satisfy the $P\star = 0.05$ criterion.

mean = S_{y+1} − LRP and variance = $\sigma^2 S + \sigma^2 LRP$ under the assumption that the S_{y+1} and LRP were independent. The S_{y+1} (and σS) was estimated by projecting forward from the terminal assessment model estimates, and the LRP (and σ_{LRP}) was obtained from the stock-recruitment analysis ($S0$). The total TAC was set by applying F_{HCR} to the population under a selectivity schedule that was averaged across fishery sectors. The commercial TAC was set as a percentage (43.1% allocation; WTG, 2009) of the overall TAC. If the interannual change in the TAC exceeded the cap on the allowable change (δTAC), then the TAC was adjusted to satisfy the constraint.

Several types of uncertainty were incorporated into the OM. Parameter uncertainty was acknowledged by applying a different set of initial conditions and system parameters used to initialize the OM for each of the 250 simulations. Each set was one Markov chain Monte Carlo (MCMC) sample of the stationary joint posterior distribution from the most recent (2012) SCA model as approximated by the Metropolis–Hastings algorithm (Gelman et al. 2004). Initial conditions included abundance and fishing mortality at age in the most recent 2 years. System parameters included stock-recruitment (α, β, p, Φ, σR), recreational effort-abundance relationship (r, $r1$, $r2$, σE), density-dependent catchability (c, $c1$, σq), vulnerability, and standard deviations for effort, catch, and CPE observations. Observation uncertainty was incorporated by distorting simulated data prior to executing assessment procedures. Fishing effort, catch, and survey CPE observation errors were lognormally distributed and errors in proportions-at-age followed a multivariate-logistic function with constant age and gear standard deviations (Schnute and Richards 1995, Cox and Kronlund 2008). Assessment uncertainty was applied by using simulated assessment-based estimates as the basis for the HCR within the OM. Implementation uncertainty was incorporated by applying lognormal deviations to the commercial fishing mortality rate.

Summary statistics for selected performance indicators from each simulation were calculated across the 25-year time horizon and compiled to produce a distribution of expected performance for each candidate HCR. Performance indicators were selected by LEPMAG. The chosen indicators were generally representative of those commonly considered when conducting MSEs (Butterworth and Punt 1999, Rademeyer et al. 2007) such as expectations relating to sustainability, risk, and industry stability. Mean walleye abundance, recreational fishery CPE, and commercial fishery harvest and yield were metrics used to evaluate long-term expected conditions. Biological risks associated with a given policy were quantified as the percentage of years the population was below 20% of the unfished spawning stock biomass. Variation in annual commercial harvest was used to quantify measures of market stability. Risks to fisheries were quantified by identifying thresholds in commercial yield and recreational catch per effort, below which industry viability would be compromised. Commercial fishery risk was assessed by computing a performance indicator that was the proportion of years in which commercial yield was less than 4 million pounds, a threshold identified by stakeholders as a minimum yield for economic viability. Risk to the recreational fishery was assessed by computing the proportion of years with an average angler catch rate less than 0.4 walleye per hour – a threshold below which stakeholders felt fishing effort and economic value from the fishery would suffer.

Results and implementation

The large number of candidate HCRs and performance measures considered, together with the stochastic nature of the model results, created significant challenges for LEPMAG when they were called upon to interpret the results, and judge which HCRs appeared to result in the best outcome. To assist with interpretation we used a variety of graphical and tabular formats for summarizing the outputs, three of which we will describe here: (1) box plots that provide comprehensive summary results for an individual performance indicator; (2) tabular summaries of median outcomes to provide a more concise summary across HCRs and performance indicators; and (3) trade-off plots.

The box plots showed the distribution of outcomes for a single performance measure across a range of candidate HCRs (e.g. see Figure 9.4). Each individual box-and-whisker plot represented the distribution of outcomes for a performance measure that resulted from a specific HCR, including medians and interquartile ranges. Boxes were arranged hierarchically, with ranges of F_t at the lowest level (groups of four adjacent boxes in Figure 9.4), alternative P^* values at the next level (contrasting shading in Figure 9.4), and alternative LRP values at the highest level. Increasing F_t resulted in a decrease in predicted abundance when the LRP and P^* were not set at risk averse levels (i.e. the set of boxes on the far left of Figure 9.4). Varying the LRP and P^* had relatively little effect on abundance, but when they were set at their most conservative levels (LRP = 40% S_0 and $P^* = 0.05$) they dominated the effect of varying F_t (i.e. set of unshaded boxes in the right panel of Figure 9.4). By examining box plots for each of the performance measures, the group concluded that within the range of HCRs examined, the LRP and P^* value had comparatively little influence on the performance of the HCR, and collectively decided to instead focus discussions on F_t levels.

Many LEPMAG members found the box plots difficult to interpret and requested a simple summary of the results. They acknowledged that a depiction of the range of possible outcomes was important, but additionally wanted to have a very clear impression of the central tendency for each HCR and performance measure. Thus, we also summarized the results in tabular form (Table 9.3), showing the median result for a preferred subset of performance measures and scenarios. The resulting table, in conjunction with the box plots, helped the group better interpret the differences among the different scenarios and among performance measures.

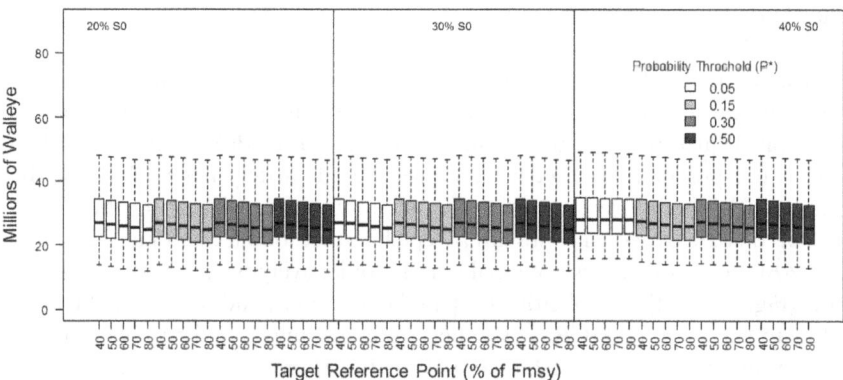

Figure 9.4 Box plot showing the results of the MSE for mean total abundance. Panels represent different limit reference points (% of unfished spawning stock biomass). Shading of the bars represents different probability thresholds (*P**). Individual bars with the same shading represent a range of F_{targ} values, as indicated on the x-axis.

Table 9.3 Tabular description of selected MSE results. For all the scenarios shown here *P** = 0.05 and TAC constraint = 20%. Values in the table represent the median across simulations. "F" is the target F (% of F_{MSY}) and "B" is the limit reference point (% of S0).

	F = 40 B = 20	F = 60 B = 20	F = 80 B = 20	F = 100 B = 20	F = 40 B = 30	F = 60 B = 30	F = 80 B = 30	F = 100 B = 30
Spawner biomass (million kg)	41.9	39.1	36.5	34.3	41.9	39.3	37.0	35.3
Commercial yield (million lbs)	4.5	6.5	8.1	9.4	4.5	6.4	7.9	8.7
Recreational catch per hour	0.44	0.43	0.42	0.41	0.44	0.43	0.42	0.42
P(Com yield < 4 mil lbs.)	0.36	0.04	0.00	0.00	0.38	0.08	0.00	0.00
— change from lowest scenario	NA%	NA%	0%	0%	NA%	NA%	0%	0%
P(Rec catch per hour < 0.4)	0.44	0.44	0.48	0.52	0.44	0.44	0.48	0.48
— change from lowest scenario	0%	0%	9%	18%	0%	0%	9%	9%

Editor's note: Two different mass units (kilogram and pounds) are used as they are more familiar to the walleye stakeholders and managers.

Many of the challenges associated with fisheries management arise from having to make decisions that try to balance conflicting objectives, often without explicit quantification of decision trade-offs. For LEPMAG, the overriding issue is the potential conflict between setting adequately large TACs for the commercial fishery to ensure economic stability, while maintaining abundance at levels that are expected to sustain acceptable recreational catch rates. To allow LEPMAG to consider this potential conflict objectively, we presented trade-off plots (Figure 9.5) that explicitly compared commercial and recreational fishery risk across a range of HCRs. As expected, increasing $F\tau$ led to lower risk to the commercial fishery, but higher risk to the recreational fishery. However, at lower $F\tau$ values (40–60% of F_{MSY}) the increase in risk to the recreational fishery as $F\tau$ increases is much less than the decrease in risk to the commercial fishery. However, at higher $F\tau$ values (80–100% of F_{MSY}), the change in risk to the commercial fishery is not as great while the risk to the recreational fishery continues to increase. The evidence that this nonlinear pattern of risk

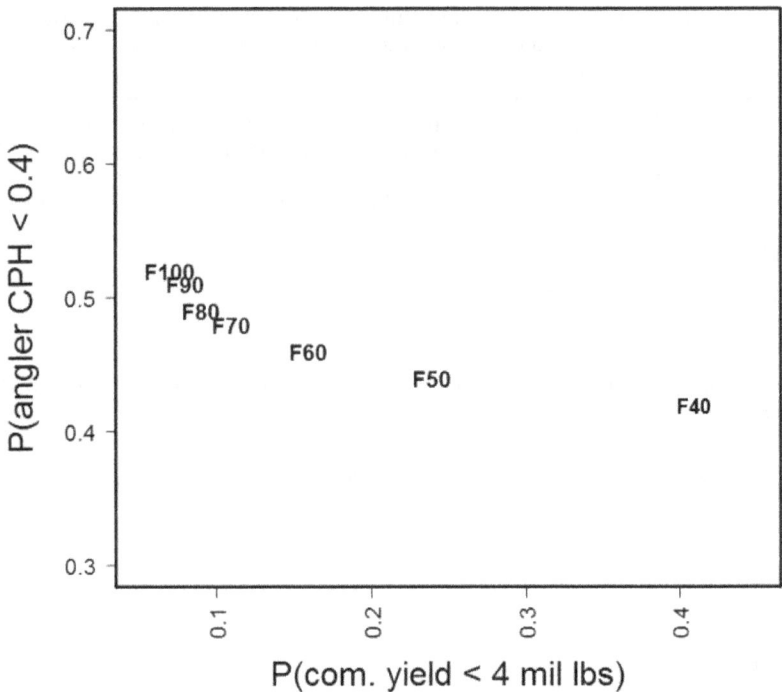

Figure 9.5 Trade-off analysis of two risk-related performance measures: probability of recreational catch per hour falling below 0.4 (y-axis) and probability of commercial yield falling below 4 million pounds of walleye (x-axis). Each point represents a different F_{targ} with the number representing the target as a percentage of F_{MSY}. All HCR results presented here used a $P*$ of 0.05 and an LRP of 20% B0.

trade-offs existed for the Lake Erie walleye fishery had a profound effect on the discussions among LEPMAG members. It eventually led to the acknowledgement by recreational fishery stakeholders that less conservative $F\tau$ values than they otherwise would have favored would be acceptable, given the benefits to the commercial fishery stakeholders. Commercial fishery stakeholders continued to favor higher $F\tau$ values (80–100% of F_{MSY}), arguing that the risk to the stock and the recreational fishery would remain quite low. However, they indicated a willingness to live with an intermediate value, at least for the next few years. As a direct consequence of the MSE stakeholder-driven process, the LEC announced at the 12th LEPMAG meeting in January 2014 adoption of a new HCR for Lake Erie walleye, with $F\tau$ at 60% of F_{MSY}, an LRP of 20% of $S0$, a $P\star$ value of 0.05, and a 20% interannual TAC constraint.

The LEPMAG experience has been widely acknowledged by both fishers and managers as enabling a substantial increase in the transparency of Lake Erie walleye management. All stakeholder groups (commercial fishers, recreational fishers, managers) indicated in a survey conducted in January 2014 that they believe "the process for making management decisions is [more] transparent" than at the start of the LEPMAG process, and that "the current harvest control rule was determined from a sound scientific analysis" (M. Jones, unpublished data). The best evidence that the participants have viewed the process favorably is provided by the fact that LEPMAG is continuing to meet to develop a similar HCR for yellow perch. Perhaps most encouraging of all is the fact that the highly technical nature of the assessment modeling and MSE process did not impair the engagement of the stakeholders. We believe this was possible because the stakeholders viewed the facilitator/modeler team as an acceptably impartial third party and were therefore willing to trust the scientific judgment of the analysts. We strongly encourage management agencies that are considering an MSE process to keep this in mind: success will depend greatly on whether the stakeholders develop a trusting relationship with the analytical team.

Acknowledgements

We would like to recognize the entire Lake Erie Percid Management Advisory Group for their engagement, and their patience, as we led them through the process described in this chapter. Travis Brenden, Allan Hicks, Owen Hamel, Charles T.T. Edwards, and Dorothy J. Dankel provided valuable comments on earlier drafts. Support for LEPMAG was provided by the Great Lakes Fishery Commission and the National Fish and Wildlife Foundation.

References

Baldwin, N.S., Saalfeld, R.W., Dochoda, M.R., Buettner, H.J., & Eshenroder, R.L. 2009. *Commercial Fish Production in the Great Lakes 1867–2006.* Great Lakes Fishery Commission, Ann Arbor, MI. Available at: www.glfc.org/databases/commercial/commerc.php (last accessed 11 January 2016).

Berger, A.M., Jones, M., Zhao, Y. & Bence, J.R. 2012. Accounting for spatial population structure at scales relevant to life history improves stock assessment: The case for Lake Erie walleye Sander vitreus. *Fisheries Research*, 115–116, 44–59.

Butterworth, D.S. & Punt, A.E. 1999. Experiences in the evaluation and implementation of management procedures. *ICES Journal of Marine Science*, 56, 985–998.

Cox, S.P. & Kronlund, A.R. 2008. Practical stakeholder-driven harvest policies for groundfish fisheries in British Columbia, Canada. *Fisheries Research*, 94, 224–237.

Deroba, J.J. & Bence, J.R. 2008. A review of harvest policies: Understanding relative performance of control rules. *Fisheries Research*, 94, 210–223.

Fournier, D.A. 2011. An introduction to AD Model Builder for use in nonlinear modeling and statistics. Version 10.0. Available at http://admb-M project.org/documentation/manuals/admb-user-manuals (last accessed 19 November 2015).

Fournier, D.A. & Archibald, C.P. 1982. A general theory for analyzing catch at age data. *Canadian Journal of Fisheries and Aquatic Sciences*, 39, 1195–1207.

Gaden, M.E. 2007. Bridging jurisdictional divides: Collective action through a Joint Strategic Plan for Management of Great Lakes Fisheries. Doctoral dissertation, University of Michigan, Ann Arbor, MI.

Gelman, A., Carlin, J.B., Stern, H.S. & Rubin, D.B. 2004. *Bayesian Data Analysis*, Chapman and Hall, Boca Raton.

GLFC (Great Lakes Fishery Commission, Editor). 2007. A joint strategic plan for management of the Great Lakes fisheries (adopted in 1997 and supersedes 1981 original). Great Lakes Fish. Comm. Misc. Publ. 2007–01.

Haeseker, S.L., Jones, M.L. & Bence, J.R. 2003. Estimating uncertainty in the stock-recruit relationship for St. Marys River sea lamprey. *Journal of Great Lakes Research*, 29 (Supplement 1), 728–741.

Hartman, W.L. 1972. Lake Erie: Effects of exploitation, environmental changes, and new species on the fishery resources. *Journal of the Fisheries Research Board of Canada*, 29, 931–936.

Hilborn, R. & Walters, C.J. 1992. Quantitative fisheries stock assessment: Choice dynamics and uncertainty, Kluwer Academic Publishers, Boston, MA.

Jones, M.L. & Bence, J.R. 2009. Uncertainty and fishery management in the North American Great Lakes: Lessons from applications of decision analysis. *In:* Krueger, C.C. & Zimmerman, C.E. (eds.) *Pacific Salmon: Ecology and Management of Western Alaska's Populations.* American Fisheries Society, Symposium 70, Bethesda, MD, pp. 1059–1081.

Locke, B.L., Belore, M., Cook, A., Einhouse, D., Kayle, K., Kenyon, R., Knight, R., Newman, K., Ryan, P. and Wright, E. 2005. *Lake Erie Walleye Management Plan.* Great Lakes Fishery Commission, Ann Arbor, MI.

Prager, M.H., Porch, C.E., Shertzer, K.W. & Caddy, J.F. 2003. Targets and limits for management of fisheries: A simple probability-based approach. *North American Journal of Fisheries Management*, 23, 349–361.

Rademeyer, R.A., Plagányi, E.E. & Butterworth, D.S. 2007. Tips and tricks in designing management procedures. *ICES Journal of Marine Science*, 64, 618–625.

Schnute, J.T., & Richards, L.J. 1995. The influence of error on population estimates from catch-age models. *Canadian Journal of Fisheries and Aquatic Sciences,* 52, 2063–2077.

Shertzer, K.W., M.H. Prager & E.H. Williams. 2008. A probability-based approach to setting annual catch levels. *Fishery Bulletin*, 106, 225–232.

Tsehaye, I.W., Jones, M.L., Brenden, T.O., Bence, J.R., Madenjian, C.P. & Warner, D.M. 2014. Assessing the balance between predatory consumption and prey dynamics in the Lake

Michigan pelagic fish community. *Canadian Journal of Fisheries and Aquatic Sciences*, 71, 1–18.

Walleye Task Group (WTG). 2009. Report of the Lake Erie Walleye Task Group to the Standing Technical Committee, Lake Erie Committee of the Great Lakes Fishery Commission.

Walters, C.J. & Martell, S.J.D. 2004. *Fisheries Ecology and Management,* Princeton University Press, Princeton, NJ.

Wang, H.-Y., Cook, H.A., Einhouse, D.W., Fielder, D.G., Kayle, K.A., Rudstam, L.G. & Höök, T.O. 2009. Maturation schedules of walleye populations in the Great Lakes Region: Comparison of maturation indices and evaluation of sampling-induced biases. *North American Journal of Fisheries Management*, 29, 1540–1554.

Wang, H., Rutherford, E.S., Cook, H.A., Einhouse, D.W., Haas, R.C., Johnson, T.B., Kenyon, R., Locke, B. & Turner, M.W. 2007. Movement of walleyes in Lakes Erie and St. Clair inferred from tag returns and fisheries data. *TAFS*, 136, 539–551.

Wilberg, M.J. & Bence, J.R. 2006. Performance of time-varying catchability estimators in statistical catch-at-age analysis. *Canadian Journal of Fisheries and Aquatic Sciences*, 63, 2275–2285.

Chapter 10

Northern Prawn Fishery: beyond biologically centred harvest strategies

Catherine M. Dichmont, André E. Punt,
Roy A. Deng, Sean Pascoe and Rik C. Buckworth

Introduction

The Northern Prawn Fishery (NPF) is a multispecies, multistock prawn fishery in tropical northern Australia (Figure 10.1), managed using an input control system. The total annual fishing effort is adjusted by changing the total length of fishing net available to the fleet, and net length can be traded between fishing vessel operators. The fishery occurs from circa April to November, with a midseason closure from roughly June to August. This acts as an additional input control measure, and the exact dates for the season, and the dates separating the first and second subseasons, depend on the assessed status of spawning stocks or in-season catch rates. In 2001, the fishery was worth over AU$160M and was one of the Australian Commonwealth's most valuable fisheries (Galeano et al., 2004). However, falling prawn prices due to cheap imports of aquaculture prawns, and the appreciation of the Australian dollar have halved this value in recent years. Increasing fuel prices have also reduced profitability of the fishery, which is now between AU$70–100M per annum (Skirtun et al., 2014). After several industry and government funded buy-back schemes, the fishery currently consists of 52 vessels and 19 operators.

The overall fishery has two parts: a banana prawn fishery and a tiger prawn fishery, each targeted by the same vessels. The tiger prawn fishery captures mainly two species of tiger prawn (*Penaeus semisulcatus, Penaeus esculentus*) and two species of endeavour prawn (*Metapenaeus endeavouri, Metapenaeus ensis*) (Venables and Dichmont, 2004), while the banana prawn fishery only targets banana prawns (*Penaeus merguiensis, Penaeus indicus*;Venables et al., 2006). Management decisions for the NPF are made by the Australian Fisheries Management Authority (AFMA) Commission (previously the AFMA Board) based on advice from the Northern Prawn Fishery Management Advisory Committee (NORMAC) and the NPF Resource Assessment Group (RAG) (Smith et al., 1999).

This chapter focuses on the management strategy evaluation (MSE) used for selecting the harvest strategy for the tiger prawn fishery only (although banana prawns are included in the simulations to account for the impact of

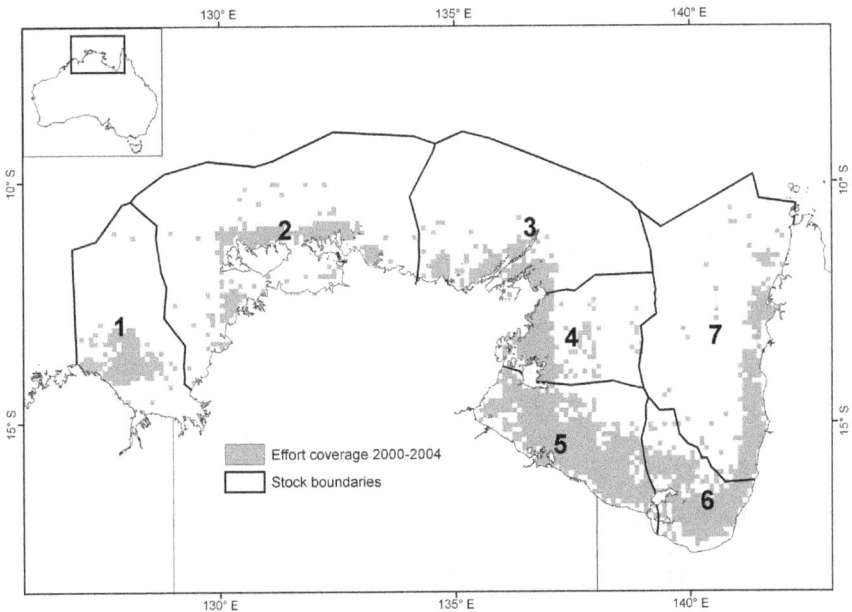

Figure 10.1 The Northern Prawn Fishery, indicating the seven stock areas considered in the operating model. Stock areas 1–3 are combined for the management strategy evaluation (MSE) analyses. The relative sizes of the seven areas are 0.11, 0.24, 0.17, 0.09, 0.13, 0.06 and 0.2, respectively.

fishing for banana prawns on the effort available for the tiger prawn fishery).[1] Tiger prawns have been the focus for quantitative stock assessments and management measures for many years (Somers, 1990, Wang and Die, 1996, Somers and Wang, 1997, Dichmont et al., 2001, 2003, Punt et al., 2010), owing to the perception that these species can be recruitment overfished (Ye, 2000). Endeavour prawns are predominantly a bycatch of targeting tiger prawns, and management measures for tiger prawns therefore also tend to impact the fishing mortality for endeavour prawns. In contrast to the tiger prawns, assessments of endeavour prawns were only conducted for the first time in 2007 (Dichmont et al., 2008). Figure 10.2 shows the time series of catches for tiger and endeavour prawns and the effort estimated to be targeted at each of the two species of tiger prawns.

In 2000, the spawning stock size of *P. esculentus* was estimated to be below S_{MSY}, the spawning stock size corresponding to MSY, which was the management target at the time (Dichmont et al., 2010). *P. semisulcatus* was also assessed to be below S_{MSY}, but to a lesser extent. In response to this, NORMAC and

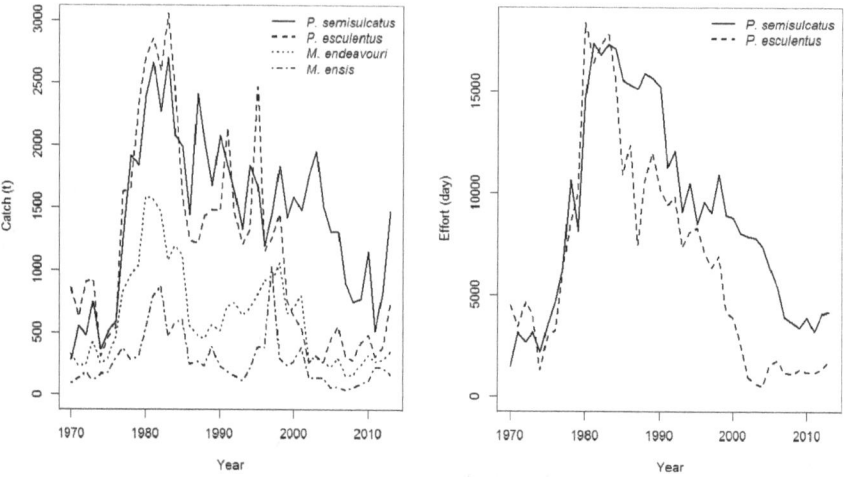

Figure 10.2 Time trajectories of catch (t) by species and effort (days) by target species. Total fishery effort is allocated to species using the approach of Venables and Dichmont (2004).

AFMA agreed to rebuild the two tiger prawn species to S_{MSY} within 5 years (i.e. by the end of 2006) and adopted a new, more conservative target reference point: "there is a 70+% chance that the spawner population at the end of 2006 will be above or at spawner level targets (S_{MSY})" (Dichmont et al., 2010). A stock rebuilding strategy to achieve this goal was implemented at that time and was successful – the 2006 stock assessment indicated that recovery had occurred, and that both species of tiger prawns were no longer overfished.

In 2005, the then Australian Federal Minister for Fisheries, Forestry and Conservation issued a formal directive to end overfishing in Commonwealth-managed fisheries through a package of buy-backs and policies. The key policy here instituted the need for formal harvest strategies (Australian Department of Agriculture Fisheries and Forestry, 2007), which defined the risk associated with fishing, as well as clear target and limit reference points. The key objective of Commonwealth fisheries management was also changed to maximizing the net economic returns from the fishery, requiring economic target reference points relating to maximum economic yield. The NPF therefore modified its harvest strategy to align with this policy – it adopted S_{MEY}, the spawning stock size corresponding to maximum economic yield (MEY) as its target reference point, and set the overfishing limit reference point to $0.5S_{MSY}$. Status relative to the limit reference point would be measured as a moving average over the most recent 5 years, to accommodate the short-lived and variable nature of the species' dynamics (Dichmont et al., 2010, Kompas et al., 2010). The objective for AFMA to achieve ecologically sustainable development also led to the need to consider the broader ecosystem impacts of the fishery when selecting harvest strategies.

MSE has been used in several ways in the NPF, and this chapter provides an overview of analyses related to a choice of harvest strategies (of which one of these has been implemented since 2010), and implications of that choice for benthic impacts. The harvest strategies are model-based and consist of combinations of assessment methods, harvest control rules to determine (1) the total annual effort in fishing days targeted at tiger prawn stocks and (2) the length of time during which the fishery is open, including the length of the midseason closure.

Effort is the currency in the MSE, but in the final implemented harvest strategy the total effort is converted to a change in the total amount of the tradeable gear units. This is done by using an algorithm based on a statistically derived relationship between headrope length (the basis of the gear unit) and total effort (Dichmont et al., 2010). Since total effort converted to gear units is the major management (input) control, the total tiger prawn effort output by the harvest strategies is not resolved by species, area or week. However, this level of detail is important for the dynamics of the different species and included in the operating model, which requires the total effort to be disaggregated to a finer scale, as described later.

The structure of the MSE has changed over time in response to changes in management objectives and their (implicit) weighting. Dichmont et al. (2006a, 2006b, 2006c) conducted the first MSE for the NPF focusing on the two tiger prawn species, with performance measures derived from biological and yield considerations. While this work provided guidance on harvest strategies for the target species, it did not adequately address the full range of management objectives for the NPF. It was much more aimed at addressing key uncertainties, including model type and spatial scale (Dichmont et al., 2006c), and therefore explored scenarios with the largest disconnect between the operating model and the harvest control rules. Consequently, Dichmont et al. (2008) extended the earlier MSE to capture a broader range of objectives, including economics, the impacts of the fishery on the benthos, and to include the two endeavour prawns; this chapter outlines this version of the MSE and subsequent developments. The MSE includes multiple, single-species models (representing four prawn species and a suite of benthic invertebrates), and harvest strategies designed to allocate a total annual fishing effort and season length to the tiger prawn fishery that is consistent with both economic and ecosystem management objectives. Given the complexity of the system being modelled, and the context provided by earlier work, it should be borne in mind that most of the key uncertainty tests that are so important in MSE modelling were undertaken by Dichmont et al. (2006c) are therefore not described in this chapter.

The harvest control rules

Since the management currency in the MSE is effort (rather than headline length or gear units), the harvest control rules tested in the MSE determine (1) the level of effort for the entire NPF, in terms of days fished, targeted at each of two tiger prawn species and (2) the length of the fishing season. Our analyses were based on a stock assessment method that involved fitting a delay-difference model to spatially aggregated catch and effort data (Dichmont et al., 2008). This delay difference model was selected based on the analyses in Dichmont et al. (2006c).

We considered two types of harvest strategy to set the total annual tiger prawn effort. The first (HS1) involved a threshold harvest control rule in which effort for each tiger prawn fishery was set according to the equation:

$$
E_t = \begin{cases} 0 & \text{if } S_t \leq 0.5S_{\text{MSY}} \\ E_{\text{target}} \dfrac{S_t - 0.5S_{\text{MSY}}}{S_{\text{target}} - 0.5S_{\text{MSY}}} & \text{if } 0.5S_{\text{MSY}} < S_t \leq S_{\text{target}} , \\ E_{\text{target}} & \text{if } S_t > S_{\text{target}} \end{cases} \tag{10.1}
$$

where E_t is fishing effort in days for year t, E_{target} is the target effort level (set as a proportion of the effort corresponding to effort at MSY, E_{MSY}), S_t is the estimate of the spawning stock size for year t, and S_{target} is the target spawning stock size (expressed as a proportion of S_{MSY}). The assessment method was used to provide the estimates of S_t, E_{MSY}, and S_{MSY}. The values of S_{target} tested were S_{MSY} and the proxy provided by the Australian federal Harvest Strategy Policy, which states that a maximum economic yield proxy should be $1.2S_{\text{MSY}}$.

The second harvest strategy type (HS2) for setting the total annual tiger prawn effort used the output from the assessment to parameterize a bioeconomic model. This model calculates the net present value (NPV) across tiger and endeavour prawns based on fishing effort for each of the two tiger prawn species. NPV is the discounted revenue by year less variable costs related to labor, capital, fuel, and other causes such as packaging:

$$
NPV = \beta_t \left(\sum_w \sum_s p_{t,w}^s H_{t,w}^s - \sum_w \sum_s \left(\alpha H_{t,w}^s + \delta E_{t,w}^s \right) \right), \tag{10.2}
$$

where β_t is the factor to discount future profit relative to present profit, $p_{t,w}^s$ is the average price per kilogram for species s during week w of future year t, $H_{t,w}^s$ is the predicted catch (kg) of prawns for species s during week w of future year t, $E_{t,w}^s$ is the effort targeted at species s during week w of future year t, α is cost-per-unit-kg landed for those costs (such as packaging) which are proportional to the catch in weight, and δ is the cost-per-unit-days fished for those costs (such as fuel) which are proportional to the amount of effort expended. The summation of costs in Equation 10.2 is taken over the two tiger prawn species while the number of species included in the summation over profit might include only the two tiger prawn species, or the two tiger prawn species and the more valuable and consistently caught of the endeavour prawn species. To implement the bioeconomic harvest strategy, a dynamic MEY is estimated in which effort for each of 50 future years is selected to maximize Equation 10.2. Effort for the first seven of these years is allowed to vary without constraint and effort for the eighth and subsequent years is set to that for the seventh year (Kompas et al., 2010). In HS2, the effort selected for the first projected year is used to set the total tiger prawn effort.

The weeks during which the fishery is open is either prespecified as part of the harvest strategy (HS1 and HS2) or is adjusted dynamically as part of the bioeconomic harvest strategy (HS2; see Figure 10.3 for an example).

In this MSE, the assessment model is similar to the operating model, but the latter explicitly accounts for spatial stock structure (Tables 10.1, 10.2 and 10.3); that is the operating model was spatially disaggregated (see later). The advantage of using a simplified model within the control rule is that a much greater range of scenarios can be explored because of computing time. On the other hand, the results from assessments are subject to structural error (because the operating model operates spatially whereas the assessments do not). Following the work presented here, a more detailed length-based assessment model has been developed and implemented for management (Punt et al., 2010). The length-based and delay difference stock assessment models provide very similar results in terms of harvest strategy testing (Dichmont et al., 2008). Therefore the MSE work presented here is still valid, and more comprehensive than it would otherwise have been had the more complicated assessment model been used.

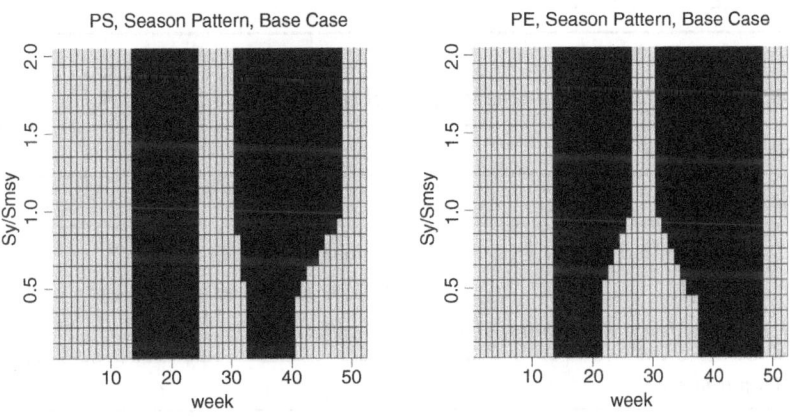

Figure 10.3 An example of the relationship between season length and S/S_{MSY}.

Table 10.1 Operating model specifications, common to each of the four species modelled.

Population dynamics

T1.1 $\quad B_{y,w+1} = (1+\rho)B_{y,w}\,e^{-Z_{y,w}} - \rho e^{-Z_{y,w}}\left(B_{y,w-1}\,e^{-Z_{y,w-1}} + w_{k-1}\,\alpha_{w-1}\,R_{\tilde{y}(y,w-1)}\right) + w_k\,\alpha_w R_{\tilde{y}(y,w)}$

T1.2 $\quad \tilde{N}_{y,w+1} = \tilde{N}_{y,w}\,e^{-Z_{y,w}} + \alpha_w\,R_{\tilde{y}(y,w)}$

Mortality and catches

T1.3 $\quad Z_{y,w} = M + F_{y,w}$

T1.4 $\quad F_{y,w} = \tilde{q}\,/\,P_s A_w Q_{y,w}\left(E_{y,w}^{T} + E_{y,w}^{B}\,/\,q_b\right)$

T1.5 $\quad Y_{y,w} = \dfrac{F_{y,w}}{Z_{y,w}}\,B_{y,w}\left(1 - e^{-Z_{y,w}}\right)$

Stock and recruitment

T1.6 $\quad S_y = \displaystyle\sum_w \beta_w\,\dfrac{1 - e^{-Z_{y,w}}}{Z_{y,w}}\,\tilde{N}_{y,w}$

T1.7 $\quad \tilde{R}_{\tilde{y}+1} = \tilde{\alpha}S_y\,e^{-\tilde{\beta}S_y + \eta_y}$

T1.8 $\quad \eta_{y+1} = \rho_r \eta_y + \sqrt{1 - \rho_r^2}\,\xi_{y+1}$

T1.9 $\quad \xi_y \sim N\left(0;\sigma_r^2\right)$

T1.10 $\quad \Omega_{i,j} = \sigma_r^2 \rho_r^{|i-j|}$

Effort dynamics

T1.11 $\quad {}_{act}E_y^{tig} = {}_{MS}E_y^{tig}\,e^{\varepsilon}$

T1.12 $\quad \varepsilon_y \sim N\left(0;\sigma_{inp}^2\right)$

T1.13 $\quad {}_{act}E_y^{tot} = {}_{act}E_y^{tig}\left({}_{act}E_{y'}^{tot}\,/\,{}_{act}E_{y.}^{tig}\right)$

Table 10.2 Notation for the operating model and estimation framework.

Symbol	Value	Description
Indices		
y	$\{1970, 1971, \ldots\}$	Annual time step
w	$\{1, 2, \ldots, 52\}$	Weekly time step within year
$\tilde{y}(y,w)$	y if $w < 40$; otherwise $y + 1$	Biological year
Model parameters		
α_w	Figure 10.4a	Fraction of annual recruitment by week
M	0.045	Natural mortality (wk^{-1}) (same for all species)

Model parameters

$R_{\tilde{y}}$	***	Annual recruitment
$\hat{R}_y$	$\alpha S_y e^{-\beta S_y}$	Recruitment predicted from the stock-recruitment relationship
ρ	0.980, 0.982, 0.968, 0.963[+]	Brody growth coefficient
w_k	$\left((1+\rho)w_{k+1}-w_{k+2}\right)/\rho$	Mass of a prawn at recruitment
w_{k-1}	15.01, 16.76, 15.39, 13.17[+]	Mass of a prawn a week before recruitment
$\tilde{q}$	0.000088	Catchability coefficient
P_s	Figure 10.1	Size of fishing covered stock area relative to the whole NPF for each of MSE areas
q_b	0.00000792, 0.00001496, 0.00006336[&]	Bycatch catchability P. semisulcatus fleet
	0.00001065, 0.00008184, 0.0[&]	P. esculentus fleet
A_w	Figure 10.4b	Availability by week
β_y	Figure 10.4c	Spawning by week
$\tilde{\alpha}$	***	Slope at the origin of the stock-recruitment relationship
$\tilde{\beta}$	***	Density-dependence parameter of the stock-recruitment relationship
ρ_r	***	Autocorrelation in recruitment deviations
σ_r	***	Standard deviation of recruitment deviations
σ_c	***	Standard deviation of the catch residuals
σ_{inp}	0.15	Standard deviation of implementation error

State variables

$\tilde{N}_{y,w}$		Numbers of recruited prawns
$B_{y,w}$		Biomass of recruited prawns
$Z_{y,w}$		Total mortality
$F_{y,w}$		Fishing morality
$Y_{y,w}$		Catch in weight
Ω		Variance-covariance matrix due to environmental factors (see Equation T1.10)

(Continued)

Table 10.2 (Continued)

Symbol	Value	Description
State variables		
$_{act}E_y^{tig}$		Effort for tiger prawns after accounting for outcome uncertainty
$_{act}E_y^{tot}$		Total fishery effort
Observations		
$E_{y,w}^T$		Targeted effort
$E_{y,w}^B$		'Bycatch' effort
$Y_{y,w}^{obs}$		Catch-in-weight
Control variables		
$_{MS}E_y^{tig}$		Effort for tiger prawns set by the harvest control rule

*** Estimated as part of the conditioning process.
+ P. semisulcatus, P. esculentus, M. endeavouri, M. ensis.
& by species in the same order as +, but ignoring the target species.

Table 10.3 Likelihood functions used in assessment and operating model.

Estimated parameters	
T3.1	$\left\{R_y : 1969, 1970, \ldots 200x\right\}, \tilde{\alpha}, \tilde{\beta}, \rho_r, \sigma_r$
Catch likelihood	
T3.2	$\displaystyle\sum_y \sum_w \left\{\ell n \sigma_C + \frac{1}{2\sigma_C^2}[\sqrt{Y_{y,w}^{obs}} - \sqrt{Y_{y,w}}]^2\right\}$
Recruitment likelihood	
T3.3	$\ell n\left(\sqrt{\det(\Omega + \mathbf{V})} + \frac{1}{2}\displaystyle\sum_{y_1}\sum_{y_2} \ell n(R_{y_1}/\hat{R}_{y_1})([\Omega + \mathbf{V}]^{-1})_{y_1,y_2} \ell n(R_{y_2}/\hat{R}_{y_2})\right)$

Note: The matrix **V** is the asymptotic variance-covariance matrix for the estimates of recruitment obtained by fitting the operating model to the data.

The evaluation process

The MSE has evolved in response to changes in management needs, and in particular the need to identify the fishing effort to achieve maximum economic returns. The evaluation structure outlined below was used by Dichmont et al.

(2008) and reflected the best system understanding at the time the analyses were conducted. The focus for the analyses was expected performance relative to management objectives, unlike Dichmont et al. (2006b) who also reported the relative errors for the estimates of various management-related quantities.

The operating model

The operating model consists of three parts: a biological prawn model, a benthic impacts model and an effort allocation model.

Biological model

It is hypothesized that there may be up to seven stocks of each of the four species in the NPF tiger prawn fishery (Figure 10.1), but three of the stock areas are combined for the operating model. Based on the resulting five spatial regions, all four species are in three of the five regions, whilst two of the regions each only contain three species. A delay-difference model was used to represent the population dynamics of the four prawn species in each area (Equation T1.1), with each area independent of the others. The data used when fitting the operating model are time series of catches in weight by week ($Y_{y,w}^{obs}$), the effort that was targeted towards the species being assessed ($E_{y,w}^{T}$) and the effort that was not targeted at the species being assessed ($E_{y,w}^{B}$). Historical changes in fishing efficiency (Bishop et al., 2008), week-dependent catchability and technical interactions by the empirically derived bycatchability parameter (Dichmont et al., 2003) were accounted for (Equation T1.4). This allowed effort to be translated into a predicted catch (Equations T1.4 and T1.5), and the model was fitted to the catch data assuming that after square-root transform it is normally distributed (Equation T3.2). The catchability parameter is based on a depletion-analysis that used logbook data to estimate biomass and catchability in a specific year, 1993 (Wang, 1999, Dichmont et al., 2003). Annual recruitments (R_y) are the only estimable parameters, which requires a strong set of assumptions regarding the biology of each species and the distribution of recruits throughout the fishing year. These assumptions take the form of parameters within the model, are listed alongside other model notation in Table 10.2 and are described in Dichmont et al. (2003). The model is initialized by the recruitments for 1969 and 1970, meaning that there is no assumption regarding the state of the stock, relative to the unfished level, in the first year for which catch and effort data are available.

Following the estimation of annual recruitments, the parameters of the stock-recruitment relationship are estimated independently using the annual spawning biomass and recruitment estimates, and accounting for estimation errors by including the variance-covariance matrix (V) in the likelihood function (Equation T3.3). This secondary fit accounts for temporal autocorrelation between recruitment deviates (Ω) that are assumed to be the result of environmental dependency (Dichmont et al., 2003). The stock-recruitment function is needed for projection purposes, but also the estimation of MSY-based reference points (Dichmont et al., 2003).

During projections, recruits are predicted using Equation T1.7, with parameter uncertainty accounted for by sampling parameter vectors from the inverse Hessian matrix. When implementing the harvest strategy, annual assessments, including estimation of annual recruitment, the stock–recruitment relationship and reference points are conducted for the whole NPF during each projection year, using historical catch and effort data alongside those generated by preceding, simulated implementations of the harvest strategy. It is assumed that catch and effort are measured without error (i.e. there is no observation error), which is realistic given the input controlled nature of the fishery.

Outcome uncertainty and effort allocation

The harvest strategies can change the level of effort targeted at each of two tiger prawn species and the weeks during the year the fishery is open. These changes are based on the estimated stock status (in the case of the MSY harvest strategies) or on the output from the bioeconomic model for the MEY harvest strategy. The required total tiger prawn effort is simply a summation of the sum of the two tiger prawn species' effort (in days). Outcome uncertainty accounts for variation in the realized effort level about the effort level recommended by the harvest control rule (Equation T1.11). This difference between $_{MS}E_y^{tig}$ and $_{act}E_y^{tig}$ captures politically motivated deviations from the outcomes from the harvest strategy, the impact of fluctuations in participation in the fishery, the difficulty of placing restrictions on a multispecies fishery and failure to fully account for changes over time in fishing efficiency.

The fishery for tiger prawns is conducted by the same fleet that targets banana prawns. The main impact of the banana prawn fishery is that it constrains the number of days available for fishing for tiger prawns. No attempt is made to model the population dynamics of banana prawns given the lack of quantification of the environmental factors that determine the success of banana prawn recruitment. Instead, an empirical approach is taken to predicting future banana prawn fishing. Specifically, a year y' is selected at random from the years 1990–2002 (excluding 1994 that had unusual seasons), giving the total fishery effort $_{act}E_y^{Tot}$, from which, after the total effort is split into week, the fraction of the total weekly effort directed at banana fishing is removed to obtain the remaining weekly tiger prawn effort (Equation T1.13).

The effort allocation model (Venables et al., 2009), which is a generalized additive modelling and Markov chain approach that uses historical, observational data from the fishery, consists of two components. The first component takes the total annual effort and allocates it to week, target species (*P. semisulcatus* and *P. esculentus*), and stock area, and the second component determines the effort applied to each of the 6-minute grid cells that constitute each stock area (Figure 10.1). The output from the first component of the effort allocation model is used to drive the biological component of the operating model while the second component is needed to compute the impact of fishing on the benthos (see later). The model explicitly takes account of fuel price and distance travelled, in addition to recent catch performance.

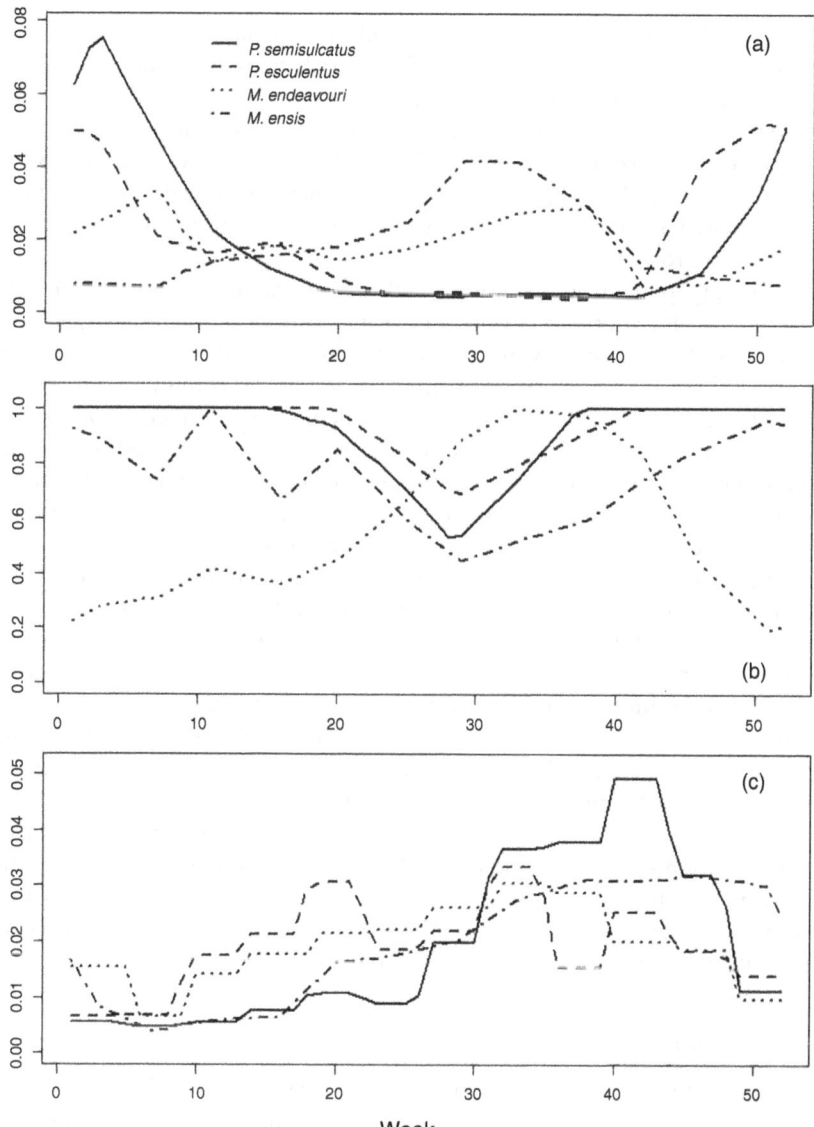

Figure 10.4 Species-specific weekly dependencies input into the operating model: (a) annual recruitment fractions, (b) weekly availability, and (c) spawning fractions. These values were obtained empirically from several years of monthly survey data (Dichmont et al., 2003).

Performance measures

A variety of performance measures were selected to capture performance relative to a range of management objectives. Stock status was evaluated relative to S_{MSY} (the original target reference point for the fishery), $0.5S_{MSY}$ (the limit reference point) and S_{MEY} (the spawning stock size at which net present value would be maximized given perfect information about the dynamics of the operating model). The impact of the choice of harvest strategy on the industry was evaluated in terms of short- and medium-term profits and interannual variation in profits.

Reporting performance measures related to the impact on the benthos required the development of a benthic impacts model. This model quantified the effects of repeated trawling on the biomass of benthic organisms. It consisted of three components given input on effort by 6-minute latitude and longitude grid cell: (1) a component that calculates the depletion of the biomass of a range of benthic taxa given repeated trawling; (2) a component that determines the rate of recovery of each taxon; and (3) a component that distributes the biomass (initially) uniformly over space. The biomass of each taxon by 6-minute grid was modelled using a continuous logistic model (Ellis and Pantus, 2001), with parameters specified based on field data (Dichmont et al., 2008). The benthos was assumed to be unfished at the start of the prawn fishery in 1970. The benthic model was restricted to the area that was fished between 2000 and 2004 (years when the fleet was smaller than 100 vessels, and the length of the fishing season was similar to the present). The assumption of random fishing at the level of 6-minute grids was conservative because trawling is somewhat aggregated at the subgrid cell level (Deng et al., 2005), which means that grid-cell scale impacts were slightly overestimated.

Although the operating model had multiple stocks of each species, the statistics used to summarize performance did not focus on results by stock but rather emphasized results by species (and aggregated over the whole fishery for catch- and profit-related performance measures). Similarly, the results for impacts on the benthos focused on measures that aggregated over species.

Simulation evaluation

The results for alternative strategies were compared in a relative sense using Zeh plots (see Figure 10.5 for an example, which shows a subsample of the full harvest strategies tested in Dichmont et al., 2008). The strategies in Figure 10.5 are setting the target reference point as

1   S_{MSY} (HS1)
2   the MEY proxy of $1.2S_{MSY}$ (HS1)
3   MEY using tiger and endeavour prawn catches in the profit function (HS2)
4   MEY for tiger prawns only (HS2).

Results were also summarized using trajectory plots of outputs for individual harvest strategies (e.g. Figure 10.6). Many of the strategies performed similarly (Dichmont et al., 2006b, 2008), although several harvest strategies could be excluded with a criterion that the spawning stock biomass should be at S_{MEY} and exceed S_{MSY}.

Unlike the MSEs applied by the International Whaling Commission and for some species in South Africa, a fairly limited number of operating model scenarios were explored as more expansive tests were undertaken in previous MSE iterations and computer time required to run scenarios. Many more were tested in the foundational work by Dichmont et al. (2006a, 2006b, 2006c). The key factors considered in sensitivity analyses were:

1   the time series of fishing efficiencies
2   the (prespecified) catchability coefficient
3   the extent of outcome uncertainty

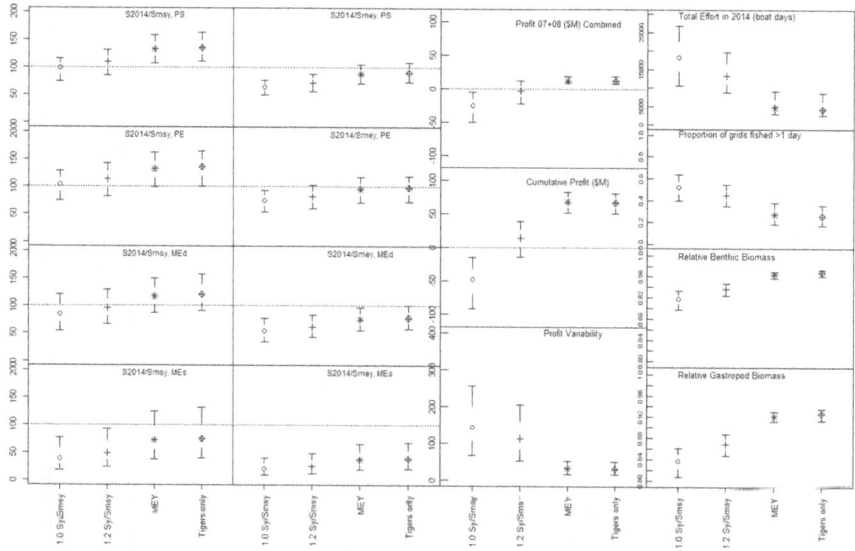

Figure 10.5 Biological, economic and ecosystem performance measures for a variety of management strategies. PS = Penaeus semisulcatus; PE = P. esculentus; MEd = Metapenaeus endeavouri; MEs = M. ensis. The symbols indicate distribution medians and the bars cover 95% of the simulation distributions. The performance statistics relate to spawning biomass relative to that at which MSY and MEY are achieved for four species (first two columns) and profit and its variability (third column). The rightmost column shows the total effort in 2014, the proportion of grids fished for more than one day in 2014, the total benthic biomass relative to unfished levels and the biomass of gastropods in 2014 relative to unfished levels.

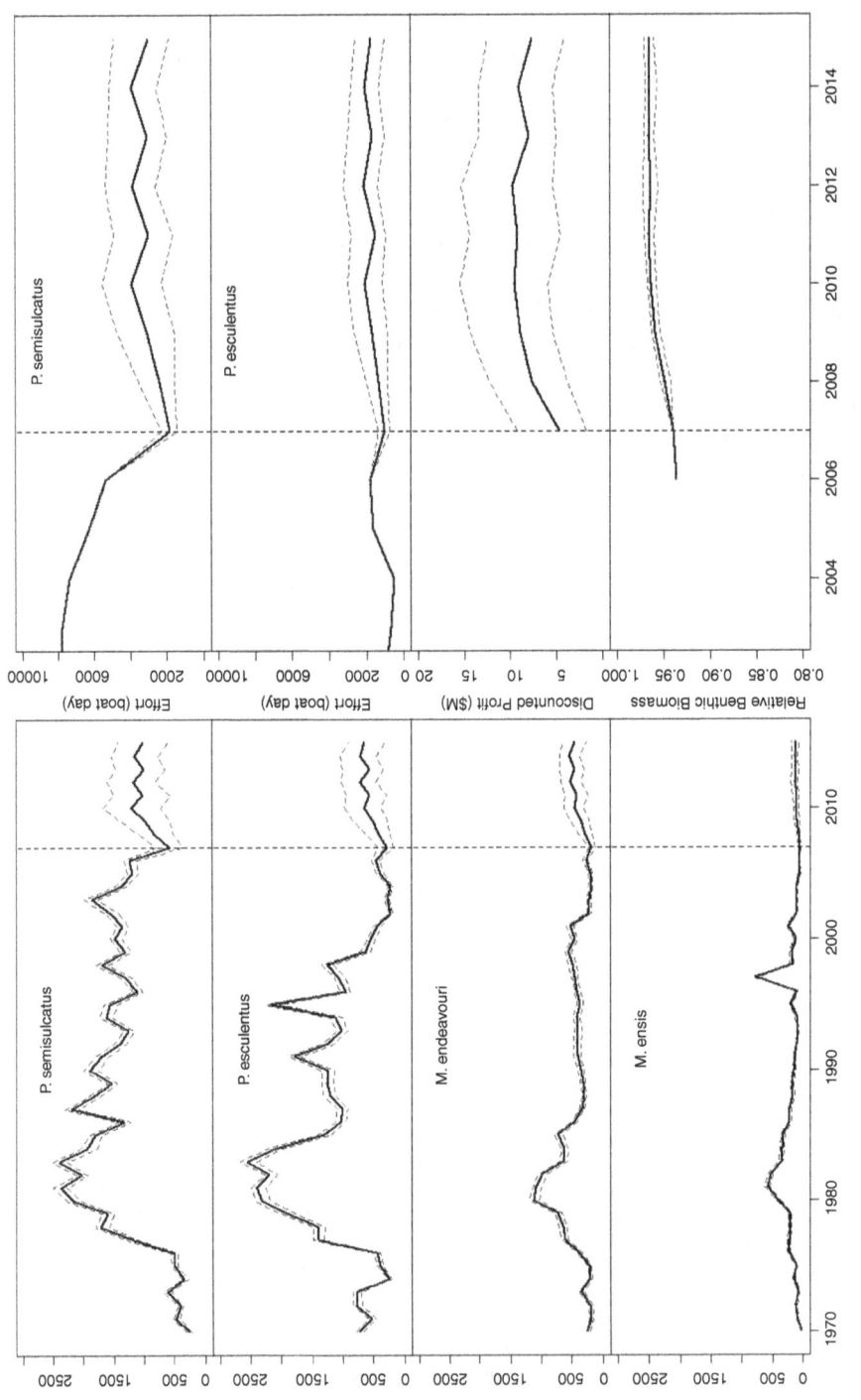

Figure 10.6 Time trajectories (medians and 90% intervals) of catch by species, effort by tiger prawn species, profit and benthic biomass for a single harvest strategy.

Source: Reproduced from Dichmont et al. (2008).

4   spatial autocorrelation in recruitment success
5   the presence of observation error associated with catches.

An uncertainty which was the focus for sensitivity evaluation was error in the assumed value for the catchability coefficient, as this directly determines fishing mortality (Equation T1.4). The factors found to have the greatest impact on the performance measures were:

1   how fishing efficiency has changed over time and whether or not the assessment is based on the correct trend in fishing efficiency
2   the catchability coefficient used to convert from fishing effort to fishing mortality
3   the difference between the intended fishing effort and the actual fishing effort expended (implementation error)
4   whether recruitment is spatially correlated among stocks.

The process of harvest strategy selection was undertaken using a co-management approach, as is typical of AFMA-managed fisheries. There was consequently no formal selection process in the way that this is undertaken in some regional fisheries management organizations, where candidate harvest strategies are developed and then tested by an MSE team. Rather, results from the MSE work were regularly and iteratively presented to the RAG and NORMAC, and both committees' members were able to influence aspects of the analyses: the RAG reviewed the technical aspects of the assessments and the MSE model, and NORMAC provided input into the harvest strategies that should be evaluated, including the weights assigned to potential management objectives. Specific combinations of OM parameterization and a harvest strategy were incrementally tested and modified with input from the RAG and the NORMAC until a harvest strategy was identified that achieved acceptable trade-offs, and conformed to the Australian government harvest policy. The final harvest strategy – the MEY target reference point bioeconomic model (HS2) that included both tiger and endeavour prawns in the profit function – became the implemented harvest strategy. Thus, the full bioeconomic model as the assessment method and the associated MEY harvest control rule (Equation 10.2) defined the official harvest strategy for the NPF and has since been used to provide management advice for the fishery (Dichmont et al., 2010).

Evaluation outcomes

The MSE was successful in defining the way management advice is given for the NPF. Some of the key conclusions from the MSE which have had management impacts are:

- Endeavour prawns should be included in calculations of MEY because they contribute to revenue, but as by-product species add little to cost.

Management advice for the fishery is now based on a harvest strategy that includes the two tiger prawn species and *M. endeavouri* in the bioeconomic model used to compute effort levels.

- A target reference point of S_{MEY} leads to higher profits and lesser impacts on other components of the ecosystem. However, inclusion of S_{MEY} in a harvest strategy increases demands for data, owing to the need for information on costs in addition to biological information (Dichmont et al., 2010). It additionally requires greater understanding of the economics of the fishery.
- Although there might have been an expectation that the fishery is having a substantial impact on the broader ecosystem, the results of the MSE suggest that the impacts are actually fairly minor (Figures 10.5 and 10.6). This conclusion was confirmed by Dichmont et al. (2013) using an MSE that included ecosystem impacts.

The MSE conducted by Dichmont et al. (2008) was extended by Dichmont et al. (2013) to evaluate spatial closure options in the NPF in terms of impacts on the fishery, the benthos and at-risk species. This highlights that MSEs which explore the broader ecosystem implications of harvest strategies for target fisheries need not be based on complex and hard-to-parameterize end-to-end models such as *Atlantis* (Fulton et al., 2011). Rather, MSEs based on intermediate complexity models (e.g. MICE models; Plagányi et al. (2014)) can be appropriate as further discussed in Plagányi (Chapter 15, this volume).

The current approach for calculating recommended levels of effort is based on a bioeconomic model which selects time series of effort to maximize the net present value of the flow of profits over time. However, unlike the harvest strategies described here, this approach involves basing the population dynamics for the two tiger prawn species on a size-structured population dynamics model (Punt et al., 2010) and those for *M. endeavouri* on the spatially structured Bayesian biomass dynamics model of Zhou et al. (2009) and Punt et al. (2011). The size structure for the tiger prawns is important as prices vary substantially by size, and hence when and where effort is applied can affect the level of fishery profitability.

The use of the bioeconomic model has increased transparency as to how decisions are made and what assumptions underlie them. Industry have become involved in providing economic data directly for use in the most recent models as well as validating key assumptions used in the analyses, further facilitating adoption of the model outcomes. The NPF is unusual in that it is mostly self-managed, with AFMA providing essentially the role of an auditor. This form of co-management has been the end result of a long history of strong relationships between industry, AFMA and scientists (Dichmont et al., 2007). Without this strong relationship, the MSE approach of iterative testing and implementation with real-time economic data would be much harder. Advice based on the harvest strategy which sets effort by tiger prawn species with the aim of maximizing net present value for the tiger prawn fishery has been provided to the NORMAC by the RAG since 2008 (based on the delay-difference model in 2008 and 2010 and on the approach of Punt et al. [2011] from 2010). The results have always been accepted as the basis for management, which is a credit to the engagement process. Deng et al. (2014) have refined the harvest strategy for the tiger prawn fishery based on its retrospective performance in terms of predicting how it operated. Thus, for this fishery, harvest strategies are adaptively developed using both MSE and other diagnostic approaches.

Conclusion

MSE has been used in this fishery to revise and refine existing approaches. In fact, a key outcome for the MSE for the NPF is that current management approaches are likely to perform adequately to satisfy objectives related to conservation of the target resource and the impacted ecosystem to a large extent. The MSE testing over a broad range of objectives was a key component of the evidence for the environmental credentials of the fishery when it successfully undertook independent eco-certification by the Marine Stewardship Council (MSC) from 2012. Prawn trawling has a poor reputation, and evidence that it can have low benthic impacts and still be profitable was important during the MSC assessment process. It is also important to note that the MSE results were unusual in that the best scenario was win–win in terms of sustainability, benthic impacts and profitability. This is because it reduced overall effort which resulted in reduced impact throughout but at the same time increased expected profits.

Note

1 Simulation tests have been applied to white banana prawns (e.g. Hutton et al., 2009, Buckworth et al., 2013).

References

Australian Department of Agriculture Fisheries and Forestry. 2007. *Commonwealth Fisheries Harvest Strategy: Policy and Guidelines*, DAFF, Canberra, ACT.

Bishop, J., Venables, W.N., Dichmont, C.M. & Sterling, D.J. 2008. Standardizing catch rates: Is logbook information by itself enough? *ICES Journal of Marine Science*, 65, 255–266.

Buckworth, R.C., Ellis, N., Zhou, S., Pascoe, S., Deng, R.A., Hill, F.G. & O'Brien, M. 2013. Comparison of TAC and current management for the White Banana Prawn fishery of the Northern Prawn Fishery. Final Report for Project RR2012/0812 to the Australian Fisheries Management Authority, June 2013.

Deng, R., Dichmont, C., Milton, D., Haywood, M.,Vance, D., Hall, N. & D. Die. 2005. Can vessel monitoring system data also be used to study trawling intensity and population depletion? The example of Australia's northern prawn fishery. *Canadian Journal of Aquatic and Fishery Sciences,* 62, 611–622.

Deng, R.A., Punt, A.E., Dichmont, C.M., Buckworth, R.C. & Burridge, C.Y. 2014. Improving catch-prediction setting for tiger prawns in the Northern Prawn Fishery. *ICES Journal of Marine Science.* doi:10.1093/icesjms/fsu033

Dichmont, C.M., Deng, A., Punt, A.E., Venables, W. & Haddon, M. 2006a. Management strategies for short-lived species: The case of Australia's Northern Prawn Fishery: 1. Accounting for multiple species, spatial structure and implementation uncertainty when evaluating risk. *Fisheries Research,* 82, 204–220.

Dichmont C.M., Deng, A., Punt, A.E.,Venables, W. & Haddon, M. 2006b. Management strategies for short lived species: The case of Australia's Northern Prawn Fishery: 2. Choosing appropriate management strategies using input controls. *Fisheries Research,* 82, 221–234.

Dichmont, C.M., Deng, A., Punt, A.E., Venables, W. & Haddon, M. 2006c. Management strategies for short lived species: The case of Australia's Northern Prawn Fishery: 3. Factors affecting management and estimation performance. *Fisheries Research,* 82, 235–245.

Dichmont, C.M., Deng, A., Punt, A.E., Ellis, N., Venables, W.N., Kompas, T., Ye, Y., Zhou, S. & J. Bishop. 2008. Beyond biological performance measures in Management Strategy Evaluation: Bringing in economics and the effects of trawling on the benthos. *Fisheries Research,* 94, 238–250.

Dichmont, C.M., Die, D., Punt, A.E.,Venables, W., Bishop, J., Deng, A. & Dell, Q. 2001. Risk Analysis and Sustainability Indicators for the Prawn Stocks in the Northern Prawn Fishery. Report of FRDC Project No 98/109. CSIRO Marine Research, Cleveland.

Dichmont, C.M., Ellis, N., Bustamante, R.H., Deng, R., Tickell, S., Pascual, R., Lozano-Montes, H. & Griffiths, S. 2013. Evaluating marine spatial closures with conflicting fisheries and conservation objectives. *Journal of Applied Ecology,* 50, 1060–1070.

Dichmont, C.M., Jarrett, A., Hill, F. & Brown, M. 2010 (updated in 2014). Northern Prawn Fishery harvest strategy under input controls. AFMA Report No R2006/828b, Available at: www.afma.gov.au/managing-our-fisheries/harvest-strategies/harvest-strategy-for-the-northern-prawn-fishery-under-input-controls/ (last accessed 11 Jan 2016).

Dichmont, C.M., Loneragan, N., Brewer, D.S. & Poiner, I. 2007. Industry, managers, and scientists as partners: Challenges, opportunities and successes in Australia's Northern Prawn Fishery. In: McClanahan, T.R. & Castilla, J.C. (eds.) *Fisheries Management: Progress Towards Sustainability.* Blackwell Publishing, Oxford, pp. 207–230.

Dichmont, C.M., Pascoe, S., Kompas, T. & Punt, A.E. 2010. On implementing maximum economic yield in commercial fisheries. *PNAS,* 107, 16–21.

Dichmont, C.M., Punt, A.E., Deng, A., Dell, Q. & Venables, W. 2003. Application of a weekly delay-difference model to commercial catch and effort data for tiger prawns in Australia's Northern Prawn Fishery. *Fisheries Research,* 65, 335–350.

Ellis, N. & Pantus, F. 2001. Management Strategy Modelling: Tools to evaluate trawl management strategies with respect to impacts on benthic biota within the Great Barrier Reef Marine Park area. CSIRO Marine Research, Cleveland, Australia.

Fulton, E.A., Link, J.S., Kaplan, I.C., Savina-Rolland, M., Johnson, P., Ainsworth, C., Horne, P., Gorton, R., Gamble, R.J., Smith, A.D.M. & Smith, D.C. 2011. Lessons in modelling and management of marine ecosystems: The Atlantis experience. *Fish and Fisheries,* 12, 171–188.

Galeano, D., Langenkamp, D., Shafron, W. & Levantis, C. 2004. *Australian Fisheries Survey Report 2003*. Australian Bureau of Agricultural and Resource Economics, Canberra.

Geromont, H.F., De Oliveira, Johnston, S.J. & Cunningham, C.L. 1999. Development and application of management procedures for fisheries in southern Africa. *ICES Journal of Marine Science*, 56, 952–966.

Hutton, T., Dichmont, C.M. & Pascoe, S. 2009. Banana Prawns and TAC. *In:* Kompas, T. & Grafton, R.Q. (eds.) *A Cost-Benefit Analysis of Alternative Management Options for the Australian Northern Prawn Fishery*. Unpublished Report, Sustainable Environment Group, 100.

Kompas, T., Dichmont, C.M., Punt, A.E., Deng, A., Che, T.N., Bishop, J., Gooday, P., Ye, Y. & Zhou, S. 2010. Maximizing profits and conserving stocks in the Australian Northern Prawn Fishery. *Australian Journal of Agricultural and Resource Economics*, 54, 281–299.

Plagányi, É.E., Punt, A.E., Hillary, R., Morello, E.B., Thébaud, O., Hutton, T., Pillans, R.D., Thorson, J.T., Fulton, E.A., Smith, A.D.M., Smith, F., Bayliss, P., Haywood, M., Lyne, V. & Rothlisberg, P.C. 2014. Models of intermediate complexity for ecosystem assessment to support tactical management decisions in fisheries and conservation. *Fish and Fisheries*, 15, 1–22.

Punt, A.E., Deng, R.A., Dichmont, C.M., Kompas, T., Venables, W.N., Zhou, S., Pascoe, S., Hutton, T., Kenyon, R., van der Velde, T. & Kienzle, M. 2010. Integrating size-structured assessment and bio-economic management advice in Australia's Northern Prawn Fishery *ICES Journal of Marine Science*, 67, 1785–1801.

Punt, A.E., Deng, R., Pascoe, S., Dichmont, C.M., Zhou, S., Plagányi, É. E, Hutton, T., Venables, W.N., Kenyon, R. & van der Velde, T. 2011. Calculating optimal effort and catch trajectories for multiple species modeled using a mix of size-structured, delay-difference and biomass dynamics models. *Fisheries Research*, 101, 201–211.

Punt, A.E. & Donovan, G. 2007. Developing management procedures that are robust to uncertainty: Lessons from the International Whaling Commission. *ICES Journal of Marine Science*, 64, 603–612.

Skirtun, M., Stephan, M. & Mazur, K. 2014. Australian fisheries economic indicators report 2013: Financial and economic performance of the Northern Prawn Fishery. ABARES, Canberra.

Smith, A.D.M., Sainsbury, K.J. & Stevens, R.A. 1999. Implementing effective fisheries-management systems – management strategy evaluation and the Australian partnership approach. *ICES Journal of Marine Science*, 56, 967–979.

Somers, I.F. 1990. Manipulation of fishing effort in Australia's penaeid prawn fisheries. *Australian Journal of Marine and Freshwater Research*, 41, 1–12.

Somers, I. & Wang, Y. 1997. A simulation model for evaluating seasonal closures in Australia's multispecies Northern Prawn Fishery. *North American Journal of Fisheries Management*, 17, 114–130.

Venables, W. & Dichmont, C.M. 2004. A generalized linear model for catch allocation: An example from Australia's Northern Prawn Fishery. *Fisheries Research*, 70, 405–422.

Venables, W.N., Ellis, N., Punt, A.E., Dichmont, C.M. & Deng, R.A. 2009. A simulation strategy for fleet dynamics in Australia's northern prawn fishery: Effort allocation at two scales. *ICES Journal of Marine Science*, 66, 631–645.

Venables, W.N., Kenyon, R.A., Bishop, J.F.B., Dichmont, C.M., Deng, R.A., Burridge, C., Taylor, B.R., Donovan, A.G., Thomas, S.E. & Cheers, S.J. 2006. Species Distribution and Catch Allocation: Data and methods for the NPF, 2002–2004. Report to AFMA No. R01/1149, CSIRO Publishers, Canberra.

Wang, Y-G. & Die, D. 1996. Stock-recruitment relationships of the tiger prawns (*Penaeus esculentus* and *Penaeus Semisulcatus*) in the Australian Northern Prawn Fishery. *Marine and Freshwater Research,* 47, 87–95.

Ye, Y. 2000. Is recruitment related to spawning stock in penaeid shrimp fisheries? *ICES Journal of Marine Science,* 57, 1103–1109.

Zhou, S., Punt, A.E., Deng, R., Dichmont, C.M., Ye, Y. & Bishop, J. 2009. Modified hierarchical Bayesian biomass dynamics models for assessment of short-lived invertebrates: a comparison for tropical tiger prawns. *Marine and Freshwater Research,* 60, 1298–1308.

Chapter 11

Incorporating technological interactions in a joint management procedure for South African sardine and anchovy

Carryn L. de Moor and Douglas S. Butterworth

Introduction

The South African pelagic fishery developed in response to a demand for canned products during World War II, initially targeting sardine (*Sardinops sagax*) and horse mackerel (*Trachurus trachurus capensis*) (De Oliveria 2002). The annual sardine catch peaked in the early 1960s around 400,000 t, but dropped sharply thereafter, likely due to a combination of prolonged high catches coinciding with poor recruitment, and remained low until the mid-1990s (Figure 11.1). With the fall in the sardine landings, the fishery began targeting anchovy (*Engraulis encrasicolus*) with the introduction of smaller-mesh nets between 1963 and 1965 (De Oliveira 2002), and anchovy dominated the pelagic fishery landings from the late 1960s to the mid-1990s, with annual catches peaking at 600,000 t in the late 1980s (Figure 11.1).

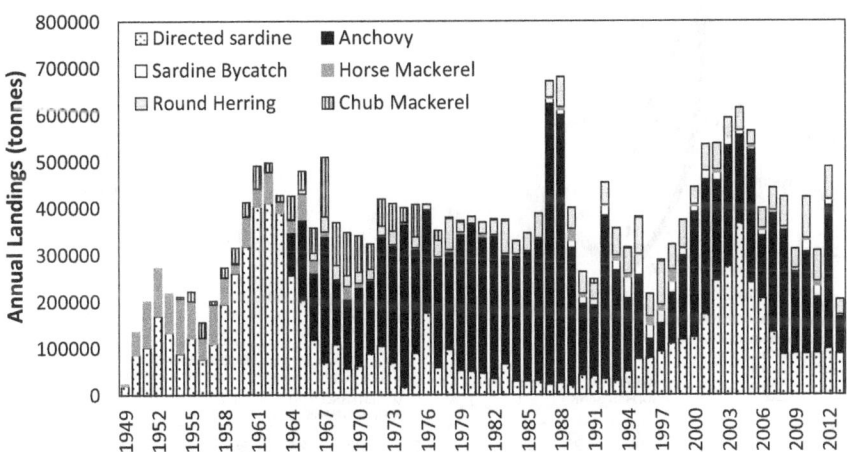

Figure 11.1 The annual landings of the main species taken by the South African pelagic fishery (1949–2013), kindly provided by the Department of Agriculture, Forestry and Fisheries (DAFF).

Today sardine and anchovy continue to form the backbone for the South African pelagic fishery, the country's second most valuable fishery in monetary terms after the demersal fishery for hake. This fishery also takes smaller quantities of other small pelagic species, now dominated by round herring (*Etrumeus whiteheadi*).

Adult sardine and anchovy spawn mainly on the Agulhas Bank in spring and summer (Figure 11.2). Eggs and larvae are transported by strong ocean currents from these southern spawning grounds to food-rich nursery areas along the west coast which result from strong upwelling in the southern Benguela current region. The recruits (0-year-olds) start to appear in the northern areas along the west coast from March/April each year, and then migrate southwards over the remainder of autumn and winter. As these fish mature into adults, they migrate further eastward onto the Agulhas Bank and generally move further offshore where they will spawn in subsequent years.

Sardine and anchovy are relatively short-lived, reaching maturity around age 1 for anchovy and age 2 for sardine. As is the norm for short-lived small pelagic species, their recruitment is highly variable. The number of anchovy recruits, for example, can fluctuate by a factor of up to five or more from one year to the next. This leads to large variations in the total abundance of these fish over relatively short time periods.

The directed sardine fishery targets primarily adult fish because these are large enough to be canned; this provides much greater revenue per ton than the alternative of fishmeal production. In contrast, the anchovy fishery (for

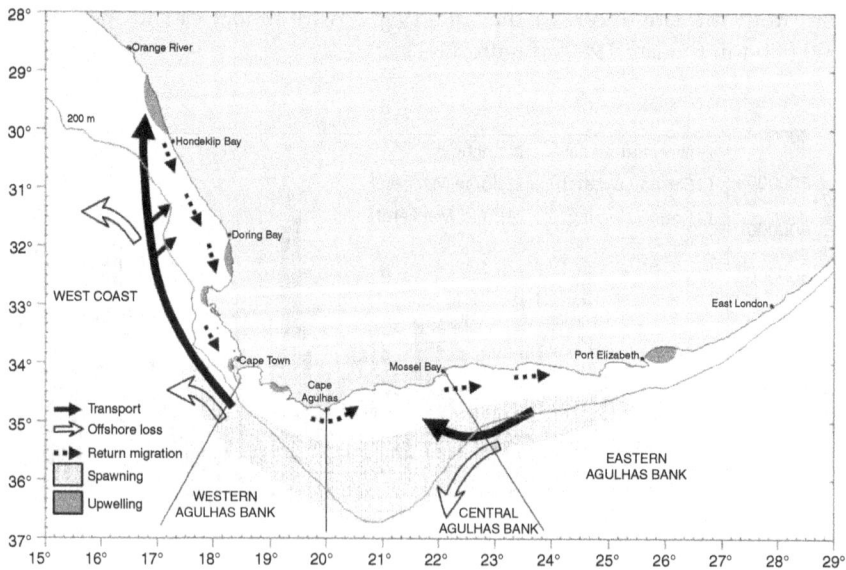

Figure 11.2 A map of the coastline of South Africa depicting the spawning areas of sardine and anchovy, the northwards transportation of eggs and larvae and the return southwards migration of the recruits.

which the catches are almost entirely converted to fishmeal) targets juveniles during their southward return migration along the west coast. This is
the only time during which anchovy are readily available for capture as
they tend to move offshore on the Agulhas Bank and disperse (rendering
them less readily catchable) as they mature. As juvenile sardine and anchovy
shoal together, catches targeting anchovy cannot avoid some juvenile sardine
bycatch. The removal of these juvenile sardines as bycatch has a negative con-
sequential effect for potential catches of adult sardine in future years. In short,
the greater the juvenile sardine bycatch this year, the fewer adult sardines
available to the directed fishery in years to come. Thus a trade-off decision
is needed: directed sardine and anchovy catches cannot be optimised simultaneously. For this reason the sardine and anchovy catch limits have been
calculated jointly since 1994.

Current management and objectives

The management procedure (MP) approach (also known as management strategy evaluation; see Rademeyer et al. 2007) has been used to develop methods for setting appropriate catch limits. An MP implements an agreed set of
rules expressed through mathematical formulae to provide management recommendations (e.g. a catch limit for a species). These rules, known as harvest
control rules (HCRs), use a prespecified set of inputs such as information on
the abundance of the resource obtained, for example either directly from surveys or through recent quantitative assessments. They are designed and tested
using mathematical and statistical models of the underlying resource(s) and
fishery to ensure that any future catches will meet pre-agreed objectives. The
MP approach thus requires the basic management objectives to be defined at
the start of any MP development process (De Oliveira et al. 2008, Punt et al. in
press), with perhaps some refinements as testing proceeds and trade-offs become
evident. The primary objective agreed for South African sardine and anchovy
has been to maximise average directed catch in the medium term, subject to
ensuring an acceptably small risk of reducing these resources to an undesirably low level (Butterworth 2008, de Moor et al. 2011). A secondary objective
has been keeping interannual TAC changes small in the interests of stability of
catches for the industry, though clearly the highly variable recruitment of these
resources places limits on the extent of stability which can be achieved without
"wasting" the resource in periods of higher abundance.

The simulation testing of the rules is the crucial aspect which sets the MP
approach apart from other processes which use pre-agreed rules: the rules
adopted for an MP to be used in practice have already been indicated through
such testing to be able to meet pre-agreed objectives (Butterworth 2007, De
Oliveira et al. 2008). In South Africa, operational management procedures
(OMPs), have been used to recommend catches for sardine and anchovy for
more than 20 years, making this one of the earliest examples of the implementation of an MP in a commercial fishery (Punt et al. in press).

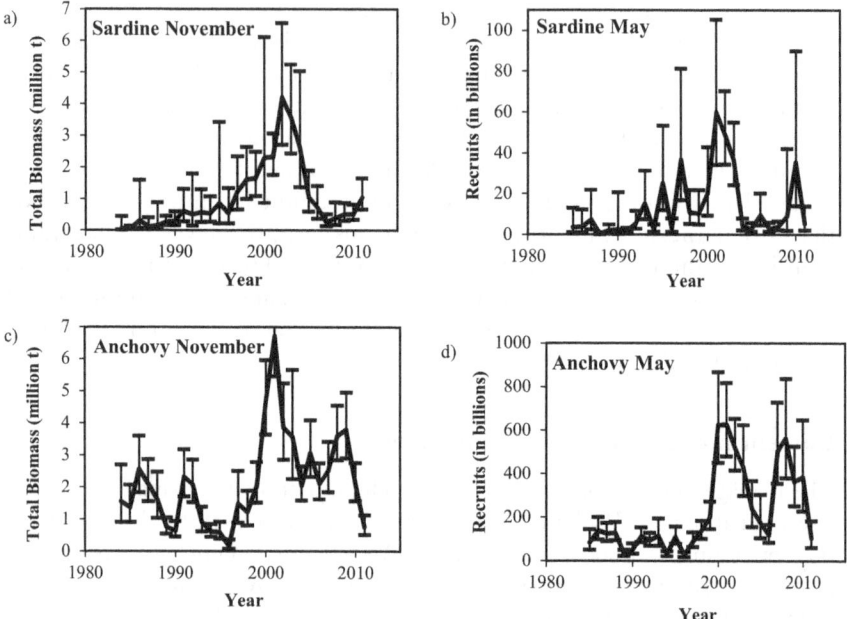

Figure 11.3 The hydroacoustic survey estimated (a) sardine total biomass, (b) sardine recruitment, (c) anchovy total biomass and (d) anchovy recruitment from the biannual surveys over 28 years, kindly provided by DAFF. The bars indicate 95% confidence intervals for the surveyed abundances.

A suite of HCRs is used to set the total allowable catches (TACs) for the directed fisheries on anchovy and adult sardine, as well as a total allowable bycatch (TAB) for sardine caught as juveniles by the anchovy fishery and as adults by the fishery for round herring. The HCRs are empirical (e.g. Butterworth 2008, De Oliveira et al. 2008), depending purely on data collected, primarily by hydroacoustic surveys. The highly variable nature of sardine and anchovy necessitates their frequent monitoring, and in South Africa hydroacoustic surveys of these two species have been conducted twice a year since 1984 (Figure 11.3). The November survey provides an estimate of the total abundance of fish and is used to calculate initial TAC/TAB values for the following year (Figure 11.4). The second survey is carried out in May/June each year to estimate the number of recruits, and this information is used to revise the anchovy TAC and sardine TAB. Since recruits commence their migration down the west coast only around March/April (which continues for some months thereafter), May is the earliest chance to reliably estimate the recruitment for the year, providing some feedback to the HCR to recommend the year's anchovy catch limit (see later).

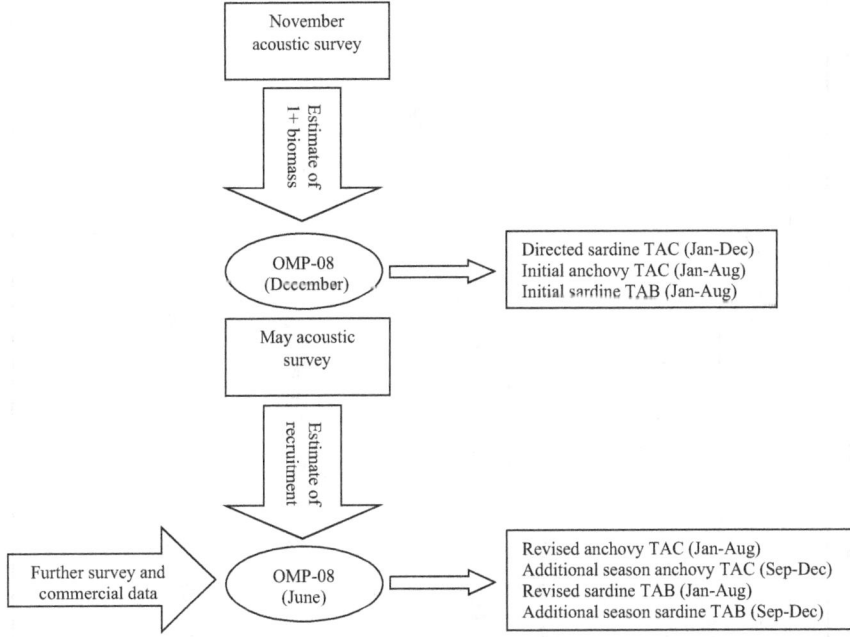

Figure 11.4 A schematic diagram indicating the annual process of applying the OMP to set sardine and anchovy TAC/Bs.

The harvest control rules

The suite of HCRs described here emerged from a process of fine-tuning the rules from the OMP used to manage sardine and anchovy for the preceding 4 years (OMP-04; de Moor et al. 2011). This was to better account for an updated understanding of the resources and their dynamics (i.e. refining the operating models [OMs], see later). In addition, precautionary changes to how TACs were recommended during periods of low resource abundance were incorporated, as well as changes to allow the fishing industry to take advantage of years with good anchovy recruitment.

The annual harvest of sardine and anchovy comprises catch and bycatch specific elements, each with an associated control rule (Tables 11.1a and 11.1b). Once the results of the November survey are available, the directed sardine TAC for the following calendar year, an initial anchovy TAC and an initial sardine TAB are calculated (Figure 11.4). The highly variable nature of anchovy recruitment means that it is not possible at this point in the year to predict accurately how many juveniles will recruit to the anchovy population during the coming year. This is problematic because the anchovy fishery is primarily

Table 11.1a The OMP-08 Harvest Control Rules for South African sardine and anchovy. In the interests of brevity, additional smoothing of the TAC for survey observations near to, but above, the Exceptional Circumstances threshold are omitted here, but can be found in de Moor and Butterworth (2008). Parameter definitions are given in Tables 11.2 and 11.3. All numbers are in billions and masses are in thousands of tonnes.

December (y – 1)

TAC/B	Basic Rules	Constraints	Equation
Directed sardine TAC	$TAC_y^S = \beta B_{y-1,N}^{obs,S}$	$\max\left\{(1-c_{mxdn}^S)TAC_{y-1}^S; c_{mntac}^S\right\} \leq TAC_y^S \leq c_{mxtac}^S$ if $TAC_{y-1}^S \leq c_{tier}^S$ $(1-c_{mxdn}^S)c_{tier}^S \leq TAC_y^S \leq c_{mxtac}^S$ if $TAC_{y-1}^S > c_{tier}^S$	11.1a.1
Initial anchovy TAC	$TAC_y^{1,A} = \alpha_{ns}\,\delta q\left(p + (1-p)\dfrac{B_{y-1,N}^{obs,A}}{B_N^{obs,A}}\right)$	$\max\left\{(1-c_{mxdn}^A)TAC_{y-1}^{2,A}; c_{mntac}^A\right\} \leq TAC_y^{1,A} \leq c_{mxtac}^A$ if $TAC_{y-1}^{2,A} \leq c_{tier}^A$ $(1-c_{mxdn}^A)c_{tier}^A \leq TAC_y^{1,A} \leq c_{mxtac}^A$ if $TAC_{y-1}^{2,A} > c_{tier}^A$	11.1a.2
Initial sardine TAB	$TAB_y^{1,S} = \gamma_y\,TAC_y^{1,A} + TAB_{rh}^S$, where $\gamma_y = 0.1 + \dfrac{0.1}{1+\exp\left(-0.0025\left(B_{y-1,N}^{obs,S} - 2000\right)\right)}$		11.1a.3

June y

TAC/B	Basic Rules	Constraints	Equation
Revised anchovy TAC	$TAC_y^{2,A} = \alpha_{ns}\, q\left(p\,\dfrac{N_{y-1,rec0}^A}{N_{rec0}^A} + (1-p)\dfrac{B_{y-1,N}^{obs,A}}{B_N^{obs,A}}\right)$	$\max\left\{TAC_y^{1,A}; (1-c_{mxdn}^A)TAC_{y-1}^{2,A}; c_{mntac}^A\right\} \leq TAC_y^{2,A} \leq \min\left\{c_{mxtac}^A; TAC_y^{1,A} + c_{mxinc}^{ns,A}\right\}$ if $TAC_{y-1}^{2,A} \leq c_{tier}^A$ $\max\left\{TAC_y^{1,A}; (1-c_{mxdn}^A)c_{tier}^A\right\} \leq TAC_y^{2,A} \leq \min\left\{c_{mxtac}^A; TAC_y^{1,A} + c_{mxinc}^{ns,A}\right\}$ if $TAC_{y-1}^{2,A} > c_{tier}^A$	11.1a.4
Revised sardine TAB	$TAB_y^{2,S} = \lambda TAC_y^{1,A} + r_y(TAC_y^{2,A} - TAC_y^{1,A}) + TAB_{rh}^S$, where $\lambda = \max\{\gamma_y, r_y\}$		11.1a.5
Final anchovy TAC[1]	$TAC_y^{3,A} = \alpha_{ads}\, q\left(p\,\dfrac{N_{y-1,rec0}^A}{N_{rec0}^A} + (1-p)\dfrac{B_{y-1,N}^{obs,A}}{B_N^{obs,A}}\right)$	$\max\{TAC_y^{2,A}; c_{mntac}^A\} \leq TAC_y^{3,A} \leq \min\{c_{mxtac}^A; TAC_y^{2,A} + c_{mxinc}^{ads,A}\}$ Additional season booster: $TAC_y^{3,A} = \min\left\{c_{mxtac}^A;\begin{cases}\dfrac{c_{mxinc}^{ads,A} - (TAC_y^{3,A} - TAC_y^{2,A})}{B_2 - B_1}(B_{y,proj}^A - B_1) + TAC_y^{3,A} & \text{if } B_1 \leq B_{y,proj}^A < B_2 \\[2ex] TAC_y^{2,A} + c_{mxinc}^{ads,A} & \text{if } B_{y,proj}^A \geq B_2\end{cases}\right\}$	11.1a.6
Final sardine TAB[2]	$TAB_y^{3,S} = TAB_y^{2,S} + \min\left\{TAB_{ads}^S; \gamma_y(TAC_y^{3,A} - TAC_y^{2,A})\right\}$		11.1a.7

[1] The anchovy TAC applied during the additional subseason is $TAC_y^{3,A} - TAC_y^{2,A}$.

[2] The sardine TAB applied during the additional subseason is $TAB_y^{3,S} - TAB_y^{2,S}$.

Table 11.1b The OMP-08 harvest control rules governing the Exceptional Circumstances provisions. Parameter definitions are given in Tables 11.2 and 11.3, and formulae for TAC_y^{S*}, TAC_y^{1A*}, TAC_y^{2A*}, and $TAC_y^{3,A*}$ are given in Table 11.1a. All numbers are in billions and masses are in thousands of tonnes.

	TAC	Exceptional Circumstances apply	Rule	Equation
December (y – 1)	Directed sardine TAC	if $B_{y-1,N}^{obs,S} < B_{ec}^S$	$TAC_{y,init}^S = 0.5\times\begin{cases}0 & \text{if } \dfrac{B_{y-1,N}^{obs,S}}{B_{ec}^S}\leq x^S\\[2mm] TAC_y^{S*}\left(\left(\dfrac{B_{y-1,N}^{obs,S}}{B_{ec}^S}-x^S\right)\Big/(1-x^S)\right)^2 & \text{if } x^S<\dfrac{B_{y-1,N}^{obs,S}}{B_{ec}^S}<1\end{cases}$	11.1b.1
	Initial anchovy TAC	if $B_{y-1,N}^{obs,A} < B_{ec}^S$	$TAC_y^{1A}=\begin{cases}0 & \text{if } \dfrac{B_{y-1,N}^{obs,A}}{B_{ec}^A}\leq x^A\\[2mm] TAC_y^{1A*}\left(\left(\dfrac{B_{y-1,N}^{obs,A}}{B_{ec}^A}-x^A\right)\Big/(1-x^A)\right)^2 & \text{if } x^A<\dfrac{B_{y-1,N}^{obs,A}}{B_{ec}^A}<1\end{cases}$	11.1b.2
June y	Directed sardine TAC	If $B_{y-1,N}^{obs,S} < B_{ec}^S$	$TAC_y^S=\begin{cases}TAC_{y,init}^S+1.2\times\dfrac{N_{y,r}^{obs,S}}{R_{crit}}\times TAC_{y,init}^S & \text{if } N_{y,r}^{obs,S}\leq R_{crit}\\[2mm] TAC_{y,init}^S+1.2\times TAC_{y,init}^S & \text{if } N_{y,r}^{obs,S}>R_{crit}\end{cases}$	11.1b.3
		if $B_{y,proj}^A < B_{ec}^S$	$TAC_y^{2,A}=\max\left\{TAC_y^{1,A};\; TAC_y^{2,A*}\left(\left(\dfrac{H(N_{y-1,rec0}^A)}{H(N_{y-1,rec0}^{A*})}-x^A\right)\Big/(1-x^A)\right)^2\right\}$ with $\begin{cases}0 & \text{if } \dfrac{H(N_{y-1,rec0}^A)}{H(N_{y-1,rec0}^{A*})}\leq x^A\\[2mm] & \text{if } x^A<\dfrac{H(N_{y-1,rec0}^A)}{H(N_{y-1,rec0}^{A*})}<1\end{cases}$ $\quad H(N)=p\dfrac{N}{\overline{N}_{rec0}^A}+(1-p)\dfrac{B_{y-1,N}^{obs,A}}{B_{Nov}^{obs,A}}$	11.1b.4
	Revised anchovy TAC		$B_{y,proj}^A=\left(\dfrac{B_{y-1,N}^{obs,A}}{\overline{w}_1^A}e^{-5*0.9/12}-C_{y,1}^A\right)e^{-7\times0.9/12}\overline{w}_2^A+B_{y,proj0}^A$ $N_{y-1,rec0}^{A*}=(\theta e^{0.5(6+r_y^A)0.9/12}+C_{y,0bs}^A)e^{[0.5(6+r_y^A)]0.9/12}$ $\theta=\dfrac{[B_{ec}^A-(B_{y,proj}^A-B_{y,proj0}^A)]}{\overline{w}_1^A}e^{5.5*0.9/12}+\dfrac{TAC_y^{2,A*}}{\overline{w}_{0c}^A}-C_{y,-}^A-C_{y,0bs}^A$ $B_{y,proj0}^A=\max\left\{0;\left(N_{y,r}^{obs,A}-\left[\dfrac{TAC_y^{2,A*}}{\overline{w}_{0c}^A}-C_{y,1}^A-C_{y,0bs}^A\right]\right)e^{-5.5*0.9/12}\overline{w}_1^A\right\}$	
	Final anchovy TAC		As for the revised anchovy TAC, except replacing TAC_y^{1A} with $TAC_y^{2,A}$ and $TAC_y^{2,A*}$ with $TAC_y^{3,A*}$	11.1b.5

Table 11.2 Parameters and constraints of OMP-08, compared to OMP-04 (de Moor et al. 2011). All numbers are in billions and masses are in thousands of tonnes.

	Control Parameter	OMP-04	OMP-08
β	directed sardine control parameter	0.14387	0.097
α_{ns}	directed anchovy control parameter for normal season	0.72858	0.78
α_{ads}	directed anchovy control parameter for additional season	1.45716	1.17

	Constraints	OMP-04	OMP-08
TAB_{rh}^{S}	fixed annual adult sardine bycatch	10	3.5
c_{mxdn}^{S}	maximum proportion by which directed sardine TAC can be annually reduced	0.15	0.20
c_{mxdn}^{A}	maximum proportion by which normal season anchovy TAC can be annually reduced	0.25	0.25
c_{mntac}^{S}	minimum directed sardine TAC	90	90
c_{mntac}^{A}	minimum directed anchovy TAC	150	120
c_{mxtac}^{S}	maximum directed sardine TAC	500	500
c_{mxtac}^{A}	maximum directed normal season anchovy TAC	600	600
c_{tier}^{S}	2-tier break for directed sardine TAC	240	255
c_{tier}^{A}	2-tier break for directed anchovy TAC	330	330
$c_{mxinc}^{ns,A}$	maximum increase in normal season anchovy TAC	200	150
$c_{mxinc}^{ads,A}$	maximum additional season anchovy TAC	150	120
TAB_{ads}^{S}	maximum sardine bycatch during the additional season	2	2
B_{ec}^{S}	threshold at which Exceptional Circumstances are invoked for sardine	250	300
B_{ec}^{A}	threshold at which Exceptional Circumstances are invoked for anchovy	400	400
B_{1}	threshold above which the anchovy additional season TAC can increase more rapidly	N/A	1,000
B_{2}	threshold above which the anchovy additional season TAC reaches a maximum	N/A	1,500
x^{s}	the proportion of the Exceptional Circumstances threshold below which sardine TAC is zero	0	0.25
x^{A}	the proportion of the Exceptional Circumstances threshold below which anchovy TAC is zero	0.25	0.25

Table 11.2 (Continued)

	Constraints	OMP-04	OMP-08
R_{crit}	sardine recruitment threshold required in order to achieve the maximum possible midyear increase in sardine TAC under Exceptional Circumstances	N/A	17.38

	Fixed Controls	OMP-04	OMP-08
δ	'scale-down' factor on initial anchovy TAC	0.85	0.85
p	weighting given to recruit survey in anchovy TAC	0.7	0.7
q	relates to average TAC under OMP99	300	300
γ_y	conservative initial estimate of juvenile sardine:anchovy ratio	0.1–0.2	0.1–0.2

based on these recruits. A TAC, once awarded, cannot subsequently be reduced (indeed it may already have been caught); thus care must be taken that the initial TAC is not too high in case poor recruitment occurs subsequently. For this reason, the anchovy TAC is initially calculated under the assumption that the forthcoming recruitment will be average; however only a portion of this TAC is recommended at the start of the year (the δ factor in Equation 11.1a.2 and Table 11.2). The May survey provides key information on the magnitude of the year's recruitment and a sound basis to adjust the anchovy TAC and associated sardine TAB midseason while the fish remain available for the industry to catch (Figure 11.4).

The HCRs for sardine and anchovy have developed into a complex set of equations as detailed management objectives proliferated over time, with each component of the rules developed to cater to some specific objective. This is most easily visualised for the HCR which determines the directed sardine TAC (Figure 11.5). The basics of how the HCR translates the survey data into a TAC can be divided into four main categories, depending on the resource abundance.

1   "Normal": At root, the TAC is calculated as a constant proportion, β, of the resource abundance estimated from the most recent November survey (this is what is known as a constant fishing mortality strategy, Equation 11.1a.1). Thus if the resource becomes more abundant the TAC will increase, and vice versa – the basic requirement for a feedback control approach. An upper limit applies to the TAC, c^S_{mxtac}, based on the industry's catch and processing capacity (Table 11.2). In order to provide some stability for the industry, the TAC may not decrease from one year to the next by more than $c^S_{mxdn} = 20\%$ for the directed sardine TAC (Table 11.2).

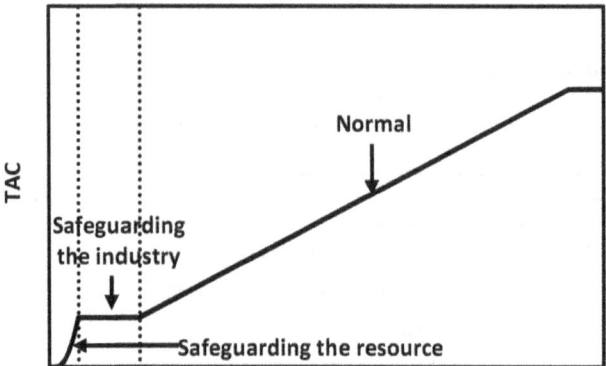

Survey Estimated Biomass

Figure 11.5 A plot showing three of the categories of the harvest control rule used to set the directed sardine TAC. In the "normal" range, the TAC is set at a proportion of the estimated resource abundance (Equation 11.1a.1). If the abundance decreases to a medium to low level, the TAC is kept constant at a minimum level ("Safeguarding the industry," c^s_{mntac} in Equation 11.1a.1). However, if the resource abundance drops further, the TAC is rapidly decreased ("Safeguarding the resource," Equation 11.1b.1). Constraints, such as the maximum interannual decrease above and below the "two-tier" threshold are not shown here. A similar figure originally appeared in de Moor and Butterworth (2009b).

2   "Boom": The constraint c^s_{mxdn} means that if the abundance increases very rapidly, the TAC could not increase as rapidly because of the need to allow for the possibility that the abundance may drop substantially in the next year or two. Were such a decrease in abundance to occur, given this constraint, the TAC could decrease only by a restricted amount, thereby increasing the risk of the resource being reduced below an acceptable level. A "two-tier" system is thus used to allow the industry to take advantage of such "booms" in this highly variable resource, without increasing risk. Once the TAC recommended increases above a "two-tier threshold," c^s_{tier} (Table 11.2), the constraint mentioned earlier on the maximum interannual decrease in the TAC is overridden, allowing the TAC to be reduced to a preset but still relatively high level the following year (Equation 11.1a.1).

3   "Safeguarding the industry": TAC decreases are also constrained by a minimum TAC, c^s_{mxtac} (Table 11.2). Thus even if the resource abundance continues to decrease below a certain level, the TAC will not decrease further

(except under (4)) (Equation 11.1a.1). This minimum is in the interests of preventing bankruptcies amongst rights holders during periods of medium to low resource abundance.

4   "Safeguarding the resource": This minimum TAC can, however, be maintained only up to a point. If the resource abundance drops below a still lower threshold, B_{ec}^S, the minimum TAC is overridden and TACs are decreased rapidly to prevent severe resource depletion. This is also known locally as "Exceptional Circumstances." As an additional precaution, only half of the OMP calculated TAC is awarded at the beginning of the season, with an increase in the TAC possible after the May survey, depending on the level of recruitment which this survey indicates (Equations 11.1b.1 and 11.1b.3) (i.e. similar to the approach for anchovy).

The HCR for anchovy can be categorised similarly, but with an important difference in the algorithm used to set the directed catch under "Normal" circumstances. Since information on anchovy recruitment is not available from the November survey, the HCR must be able to accommodate subsequent recruitment survey data in a consistent manner. This is achieved by setting the TAC using a weighted average of the relative November biomass and relative recruitment strength, with the initial TAC set using a simple assumption that relative recruitment will be equal to one (i.e. recruitment will be equal to the historic average; Equation 11.1a.2). The anchovy TAC is split into two seasons using the control parameters α_{ns} and α_{ads} (Table 11.2), with the later "additional" September to December season (Equation 11.1a.6) designed to enable the targeting of anchovy during a period when the associated sardine bycatch should be minimal. As the year progresses, recruiting sardine grow faster than anchovy so that the juveniles of the two species no longer shoal together to the same extent. This second season therefore mitigates the underutilisation of anchovy that would otherwise arise because of the need to limit sardine bycatch.

The sardine TAB has two parts. The first is a fixed amount, TAB_{rh}^S, which is primarily intended to provide an allowance for the unavoidable bycatch of adult sardine when round herring is targeted. The second is proportional to the anchovy TAC (Equations 11.1a.3 and 11.1a.5), and is intended to make allowance for mainly juvenile sardine bycatch in the anchovy directed fishery. An estimate of the ratio of juvenile sardine to anchovy occurring in catches in May each year (r_y, Table 11.3) is used when revising the sardine TAB at midyear. High juvenile sardine catches in the anchovy directed fishery will impact the targeted adult sardine fishery negatively in subsequent years, and this trade-off needs to be quantified to be able to set the sardine TAB as a function of the anchovy TAC. Because the system is complex and in a nonequilibrium state, the only viable method for doing this is as part of a simulation-based evaluation, which is described in detail later.

Table 11.3 Operating model, implementation model and harvest control rule notations, where $i = S, A$ denotes sardine or anchovy, respectively. All numbers are in billions and masses are in thousands of tonnes.

Parameter	Definition
$N^i_{y,a}$	Predicted numbers-at-age a of species i at the beginning of November in year y
M^i_j	Natural mortality rate (year^{-1}) of juvenile (age 0) fish of species i
M^i_{ad}	Natural mortality rate (year^{-1}) of adult (age 1+) fish of species i
$C^i_{y,a}$	Predicted catches-at-age a in year y of species i
$B^i_{y,N}$	Predicted November 1+ biomass of species i in year y
$SSB^i_{y,N}$	Predicted November spawner biomass of species i in year y
$\bar{w}^i_a$	Average historical November survey weights at age a of species i
a^i	Median maximum recruitment of species i
b^i	Spawner biomass below which median recruitment is impaired for species i
σ^i_r	Standard deviation of the recruitment residuals for species i
s^i_{cor}	Recruitment serial correlation for species i
$B^{obs,i}_{y,N}$	November survey estimate of 1+ biomass of species i in year y
k^i_N	Multiplicative bias between November survey estimated and model predicted 1+ biomass
$N^{obs,i}_{y,r}$	May survey estimate of recruitment of species i in year y
k^i_r	Multiplicative bias between May survey estimated and model predicted recruitment
$N^i_{y,r}$	Predicted recruits of species i at the time of the May survey in year y

Table 11.3 (Continued)

Parameter	Definition
$C^i_{y,Obs}$	Catches of 0-year-old fish of species i taken before the May survey in year y
k_{sur}	Slope of the linear regression of the historical ratios of juvenile sardine bycatch to anchovy catch and the ratio of juvenile sardine to anchovy in the May survey
σ^i_{sur}	November ($sur = Nov$) and May ($sur = rec$) survey sampling error standard deviation for species i
$\omega^i_y \sim N(0,1)$	Random error
$\eta^i_{y,sur} \sim N(0,1)$	Random error
$\eta^i_{y,m} \sim N(0,1)$, $m = jan: may, may, jun, aug$	Random error
S^s_a	Sardine fishing selectivity-at-age a
$\overline{w}^i_{ac}$	Average historical catch weights-at-age a of species i
$C^A_{y,m}$	Anchovy catch in month m of year y from landings that have targeted anchovy
$C^{S,byc}_{y,m}$	Sardine bycatch with targeted anchovy in month m of year y
$\overline{B}^{obs,A}_{Nov}$	Average historical November survey estimates of anchovy 1+ biomass between 1984 and 1999
$\overline{N}^{obs,A}_{rec0}$	Average historical May survey estimates of anchovy recruitment between 1985 and 1999, back-calculated to 1 November
$r_y = \frac{1}{2}\left(r_{y,sur} + r_{y,com}\right)$	Ratio of juvenile sardine to anchovy in the sea during May of year y
$N^A_{y-1,rec0} =$ $\left(N^{obs,A}_{y,r}e^{0.5(6+t^A_y)0.9/12} + C^A_{y,Obs}\right)e^{0.5(6+t^A_y)0.9/12}$	May survey estimate of anchovy recruitment in year y, back-calculated to 1 November of $y-1$

Evaluation

One of the advantages of the MP development process is that it draws together a wide range of stakeholders, from fish ecologists to representatives of industry and of environmental nongovernmental organisations (ENGOs). This allows for greater transparency for all stakeholders, so that it is not only the scientists involved in the OMP development who have a broad understanding of the process (Butterworth 2008). In South Africa, this has been primarily facilitated through discussions amongst members and observers in the Small Pelagics Scientific Working Group of the Department of Agriculture, Forestry and Fisheries (DAFF).

Table 11.4 Key performance statistics: the probability that adult sardine biomass falls below the average adult sardine biomass over November 1991 to November 1994 (the "risk threshold," $Risk^S$) at least once during the projection period of 20 years, $risk_S$; the probability that adult anchovy biomass falls below 10% of the average adult anchovy biomass between November 1984 and November 1999 at least once during the projection period of 20 years, $risk_A$; the average directed catch over 20 years (in thousands of tonnes), $\bar{C}^{A/S}$; the average proportional annual change in directed catch over 20 years, $AAV^{A/S}$; the average biomass at the end of the projection period ($\bar{B}_{2027}^{A/S}$) as a proportion of carrying capacity, ($K^{A/S}$), as a proportion of the risk threshold and as a proportion of biomass at the beginning of the projection period; the average minimum biomass over the projection period ($B_{min}^{A/S}$) as a proportion of carrying capacity and as a proportion of the risk threshold; the proportion of times Exceptional Circumstances are declared, $EC_{declare}^{A/S}$; and the average number of years for which Exceptional Circumstances, if declared, are declared consecutively, $EC_{consec}^{A/S}$. Performance statistics are compared between OMP-08 and a control rule with alternative constraint values (see Table 11.2).

	Sardine			Anchovy	
	OMP-08	$c_{mxdn}^S = 0.15$ and $B_{ec}^S = 250$		OMP-08	$c_{mxdn}^S = 0.15$ and $B_{ec}^S = 250$
β	0.097		α_{ns}	0.780	
$risk_S$	0.178	0.175	$risk_A$	0.097	0.099
$\bar{C}^S$	190	175	$\bar{C}^A$	381	383
AAV^S	0.24	0.20	AAV^A	0.30	0.30
$\bar{B}_{2027}^S/K^S$	0.68	0.69	B_{2027}^A/K^A	0.61	0.61
$\bar{B}_{2027}^S/Risk^S$	10.45	10.75	$B_{2027}^A/Risk^A$	1.81	1.80

Table 11.4 (Continued)

	Sardine			Anchovy	
$\overline{B_{2027}^S/B_{2007}^S}$	5.66	5.82	$\overline{B_{2027}^A/B_{2007}^A}$	0.84	0.83
B_{min}^S/K^S	0.26	0.26	B_{min}^A/K^A	0.14	0.14
$B_{min}^S/Risk^S$	1.78	1.78	$B_{min}^A/Risk^A$	0.39	0.39
$EC_{declare}^S$	0.05	0.04	$EC_{declare}^A$	0.08	0.16
EC_{consec}^S	4	4	EC_{consec}^A	4	4

Each objective is linked to a quantitative performance statistic (Table 11.4). For example, pre-agreed objectives include ensuring that Exceptional Circumstances are not declared too often, and that when they are declared the resource is able to recover quickly. These are measured by the performance statistics obtained from MP simulation trials, specifically the proportion of times Exceptional Circumstances are declared ($EC_{declare}^S$) and the average number of years for which Exceptional Circumstances continue once declared (EC_{consec}^S).

By considering this range of evaluation outputs, the stakeholders can compare the trade-offs evident in the ability of different candidate MPs to meet the various objectives. Naturally, the primary objectives of different stakeholders can conflict, so that the final MP adopted may need to reflect some compromises amongst these groups, as reached at meetings of stakeholders convened by DAFF.

In line with international practice concerning the development of OMPs (Punt et al. in press), candidate MPs (which are defined entirely by the HCRs and data inputs when empirical rules like those described earlier are used) have been subject to rigorous testing prior to implementation to ensure these objectives can be met. This testing is conducted using the simulation models described later.

Operating model (OM)

The OM is a model of the underlying "true" dynamics of the resources and fisheries of interest (e.g. Schnute et al. 2007, De Oliveira et al. 2008, Punt et al. in press) and is used to project the resource into the future to test the performances of candidate MPs. There are many alternative OMs, each necessarily consistent with the data available. Typically a baseline OM is defined with alternatives being specified in terms of their differences from this baseline; the baseline population dynamics model either is, or is close to, the preferred stock assessment model for the resource and is generally chosen on the basis of model selection criteria. Both sardine and anchovy are modelled in annual steps, incrementing at the beginning of November each year (Table 11.5a), and model parameters such as those determining the relationship of recruitment to spawner biomass or the initial numbers-at-age are determined by "fitting" such a model to available historical data and survey estimates of abundance (e.g. de Moor and Butterworth 2009a).

Future management recommendations will arise from the continued collection of existing time series of data for sardine and anchovy. Thus when projecting into the future, hydroacoustic survey estimates of abundance also need to be generated. One reason for the simulation testing is to ensure that a candidate MP is sufficiently flexible to cope with uncertainties regarding the resource monitoring data. An example of such uncertainty includes imprecision in the survey estimates of abundance – the surveys for sardine and anchovy typically have sampling errors of about 20–40% (Figure 11.3). In addition, the survey estimates are a biased measure of the "true" abundance; this bias arises from various individual hydroacoustic error factors such as target identification and calibration (de Moor and Butterworth 2015, Supplementary Material S3). The OM thus also takes this survey bias into account, as well as the associated sampling errors (see Table 11.5a).

Implementation model

The implementation model forms part of the operating model and consists of equations for determining how the annual catch limits – which are recommended in terms of weight – are removed from the age-structured population each year (Table 11.5b). For this resource, the implementation model forms a more important component of the MP development process than is normally the case, as it must take into account the dynamically changing juvenile sardine bycatch associated with targeted anchovy catches. This sardine bycatch is expected to decrease during the second half of the year once faster-growing sardine begin to separate from the juvenile anchovy shoals. The TAB is designed to be set high enough so as not to needlessly hamper the anchovy directed fishery. However, if the TAB is set too high it will result in an unnecessary decrease in the directed sardine TAC in order to maintain the same perceived risk of unintended depletion of the sardine resource. The final TAB is thus determined by the ratio of juvenile sardine to anchovy in the recruitment survey and in catches in May. However, since this ratio often decreases during the year, it is expected that in most years the sardine TAB will not be reached. Accordingly, the implementation model does not simply assume that the full sardine TAB is removed from the sardine resource each year. Rather, the numbers of juvenile sardine caught is based on the anchovy catch and the ratio of juvenile sardine to anchovy expected (see Table 11.5b). However, in the few cases that the sardine bycatch limit is reached under this model, the closure of the anchovy fishery without the full anchovy TAC being taken is also simulated. The parameters governing these simulations are set based on what has occurred in the fishery in the most recent decade (e.g. ratios of juvenile sardine bycatch

Table 11.5a The baseline operating model for generating age-structured population dynamics for sardine (superscript S) and anchovy (superscript A). Parameter definitions are given in Table 11.3. Further details and parameter values are given in the supplementary material of de Moor et al. (2011) and in Cunningham and Butterworth (2008).

Type	Equations	Residuals and correlations
Sardine state dynamics	$N_{y,1}^{S} = \left(N_{y-1,0}^{S} e^{-M_{j}^{S}/2} - C_{y,0}^{S}\right) e^{-M_{j}^{S}/2}$ $N_{y,a}^{S} = \left(N_{y-1,a-1}^{S} e^{-M_{ad}^{S}/2} - C_{y,a-1}^{S}\right) e^{-M_{ad}^{S}/2} \quad a = 2,3,4$ $N_{y,5+}^{S} = \left(N_{y-1,4}^{S} e^{-M_{ad}^{S}/2} - C_{y,4}^{S}\right) e^{-M_{ad}^{S}/2} + \left(N_{y-1,5+}^{S} e^{-M_{ad}^{S}/2} - C_{y,5+}^{S}\right) e^{-M_{ad}^{S}/2}$ $B_{y,N}^{S} = \sum_{a=1}^{5+} N_{y,a}^{S} \overline{w}_{a}^{S}$ $SSB_{y,N}^{S} = \sum_{a=2}^{5+} N_{y,a}^{S} \overline{w}_{a}^{S}$	
Anchovy state dynamics	$N_{y,1}^{A} = \left(N_{y-1,0}^{A} e^{-8.5 M_{j}^{A}/12} - C_{y,0}^{A}\right) e^{-3.5 M_{j}^{A}/12}$ $N_{y,2}^{A} = \left(N_{y-1,1}^{A} e^{-5 M_{ad}^{A}/12} - C_{y,1}^{A}\right) e^{-7 M_{ad}^{A}/12}$ $N_{y,3}^{A} = N_{y-1,2}^{A} e^{-M_{ad}^{A}}$ $N_{y,4+}^{A} = N_{y-1,3}^{A} e^{-M_{ad}^{A}} + N_{y-1,4+}^{A} e^{-M_{ad}^{A}}$ $B_{y,N}^{A} = SSB_{y,N}^{A} = \sum_{a=1}^{4+} N_{y,a}^{A} \overline{w}_{a}^{A}$	

(Continued)

Table 11.5a (Continued)

Type	Equations	Residuals and correlations
Stock-recruitment relationship	$$N_{y,0}^i = \begin{cases} d\,e^{\xi_r^i \sigma_r^i}, & \text{if } SSB_{y,N}^i \geq b^i \\[2mm] \dfrac{d}{b^i}\,SSB_{y,N}^i e^{\xi_r^i \sigma_r^i}, & \text{if } SSB_{y,N}^i < b^i \end{cases}$$	$$\varepsilon_y^i = s_{cor}^i\,\varepsilon_{y-1}^i + \sqrt{1-\left(s_{cor}^i\right)^2}\;\omega_y^i$$
Simulated survey data	$$B_{y,N}^{obs,i} = k_N^i B_{y,N}^i e^{\xi_{y,Nov}^i}$$ $$N_{y,r}^{obs,i} = k_r^i N_{y,r}^i e^{\xi_{y,rec}^i}$$ $$N_{y,r}^i = \left(N_{y-1,0}^i e^{-3.25 M_i^i/12} - C_{y,0bs}^i\right) e^{-3.25 M_i^i/12}$$	$$\varepsilon_{y,sur}^S = \eta_{y,sur}^S\,\bar{\sigma}_{sur}^S$$ $$\varepsilon_{y,sur}^A = \left(\rho_{sur}\,\eta_{y,sur}^S + \sqrt{1-\rho_{sur}^2}\;\eta_{y,sur}^A\right)\bar{\sigma}_{sur}^A \text{ where } sur = Nov,rec$$ $$\rho_{Nov} = \left(\sum_{y=1984}^{2006} 1\right)^{-1} \sum_{y=1984}^{2006} \frac{\varepsilon_{y,N}^S \varepsilon_{y,N}^A}{\sigma_{Nov}^S \sigma_{Nov}^A}$$ $$\rho_{rec} = \left(\sum_{y=1985}^{2006} 1\right)^{-1} \sum_{y=1985}^{2006} \frac{\varepsilon_{y,r}^S \varepsilon_{y,r}^A}{\sigma_{rec}^S \sigma_{rec}^A}$$ $$\varepsilon_{y,N}^i = \ln B_{y,N}^{obs,i} - \ln\left(k_N^i B_{y,N}^i\right)$$ $$\varepsilon_{y,r}^i = \ln N_{y,r}^{obs,i} - \ln\left(k_r^i N_{y,r}^i\right)$$

Simulated juvenile sardine: anchovy in May survey and commercial catch

$$r_{y,sur} = k_{sur}\,\frac{N_{y,r}^{obs,S}}{N_{y,r}^{obs,A}}$$

$$r_{y,com} = k_{may}\,\frac{N_{y,r}^{S}}{N_{y,r}^{A}}\,e^{\sigma_{may}\,\varepsilon_{y,may}}$$

$$\sigma_{Nov}^{j} = \sqrt{\left.\sum_{y=1984}^{2006}\left(\dot{\varepsilon}_{y,N}^{j}\right)^{2}\right/\sum_{y=1984}^{2006}1}$$

$$\sigma_{rec}^{j} = \sqrt{\left.\sum_{y=1985}^{2006}\left(\dot{\varepsilon}_{y,r}^{j}\right)^{2}\right/\sum_{y=1985}^{2006}1}$$

$$\varepsilon_{y,may} = \rho_{may}\,\eta_{y,jan:may} + \sqrt{1-(\rho_{may})^{2}}\;\eta_{y,may}$$

$$\rho_{may} = \frac{\displaystyle\sum_{y=1987}^{2006}\varepsilon_{y,jan:may}^{l}\,\varepsilon_{y,may}^{l}}{\left(\displaystyle\sum_{y=1987}^{2006}1\right)\sigma_{jan:may}\,\sigma_{may}}$$

k_{may}, $\varepsilon_{y,jan:may}^{l}$, $\varepsilon_{y,may}^{l}$, $\sigma_{jan:may}$ and σ_{may} are given in Table 11.5b

Table 11.5b The implementation model for generating catches-at-age (in billions) for sardine (superscript S) and anchovy (superscript A) from HCR generated TAC/Bs (in thousands of tonnes; see Table 11.1). Parameter definitions are given in Table 11.3. Further details and parameter values are given in the supplementary material of de Moor et al. (2011) and in Cunningham and Butterworth (2008). In addition to the below, catch is limited to a maximum of 95% of the exploitable stock.

Type	Equations
Annual sardine adult catch	$C_{y,a}^{S} = N_{y-1,a}^{S} S_{a}^{S} F_{y} e^{-M_{ad}^{S}/2}, \quad a = 1,\ldots,5+ \qquad F_{y} = \left(\dfrac{TAC_{y}^{S} + TAB_{rh}^{S}}{\left(\sum\limits_{a=1}^{5+} N_{y-1,a}^{S} S_{a}^{S} \bar{w}_{\alpha c}^{S}\right) e^{-M_{ad}^{S}/2}} \right)$
Annual anchovy 1-year-old catch	$C_{y,1}^{A} = \dfrac{1}{\bar{w}_{k}^{A}}\left(0.36 \times TAC_{y}^{1,A} \right)$
Anchovy 0-year-old catch prior to the May survey	$C_{y,0bs}^{A} = 0.28 \dfrac{TAC_{y}^{1,A}}{\bar{w}_{0c}^{A}}$
Anchovy 0-year-old catch for the normal season (Jan–Aug)	$C_{y,0}^{A*} = \dfrac{1}{\bar{w}_{0c}^{A}}\left(TAC_{y}^{2,A} - C_{y,1}^{A} \times \bar{w}_{k}^{A} \right)$
Sardine 0-year-old catch prior to the May survey	$C_{y,0bs}^{S} = k_{jan:may} \dfrac{N_{y-1,0}^{S}}{N_{y-1,0}^{A}} e^{\sigma_{jan:may} \eta_{y,jan:may}} \dfrac{(C_{y,0bs}^{A} \bar{w}_{0c}^{A} + C_{y,1}^{A} \bar{w}_{k}^{A})}{\bar{w}_{0c}^{S}}$ $k_{jan:may} = \exp\left\{ \sum\limits_{y=1987}^{2006} \left[\ln(C_{y,jan:may}^{S,byc} / C_{y,jan:may}^{A}) - \ln(N_{y-1,0}^{S} / N_{y-1,0}^{A}) \right] \Big/ \sum\limits_{y=1987}^{2006} 1 \right\}$ $\sigma_{jan:may} = \sqrt{ \sum\limits_{y=1987}^{2006} \left(\varepsilon'_{y,jan:may} \right)^2 \Big/ \sum\limits_{y=1987}^{2006} 1 }$ $\varepsilon'_{y,jan:may} = \ln(C_{y,jan:may}^{S,byc} / C_{y,jan:may}^{A}) - \ln(k_{jan:may} N_{y-1,0}^{S} / N_{y-1,0}^{A})$

Sardine 0-year-old catch for the normal season (Jan–Aug)

[This allows for a decrease in the bycatch ratio r_y during the year]

$$C_{y,0}^{S*} = 1.404 \times C_{y,0bs}^{S} + \frac{1}{\overline{w}_{0c}^{S}} \left(r_{y,jun} C_{y,jun}^{A} + r_{y,jul} C_{y,jul}^{A} + r_{y,aug} C_{y,aug}^{A} \right)$$

$$C_{y,jun}^{A} = 0.39 \times TAC_{y}^{1A} \quad C_{y,jul}^{A} = 0.55(TAC_{y}^{2A} - TAC_{y}^{1A}) \quad C_{y,aug}^{A} = (1 - 0.55)(TAC_{y}^{2A} - TAC_{y}^{1A})$$

$$r_{y,m} = k_m \frac{N_{y,r}^{S}}{N_{y,r}^{A}} e^{\sigma_m \varepsilon_{y,m}}, \quad m = jun, jul, aug$$

$$\varepsilon_{y,jun} = \rho_{byc} \varepsilon_{y,may} + \sqrt{1 - (\rho_{byc})^2} \, \eta_{y,jun}$$

$$\varepsilon_{y,jul} = (\rho_{byc})^2 \varepsilon_{y,may} + \rho_{byc} \sqrt{1 - (\rho_{byc})^2} \, \eta_{y,jun} + \sqrt{1 - (\rho_{byc})^2} \, \eta_{y,jul}$$

$$\varepsilon_{y,aug} = (\rho_{byc})^3 \varepsilon_{y,may} + (\rho_{byc})^2 \sqrt{1 - (\rho_{byc})^2} \, \eta_{y,jun} + \rho_{byc} \sqrt{1 - (\rho_{byc})^2} \, \eta_{y,jul} + \sqrt{1 - (\rho_{byc})^2} \, \eta_{y,aug}$$

ρ_{byc} is estimated from the actual correlations between residuals in successive months from past data

$$k_m = \exp \left\{ \sum_{y=1987}^{2006} \left[\ln(C_{y,m}^{S,byc} / C_{y,m}^{A}) - \ln(N_{y,r}^{S} / N_{y,r}^{A}) \right] \Big/ \sum_{y=1987}^{2006} 1 \right\} \quad \sigma_m = \sqrt{ \sum_{y=1987}^{2006} (\varepsilon_{y,m}')^2 \Big/ \sum_{y=1987}^{2006} 1 }$$

$$\varepsilon_{y,m}' = \ln(C_{y,m}^{S,byc} / C_{y,m}^{A}) - \ln(k_m N_{y,r}^{S} / N_{y,r}^{A}), \quad m = may, jun, jul, aug$$

(Continued)

Table 11.5b (Continued)

Type	Equations
Closure of anchovy fishery if normal season sardine TAB reached	$$C_{y,0}^{A**} = \begin{cases} C_{y,0}^{A*} & \text{if } C_{y,0}^{S*}\,\overline{w}_{0c}^{S} \leq TAB_y^{2,S} - TAB_{rh}^{S} \\[2ex] \dfrac{1}{\overline{w}_{0c}^{A}}\left(TAC_y^{2,A}\left[\dfrac{TAB_y^{2,S} - TAB_{rh}^{S}}{C_{y,0}^{S*}\,\overline{w}_{0c}^{S}}\right] - C_{y,1}^{A*}\,\overline{w}_{lc}^{A}\right) & \text{if } C_{y,0}^{S*}\,\overline{w}_{0c}^{S} > TAB_y^{2,S} - TAB_{rh}^{S} \end{cases}$$ $$C_{y,0}^{S**} = \min\left\{ C_{y,0}^{S*},\ \frac{TAB_y^{2,S} - TAB_{rh}^{S}}{\overline{w}_{0c}^{S}} \right\}$$
Additional season	$$C_{y,0}^{S} = C_{y,0}^{S**} + \frac{1}{\overline{w}_{0c}^{S}}\min\left\{ TAB_{ads}^{S};\, r_y\left(TAC_y^{3,A} - TAC_y^{2,A}\right)\right\}$$ $$C_{y,0}^{A} = C_{y,0}^{A**} + \frac{1}{\overline{w}_{0c}^{A}}\left(TAC_y^{3,A} - TAC_y^{2,A}\right)$$

to targeted anchovy catch in relation to their estimated relative abundances) and through estimation by fitting the OM to available data (e.g. to estimate the selectivities-at-age).

Sardine-anchovy trade-off

Given the primary objective of maximising catch subject to an acceptably small risk of reducing the resources to an undesirably low level, the TAC recommendation reflects, in essence, the amount of fish that can be caught while still maintaining the potential for reasonable levels of resource productivity in the future.

Figure 11.6 shows the average directed sardine and anchovy catches projected to result over 20 years of simulation, while satisfying the primary objective in different ways. Every point on the curve in Figure 11.6 corresponds to a specific set of control parameters, β, α_{ns} and α_{ads} (Table 11.2), which were identified through simulation over a grid of the control parameters as being able to maximise the average directed sardine and anchovy catches, $\bar{C}^S$ and $\bar{C}^A$, while still keeping the risk of both of the resources dropping to an undesirably low level to less than a threshold set separately for each resource ($risk_S < 0.18$ and $risk_A < 0.10$; Table 11.4). If the right-hand part of the curve was shifted to the right, the risk to the anchovy resource would increase, while if the top part of the curve was shifted upwards, the risk to the sardine resource would increase. The top part of the curve shows that an increase in average directed sardine catch requires a (larger) decrease in anchovy catch with its associated sardine bycatch in order to maintain the same risk to the sardine resource.

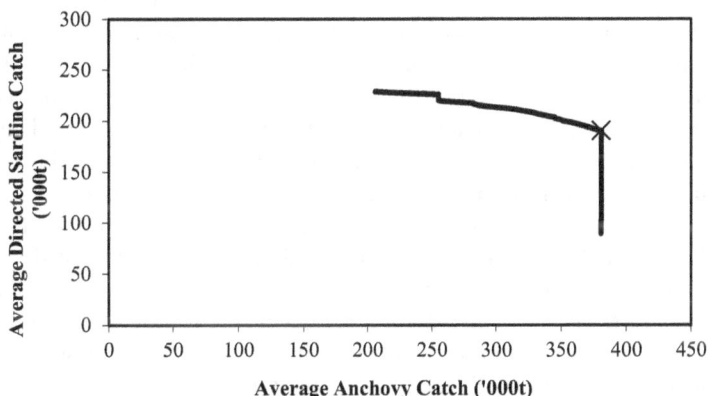

Figure 11.6 The trade-off curve between directed sardine and anchovy catch, the latter having an associated juvenile sardine bycatch. Every point on the curve maximises the catches subject to satisfying the risk criteria for both sardine and anchovy. The combination of control parameters chosen for OMP-08 is shown by an X.

Robustness tests

One also needs to ensure that the candidate MPs do not result in substantially different performance statistics if different key assumptions are made from those in the baseline OM (Table 11.5a). Candidate MPs were thus also simulation tested against alternative plausible OMs, called robustness tests. The uncertainties considered included natural mortality of the fish and the relationship between spawner biomass and recruitment, which in reality will differ by an unknown amount from the values assumed by the baseline OM. Another uncertainty considered was the bias associated with the survey estimates of abundance, that is whether they were greater or less than the bias assumed or estimated when fitting the OM (Cunningham and Butterworth 2007).

Final choice

The MP development process for South African sardine and anchovy has tended to flow from an initial update of the previous OMP taking into consideration the new OM, to then testing possible changes to the rules in combination with changes to the constraints. All these alternatives are considered in light of the primary objectives. The OMP that was finally recommended by the Small Pelagics Scientific Working Group and adopted by decision-makers, termed OMP-08, included decisions on the following, particularly in relation to the previous OMP-04.

1   New rules: The additional season anchovy TAC under OMP-04 was calculated using an equation similar to that for the normal season only (first part of Equation 11.1a.6). A new "booster" rule was introduced in OMP-08 to enable the additional season anchovy TAC to increase rapidly towards the maximum if the projected biomass is above a threshold (Equation 11.1a.6). This facilitated better use of the anchovy resource during periods of high abundance without increasing the risk to the sardine resource. Changes to the Exceptional Circumstances rules included ensuring that the directed sardine TAC reaches zero before observed abundance reaches zero, as previously applied to anchovy (Table 11.1b, 11.2) and recommending that only half of the sardine TAC be awarded at the beginning of the year if Exceptional Circumstances are declared (Equations 11.1b.1 and 11.1b.3).

2   Constraints: The most important changes in constraints were an increase in the maximum proportion by which directed sardine TAC could be reduced interannually, c^S_{mxdn}, and an increase in the threshold below which sardine Exceptional Circumstances are declared, $B^{obs}_{y-1,N}$ (Table 11.2). Both of these changes were prompted by the lower abundance of the sardine resource, compared to that when OMP-04 was developed, in order to maximise the average projected catch at a risk level that was deliberately set lower (Table 11.4). Less substantial changes to other constraints, some of

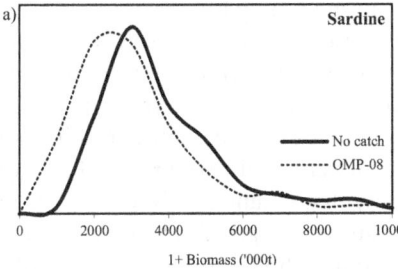 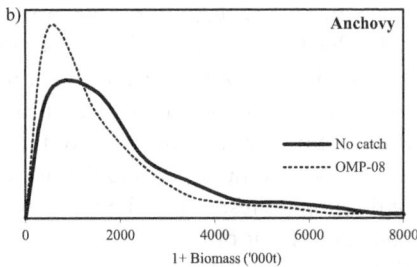

Figure 11.7 The (a) sardine and (b) anchovy distributions of 1+ biomass after 20 years of projection assuming either no future catch or management according to OMP-08.

which were designed based on the past decade, were to closer match what the pelagic industry expected to attain (Table 11.2).

3   Control parameters: The values for the control parameters of OMP-08, β, α_{ns} and α_{ads}, were then chosen to match the corner point of the trade-off curve (Figure 11.6) in order to maximise the expected average directed sardine catch without decreasing the expected average anchovy catch.

Figure 11.7 shows the difference in the resource biomass after 20 years of projection under OMP-08 compared to a no-catch scenario.

Implementation and conclusions

South Africa has a long history of fisheries management using the MP approach. Prior to the development of OMP-08, the MP-recommended TAC/Bs for the sardine and anchovy fisheries had been endorsed unchanged by the Minister responsible for well over a decade. OMP-08 was adopted in 2007, and resulted in a continuation of this endorsement from 2008 through to 2012. The HCRs developed for South African sardine and anchovy, although complex, have shown that the MP approach can be applied successfully to manage highly variable, short-lived species (de Moor et al. 2011). Having an agreed set of rules to calculate a TAC/B recommendation provides increased transparency and removes the tendency for the annual TAC recommendation process to become a "lottery" (Butterworth 2007). Although the inclusion of all stakeholders in OMP development probably results in a slower process, this disadvantage is far outweighed by the benefit of all the stakeholders having a clear idea of what to expect in terms of future TACs and resource safeguards (Butterworth 2008). One of the key benefits to industry from the MP approach has been the ability to safely incorporate constraints on variations in the TAC from year to year (Butterworth 2008, de Moor et al. 2011). This facilitates planning by the industry and allows for a more socioeconomically stable environment, providing better job security for those employed in the fishing industry.

Understanding of the dynamics of these resources is continually advancing as new data become available. For this reason, OMP revisions for South African fisheries are planned to take place every 4 years (Rademeyer et al. 2008), though this is not always achieved due to unforeseeable complications in data analysis and limited personnel. For South African small pelagics, one key area of developing research is the likely possibility that sardine consist of more than one subpopulation (Coetzee et al. 2008, van der Lingen 2011, de Moor and Butterworth 2015). Management of this resource should thus ensure that fishing is not detrimental to either of these individual populations, rather than only to the resource as a whole. A new OMP that takes account of this concern is expected to be agreed upon soon.

Acknowledgements

The authors gratefully acknowledge financial support from DAFF. We thank scientists and technicians from DAFF for their collection of data upon which these analyses depend and Dagmar Merkle (DAFF) for her help in producing Figure 11.2. Thanks are also extended to the editors for their guidance and Andrea Ross-Gillespie (MARAM) for checking the technical aspects of this chapter.

References

Butterworth, D.S. 2007. Why a management procedure approach? Some positives and negatives. *ICES Journal of Marine Science,* 64, 613–617.

Butterworth, D.S. 2008. Some lessons from implementing management procedures. *In:* Tsukamoto, K., Kawamura, T., Takeuchi, T., & Beard, Jr., T.D. (eds.) *Fisheries for Global Welfare and Environment.* 5th World Fisheries Congress 2008, pp 381–397. Tokyo: Terrapub.

Coetzee, J.C., van der Lingen, C.D., Hutchings, L., & Fairweather, T.P. 2008. Has the fishery contributed to a major shift in the distribution of South African sardine? *ICES Journal of Marine Science,* 65, 1676–1688.

Cunningham, C.L., & Butterworth, D.S. 2007. Development and Testing of OMP-08. Report No. MCM/2007/26NOV/SWG-PEL/01. Cape Town: Marine and Coastal Management. Available at www.mth.uct.ac.za/maram/pub/2007/MCM_2007_NOV_SWG-PEL_01. pdf (last accessed 12 January 2016).

Cunningham, C.L. & Butterworth, D.S. 2008. Proposed Final OMP-08. Report No. MCM/2008/SWG-PEL/09. Cape Town: Marine and Coastal Management. Available at www.mth.uct.ac.za/maram/pub/2008/MCM_2008_SWG-PEL_09.pdf (last accessed 12 January 2016).

de Moor, C.L. & Butterworth, D.S. 2008. OMP-08. Report No. MCM/2008/SWG-PEL/23. Cape Town: Marine and Coastal Management. Available at www.mth.uct.ac.za/maram/pub/2008/MCM_2008_SWG-PEL_23.pdf (last accessed 12 January 2016).

de Moor, C.L. & Butterworth, D.S. (2009a) The 2004 re-assessment of the South African sardine and anchovy populations to take account of revisions to earlier data and recent record abundances. *African Journal of Marine Science,* 31, 333–348.

de Moor, C.L. & Butterworth, D.S. (2009b). Setting the TAC for sardine and anchovy. *Maritime Southern Africa*, Nov/Dec 2009, 29–31.

de Moor, C.L., & Butterworth, D.S. 2015. Assessing the South African sardine resource: two stocks rather than one? *African Journal of Marine Science*, 37, 41–51.

de Moor, C.L., Butterworth, D.S., & De Oliveira, J.A.A. 2011. Is the management procedure approach equipped to handle short-lived pelagic species with their boom and bust dynamics? *ICES Journal of Marine Science*, 68, 2075–2085.

De Oliveira, J.A.A. 2002. *The Development and Implementation of a Joint Management Procedure for the South African Pilchard and Anchovy Resources*. Ph.D. Thesis, University of Cape Town.

De Oliveira, J.A.A., Kell, L.T., Punt, A.E., Roel, B.A., & Butterworth, D.S. 2008. Managing without best predictions: The Management Strategy Evaluation framework. *In:* Payne, A.I.L., Cotter, A.J.R., & Potter, T. (eds.) *Advances in Fisheries Science. 50 Years on from Beverton and Holt*, pp. 104–134. Blackwell Publishing, Oxford.

Punt, A.E., Butterworth, D.S., de Moor, C.L., De Oliveira, J.A.A., & Haddon, M. in press. Management strategy evaluation: Best practices. *Fish and Fisheries,* doi:10.1111/faf.12104

Rademeyer, R.A., Butterworth, D.S., & Plagányi, É.E. 2008. A history of recent bases for management and the development of a species-combined Operational Management Procedure for the South African hake resource. *African Journal of Marine Science*, 30, 291–310.

Rademeyer, R.A., Plagányi, É.E. & Butterworth, D.S. 2007. Tips and tricks in designing management procedures. *ICES Journal of Marine Science*, 64, 618–625.

Schnute, J.T., Maunder, M.N., & Ianelli, J.N. 2007. Designing tools to evaluate fishery management strategies: Can the scientific community deliver? *ICES Journal of Marine Science*, 64, 1077–1084.

van der Lingen, C.D. 2011. The biological basis for hypothesizing multiple stocks in South African sardine *Sardinops sagax*. Report No. MARAM IWS/DEC11/P/OMP11/P7. Cape Town: 2011 International Fisheries Stock Assessment Workshop. Available at www.mth.uct.ac.za/maram/workshop/workshop2011.php (last accessed 12 January 2016).

Chapter 12

North Sea haddock

The EU-Norway management procedure evaluation

Coby L. Needle

Introduction

Haddock (*Melanogrammus aeglefinus*) is a medium-sized gadoid species that can be found across a wide area of the northern Atlantic. Important commercial fisheries of haddock are located on Georges Bank off New England (Brooks et al. 2012), Iceland (ICES-NWWG 2014), the Barents Sea north of Norway (ICES-AFWG 2014) and the North Sea (ICES-WGNSSK 2013). In the latter area, haddock is a key target for the demersal fleets of a number of northern European countries, particularly Scotland for which it is the second most valuable demersal stock (after monkfish), and was worth around £30 million in 2011, or around 8% of the value of all seafood landed into Scotland (Scottish Government 2012). Haddock have been exploited by Scottish fleets for many years, and in the past formed the basis of economic growth for many coastal communities (particularly in the north-east of Scotland: see Coull 1997). In my home town of Aberdeen, for example, a portion of fish and chips would always feature haddock, while the same meal might be whiting (*Merlangius merlangus* L.) in Glasgow or cod (*Gadus morhua* L.) in much of England. However, fishing has declined in many small northeast villages, and most haddock are now landed into the large whitefish markets in Peterhead, Fraserburgh and Shetland (Scottish Government 2012).

In the North Sea, haddock are found mostly in the central and northern areas, and are very seldom encountered south of a line joining the north-east of England and southern Norway through the Dogger Bank (see Figure 12.1). For this reason, and following historical fishing patterns, the bulk (around 84%) of the European Union (EU) annual landings quota for haddock is allocated to the UK, with around 66% usually being assigned to Scotland. Scotland therefore has a keen interest in maintaining a sustainable and profitable fishery for haddock.

Like all haddock stocks, the North Sea stock is characterised by an episodic recruitment pattern, with occasional large year-classes interspersed with several years of very poor recruitment (see Figure 12.2). During the so-called "gadoid outburst" in the 1960s and 1970s the productive years produced very large

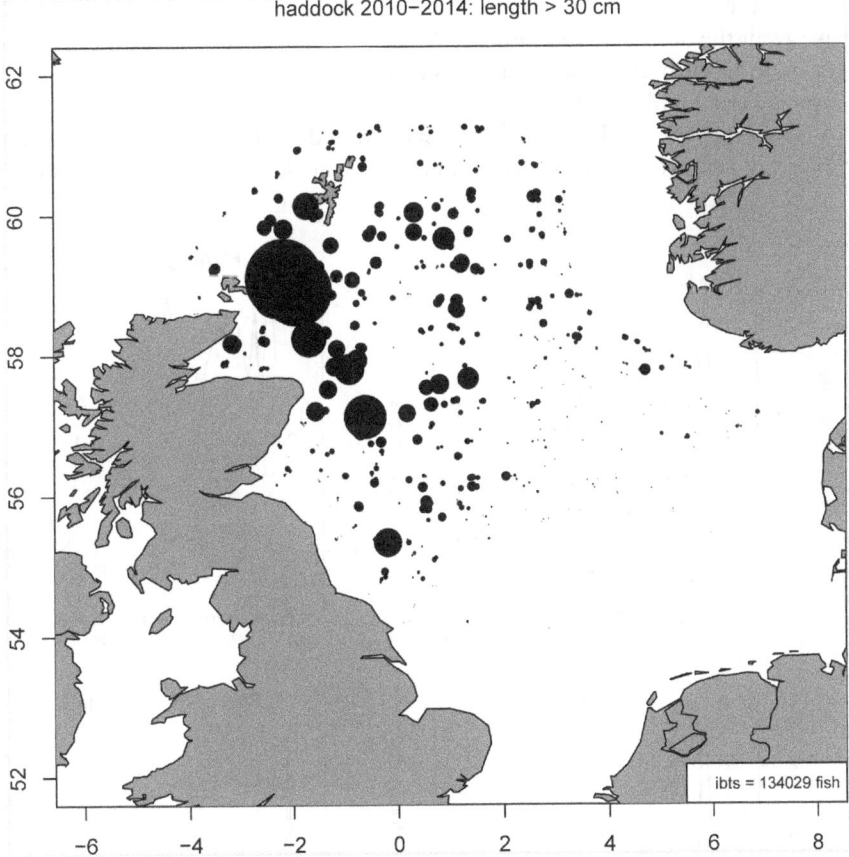

Figure 12.1 Distribution map for haddock in the North Sea, using observations during 2010–2014 from the IBTS Q1 survey (ICES-WGIBTS 2014). The size of the circles is proportional to the observed abundance per hour of haddock greater than 30 cm long, and the total number of fish observed in this category is indicated in the legend.

recruitments (Hislop 1996), and the fishery was able to exert relatively high fishing mortalities (F) that appeared to be sustainable. More recently, while the frequency of the recruitment events has remained about the same, the magnitude of the large year-classes has diminished (for reasons unknown) and the stock is no longer able to withstand the fishing pressure experienced historically. Spawning stock biomass (SSB) appears to be closely related to previous recruitment, rather than the other way round: there is no clear relationship between parental SSB and subsequent recruitment. Haddock therefore presents an interesting management problem: we cannot predict when a large year-class will emerge, and when this does happen those fish must be allowed to contribute to the fishery yield and spawning population for as long as possible while we await the next large recruitment. While this is a feature of many marine fisheries, the North Sea haddock is unusual in the scale of the difference between average and good year-classes.

The North Sea haddock stock is exploited both by EU member states and by Norway, and is therefore managed as a shared stock along with species such as

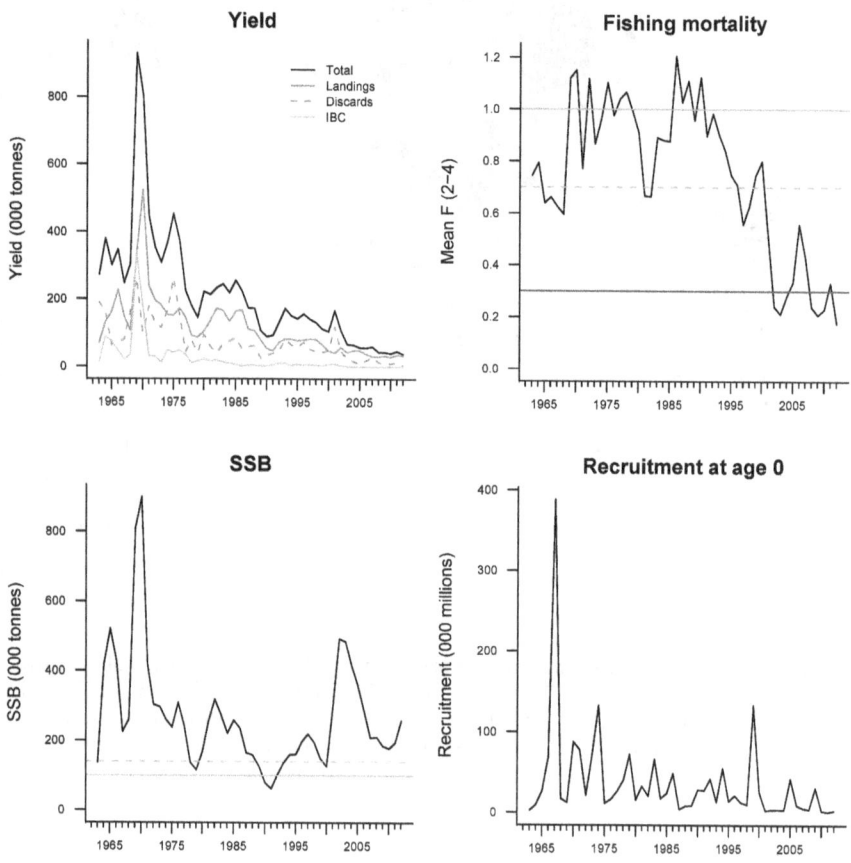

Figure 12.2 Summary plots for final assessment from the 2013 ICES Working Group for the Assessment of Demersal Stocks in the North Sea and Skagerrak (ICES-WGNSSK 2013). Dotted horizontal grey lines indicate precautionary reference points F_{pa} (top right plot) and B_{pa} (bottom left plot), while solid horizontal grey lines indicate limit reference points F_{lim} and B_{lim} in the same plots. The solid horizontal black line in the top right plot represents the target $F = 0.3$ in the EU-Norway management plan, which is also considered to be a proxy for F_{MSY}.

cod, whiting, saithe (*Pollachius virens* L.) and plaice (*Pleuronectes platessa* L.). During the late 1990s there was a political move to formalise the way in which such stocks are managed, and a series of joint EU-Norway management procedures (MPs) started to be developed through collaborative consultations between scientists, fisheries managers and stakeholders. One of the first of these was for haddock, for which the broad terms were agreed in 2005 with two main elements:

1   Management regulations should seek to achieve a target fishing mortality F_{target} of 0.3.
2   Spawning stock biomass (SSB or *B*) should be kept above the precautionary level (B_{pa} = 140,000 t). If *B* falls below B_{pa}, additional measures should be taken to ensure that *B* increases.

Here, when assessing stock status, F_{target} is compared with $F_{a,y}$ averaged across a predefined range of ages, *a* (2 to 4 for North Sea haddock) for a given year, *y*. B_{pa} is defined as the precautionary level of biomass below which the stock should not fall. These definitions were retained for subsequent updates to the procedure discussed here.

The management procedure contained a clause specifying that an evaluation should be carried out by the end of 2006. In April of that year, the EU and Norway approached the International Council for the Exploration of the Seas (ICES) to inquire about the feasibility of addressing this review through ICES assessment working group channels. It was agreed that the ICES Working Group for the Assessment of Demersal Stocks in the North Sea and Skagerrak (ICES-WGNSSK) would coordinate the evaluation of the existing procedure and any proposed modifications, and that the review would be prepared subsequently during the October 2006 meeting of the ICES Advisory Committee for Fisheries Management (ACFM).

Initial analyses (Needle 2006a) were presented in June 2006 at the ICES Working Group on Methods of Fish Stock Assessment (ICES-WGMG 2006), at which improvements and modifications were suggested. Consultations with managers and stakeholders followed, and updated results (Needle 2006b) were discussed at the October meeting of ACFM (ICES-ACFM 2006). The new analyses formed the basis of ICES advice to the EU and Norway which I presented at their annual bilateral negotiations in November 2006. In the margins of these meetings further discussions were held on the likely sustainability of the proposed procedure, and this led to modifications (the addition of a sliding-*F* rule, and the clarification of the time when biomass should be measured – see later). Following this process, the revised procedure came into force on 1 January 2007 and was first used as the basis for advice for landings in 2008. In this chapter, we summarise the procedure and its evaluation, and discuss whether or not it proved to be successful.

The decision rule

The text of the EU-Norway MP as implemented in 2007 is as follows:

1   Maintain a minimum level of spawning stock biomass (SSB) greater than 100,000 ts (B_{lim}).
2   For 2007 and subsequent years, restrict fishing on the basis of a total allowable catch (TAC) consistent with a fishing mortality rate of no more than 0.3 for appropriate age groups, when the SSB in the end of the year in which the TAC is applied is estimated above 140,000 t (B_{pa}).
3   Where the rule in paragraph 2 would lead to a TAC which deviates by more than 15% from the TAC of the preceding year, establish a TAC that is no more than 15% greater or 15% less than the TAC of the preceding year.
4   Where the SSB referred to in paragraph 2 is estimated to be below B_{pa} but above B_{lim} the TAC shall not exceed a level which will result in a fishing mortality rate equal to $0.3 - 0.2 \times \left(\dfrac{B_{pa} - SSB}{B_{pa} - B_{lim}} \right)$. This consideration overrides paragraph 3.
5   Where the SSB referred to in paragraph 2 is estimated to be below B_{lim} the TAC shall be set at a level corresponding to a total fishing mortality rate of no more than 0.1. This consideration overrides paragraph 3.
6   In order to reduce discarding and to increase the spawning stock biomass and the yield of haddock, the exploitation pattern shall, while recalling that other demersal species are harvested in these fisheries, be improved in the light of new scientific advice from *inter alia* ICES.
7   In the event that ICES advises that changes are required to the precautionary reference points B_{pa}(140,000 t) or B_{lim}(100,000 t), paragraphs 1–5 should be reviewed.
8   No later than 31 December 2009, the arrangements in paragraphs 1 to 7 should be reviewed to ensure that they are consistent with the objective of the procedure. This review shall be conducted after obtaining inter alia advice from ICES concerning the performance of the procedure in relation to its objective.

The text of the procedure refers throughout to catch as the quantity that is being managed. In reality, throughout the lifetime of the procedure it was landings that were being managed. No restrictions on over-quota discarding in the relevant fisheries were in place (indeed, such discarding was mandatory as over-quota landings were strictly forbidden). TAC therefore in effect refers to the total allowable *landings*. The management procedure described earlier is limited to use of quotas as a management tool, rather than effort restrictions, and in essence can be simplified to two key points:

* F_{target} and TAC constraint: In assessment year y, set the TAC in the quota year $y + 1$ so that the expected fishing mortality rate, averaged over ages

2–4, is equal to the target reference point $F_{target} = 0.3$. If this results in spawning stock biomass $B < B_{pa}$ in year $y + 2$, then implement the sliding-F rule instead (below). Modify the TAC to ensure that the maximum interannual change is within $\pm15\%$. This means that management decisions are based on predictions of the results of management actions, rather than purely on historical observations.

- Sliding-F rule: If, following application of the F_{target} mortality rate in years $y + 1$ and $y + 2$, $B < B_{pa}$ in year $y + 2$, apply the sliding-F rule to set the TAC in year $y + 1$ with no constraint (see paragraph 4 of the procedure and Figure 12.3).

These points constitute a harvest control rule that is analogous to rules commonly used in salmon fisheries, and can be seen as a simplified version

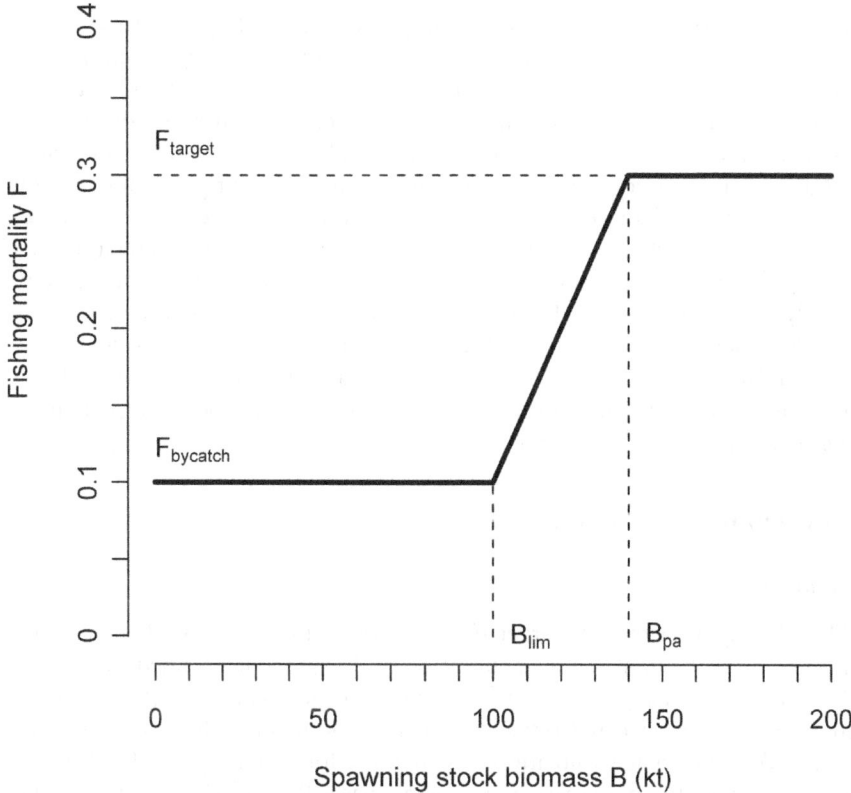

Figure 12.3 The sliding-F rule (thick line) for specifying the intended fishing mortality rate $\bar{F}$, based on the expected spawning stock biomass B remaining after the corresponding quota has been taken. For North Sea haddock: $F_{target} = 0.3, F_{bycatch} = 0.1, B_{lim} = 100,000$ t and $B_{pa} = 140,000$ t.

of the generic rule subsequently proposed by Froese et al. (2011). This class of rule is characterised by a nonlinear relationship between stock status (measured as either spawning biomass or recruits) and the target fishing mortality. It is justified on the grounds that a constant harvest rate, rather than a constant catch, is most likely to lead to long-term sustainability of a randomly fluctuating resource (Beddington and May 1977, Rosenberg et al. 1993), but that additionally a minimum biomass must be retained so as to allow recovery following depletion of the stock to low levels (Walters 1975, Quinn et al. 1990). This need for a minimum escapement level requires that fishing mortality be reduced at low stock sizes. Approaches analogous to the sliding-F rule are widely used in the management of European fish stocks (examples include North Sea cod and saithe), and the ICES maximum sustainable yield (MSY) approach now also features a linearly declining F_{target} mortality rate below a given biomass level.

The agreed F_{target}, TAC constraint and sliding-F components of the rule reflect many of the concerns expressed by fisheries managers, scientists and stakeholders following a series of meetings during 2006. In particular, the TAC constraint was included at the behest of the fishing industry to ensure stability of opportunity, while the sliding-F rule was requested by scientists to ensure that the target F used would be appropriate to the current biomass level. With reference to the terminology used globally: the data inputs, assessment model and harvest control rule constitute the management procedure (or MP, as defined in Rademeyer et al. 2007, and Edwards, Chapter 2, this volume) for North Sea haddock. It is widely referred to in the relevant literature as a management "plan," but this is something of a misnomer as it does not specify clearly (save for mention of quota setting) how the fishery is to be managed. The MP specifies a harvest control rule for determining quota, but says nothing about related management measures such as gear regulations, the use of closed areas or seasons, how the permitted quota is to be divided between participating nations, or effort regulations.

The evaluation process

Structure

The MP stipulates that the target F is a function of the biomass B as measured at the start of the year *after* the application of the quota determined by the MP. The biomass at this point is itself a function of the assumed target F, and the interaction between F and B makes evaluation of the procedure quite complicated. We can envisage a situation in which the forecast at $F_{target} = 0.3$ leads to a biomass after the quota year which is between B_{lim} and B_{pa}. In this case the sliding-F rule stipulates a different F_{target} (Figure 12.3), so the forecast must be performed again, which leads to another different $F_{target,}$ and so on. This cycle converges to a single solution, but only at the cost of computational complexity and potentially long run times. The code to carry out the evaluations was written in R (R Development Core Team 2011, v.2.8.1) using modules and functions from the FLR library (FLR Team 2006, Kell et al. 2007, Hillary 2009, http://flr-project.org/). Subsequent improvements to the core FLR code mean that it can now carry out the analysis much more quickly, but there are elements of the evaluation (such as the existence of distinct landings, discards and industrial bycatch catch components) that still render the standard FLR code insufficient.

In the evaluation process, 50 simulation-based evaluations of the management procedure were run (Needle 2008b), each for 25 years into the future, and for a series of scenarios with different F_{target} values (Equation T12.19 in Table 12.3). The time period for the simulations was chosen to ensure that the conclusions were not wholly dependent on the particular starting values: the oldest age in the historical assessment was 10, so at least 10 simulation years would be required to ensure that the influence of the starting year was minimised. Simulations of 25 years ensured a reasonable year span after this point, while not requiring unfeasibly long run times. The limit on interannual quota variability was set to ±15% (Equation T12.18). Through each simulation, two separate stock data streams were maintained: one containing *true* information on stock dynamics (abundance, mortality and so on), and one giving the *assessed* stock information as would be available to managers in reality. These two stocks (true and assessed) can be quite different, and the differences between them can be very influential on performance of the MP during the evaluations. Outcomes were expressed in terms of the number of years in each simulation for which spawning stock biomass B fell below the defined precautionary (B_{pa}) or limit (B_{lim}) biomass reference points.

Historical data and parameter setting

An initial FLXSA stock assessment (Fish Lab eXtended Survivors' Analysis; Shepherd 1992, Darby and Flatman 1994, Kell 2011) was run, using data up to and including year $y - 1$ (which was 2006 in this case study). Input data on the estimated numbers-at-age and mean weights-at-age for three catch components (landings, discards and industrial bycatch), along with time-invariant natural mortality and proportion mature values (Equations T12.10 and T12.11), were taken from the relevant ICES assessment working group reports (ICES-WGNSSK 2006, ICES-WGNSSK 2007). This assessment generated estimates of abundance at age $\hat{N}_{a,y}$, recruitment $\hat{R}_y = \hat{N}_{0,y}$ (age 0 is the first age in the North Sea haddock data set), and fishing mortality at age $\hat{F}_{a,y}$. These were used to initialise the projections. Assessment model outputs further enabled estimation of the following components necessary for simulation of the system into the future.

1   Catchability at age $q_{i,a}$ for each survey. This was assumed to be related to estimated abundance via a power relationship, and was estimated for each survey i and age a by minimising the sum-of-squares:

$$SSQ_q = \sum_{i,a,y} \left(I_{i,a,y} - \hat{q}_{i,a}\hat{N}_{a,y}^{\hat{p}_{i,a}}\right)^2,$$

where $\hat{p}_{i,a}$ is the estimated power term in the catchability relationship for each age a and survey i.

2   Recruitment parameters. These were the mean and variance of the low-to-moderate recruitments observed during 1995–1998 and 2000–2006, and the estimate of the high 1999 year-class recruitment. A key requirement for applying the model to North Sea haddock is to ensure that it encapsulates adequately the sporadic nature of recruitment for the stock. Historically, North Sea haddock recruitment has followed a pattern of occasional large year-classes (the size of which seems unrelated to parental stock size, at least directly), interspersed with years of low-to-moderate recruitment (Figure 12.2). In the model, this pattern is replicated by stipulating one large recruitment (of the order of the 1999 year-class; around 100 billion fish) in a random year within each 10-year simulation period. As these simulations are 25 years long, there will therefore be at most two or three large year-classes within each iteration; this seems to be consistent with historical observations. A further proviso is to ensure that the large year-classes are separated by at least two years, as North Sea haddock have never been observed to produce two large year-classes in succession. This may be due to cannibalism (age-1 fish predating on age-0 juveniles from the following year-class) or other density-dependent effects (Fogarty et al. 2001). Recruitment for the remaining years in the simulation is given by a log-normal distribution about the geometric mean (around 10 billion fish) of the 10 years prior to 2006, not including 1999 (Equation T12.13).

3   Selection at age ζ_a for the fishery. This is a measure of how fishing mortality F varies with age a, and was given for each age by the mean of the last three historical F estimates (Equation T12.16). Selection ζ_a must be assumed to be known throughout the simulation period: if this were not the case, there would be no unique solution to the estimation of F that would result in the required catch.

4   The proportions of the total catch numbers in each of the catch components (landings ρ_a^l and discards ρ_a^d) were calculated using Equations T12.6 and T12.7 (Table 12.1), also fixed through the simulation period.

More generally, it was also assumed that the fishing fleet would remain at the same capacity, and fish in the same way as before. As mentioned, discarding was known to be a very big issue for this stock, due to stringent landing regulations. The modelling of discarding of haddock from the commercial North Sea fishery should therefore be very important for the effective simulation of the fishery and its management. However, this is a difficult issue to address, as discarding activity is a function not only of stock abundance, but also of prices, costs, quota constraints, gear regulations, fishing distribution, and the availability of other fishing opportunities (amongst other factors). There was no attempt here to model discards in this way (because appropriate models do not yet exist), and fixed proportions of discarding at each age were used throughout the simulations. As stipulated in Equations T12.6 and T12.7, the proportion discarded at age a in the simulations is the mean of the proportions discarded at age a for the years $\gamma - 3$ to $\gamma - 1$ (here γ denotes the first assessment year, which is 2007 for this case study). The remaining input settings used in the North Sea haddock case study are summarised in Equations T12.14 to T12.22.

The analysis algorithm

The North Sea haddock MSE analysis consisted of three concentric loops: the F_{target} loop, which considers different values of F_{target}; the k or iteration loop, which loops over different randomly generated recruitments, and the inner y or year loop. Although the inclusion of a stochastic recruitment term has been decried as "paradoxical" by a minority of writers (e.g. Rochet and Rice 2009), it is a very widespread technique that seems intrinsically reasonable when precise estimates of recruitment are impossible to achieve. In addition, there has as yet been no quantitative demonstration that alternatives (such as expert judgement or comparative approaches) produce results that are any more reliable. For these reasons, we use this stochastic approach here.

ICES advice generally refers to the "intermediate year," which is the year after the final historical year (and is usually the year *in which* the advice is being prepared), and the "forecast year," which is the year following (and is the year *for which* the advice is being prepared). In the first year of the projection period ($y = \gamma + 1$), the intermediate year fishing mortality (i.e. $F_{a,y}$ in the first year after the assessment) is set to the mean of the last three historical years. Thereafter, for $y > \gamma + 1$, the intermediate year F is determined by previous applications of the HCR, and the y-loop dynamics proceed as follows:

1   Recruitment in year y is generated using Equation T12.1, and numbers at age by Equation T12.2.
2   Catch C^c, landings C^l and discard C^d numbers (as well as associated yields Y^c, Y^l, and Y^d) in year y are now calculated, using Equations T12.3, T12.6 and T12.7. Note that $\tilde{F}_{a,y}$ here is the *intended* fishing mortality produced by a previous management decision. Under catch–based management, Equation T12.3 must be solved for $\tilde{F}_{a,y}$ because the realised fishing mortality will differ from that expected when the management decision was made. In other words it is the TAC (or intended landings yield), $\tilde{Y}_y^l$, that is determined by a previous management decision, not the realised fishing mortality $F_{a,y}$, and the simple removal of fish equivalent to the TAC may result in negative population numbers if the population has changed in the time period

between the determination of the TAC and its implementation. Therefore a multiplier λ_y must be determined such that the application of $F_{a,y} = \lambda_y \zeta_a$ results in positive numbers at age and a yield less than or equal to the TAC. Here we are modelling a fishery which will take the predetermined TAC if possible, but not if doing so would result in negative abundance at any age. λ_y is estimated by minimising the sum-of-squares between intended yield $\tilde{Y}_y^l$ and the actual yield Y_y^l, constrained so that $N_{a,y} > 0$ for all ages. Once this is done, realised fishing mortality is given by $F_{a,y} = \lambda_y \zeta_a$ and catch by Equation T12.3.

3   Observational data are generated for the assessment in year $y + 1$ using Equations T12.8 and T12.9. Survey indices are generated assuming catchability $q_{i,a}$, abundance $N_{a,y}$ and a random term $\varepsilon_{a,y}^i \sim N(0, \sigma_i^2)$ (Equation T12.8). Catch, landings and discards data for assessments are produced by applying random noise to true values. Given an assumed measurement error variance on catch data σ_c^2, assessment catch data is given by Equation T12.9, where ρ_a^l and ρ_a^d are given by Equations T12.6 and T12.7 respectively. Measured yields $\hat{Y}^c$, $\hat{Y}^l$ and $\hat{Y}^d$ are calculated in a similar way to that given in Equation T12.5.

4   An assessment is carried out, using data (up to and including year $y - 1$) for $\hat{C}^c$, $\hat{Y}^c$, W^a, **I**, **M**, **Mat**, prop F and prop M (see Equations T12.10 and T12.11). The FLXSA function of FLR is used for this purpose, and returns assessment estimates of abundance $\hat{N}$ and fishing mortality $\hat{F}$.

5   At this point we consider whether the sliding-F management rule needs to be applied, based on estimated stock status. An F-multiplier (F^m) is estimated that results in $F_{y+1} = F_{target}$ when applied to $\hat{F}_{y-1}$. A short-term (3-year) forecast is carried out on the basis of this value of fishing mortality, using year $y - 1$ as the starting point, and the resultant spawning stock biomass $\hat{B}_{y+2}$ in the year following the quota year is generated. If $\hat{B}_{y+2} < B_{pa}$, the sliding-F rule is applied to generate a new F_{target} and the forecast procedure is repeated to produce a new $\hat{B}_{y+2}$. This may imply a different F_{target}, in which case the procedure is repeated until the difference between subsequent values of F_{target} is less than a prespecified iteration tolerance. This iteration nearly always converges: if $\hat{B}_{y+2}$ flips between a value above B_{pa} and a value below B_{lim} (which can happen if B_{pa} and B_{lim} are close together), then the average of F_{target} and $F_{bycatch}$ is used.

If the final $\hat{B}_{y+2} > B_{pa}$, the TAC constraint is applied ($\Delta TAC = \pm 15\%$ in this case). The implied intended landings yield $\tilde{Y}_{y+1}^l$ is compared with $\tilde{Y}_y^l \pm \Delta TAC$. If $\tilde{Y}_{y+1}^l$ is within this range, then the intended yield is set to that which is implied by $\tilde{F}_{y+1} = \hat{F}_{y-1} \times F^m$. On the other hand, if the implied yield $\tilde{Y}_{y+1}^l$ from the original forecast is *not* within the bounds specified by ΔTAC, then $\tilde{Y}_{y+1}^l$ is set to $\tilde{Y}_y^l \pm \Delta TAC$ as required (thus implementing the TAC constraint).

A series of low recruitments can lead to an exponential (and irreversible) increase in F as our virtual managers try to ensure that the full TAC is taken.

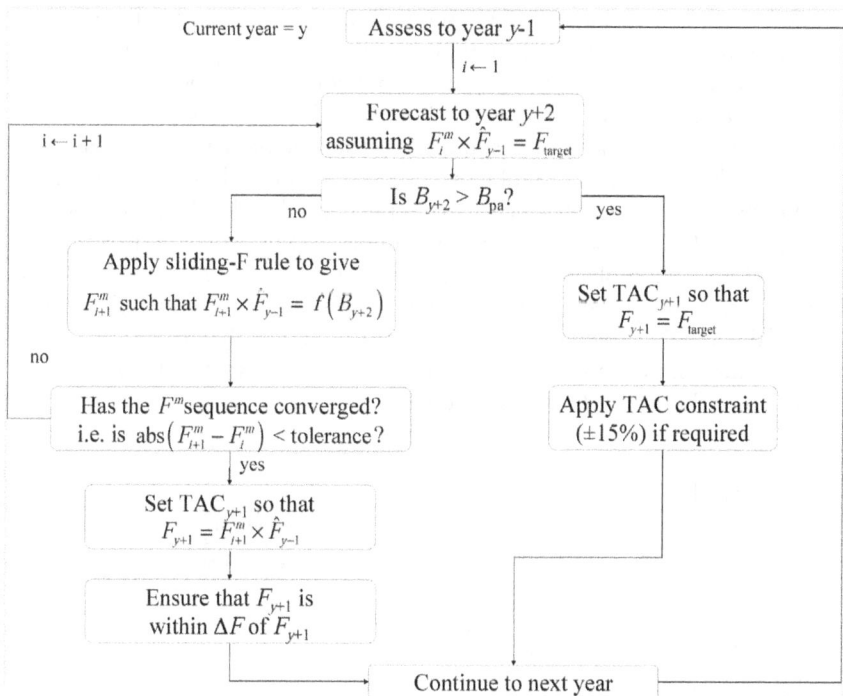

Figure 12.4 Flowchart outlining the key points of the North Sea haddock management plan. F_i^m is the F-multiplier applied during the ith iteration of the
sliding-F rule; $\hat{F}_y$ a limit on the permitted interannual change in F (currently set to ±25%); and $f(B_{y+2})$ is shorthand for "the sliding-F rule applied
to biomass in the year $y + 2$."

To prevent this, a limit ΔF is stipulated on interannual change in F. If $F^m > 1.0$
$+ \Delta F$ then F^m is set to $1.0 + \Delta F$. Similarly, if $F^m < 1.0 - \Delta F$ then F^m is set to
$1.0 - \Delta F$.

The process is summarised in Figure 12.4. Although it is quite complicated, the
output from the evaluation is simple: the intended landings yield (or TAC), $\tilde{Y}_{y+1}^l$.

6   With management decisions now determined, the y-loop carries on to the
     start of the next year. The quota year from the previous y-iteration now
     becomes the intermediate year, and effect of fishing on the stock is now
     largely determined by the intended yields. Once the y-loop is completed,
     the simulation begins again with the next k-loop and a different time series
     of recruitments, and subsequently the next F_{target} loop.

Results

Figure 12.5 summarises the model output for a single iteration for $F_{target} = 0.3$
(50 such iterations comprised the full analysis for a particular F_{target}). In this

iteration, the effect of the ±15% TAC constraint is clear in the time series of intended landings (i.e. quota or TAC; see Figure 12.5a). Following the sliding-F rule, the interannual change in TAC deviates from the ±15% limits only for 5 years between 2018 and 2025 (Figure 12.5f), which correspond to years for which *assessed* biomass falls below B_{pa} (Figure 12.5c). However, in these years (and, indeed, in most years in the simulation), the true biomass was higher, and this reveals the effect of the problem with discard modelling that was mentioned earlier. As fishing mortality is maintained at a low level, biomass begins to rise; but the TAC constraint means that landings do not rise commensurately. In reality this would probably lead to increased discarding rate for younger ages, all else being equal, but that cannot happen in this model for which fixed proportions discarded at each age have been stipulated. For this reason, the catch (landings plus discards) on which the assessment is based is lower than it should be, and hence the assessed biomass is lower.

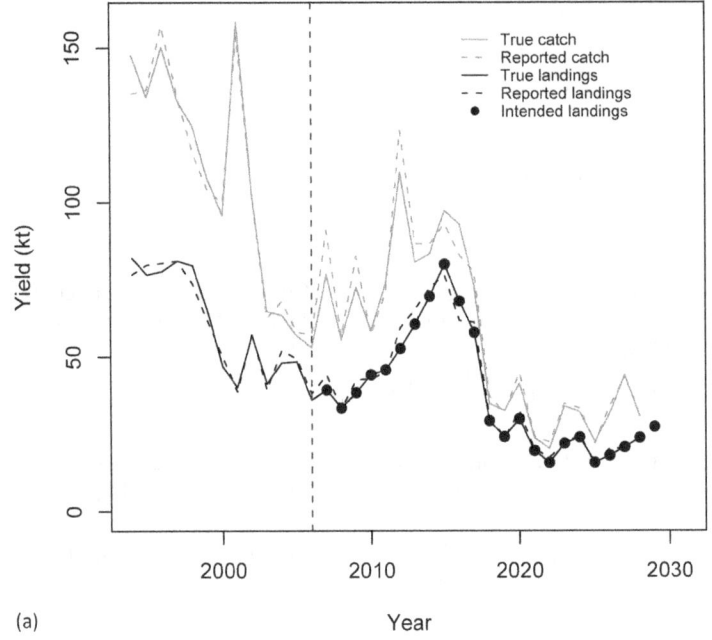

(a)

Figure 12.5 Summary plots of a single simulation iteration for North Sea haddock, with F_{target} = 0.3. In each plot the vertical dashed line delineates the last historical year. Each grey line in the top four plots shows the assessment result for 1 year in the future simulation. (a) Yield. (b) True (black) and assessed (grey) fishing mortality F, with dashed horizontal lines indicating F_{lim} (upper) and F_{pa} (lower). (c) True (black) and assessed (grey) spawning stock biomass B, with dashed horizontal lines indicating B_{pa} (upper) and B_{lim} (lower). (d) True (black) and assessed (grey) recruitment to the fished stock. (e) Comparison of frequency distribution of historical and simulated recruitment. (f) The percentage interannual change in quota (TAC), with dashed horizontal lines showing the ±15% level.

Source: Needle (2008b).

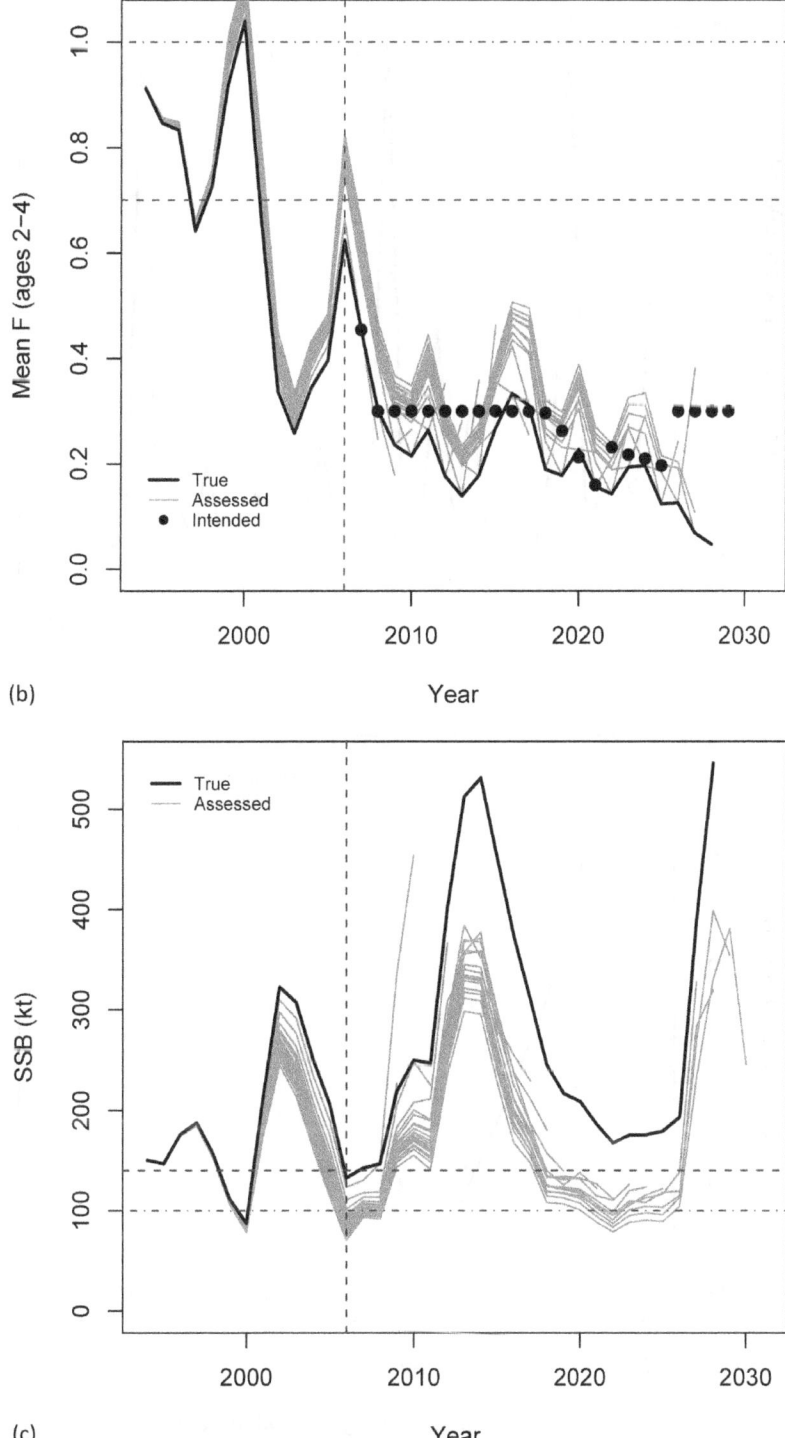

(b)

(c)

Figure 12.5 (Continued)

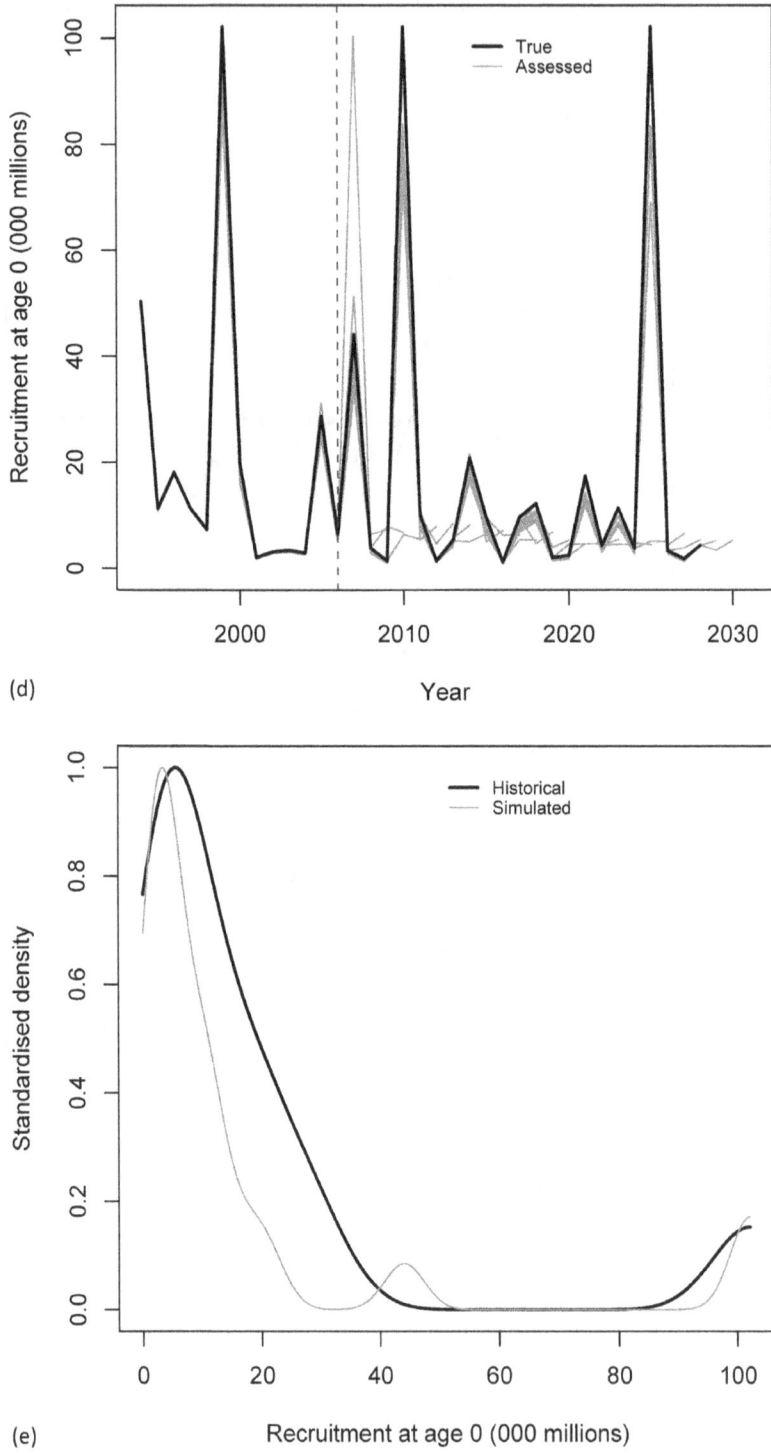

(d)

(e)

Figure 12.5 (Continued)

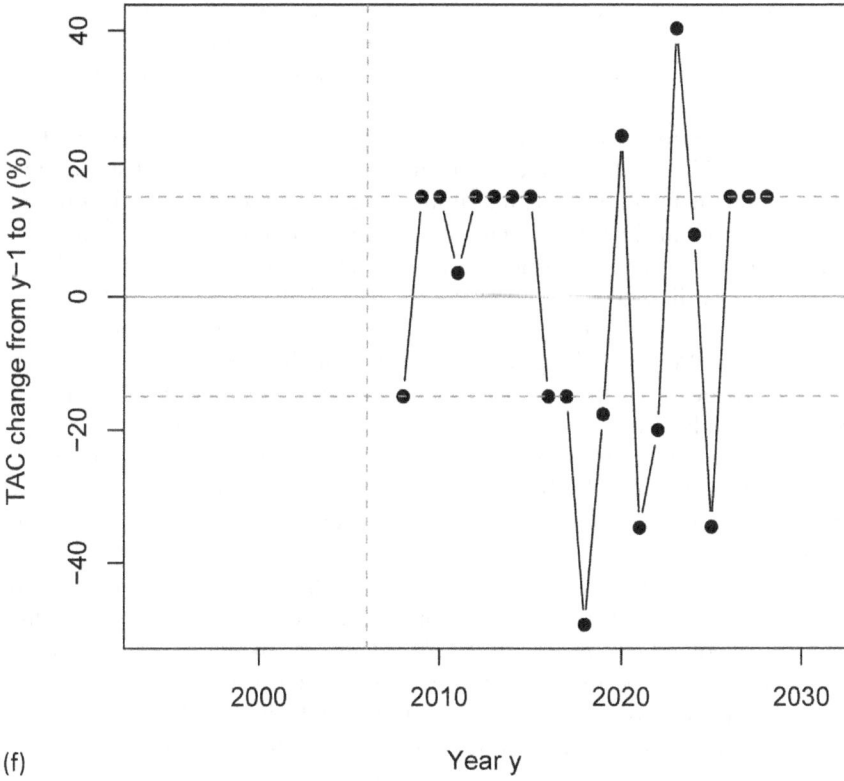

(f)

Year y

Figure 12.5 (Continued)

The data from the simulated survey series did not have this bias, but the assessment model used (FLXSA, an implementation of XSA) is largely driven by catch data with surveys playing a calibrative role (Darby and Flatman 1994). Extensive testing during model development showed that none of the FLXSA settings (shrinkage and so on) could ameliorate the effect. A higher F_{target} or the removal of the TAC constraint did reduce the problem: however, these were not part of the procedure under evaluation. The underestimation also has consequences for fishing mortality, which was mostly estimated to be higher than it really was. The result was a management procedure which was actually more conservative than it needed to be.

TAC increases of greater than 15% are possible (Figure 12.5f). This will happen when a low assessment of spawning stock biomass is combined with a large incoming year-class. The management procedure operates on the basis of assessed spawning stock biomass, which remains low while the fish are young, so the constraint on interannual TAC variation does not apply. At the same time, the abundant young fish contribute to a higher quota forecast. The TAC must therefore increase by more than 15% if the management procedure is to be followed. This result seems contradictory in a situation of low biomass, but is inevitable if the management procedure is implemented as written.

The 50 iterations carried out with target F_{target} = 0.3 are summarised in Figure 12.6. The median values from these plots are the result of smoothing across different realisations of recruitments, and are therefore useful only as an *indication* of likely future events. The median outcome itself is not at all likely, given that each recruitment time series always has two large year-classes which are not directly reflected in the median. Median landings yield falls to a low level as the 1999 year-class is exhausted, before rising to a steady state of around 45 kt. Median fishing mortality is, for much of the time series, much lower than the F_{target} that the management procedure should provide; this is due to a combination of the ±15% TAC constraint and the simple discard model mentioned earlier. Median biomass remains stable for 4 years or so, before rising swiftly and rebounding to a steady state well above B_{pa}. Finally, the median recruitment is low, but the occasional large year-classes are also evident as outliers.

As well as medians, Figure 12.6 also indicates the spread of possibilities, and on this basis we can examine the risk of occurrence of unwanted events. One such event would be biomass falling below B_{pa} or B_{lim} which is what the management procedure is attempting to avoid. We can estimate this risk by counting the number of years in a given iteration for which $B < B_{pa}$ or $B < B_{lim}$. If we denote the spawning stock biomass in year y in iteration k of a simulation using $F = F_{target}$ by $B_{y,k,F}$, allow B_{ref} to stand for B_{pa} or B_{lim} as appropriate, and use an indicator function $K_{y,k,F}$ such that

$$K_{y,k,F} = \begin{cases} 1, & B_{y,k,F} < B_{ref} \\ 0, & B_{y,k,F} \geq B_{ref} \end{cases}$$

then the required risk is given by

$$\text{Risk}_{k,F} = \sum_{y} K_{y,k,F}.$$

The distributions and central tendencies of $\text{Risk}_{k,F}$ are then used to indicate the degree of risk associated with each management measure (which in this case means each value of F_{target}).

The risk estimates for each F_{target} are summarised in Figure 12.7, which considers the risk of both $B < B_{pa}$ and $B < B_{lim}$. The distributions of $\text{Risk}_{k,F}$ for both have had loess smoothers passed through them, to give an indication of the central tendency of risk. On the basis of these smoothers, the number of years for which $B < B_{lim}$ ranges from 0.26 years (which is 1.18% of the total) to 1.90 years (8.64%), while the values for $B < B_{pa}$ range from 1.73 years (7.86%)

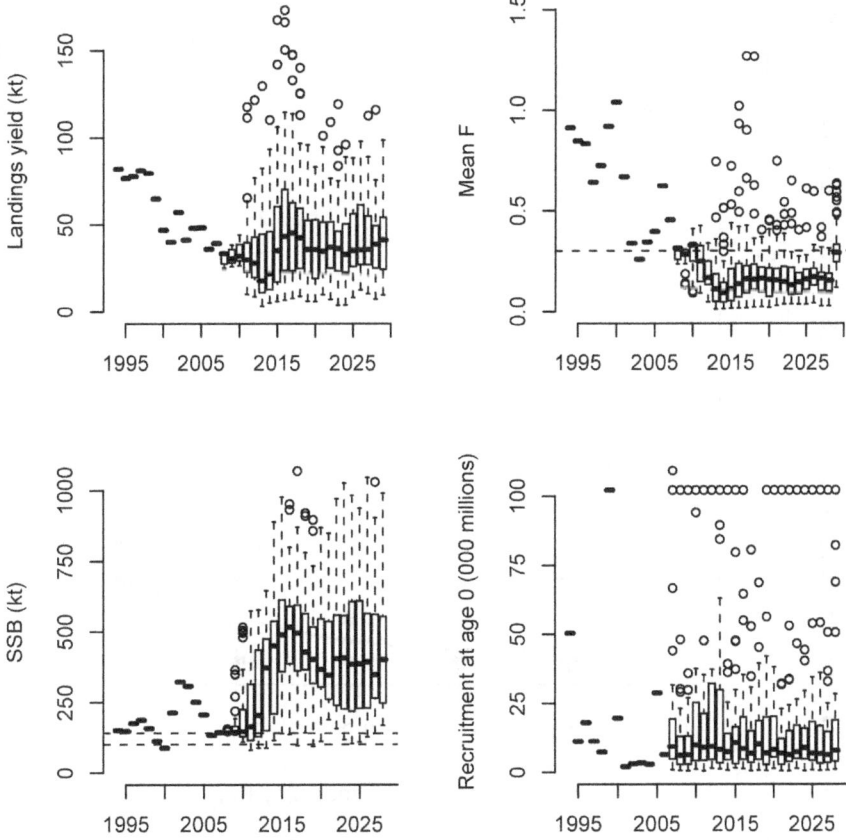

Figure 12.6 Summary plots of 50 simulation iterations, with F_{target} = 0.3. The short horizontal lines indicate the medians, and the boxes the quartiles (25%ile and 75%ile). The lower whisker gives the value of 25%ile − (1.5 × (75%ile − 25%ile)) and the upper whisker gives 75%ile + (1.5 × (75%ile − 25%ile)). Outliers beyond this range are shown by open circles. The dashed horizontal line on the top right plot shows F_{target}, while those on the bottom left plot show B_{pa} (upper) and B_{pa} (lower). Historical estimates (pre-2007) are shown as short horizontal lines only.

Source: Needle (2008b).

to 4.32 years (19.64%). That is, the risk of spawning biomass being below the limit reference point for the next 22 years, given the assumptions of the model, remains less than 10% for values of F_{target} as high as 0.5. The results for F_{target} = 0.3, the value stipulated in the management procedure, are 0.46 years (2.10%) for $B < B_{lim}$ and 2.35 years (10.70%) for $B < B_{pa}$.

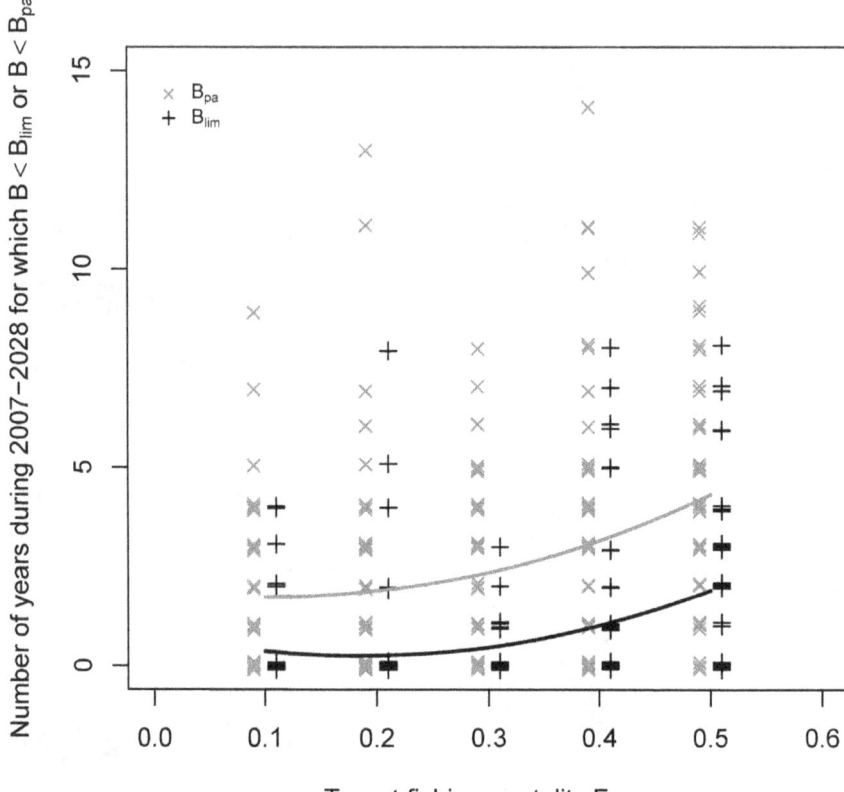

Figure 12.7 Summary of risk of $B < B_{pa}$ (grey) and $B < B_{lim}$ (black) for different values of F_{target} (over 50 iterations for each value). The correspondingly coloured solid lines show the fits of loess smoothers to the full time series of risk estimates. Small random perturbations have been applied to the vertical position of each cross to improve visualisation. Risk is defined as the number of years in each simulation for which spawning stock biomass of $B < B_{pa}$ or $B < B_{lim}$ as appropriate.

Source: Needle (2008b).

Fishery implementation of feedback control

The 2006 and subsequent reports to ICES (Needle 2006a, 2006b, 2008b) con-cluded that the EU-Norway management procedure for North Sea haddock was likely to be sustainable – that is, there was a relatively low risk of bio-mass falling below specified biomass reference and limit points, given reason-able assumptions about growth, fishery selectivity and recruitment. The advice provided by ICES in 2007 (for the fishery in 2008) was based on the management procedure, and this continued to be the case until 2013. During a benchmark assessment meeting early in 2014 (ICES-WKHAD 2014), ICES determined that the haddock stocks in the North Sea and Skagerrak (Subarea IV and Division IIIa; see Figure 12.1) could not be considered to be biologically distinct from the stock in the west of Scotland (Division VIa). One consequence of this is that the extant North Sea haddock management procedure could no longer be applied, as it was designed for the management of a stock that (de facto) no longer existed. Catch advice for 2015 has therefore been based on MSY considerations (ICES-ACOM 2014), and a new evaluated management procedure will be required for the future.

In the meantime, a joint EU-Norway request for advice was made during September 2014 to ask to "evaluate the performance of the [procedure] in achieving its stated objective of providing for sustainable fisheries with high and stable yields in conformity with the precautionary approach" (ICES-ACOM 2014, section 6.2.3.5, page 1). This did not require a formal simulation analysis, but gave an opportunity to consider how successful the procedure had been in achieving the stated objectives. We concluded (Needle 2014, ICES-ACOM 2014) that:

- During 2008–2013, SSB had been maintained above the B_{pa} reference point.
- Fishing mortality F_{2-4} declined strongly after 2000, preceding the MP. However, the average of $\bar{F}_{2-4}$ during 2008–2013 was 0.27, which is below the target F in the MP, and at the lower end of F values considered consistent with F_{MSY}.
- The ICES advice followed the MP in each of the years 2008–2013. However, never more than 75% of the advised TAC was actually taken during those years, so it is clear that factors other than the MP itself helped to maintain a sustainable fishery.
- The interannual change in advised landings never exceeded ±15%, and the change in realised landings was also limited to ±20%, so yield stability was achieved.
- The MP also called for improvement in selectivity and exploitation patterns. Discard rates did decline significantly during the application of the MP, from 48% of the catch in weight in 2007 to 12% in 2013. However, recruitment was poor following the reasonable 2009 year-class, and it seems likely that low numbers of young fish contributed more to reduced discarding than did the MP itself.

The analysis presented in various papers from 2006 onwards (Needle 2006a, 2006b, 2008b) was one of the first FLR-based management strategy evaluations, and as such exemplified how procedures could be evaluated in this way. Subsequent work included the application of the approach to interannual quota flexibility (Needle 2008a), as well as MP evaluations for haddock in the west of Scotland (Division VIa; Needle 2009, 2010a, 2010b) and Rockall (Division VIb; Needle and Mosqueira 2011): a further overview is given in Needle (2012). These analyses improved in efficiency and interpretability as the functions available in FLR were developed, so that many more iterations are now possible than were available for the North Sea MSE, with consequent improved coverage of the solution space. There will be a requirement in the near future for an evaluation of a new EU-Norway management procedure for the new Northern Shelf haddock stock (ICES-WGNSSK 2014), and the approach outlined here is likely to continue to prove its worth for scientists, stakeholders and policy-makers.

Table 12.1 Operating model dynamics.

Recruitment	$R_y = \begin{cases} R_y^{high} & y = y_1, y_2 \\ R_y^{low} & \text{otherwise} \end{cases}$	T12.1
Numbers at age	$N_{0,y} = R_{y-1}$ $N_{a,y} = N_{a-1,y-1} \exp\left(-F_{a-1,y-1} - M_{a-1,y-1}\right)$ for $0 < a < 10$ $N_{a,y} = N_{a-1,y-1} \exp\left(-F_{a-1,y-1} - M_{a-1,y-1}\right) + N_{a,y-1} \exp\left(-F_{a,y-1} - M_{a,y-1}\right)$ for $a = 10$	T12.2
Catch at age	Total catch: $\mathbf{C}^c = C_{a,y}^c = \dfrac{\tilde{F}_{a,y} N_{a,y} \left(1 - \exp\left(-\tilde{F}_{a,y} - M_{a,y}\right)\right)}{\tilde{F}_{a,y} + M_{a,y}}$ Landings: $\mathbf{C}^l = C_{a,y}^l = \rho_a^l C_{a,y}^c$ Discards: $\mathbf{C}^d = C_{a,y}^d = \rho_a^d C_{a,y}^c$	T12.3
Biomass	$B_y = \sum_a N_{a,y} W_{a,y} \text{Mat}_{a,y}$	T12.4
Yield	$\mathbf{Y}^c = Y_y^c = \sum_a C_{a,y}^c W_{a,y}^c$ $\mathbf{Y}^l = Y_y^l = \sum_a \rho_a^l C_{a,y}^c W_{a,y}^l$ $\mathbf{Y}^d = Y_y^d = \sum_a \rho_a^d C_{a,y}^c W_{a,y}^d$	T12.5
Proportionate landings	$\rho_a^l = \dfrac{1}{3} \sum_{y=2004}^{2006} \dfrac{C_{a,y}^l}{C_{a,y}^l + C_{a,y}^d}$	T12.6
Proportionate discards	$\rho_a^d = 1 - \rho_a^l$	T12.7

Table 12.2 Simulated observational data for assessment model fit.

Survey	$I_{i,a,y} = q_{i,a} N_{a,y}^{p_{i,a}} \exp\left(\varepsilon_{a,y}^i\right)$	T12.8
Catch	$\hat{C}^c = \hat{C}_{a,y}^c = C_{a,y}^c \exp(\varepsilon_{a,y}^f)$	T12.9
	$\hat{C}^l = \hat{C}_{a,y}^l = \rho_a^l \hat{C}_{a,y}^c$	
	$\hat{C}^d = \hat{C}_{a,y}^d = \rho_a^d \hat{C}_{a,y}^c$	

Table 12.3 Parameter inputs for simulated operating model dynamics.

Natural mortality	$\mathbf{M} = M_{a,y} = (2.05, 1.65, 0.40, 0.25, 0.25, 0.20, \ldots)$ for all y and $a = 0, \ldots, 10$	T12.10
Maturity	$\mathbf{Mat} = \text{Mat}_{a,y} = (0.00, 0.01, 0.32, 0.71, 0.87, 0.95, 1.00, \ldots)$ for all y and $a = 0, \ldots, 10$	T12.11
Proportion F and M before spawning	Prop $F = 0.0$ Prop $M = 0.0$	T12.12
Recruitment	$\ln\left(R_y^{low}\right) \sim N\left(\bar{R}^{low}, CV\left(R_y\right)\right),$	T12.13
	$\bar{R}^{low} = \exp\left(\dfrac{1}{10} \displaystyle\sum_{y=95,\ldots,98,00,\ldots,05} \ln\left(R_y\right)\right)$	
	$CV\left(R_y\right) = \text{sd}\left(\ln R_{95}, \ldots, \ln R_{98}, \ln R_{00}, \ldots, \ln R_{05}\right).$	
	$y_1 \sim U(2006, 2013)$	
	$y_2 \sim U(2016, 2023)$	
	$R_{y_1, y_2}^{high} \sim N\left(R_{99}, 0.1 R_{99}\right)$	
Survey standard error	$\sigma_i = 0.2$	T12.14
Catch standard error	$\sigma_c = 0.1$	T12.15
Selectivity	$\zeta_a = \dfrac{1}{3}\left(\hat{F}_{a,y-3} + \hat{F}_{a,y-2} + \hat{F}_{a,y-2}\right)$	T12.16
First assessment year	$\gamma = 2007$	T12.17
Limit in TAC change	$\Delta TAC = 0.15 (\Rightarrow \pm 15\%)$	T12.18
Target F	$F_{target} \; \varepsilon \; (0.1, 0.2, 0.3, 0.4, 0.5)$	T12.19
Bycatch F	$F_{bycatch} = 0.1$	T12.20
Limit in F change	$\Delta F = 0.25 (\Rightarrow \pm 25\%)$	T12.21
Biomass reference points	$B_{lim} = 100kr, B_{pa} = 140 \text{ kt}$	T12.22

References

Beddington, J.R. & May, R.M. 1977. Harvesting natural populations in a randomly fluctuating environment. *Science*, 197, 463–465.

Brooks, E.N., Sutherland, S.J., Van Eeckhaute, L. & Palmer, M. 2012. Groundfish Assessment Updates 2012: Georges Bank haddock. Available at www.nefsc.noaa.gov/publications/crd/crd1206/gbhaddock.pdf (last accessed 10 November 2014).

Butterworth, D.S. & Punt, A.E. 1999. Experiences in the evaluation and implementation of management procedures, *ICES Journal of Marine Science,* 56, 985–998.

Coull, J.R. 1997. *The Sea Fisheries of Scotland: A Historical Geography,* John Donald Publishers, Edinburgh.

Darby, C.D. & Flatman, S. 1994. *Lowestoft VPA Suite Version 3.1 User Guide,* Lowestoft, MAFF.

FLR Team 2006. *FLR: Fisheries modelling in R. Version 1.2.1. Initial design by L.T. Kell and P. Grosjean.* Available at www.flr-project.org/. Last accessed 10 November 2014.

Fogarty, M.J., Myers, R.A. & Bowen, K.G. 2001. Recruitment of cod and haddock in the North Atlantic: A comparative analysis, *ICES Journal of Marine Science,* 58, 952–961.

Froese, R., Branch, T.A., Proelß, A., Quaas, M., Sainsbury, K. & Zimmermann, C. 2011. Generic harvest control rules for European fisheries, *Fish and Fisheries,* 12(3), 340–351.

Hillary, R. 2009. An introduction to FLR fisheries simulation tools. *Aquatic Living Resources,* 22, 225–232.

Hislop, J.R.G. 1996. Changes in the North Sea gadoid stock. *ICES Journal of Marine Science,* 53, 1146–1156.

Holt, C.A. & Peterman, R.M. 2006. Missing the target: Uncertainties in achieving management goals in fisheries on Fraser River, British Columbia, sockeye salmon (*Oncorhynchus nerka*), *Canadian Journal of Fisheries and Aquatic Science,* 63, 2722–2733.

ICES 2014. EU–Norway request to evaluate the performance of the long-term management plan for North Sea haddock. Special request from November 2014. ICES Advice 2014, Book 6.

ICES-ACOM 2006. Report of the ICES Advisory Committee on Fishery Management and Advisory Committee on Ecosystems (October). ICES Advice, Books 1–10.

ICES-ACOM 2014. ICES Advice 2014. Available at www.ices.dk/community/advisory-process/Pages/Latest-Advice.aspx (last accessed 20 November 2014).

ICES-AFWG 2014. Report of the Arctic Fisheries Working Group (AFWG), 2014, Lisbon, Portugal. ICES CM 2014/ACOM:05.

ICES-NWWG 2014. Report of the North-Western Working Group (NWWG), 24 April–1 May 2014, ICES HQ, Copenhagen, Denmark. ICES CM 2014/ACOM:07.

ICES-WGIBTS 2014. Report of the International Bottom Trawl Survey Working Group (IBTSWG). ICES Headquarters, Copenhagen.

ICES-WGMG 2006. Report of the Working Group on Methods of Fish Stock Assessment, Galway, June 2006. ICES CM 2006/RMC:07.

ICES-WGNSSK 2006. Report of the ICES Working Group on the Assessment of Demersal Stocks in the North Sea and Skagerrak. ICES CM 2006/ACFM:35.

ICES-WGNSSK 2007. Report of the ICES Working Group on the Assessment of Demersal Stocks in the North Sea and Skagerrak. ICES CM 2007/ACFM:18.

ICES-WGNSSK 2013. Report of the Working Group on the Assessment of Demersal Stocks in the North Sea and Skagerrak (WGNSSK), 24–30 April 2013, ICES Headquarters, Copenhagen. ICES CM 2013/ACOM:13.

ICES-WGNSSK 2014. Report of the Working Group on the Assessment of Demersal Stocks in the North Sea and Skagerrak. ICES CM 2014/ACOM:13.

ICES-WKHAD 2014. Report of the ICES Benchmark Workshop for Northern Haddock Stocks. Marine Laboratory, Aberdeen, 27–29 January 2014 and ICES Headquarters, Copenhagen, 24–28 February 2014. ICES CM 2014/ACOM:41.

Kell, L. 2011. *FLXSA: eXtended Survivor Analysis for FLR.* R package library, version 2.0.

Kell, L.T., Mosqueira, I., Grosjean, P., Fromentin, J.-M., Garcia, D., Hillary, R., Jardim, E., Mardle, S., Pastoors, M., Poos, J.J., Scott, F. & Scott, R. 2007. FLR: An OpenSource framework for the evaluation and development of management strategies, *ICES Journal of Marine Science,* 64, 640–646.

Needle, C.L. 2006a. Evaluating harvest control rules for North Sea haddock using FLR. Working Paper for the ICES Working Group on Methods of Stock Assessment, Galway, Ireland, 21–26 June 2006.

Needle, C.L. 2006b. Revised FLR-based evaluation of candidate harvest control rules for North Sea haddock. Working paper for the ICES Advisory Committee for Fisheries Management, Copenhagen, October 2006.

Needle, C.L. 2008a. Evaluation of interannual quota flexibility for North Sea haddock: Final report. Working paper for the ICES Advisory Committee (ACOM), September 2008.

Needle, C.L. 2008b. Management strategy evaluation for North Sea haddock. *Fisheries Research,* 94(2), 141–150.

Needle, C.L. 2009. An evaluation of a proposed management plan for haddock in Division VIa. Working paper to ICES ACOM.

Needle, C.L. 2010a. An evaluation of a proposed management plan for haddock in Division VIa (2nd edition). Working paper to ICES ACOM.

Needle, C.L. 2010b. Management strategy evaluation for VIa haddock: An additional TAC constraint option. Working paper to ICES ACOM and EC STECF.

Needle, C.L. 2012. *Fleet Dynamics in Fisheries Management Strategy Evaluations,* PhD thesis, University of Strathclyde, Glasgow.

Needle, C.L. 2014. Response to Joint EU-Norway request to ICES to evaluate the performance of the long-term management plan for North Sea haddock. Working paper to ICES Advisory Committee. Available from the author (needlec@marlab.ac.uk).

Needle, C.L. & Mosqueira, I. 2011. An evaluation of a proposed management plan for haddock in Division VIb (Rockall). Working paper to the ICES Advisory Committee, August 2011.

Quinn, T.J., II, Fagen, R. & Zheng, J. 1990. Threshold management policies for exploited populations. *Canadian Journal of Fisheries and Aquatic Science,* 47, 2016–2029.

R Development Core Team 2011. *R: A language and environment for statistical computing,* R Foundation for Statistical Computing, Vienna, Austria. Available at www.R-project.org (last accessed 20 November 2014).

Rademeyer, R.A., Plagányi, É.E. & Butterworth, D.S. 2007. Tips and tricks in designing management procedures. *ICES Journal of Marine Science,* 64, 618–625.

Rochet, M.-J. & Rice, J.C. 2009. Simulation-based management strategy evaluation: Ignorance disguised as mathematics?, *ICES Journal of Marine Science,* 66, 754–762.

Rosenberg, A.A., Fogarty, J., Sissenwine, M.P., Beddington, J.R. & Shepherd, J.G. 1993. Achieving sustainable use of renewable resources. *Science,* 262, 828–829.

Scottish Government 2012. Scottish Sea Fisheries Statistics 2011. Available at www.scotland. gov.uk/Publications/2012/09/1840/downloads (last accessed 14 October 2014).

Shepherd, J.G. 1992. Extended Survivors' Analysis: An improved method for the analysis of catch-at-age data and catch-per-unit-effort data. Working Paper to the ICES Multispecies Working Group, June 1992.

Walters, C.J. 1975. Optimal harvest strategies for salmon in relation to environmental variability and uncertain production parameters. *Journal of the Fisheries Research Board of Canada*, 32(10), 1777–1784.

Part III

Perspectives on fisheries management science

Fisheries management science continues to evolve rapidly. The case studies in Part 2 provide an introduction to how it is currently practiced. Part 3 provides some perspectives on how management science is developing and what aspects are likely to be at the forefront of future work.

To begin, Chapter 13 addresses an important dichotomy between model-based and empirical (model-free) management procedures (MPs), which can be differentiated by whether or not they include an estimator of stock status based on an assessment model. Both approaches have advantages, which are discussed, before the chapter explores how a fusion of the best features of each might prove a fruitful area of future research.

Chapter 14 extends management strategy evaluation (MSE) to an unusual context, providing an example of how a harvest control rule can be used to limit bycatch of an endangered turtle species by the Californian drift gillnet fishery off the west coast of the United States. Because bycatch must be estimated from observer data, this evaluation framework includes an additional layer of complexity that extends the standard MSE approach, and could make it applicable to a variety of other applications in which we cannot assume that the catch is known.

One persistent criticism of fisheries science as it is currently practiced is its failure to account for the wider ecosystem. This limitation is reviewed in Chapter 15 within the context of fisheries management. The chapter gives examples of how ecosystem and multispecies considerations can be included in MSE, primarily through modifications to the operating model (OM). Realism of the OM is a cornerstone of MSE, and Chapter 16 continues this theme by reviewing the use of spatially explicitly models in this context. Because most stock assessment models that inform fishery management are not spatially explicit, management actions are often not matched to the spatial scale of assessment model outputs. This can have unintended and potentially deleterious consequences. Spatially explicit OMs that simulate the impacts of management measures at finer scales can be used to mitigate unforeseen localized depletions or to achieve optimum yield from substock abundance.

Fisheries management science is performed at the interface of science with policy. This represents a challenge for both scientists and policy-makers, and Chapter 17 deals with an important aspect of this interface, namely the communication of risk. The risk associated with different management options is generally of fundamental concern for fisheries managers, and there has been a concerted effort by the international tuna management organisations to develop a consistent method for communicating this result. This chapter reviews the concept of risk in fisheries management and presents a framework for how it can be communicated effectively.

Finally, in a book that has focused on simulation-based management science, Chapter 18 describes some of the fundamental limitations of this approach, proposes some alternatives, and gives steps to ensure that fisheries management science is always conducted in a rigorous and policy-relevant manner.

## Chapter 13

Empirical and model-based control rules

Richard M. Hillary

Introduction

From their first incarnation within the International Whaling Commission (IWC), management procedures (MPs) were model based, essentially being stock assessment models with associated harvest control rules (HCRs) (de la Mare 1986). The term *model based* means that observed data are fitted within a statistical framework that assumes a population dynamics model with associated parameters. The outputs of that fitting process (e.g. abundance, productivity, fishing mortality and trends thereof) are then used in an HCR (as either parameters or driving indices). However, more recently there has been a move towards empirical MPs (Geromont et al. 1999, De Oliveira and Butterworth 2004, de Moor et al. 2011) where derived statistics and trends from the raw observations are used in the HCR. In this chapter I will briefly cover the history of model-based and empirical MPs, in both the literature and actually implemented, as well as the pros and cons of their utility within a full MSE framework. Finally, including worked examples, I will discuss how recent advances in both stock assessment and MSE suggest that a fusion of both approaches has real potential for future MPs.

A brief history of management procedures and management strategy evaluations

The first MP – adopted in 1974, though not explicitly evaluated until much later – was model based and developed in the IWC, called the New Management Procedure (NMP) (Punt and Donovan 2007). It included an HCR designed to set quotas according to the biomass depletion status of the stock as estimated by a Pella-Tomlinson production model. Concerns around how exactly the parameters required to run the MP were to be estimated, and other issues around defining stock units, led to the first instance of what is now described as MSE by de la Mare (1986), where an example implementation of the NMP was found to perform poorly.

This initial evaluation of the NMP led to the IWC, alongside the already declared moratorium on whaling in 1982, initiating a process of MP testing and development in the first fully described MSE framework (including the setting of objectives and a wide stakeholder engagement base). Development of the Revised Management Procedure (RMP) has, arguably, set the general framework for all subsequent MSE work. After an extensive period of testing of a number of candidate MPs a final Catch Limit Algorithm (CLA), at the core of the RMP, was selected and adopted (Cooke 1999). The CLA is model based and structured as follows:

- *Stock assessment model*: modified Schaefer production model with a quadratic surplus production term.
- *Input data*: abundance indices (both relative and absolute) combined into one representative index of stock abundance.
- *Estimation framework*: a Bayesian approach is used with informative priors for the assessment model parameters alongside the information from the (combined) abundance index.
- *HCR*: the TAC is basically a function of the HCR parameters plus the biomass and parameter estimates from the stock assessment model. The HCR parameters could be adjusted (tuned) during the MP development phase.

While there are other elements to the RMP, such as stock definition (Punt and Donovan 2007), the CLA structure defined earlier is general enough to define the main elements of all model-based MPs.

Given the involvement of South African scientists in the IWC RMP development process, a model-based MP approach was explored for a number of South African fisheries such as anchovy (Butterworth and Bergh 1993) and hake (Punt 1993). One of the more interesting and perhaps pivotal findings from Punt (1993) was that, contrary to the prevailing wisdom of the time, more complex and realistic models did not necessarily perform better than simpler models when you actually simulated their implementation. The work summarised in the paper found that VPA methods that included more biological realism actually performed poorly in comparison to a simple production model in relation to variability in the TAC, even for similar biomass depletion levels.

Towards the end of the 1990s a number of empirical MPs progressed from the simulation testing arena to the real world as actively adopted MPs for South African fisheries (Geromont et al. 1999). The empirical MPs' relative ease of implementation, allowing more extensive simulation testing, was one of the prime arguments given in their favour, as well as the fact that stakeholders found it easy to understand and interpret how the MPs worked. In addition, it was found that empirical MPs could more often than not achieve the same conservation and economic goals as model-based MPs. This did not mean that work on model-based MPs rapidly disappeared, but it did demonstrate that relatively simple rules can perform as well as – if not better than – more complex model-based stock assessment-type MPs in managing often complex populations and fisheries.

Moving towards the present let me first provide three useful distinctions between different types of MP:

1   Model-based with an agreed HCR which has been evaluated using the MSE approach and adopted by the relevant management authority.
2   Model-based with an agreed HCR which has not been evaluated but is implemented in terms of setting management advice (also termed an implicit MP by Rademeyer et al. [2007]).
3   Empirical with an agreed HCR which has been evaluated and adopted using the MSE approach and adopted by the relevant management authority.

While definitions 1 and 3 are "proper" instances of an MP tested using MSE, number 2 is included to cover what are effectively stock assessments, with clearly defined monitoring data inputs, and that use an agreed HCR (and are distinct from situations where stock assessment results are considered along with other factors in setting management advice). This additional category covers a lot of fisheries that are using a de facto (or implicit) MP, albeit one that has not been tested. Classification number 2 currently covers most North American and (assessed) European fisheries, but also a number of high profile/value fisheries in Australia and New Zealand. In South Africa, it is empirical MPs that dominate (Geromont et al. 1999, De Oliveira and Butterworth 2004, de Moor et al. 2011; see also de Moor and Butterworth (2015) and Chapter 11, this volume) but in Australia there are a number of data-poor stocks that are managed via both empirical and model-based MPs, some of which have been formally evaluated (Smith et al. 2008).

In the context of regional fisheries management organisations (RFMOs), which oversee the management of fisheries in international waters): Antarctic fisheries assessed by the Commission for the Conservation of Antarctic Marine Living Resources (CCAMLR) fall into category number 2; the North Atlantic Fisheries Organisation (NAFO) has implemented an empirical HCR for Greenland halibut (Anonymous 2010); in the Southern Hemisphere the Commission for the Conservation of Southern Bluefin Tuna (CCSBT) implemented what is probably classified as a model-based fully evaluated MP in 2011 (see Hillary et al., Chapter 8, this volume; more on this later in this chapter). But outside of these examples, the dominant approach for RFMOs is stock assessment and management advice via an ad hoc consensus-based approach. Table 13.1 summarises some of the different MP categories for a selection of fisheries.

In summary, fully tested MPs came initially from the model-based world view but, gradually, the use of empirical MPs has become more and more common and arguably is the most populous form of fully evaluated and implemented MP. However, and this was the main reason for including category number 2, the dominant form of MP is both model based and almost exclusively untested: stock assessments with agreed HCRs (presumably because that

Table 13.1 Indicative summary of MP types from across the world's fisheries.

Location	Fishery	MP type(s)
Antarctic (CCAMLR)	toothfish/icefish	2
Australia	SE scalefish and shark	1, 2, 3
Australia	tuna and billfish	3
Europe	assessed fisheries	2
Global (IWC)	whales	1
North Atlantic (NAFO)	Greenland halibut	3
New Zealand	hoki, snapper, roughy	2
New Zealand	lobster	3
South Africa	hake, sardine, anchovy	3
Southern Hemisphere (CCSBT)	southern bluefin tuna	1
US	assessed fisheries	2

is what was always done – in some fashion – and that this approach is considered to be intrinsically "correct"). It is worth reflecting a little on this point, without opening the MP versus assessment debate wider, as I outline the pros and cons of both MP approaches, and whether a middle ground can be explored that allows for the best of model-based approaches whilst still ensuring the models themselves can be properly evaluated.

The arguments for and against

One of the best concepts encompassed by MSE is this: evaluation is not dependent on the form of the candidate MP but how it *actually* performs in the simulations. Nevertheless, while cognizant of this, it is worth exploring some of the potential positives and negatives of both the empirical and model-based approaches, from a broader perspective.

Model-based

Raw observations, even both varied and informative, rarely – if ever – allow direct inference on variables such as absolute biomass, depletion, mortality rates and year-class strength. A model-based approach therefore has advantages related to the inference of stock status. Specifically it can:

- statistically combine different observations;
- provide additional parameters and variables for use within an MP beyond the observations (or quantities derived from them);

- reduce uncertainty in the underlying index or indices by accommodating observation and process errors explicitly;
- provide estimates of uncertainty in the variables used as inputs to the MP.

Consider a very simple MP using mean length as measured from a survey: in the absence of any other information, a decrease in mean length can occur from either a reduction in the larger fish from fishing mortality, or an influx of smaller fish from a larger recruitment event. A model with the requisite structure could tell us which case is likely to be true, and a model-based MP will very likely have better information in terms of what to do about the mean length change (i.e. whether it is a positive or negative indication of future productivity). Another positive is noise reduction: models that actively (or otherwise) account for observation and process error can reduce the variability in the inputs to an MP, relative to the observations. One of the key performance measures in MSE is usually catch stability, so an MP input that is less variable but still follows the right "signal" embedded in the raw index can perform better. Finally, models can deal with missing data quite naturally, whereas empirical approaches have to define ad hoc solutions to, for example, a survey not being done in a given year.

For MP models that are effectively stock assessment models (such as a surplus production or age-structured models), a big issue is our ability to actually test them properly in an MSE context. As computational power has rapidly increased it is now possible to run lower to medium complexity stock assessment models within an MSE framework (see e.g. Hicks et al., Chapter 4, and Cox and Kronlund, Chapter 5, this volume). However, some factors remain, likely permanently, untestable. Anyone who has run an assessment, or at least attended meetings that run and review assessments, will probably recognise the following steps:

- The weighting of various data sets is often done via a mixture of more statistically based arguments and more subjective information from perhaps anecdotal sources or based on expert judgement.
- Numerous sensitivity runs are performed or requested that, while not perhaps becoming base cases themselves, do often alter the settings and structure of the base case.
- Often multiple models are run and results presented as a summary across models with different weightings applied.
- Sometimes, a model may be thrown out due to the inability of obtaining a statistically defensible result.

Realistically, none of the preceding scenarios can be adequately "simulated" within an MSE context. So, while one can run a stock assessment model with a fixed structure (and some approach to dealing with convergence issues) within an MSE framework, an MP that is basically a realistic stock assessment with an agreed HCR – even if the HCR is rigorously adhered to – is, arguably, not fully testable via MSE.

Empirical

A number of the advantages of the empirical approach come from their (generally) innate simplicity, relative to model-based approaches. First, they are easier to understand by a much wider range of stakeholders. In contrast, model-based approaches – especially de facto stock assessments – are often only really understood by a small cadre of scientists involved in the wider assessment group. That simplicity also means empirical MPs are easier to simulate, which makes them faster to test and also permits a wider range of candidate MPs and robustness tests to be explored. Simplicity also makes them applicable to a much wider range of potential stocks – a large number of the world's fisheries still remain unassessed due to both a lack of assessment capacity or just not enough informative data. Empirical MPs explicitly designed and tested for data and capacity poor scenarios such as these offer very tempting ways to manage such fisheries that are often just as valuable as their data-rich, assessed counterparts.

But there are also drawbacks. One problem is noise – in fisheries almost all observations are quite noisy, and with correlated error structures. Bias in an index or observation is a problem for both empirical and model-based approaches, but the innate variability in the observation process cannot easily be dealt with when taking an empirical approach. It's not just observation error either, as there will likely be additional process error contributing to the overall variance in the observations as well. Another issue is missing data – models can deal with instances of missing data, as long as it is not too long. The empirical approach either relies on the continued existence of the key MP input observations, or has to define some kind of mathematical procedure for dealing with missing data that is both sensible and testable.

How about a little of both?

In this section I make the case for a middle ground between what I consider model-based and empirical approaches. Previous model-based approaches have basically been stock assessment models of differing levels of complexity, estimating absolute biomass, fishing mortality rates, year-class strength and so on. Recently, in response to challenges around assessing data-poor stocks, there has been a number of assessment models that are more like biologically plausible filters (Trenkel 2008). They tend to estimate absolute abundance only if the observations are absolute themselves (Trenkel 2008), and impose a very general structure such as a simple Markovian process (Spencer et al. 2013) or more structured delay-difference or autoregressive models (Trenkel 2008, Thompson 2013). These models can take a nonparametric (Hillary 2012) or random-walk (Spencer et al. 2013) type approach to the dynamics with model flexibility rigorously estimated. I use the term "flexibility" to refer specifically to the ability of a model to adjust *at run time* the effective degrees of freedom ("number" of free parameters) which, in turn, influences how well the model can fit the data. The degree of fitting ability is controlled by a variance parameter, which is characteristically estimated by the type of model being discussed here. In terms of how the dynamics are described, while they are nowhere near as complex as many traditional assessment models, they do tend to tackle the problem of model flexibility more rigorously than a lot of the larger integrated assessments.

Within a simple, flexible, but statistically estimated model structure, these kinds of hybrid approaches are looking to find the signal amid the noise. Their simplicity makes them well suited to an MSE context. They can be run efficiently and compactly – in the sense that traditional "human" decision-making processes (discussions on model settings, weightings and what data should be included) in stock assessments don't really apply – meaning they can be implemented within the MSE loop as they would be in the real world. The outputs are mostly just filtered versions of the observations used as inputs. In this sense, and at least for unbiased observations, we can obtain less variable indices that follow the same signal as the raw observations. This means that the algorithmic form of simple empirical HCRs can still be used to convert the filtered model output into a management decision. We may lose some of the wider stakeholder understanding by including statistical estimation into the MP, but the outputs and how they are used are just as easy to understand as their empirical counterparts.

At least for one example, these theoretical benefits proved to be true. For the southern bluefin tuna MP (Hillary et al. (2015), Hillary et al., Chapter 8, this volume) an integrated delay-difference random-effects model of relative abundance was used to filter the recruitment and subadult abundance indices prior to using them as inputs to an empirical-type HCR. In the MSE process it was observed that the model-based approach could meet the same risk criteria as purely empirical MPs with the same input data and very similar HCRs, but achieved higher average catches. This was due to noise reduction: the filtered indices were less variable, and the associated TACs from similar HCRs were therefore also less variable. As a consequence, when tuning to the same rebuilding criteria, higher average catches could be taken as the variability introduced into the stock biomass from TAC changes was lower. In the MSE process model convergence was always achieved, and the settings were always fixed, so the model was easily testable alongside the other empirical MPs. In summary, while the final adopted MP incorporated HCR elements from some of the empirical MPs, the model-based approach was retained since it had demonstrated good performance on the reference operating model (OM) and several robustness tests.

A worked example using Indian Ocean bigeye tuna

To demonstrate the kind of MP that could bridge the gap between traditional model-based and empirical procedures I will use the example of catch biomass and long-line catch per unit effort (CPUE) data for Indian Ocean bigeye tuna. Figure 13.1 shows the bigeye tuna data, illustrating the kind of one-way trip dynamics fairly common in tropical tuna stocks. A potentially useful exercise is an initial rapid appraisal of what kinds of model-based approaches are likely to be useful given the input data available for a candidate MP. Given the data (catch biomass and CPUE) the most likely traditional model-based candidates are a biomass surplus production model or perhaps even an age-structured production model (ASPM), though this would require a lot of assumed values.

One thing we do know fairly well is that, with the one-way trip trends in these data (catches going up, CPUE going down, both fairly monotonically), we cannot really separate absolute abundance from resilience/productivity. From a surplus production model perspective, this makes joint estimation of the intrinsic growth rate r, and carrying capacity K, virtually impossible; for an ASPM, estimating unfished abundance and steepness of the stock-recruit

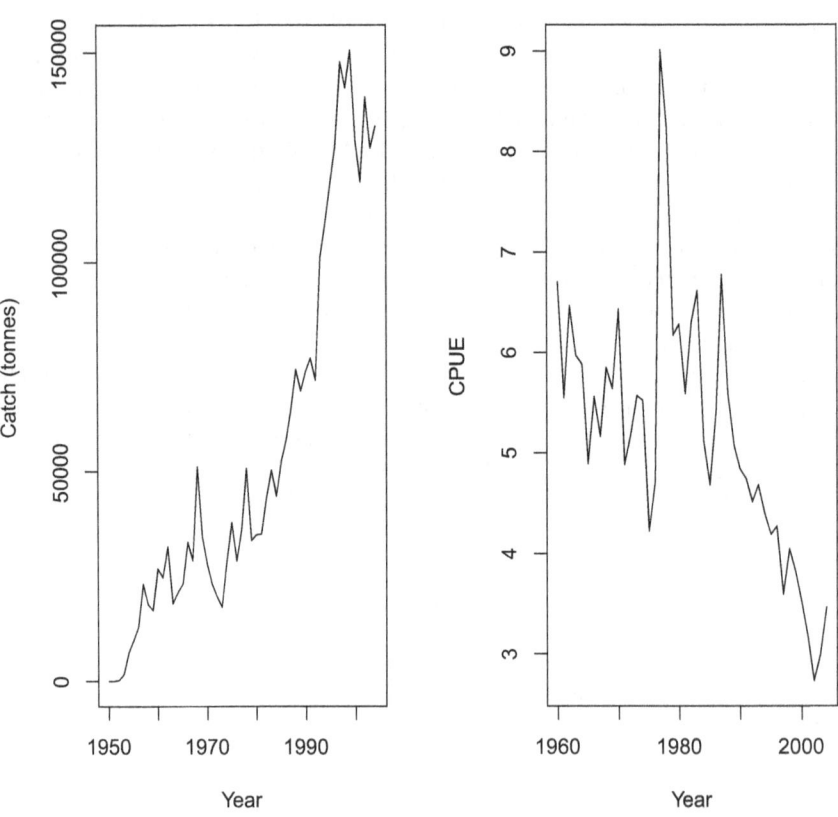

Figure 13.1 Catch biomass (left) and Japanese long-line CPUE (right) for Indian Ocean bigeye tuna.

curve – even with a fixed value of natural mortality – would be equally quix-otic. Using these models requires either fixing or defining strongly informative priors for dimensionless parameters like *r* or steepness – in fact Hillary (2008) demonstrated for this very data set that the prior for *r* is not updated when esti-mating the joint posterior of *r* and *K*. This means that any absolute abundance and maximum sustainable yield (MSY) type information one may obtain from the estimation model will be either totally or very strongly driven by the speci-fied prior – for the logistic surplus production model for example, the fishing mortality and catch at MSY, $F_{MSY} = r/2$ and $C_{MSY} = rK/4$ respectively, both of which are functions of *r*. If one wanted to use these kinds of variables in an HCR, what additional information does the estimation model supply? Given any absolute information on abundance will be equally highly conditional on what we (have to) assume about resilience, one might argue that – for cases such as these – a relative approach that aims to simply extract a trend signal from the input data is more sensible.

The aim here is to implement the same simple MP with input indices defined as follows:

1   The raw standardised long-line CPUE index: I_y;
2   A filtered index estimated using a random–walk model: I_y^{rw};
3   A filtered index estimated using a nonparametric state-space model: I_y^{np}.

The HCR itself is the commonly used log-scale slope in the abundance index (Magnusson and Stefánsson 1989, Geromont et al. 1999, Anonymous 2010):

$$TAC_{y+1} = TAC_y \times (1 + k\lambda),$$ (13.1)

where TAC_y is the catch quota, *k* is a gain parameter controlling how reactive the HCR is, and λ is the slope of the linear trend in the log-scale abundance index for a given time frame of length τ, ending in year *y*. To investigate how each MP might alter the catch I began the process by calculating the catch for the most recent 10 years (1996 to 2007) according to the HCR defined in Equation 13.1, for purely example settings of $k = 1.5$ and $\tau = 7$.

The abundance filter models

I briefly define the underlying estimation models that filter the CPUE index prior to input into the HCR. The random–walk approach assumes that the log-scale "true" index Θ_y evolves according to the following dynamics:

$$\theta_y = \theta_{y-1} + u_y,$$

$$u_y \sim N(0, \sigma_u^2),$$

with initial value θ_0. Using a normal likelihood for θ_y given the observed $\ln(I_y)$, the estimated parameters are θ_0, u_y, and σ_u^2. In the maximum likelihood context this is a simple random-effects model, or in the Bayesian paradigm a hierarchical model – the important factor is that model flexibility, controlled by σ_u^2, is explicitly estimated.

The second model, a nonparametric state-space model first explored in Hillary (2012), also models the log-scale "true" index θ_y but now asserts that there is a functional relationship (via a neural network) between the index and catch biomass in the previous year and the index in the following year:

$$\theta_{y+1} = \theta_y + \sum_{i=1}^{N} \omega_i \varphi\left(\theta_y, c_y, \beta_i, \gamma\right),$$

$$\omega_i \sim N\left(0, \sigma_\omega^2\right),$$

where c_y is the log-scale catch biomass; the number of neurons in the network is N; ω is the vector of weights in the network; and β_i are sensibly distributed nodes in the paired CPUE and catch data at which the network basis function is calculated:

$$\varphi\left(x = \{\theta, c\}, \beta, \gamma\right) = \left(\| x - \beta \|^2 + \gamma^2\right)^{1/2},$$

where γ is a scaling parameter (mean Euclidean distance between the nodes). The model is basically a nonparametric biomass dynamic model with estimated parameters θ_0, ω and σ_ω^2. The same normal likelihood as defined for the random-walk models is used and the model is, in the maximum likelihood sense, a random-effect model or for the Bayesian a hierarchical model with flexibility controlled by the variance parameter σ_ω^2. Given 39 observations, the number of network nodes is set at $N = 10$, with the procedure to select the basis centres β_i and the scaling parameter γ defined as in Hillary (2012).

For both models efficient Markov chain Monte Carlo (MCMC) routines were developed to fit them to the CPUE data. The posterior mean of the fitted index could then be used as input to the HCR defined in Equation 13.1: $I_y^{nw,sp} = \mathbb{E}\left[\exp\left(\theta_y\right)\right]$. Figures 13.2 and 13.3 summarise the fits to the observed CPUE for the random-walk and nonparametric models, respectively. The random-walk model clearly fits the data more closely, but with more variability in the filtered index; the nonparametric model estimates the same general trend as the observed index, but does not fit to the highly variable data in the middle and towards the end of the CPUE index. From the posterior predictive plots both models clearly explain the data fairly well. Posterior predictive methods simulate data from the posterior distribution and can be used to evaluate model fit through a comparison with the observed data. In this case the median absolute deviation in the residuals was calculated for both the simulated and observed data as a measure of the residual variation and p-values (which are probabilities of the simulated residual variation being greater than the observed) close to 0.5 show that, on balance, the simulated data possesses very similar variability to the observed data.

Simulating alternative TAC histories with the MPs

Figure 13.4 shows the TAC levels predicted by the empirical and two model-based MPs. Given the declining CPUE in the initial period, all the MPs act to reduce the TAC below the previous levels. After the CPUE decline seems to become an increase in CPUE into the 2000s, both the empirical and random-walk MPs give stronger increases in TAC than the nonparametric MP, since the nonparametric model estimates a flatter biomass trend in this period.

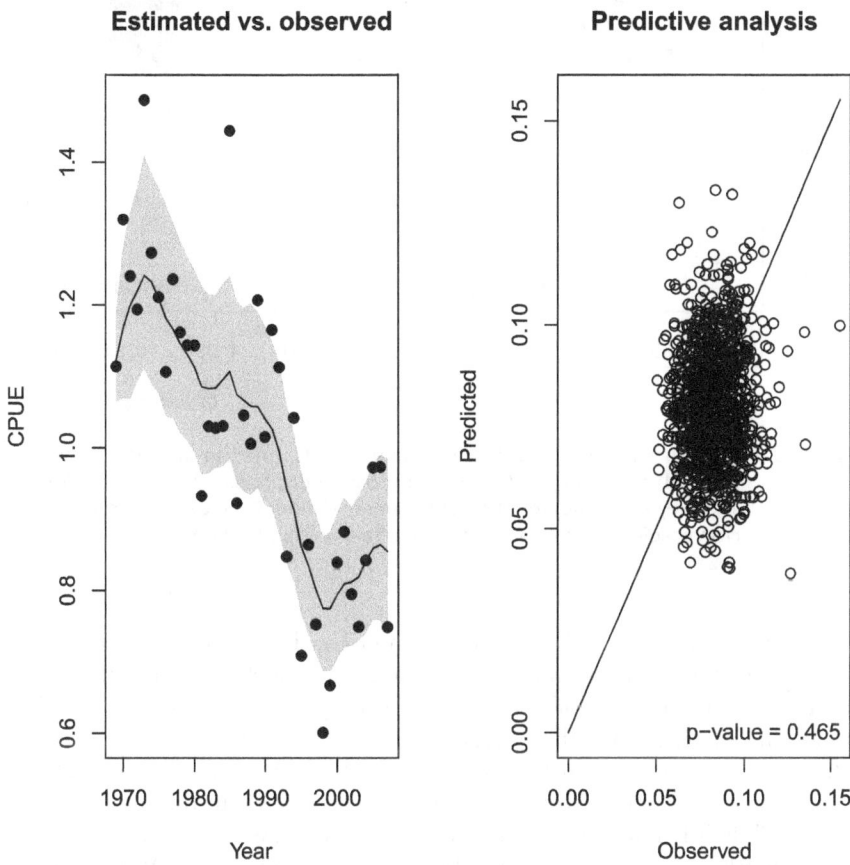

Figure 13.2 On the left are the posterior summary fits (median and 95% credible interval) to the bigeye CPUE data (observed, circles; predicted, black line and grey envelope) for the random-walk model. On the right the posterior predictive summary, including the Bayesian *p*-value.

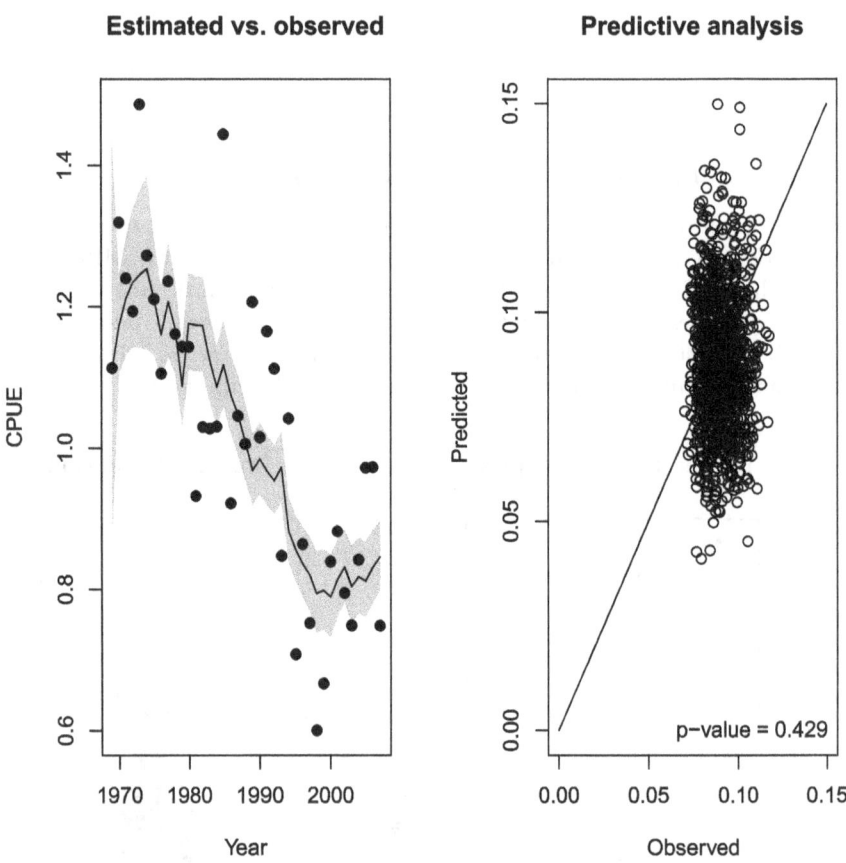

Figure 13.3 On the left are the posterior summary fits (median and 95% credible interval) to the bigeye CPUE data (observed, circles; predicted, black line and grey envelope) for the nonparametric model. On the right the posterior predictive summary, including the Bayesian *p*-value.

In terms of average catch and average annual variation (AAV) in catch, the year-averaged percentage difference in successive catches, Table 13.2 summarises the performance of the MPs. Comparing among the candidates, the nonparametric MP gives the highest average catch for the lowest variability. The empirical and random-walk MPs, driven by very similar biomass trends in both, give very similar average catch and AAV levels (but the model-based MP is still slightly less variable than the empirical one). Without a full MSE, we are obviously not in a position to select between these options. What it does hopefully show is that, for the same data requirements and HCR structure as empirical MPs, model-based MPs that give us fully probabilistic estimates of the trends used in the HCR can easily be implemented, and have the potential to reduce variability in the TAC and thereby increase MP performance.

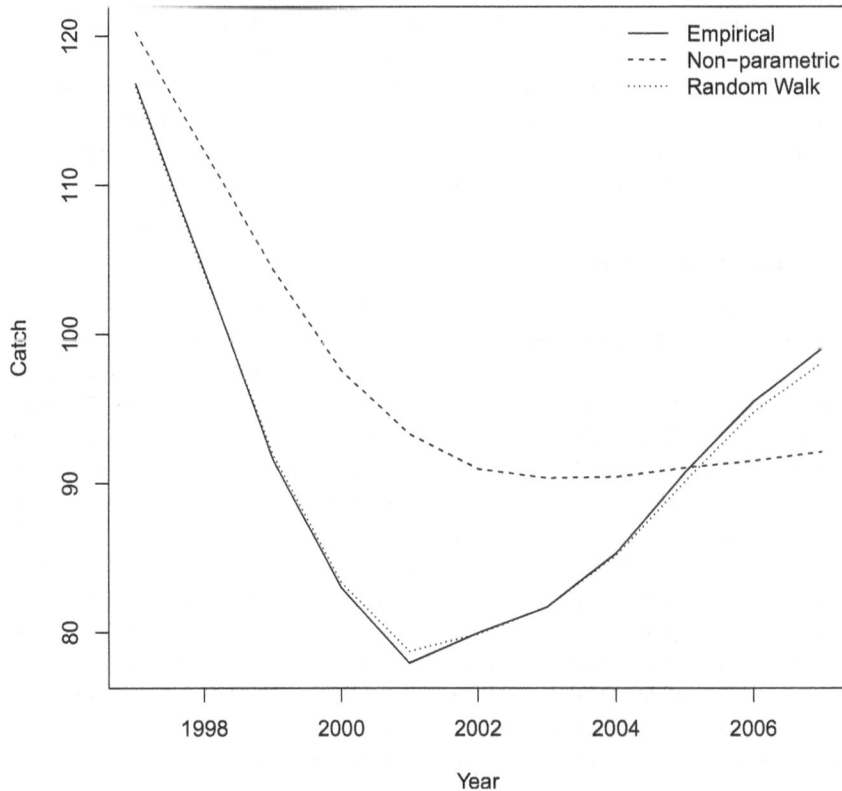

Figure 13.4 Indian Ocean bigeye tuna catch that the simple empirical and model-based MPs would have predicted for the 1996–2007 period.

Table 13.2 Hypothetical historical TAC summaries for the bigeye tuna example.

Scenario	Mean catch ($t \times 10^{-3}$)	AAV (%)
Empirical	91	6.3
Random walk	93	5.1
Nonparametric	98	2.9

Two contrasting full MSE examples

The current canon of MSE work has demonstrated one observation very well: not all apparently good ideas actually work in many instances, and not all additional complexity – however small – results in performance improvement. To be balanced about the potential of the intermediate complexity model-based MPs discussed in this chapter, I will outline two contrasting MSE examples: one where the empirical and model-based performance is practically identical (and one would probably use the empirical MP), and another where the model-based MPs do outperform the empirical MP and would be a better option. These simple examples will demonstrate that context is very important: the reasons for the performance differences (or lack thereof) are not complicated, but are totally driven by the peculiarities of the simulation problem.

The operating model

A simple age-structured annual time step OM is parameterised, and see Table 7 in Edwards et al. (2012) for specifics. In terms of life history I assume a natural mortality rate of $M = 0.2$, a steepness of $h = 0.75$, an unfished spawning stock biomass (B_0) of 1 (making it interpretable as a relative model), and independent and identically distributed (i.i.d.), log-normal recruitment variability with a coefficient of variation (CV) of 0.5. Ages range from 1 to a plus group of 10, with a von Bertalanffy growth curve (assuming $\kappa = 0.25$, $L_\infty = 100$ cm, and $L_0 = 10$ cm), and weight-at-age with a cubic power coefficient relative to length (but arbitrary units). In terms of the fishery a single selectivity is defined and assumed to be a logistic function of age a, described by the ages at which 50% and 95% of the population are vulnerable (Haddon 2011), with $a_{50\%}^s = 3$ and $a_{95\%}^s = 6$, respectively. Maturity is also logistic in terms of age, with $a_{50\%}^m = 4$ and $a_{95\%}^m = 7$. The key MP observations simulated are a relative biomass survey with the same selectivity as the fishery and independent and identically distributed (i.i.d.) log-normal observation errors with a prespecified CV.

The harvest control rule

The base HCR of the candidate MPs is a modified version of the slope-in-the-index rule explored in the previous section:

$$\text{TAC}_{y+1} = \text{TAC}_y \times \begin{cases} 1 + k_c \lambda & \text{for } \lambda < 0 \\ 1 + \delta_c k_c \lambda & \text{for } \lambda \geq 0 \end{cases} \tag{13.2}$$

In Equation 13.2 k_c is (again) the gain parameter of the HCR, δ_c is an asymmetry parameter that allows differential HCR reactions to positive and negative trends, and λ is the slope in the log-scale survey index (either raw or filtered) taken over a 7-year period ending in year y. Three variants on this MP are simulated, making use of the models defined earlier: (1) empirical using the raw survey indices (Case 1), (2) model-based using the random-walk model to filter the index (Case 2) and (3) model-based using the neural network model to filter the index (Case 3).

Scenarios and performance criteria

Two scenarios are explored to test the three candidate MPs. The first is explicitly defined as a stock recovery example: any MP has to increase the spawning stock biomass (SSB) from a starting (stochastic equilibrium) SSB of 0.2 B_0 to the stochastic estimate of spawning biomass at MSY (B_{MSY}) with probability 0.5 and within 10 years. It is assumed the survey has existed for 20 years prior to the MP being implemented with an observation error CV of 0.2. Given the recovery nature of the objectives, the asymmetry parameter is set at $\delta_c = 0.5$, otherwise it would be very hard for any such relative trend-based MP to adequately achieve the objectives for any sensible range of the gain parameter. The tuning parameter is the gain parameter k_c and the same seeds are used for each simulation to ensure the tuning algorithm (a simple but robust bisection root finder) can sensibly estimate the target level of. k_c The second scenario concerns a healthy fishery with objectives that are less well specified. A starting depletion of 0.6 B_0 is assumed, all three candidate MPs are given the same HCR parameters ($k_c = 2, \delta_c = 1$) and the survey is now assumed to have an observation error CV of 0.3. The survey is similarly assumed to exist for 20 years prior to the implementation of the MP and the projection period is also 10 years.

In terms of fishery performance criteria, year-averaged catch biomass (C_{fut}) and average annual variation (AAV) are used, calculated for each iteration and summarised as a median value with 95% probability intervals. To assess conservation of the stock, the future SSB is measured against a limit value (B_{lim}). For the first scenario, SSB performance – because the MPs are tuned they all meet the MSY target specified – focusses only on the probability that the SSB goes below $B_{lim} = 0.1(B_0)$ in the projection period. For the second scenario, the stock is started above MSY so biomass performance focusses on the probability that the SSB goes below $B_{lim} = B_{MSY}$.

Results

Table 13.3 details a performance summary for the three candidate MPs under the two scenarios defined. For the rebuilding scenario all three MPs have virtually identical performance, in terms of expected future catches, levels of AAV and risk to future SSB. Indeed the only factor that really differentiates the empirical and model-based MPs is the tuned gain parameters. The tuned value for the empirical MP ($k_c = 1.98$) is noticeably lower than the tuned value for the neural network ($k_c = 2.7$) or random-walk ($k_c = 3.6$) MPs, meaning that the HCR is less sensitive to changes in perceived abundance. To attain the rebuilding objectives the model-based MPs must become much more reactive, per unit change in the input index, than the empirical MP. What is interesting is that, even in spite of being far more reactive, they result in practically identical levels of AAV when compared to the less reactive empirical MP. This effect demonstrates their noise-reduction performance, even though in this case this effect is balanced out by their higher gain parameters. For this scenario, one would probably choose the empirical MP because it is the simplest of the three and achieves the same performance. The results, however, should not be so surprising. The scenario is well specified (clear objectives with enough time allowed to meet them), with no assumed bias and an observation error CV of

Table 13.3 Performance summary for the three candidate MPs and the two scenarios defined (for average future catch and AAV the estimates are the median with the brackets covering the 95% probability interval).

MP	Scenario	Tuned	k_c	B_{lim}	$E(C_{fut})$	AAV	$p(B_{fut} < B_{lim})$
Case 1	Rebuild	Yes	1.98	$0.1B0$	0.09 (0.05–0.16)	10 (3–25)	0.05
Case 2	Rebuild	Yes	2.7	$0.1B0$	0.09 (0.04–0.17)	10 (3–26)	0.06
Case 3	Rebuild	Yes	3.6	$0.1B0$	0.09 (0.04–0.16)	11 (1–26)	0.06
Case 1	Status quo	No	2	BMSY	0.08 (0.04–0.17)	13 (5–26)	0.04
Case 2	Status quo	No	2	BMSY	0.09 (0.04–0.18)	9 (4–21)	0.03
Case 3	Status quo	No	2	BMSY	0.09 (0.05–0.15)	2 (1–16)	0.02

only 0.2. Even in the raw index, a log-scale slope taken over 7 years captures the dynamics fairly well.

The second scenario, which can be thought of as a *status quo* management problem, is more revealing. There are no well-defined objectives or targets – each MP is given the same HCR parameters and time frame over which to operate from a starting SSB level of $0.6 B_0$. Here the empirical MP underperforms relative to the model-based MPs: slightly lower average catches, noticeably higher levels of AAV, and with a slightly elevated probability of exceeding B_{MSY} in the future. This is a particular scenario where one does indeed see the benefits of the noise-reducing properties of the model-based filter-type MPs. They reduce catch variability (considerably in this case), and are able to achieve (slightly) higher catch levels for (slightly) lower biomass risk, as they are less likely to track spurious observation-error driven trends in the input index. Also of note is that the random-walk model is a clear winner when comparing the model-based MPs. This is because of the (quasi) steady-state nature of the scenario: the MPs are basically being tasked with maintaining the current stock level. This results in little information relating to surplus production levels in the survey index, and the nonlinear nature of the neural network model is trying to estimate this very process; as a result, we see lower precision in the estimated indices relative to the random-walk model and poorer performance in terms of AAV.

Discussion

The model-based approach was taken at the genesis of the MP approach to managing fisheries in the IWC, where the model was a fairly simple stock assessment model (Cooke 1999). As time went on further examples appeared, with more complex age-structured models being tested (Butterworth and Bergh 1993, Punt 1993), but were shown to occasionally perform worse than simpler approaches (Punt 1993). Alongside a need to better communicate how MPs actually work, the idea that simpler could be better led to the development of more empirical MPs (Geromont et al. 1999, De Oliveira and Butterworth 2004, de Moor et al. 2011), which now arguably account for the majority of the fully tested and implemented MPs currently used around the world.

In the end, the MP that performs best (however that is decided) should be the one to use. So, ultimately, the argument for model-based vs. empirical MPs can only take place within the context of a specific fishery. However, in this chapter I have tried to explore other issues around this subject such as noise reduction, wider stakeholder understanding and making the best use of the available data in providing HCR input variables.

One constant in the empirical vs. model-based debate has been the general form of the model-based MP: it is almost always some form of stock assessment (e.g. a production, delay-difference or age-structured model) combined with an F or vaguely MSY-related HCR (Punt 1993, Punt and Donovan 2007). In this chapter I have introduced a recently developed class of model for which there is currently only one example of implementation: southern bluefin tuna (see Hillary et al., Chapter 8, this volume). Inspired by the bluefin tuna work, some simple worked examples have been explored here to illustrate the points being made. I have based these examples on two model-based filters and a simple but popular slope-in-the-index type HCR. Using CPUE data for Indian Ocean bigeye tuna as input, it was shown that model-based MPs of the type being explored could yield a more stable catch. The MPs were neither tuned nor evaluated, but the example does demonstrate that such approaches can easily be included in empirical-like MPs but also provide the advantages of model-based methods.

To explore this claim in a more formal MSE sense, the empirical, neural network and random-walk slope-in-the-index MPs were tested using a simple age-structured OM. Two main scenarios were explored: the first was a well-specified problem with clear and achievable objectives and an informative survey index with the gain parameters tuned to the primary objectives; the second was less well specified, essentially starting from a midrange SSB depletion level with the same gain parameters for each MP, and with a more variable survey index. For the first scenario, MP performance for both catch and SSB was practically identical, and one would probably (on the basis of model parsimony) choose the empirical MP. For the second scenario, the model-based MPs clearly outperformed the empirical MP, attaining slightly higher average catches, for slight lower biomass risk, and with noticeably lower levels of AAV (with the random-walk MP the overall winner).

This simple example demonstrated what might be characterised as something close to emergent properties of MSE work: (1) the context of the problem is very important, and has a strong effect on the results; and (2) it is *never* to be assumed that additional complexity, however gradually it is included, will always improve MP performance. It is also important to resist inferring general principles from specific examples. The result from this example evaluation was that, for a well-specified problem, empirical approaches appeared to outperform model-based ones; but for a simpler but more noisy status quo type problem the model-based MPs clearly outperformed the empirical MP. The first result is the opposite of what was found for the CCSBT MP covered by Hillary et al. (Chapter 8, this volume), which is also a recovery problem. For that example, which includes a well-specified data-rich problem with clear objectives and an expansive suite of robustness trials, the model-based approach delivered superior performance when compared to empirical ones. In summary, the type of simpler model-based approaches explored in this chapter *can* deliver better performance relative to empirical MPs using the same HCR, but the circumstances for which this is likely to be the case will be highly dependent on the specifics of the problem at hand.

In this chapter, I have put forward the case for a simpler approach to model-based MPs. The use of simpler assessment models has seemingly increased over the last few years as fisheries scientists have more and more data-limited stocks and species to deal with. I would also hope to see their uptake within future MSE work, given they fit well within the kind of full feedback simulation frameworks used in MSE, and have the potential to deliver performance improvements on empirical approaches without sacrificing their simplicity.

References

Anonymous 2010. Report of the FC Working Group on Greenland Halibut Management Strategy Evaluation (WGMSE), Technical report, NAFO.

Butterworth, D.S. & Bergh, M.O. 1993. The development of a management procedure for the South African anchovy resource. *In:* S.J. Smith, J.J. Hunt & D. Rivard (eds.) 'Risk Evaluation and Biological Reference Points for Fisheries Management', Canadian Special Publication in Fisheries and Aquatic Sciences, pp. 83–99. Ottawa, National Research Council of Canada.

Cooke, J.G. 1999. Improvement of fishery-management advice through simulation testing of harvest algorithms. *ICES Journal of Marine Science,* 56, 797–810.

de la Mare, W.K. 1986. Simulation studies on management procedures. *Reports of the International Whaling Commission,* 36, 429–450.

de Moor, C.L., Oliveira, J.A.A.D. & Butterworth, D.S. 2011. Is the management procedure approach equipped to handle short-lived pelagic species with their boom and bust dynamics? The case for the South African fishery for sardine and anchovy, *ICES Journal of Marine Science,* 68, 2075–2085.

De Oliveira, J.A.A. & Butterworth, D.S. 2004. Developing and refining a joint management procedure for the multispecies South African pelagic fishery. *ICES Journal of Marine Science,* 61, 1432–1442.

Edwards, C.T.T., Hillary, R.M., Levontin, P., Blanchard, J.L. & Lorenzen, K. 2012. Fisheries assessment and management: A synthesis of common approaches with special reference to deepwater and data-poor stocks. *Reviews in Fisheries Science,* 20, 136–153.

Geromont, H.F., Oliveira, J.A.A.D., Johnston, S.J. & Cunningham, C.L. 1999. Development and application of management procedures for fisheries in southern Africa. *ICES Journal of Marine Science*, 56, 952–966.

Haddon, M. 2011. *Modelling and Quantitative Methods in Fisheries*, second edn, New York, Chapman and Hall/CRC.

Hillary, R.M. 2008. Surplus production analyses for Indian Ocean yellowfin and bigeye tuna, Technical Report IOTC-2008-WPTT-12, IOTC.

Hillary, R.M. 2012, Practical uses of non-parametric models in fisheries assessment modeling. *Marine and Fresh Water Research*, 63, 606–615.

Magnusson, K.G. & Stefánsson, G. 1989. A feedback strategy to regulate catches from a whale stock. *Report of the International Whaling Commission*, Special Issue II, 171–191.

Punt, A.E. 1993. The comparative performance of production-model and ad hoc tuned VPA based feedback-control management procedures for the stock of Cape hake off the west coast of South Africa. *In:* S.J. Smith, J.J. Hunt & D. Rivard (eds.) *Risk Evaluation and Biological Reference Points for Fisheries Management*, Canadian Special Publication in Fisheries and Aquatic Sciences, pp. 283–299, Ottawa, Council of Canada.

Punt, A.E. & Donovan, G.P. 2007. 'Developing management procedures that are robust to uncertainty: lessons from the International Whaling Commission', *ICES Journal of Marine Science*, 64, 603–612.

Rademeyer, R.A., Plagányi, É.E. & Butterworth, D.S. 2007. 'Tips and tricks in designing management procedures', *ICES Journal of Marine Science*, 64, 618–625.

Smith, A.D.M., Smith, D.C., Tuck, G.N., Klaer, N., Punt, A.E., Knuckey, I., Prince, J., Morison, I., Kloser, R., Haddon, M., Wayte, S., Day, J., Fay, G., Pribac, F., Fuller, M., Taylor, B. & Little, R.L. 2008. Experience in implementing harvest strategies in Australia's southeastern fisheries. *Fisheries Research*, 94, 373–379.

Spencer, P., Thompson, G., Ianelli, J. & Heifetz, J. 2013. Random walk models for estimating abundance from a series of resource surveys. *In: World Conference on Stock Assessment Methods, Copenhagen, International Council for the Exploration of the Sea*.

Thompson, G. 2013. A survey/exploitation vector autoregressive model for use in marine fishery stock assessment. *In:* World Conference on Stock Assessment Methods, *Copenhagen, International Council for the Exploration of the Sea*.

Trenkel, V. 2008. 'A two-stage biomass random effects model for stock assessment without catches: What can be estimated using only biomass survey indices?', *Canadian Journal of Fisheries and Aquatic Sciences*, 65, 1024–1035.

## Chapter 14

Developing control rules for threatened bycatch species

Jeffrey E. Moore and K. Alexandra Curtis

Introduction

Bycatch – the incidental catch of nontarget species, or any unregulated or unmanaged catch – is a well-known and globally ubiquitous fisheries issue (Hall et al. 2000, Davies et al. 2009, Lewison et al. 2014), reproached on the basis of wastefully harming animals and ecosystems. Bycatch mortality has caused population declines for diverse species, especially marine "megafauna" (e.g. elasmobranchs, Dulvy et al. 2014; marine mammals, Read 2008; seabirds, Croxall et al. 2012), and imminently threatens the very existence of some populations. Bycatch was a key driver of the likely extinction of Yangtze River dolphin or baiji (*Lipotes vexillifer*), for example (Turvey et al. 2007), and has pushed other species, such as the vaquita porpoise (*Phocoena sinus*; Rojas-Bracho et al. 2008, CIRVA 2014), Maui dolphin (*Cephalorhynchus hectori*; Reeves et al. 2013) and eastern and western Pacific populations of leatherback sea turtles (*Dermochelys coriacea*; Wallace et al. 2013, Tiwari et al. 2013) to critical states. Species most vulnerable to impacts of bycatch tend to have "slow" life histories, with long lives and delayed maturity, for which even small reductions in older juvenile and adult survival rates can be unsustainable (Heppell et al. 1999, 2005). Attempts to mitigate bycatch impacts can include various forms of bycatch reduction technologies (e.g. Werner et al. 2006, Gilman 2011) or regulatory measures such as time-area closures (Senko et al. 2014). The question is: *how much* bycatch reduction is needed to ensure populations do not suffer serious demographic impacts? The answer to this is typically unknown because assessments to estimate acceptable removal levels are rarely conduced for nontarget populations. This is due to a variety of factors, including lack of data or scientific capacity for doing assessments, vague or missing management objectives, operational challenges to implementing the monitoring schemes needed to support a control rule management framework and lack of political will to manage bycatch (Moore et al., 2013).

Biological reference points can be effective for limiting bycatch to appropriate levels (e.g. Taylor et al. 2000, Hall and Mainprize 2004, Reuter et al. 2010) and supporting an ecological risk assessment approach to management

(Zhou and Griffiths 2008, Hobday et al. 2011, Zhou et al., 2011, Moore et al., 2013). Reference points are the norm for managing target catch in many fisheries (Caddy and Mahon 1995, Quinn and Deriso 1999, Garcia and Staples 2000) but are rarely used to manage bycatch of threatened megafauna (Curtis et al. 2015a). One exception is potential biological removal (PBR), which limits marine mammal bycatch under the US Marine Mammal Protection Act (MMPA; Wade 1998, Taylor et al. 2000). It represents an estimate of the number of individuals that can be removed annually from the population while still achieving management objectives set forth by the MMPA, and is calculated as: $PBR = 0.5R_{max}N_{min}F_R$, where R_{max} is the maximum potential annual net growth rate for the population, N_{min} is a minimum estimate of population abundance, and F_R is a recovery factor (valued between 0.1 and 1) that may be adjusted to address various uncertainties or management considerations. PBR is a limit reference point (LRP) corresponding to a population threshold (maximum net productivity level, or MNPL) below which marine mammal populations are considered depleted under the MMPA. A version of PBR (called FRML for fishery related mortality limit) is used to limit bycatch mortality of New Zealand sea lions (*Phocartos hookeri*) in squid trawl fisheries pursuant to the New Zealand MMPA (Chilvers 2008), and bycatch limits are used for skates and rays in the Southern Ocean by CCAMLR (Kock et al. 2007). But in general there are few formal management frameworks that base bycatch management of megafaunal species on LRPs. Nevertheless LRPs are increasingly being used by researchers to evaluate bycatch sustainability and recommend bycatch limits in fisheries around the world (e.g. Żydelis et al. 2009, MRAG 2011, Zhou et al. 2011, Richard and Abraham 2013, Diffendorfer et al. 2015).

We describe a potential management framework based on the use of LRPs to determine maximum sustainable bycatch levels, and we use Bayesian inference to evaluate the probability of exceeding the LRPs given past, current and future fishing effort levels. Probabilistic inference is needed because bycatch must be estimated from sample observer data; it cannot be treated as known as is commonly the case for target species. This is a newly proposed approach, not formally adopted by managers in the United States, but developed in the context of a particular US management issue: bycatch of Pacific leatherback turtles in the California drift gillnet (CDGN) fishery off the west coast of the United States. This case study analysis highlights two issues that our work is particularly designed to address: improving inference about rare-event bycatch (e.g. ranging from zero to a few animals annually) estimated from incomplete observer coverage, and estimating LRPs for a management area that encompasses only a fraction of a population that incurs mortality from fisheries and other anthropogenic sources (e.g. from different countries) throughout its range. Key features of our approach are: (1) informing domestic management with local LRPs based on the fraction of the population in a management jurisdiction; (2) using LRP estimators that have been evaluated and tuned using simulation-based methods to ensure high probability of management success; and (3) making model-based probabilistic inference about whether bycatch exceeds the LRP over prolonged (multiyear) periods. This framework has the potential for widespread application to other bycatch species issues and fisheries, particularly for highly mobile or transboundary species or where bycatch is a rare event and thus difficult to estimate well when observer coverage is limited.

The California drift gillnet fishery

Martin et al. (2015) provided a historical overview of the CDGN fishery with respect to leatherback interactions. Briefly, the fishery has operated since the 1970s, primarily from August through December each year. Large-mesh (> 36 cm) nets, up to 1,800 m long, are used to target large pelagic species such as swordfish and thresher shark throughout the US West Coast Exclusive Economic Zone (EEZ) (Figure 14.1; also see Carretta et al. 2014, appx. 1). Fishing effort peaked during the mid-1980s at about 10,000 sets per year (NOAA logbook data, compiled by Dewar et al. 2013). An observer program has been in place since 1990. During this time, effort has declined from over 4,000 sets annually to fewer than 500 (Figure 14.2).

Documented bycatch in the fishery includes numerous protected species, including white sharks, marine mammals and sea turtles (Hanan et al. 1993, Moore et al. 2009). Leatherbacks interacting with the CDGN fishery are from the Western Pacific (WP) genetic stock (Dutton et al. 2007, Benson et al. 2011), which nests primarily in West Papua, Indonesia, Papua New Guinea, Vanuatu and the Solomon Islands (Dutton et al. 2007), and is listed as Critically Endangered on the IUCN Red List due to precipitous population decline (Tiwari et al. 2013). The annual number of female turtles nesting at primary nesting beaches in West Papua has declined by 6% per year since 1984 (Tapilatu et al. 2013). The population suffers from a multitude of impacts, including mortality from diverse fisheries throughout their range (Kinan 2005, Wallace et al. 2011, Tiwari et al. 2013). The WP stock can be further subdivided demographically into boreal summer–nesting and boreal winter–nesting groups, with distinct foraging regions (Benson et al. 2011, Gaspar et al. 2012). The CDGN fishery interacts with the former.

Leatherback bycatch in the CDGN is a rare event, with observed catches per year historically ranging from zero in most years to five (Figure 14.2). Since 2001, NOAA has implemented a large seasonal-area closure of the CDGN fishery called the Pacific Leatherback Conservation Area (PLCA) (Figure 14.1), in effect from August 15 to November 15 each year, to reduce leatherback mortality. The PLCA, possibly in combination with set-depth regulations implemented in 1997 to reduce marine mammal mortality (Moore et al. 2009), has been effective in reducing observed bycatch levels (Martin et al. 2015), but concern for leatherbacks remains. The Pacific Fishery Management Council ("the Council"), the primary advisory body to NOAA for managing US west coast fisheries, adopted in September 2015 a "hard cap" control rule to limit mortality of endangered leatherback sea turtles and other protected species in

DGN Leatherback Bycatch Events

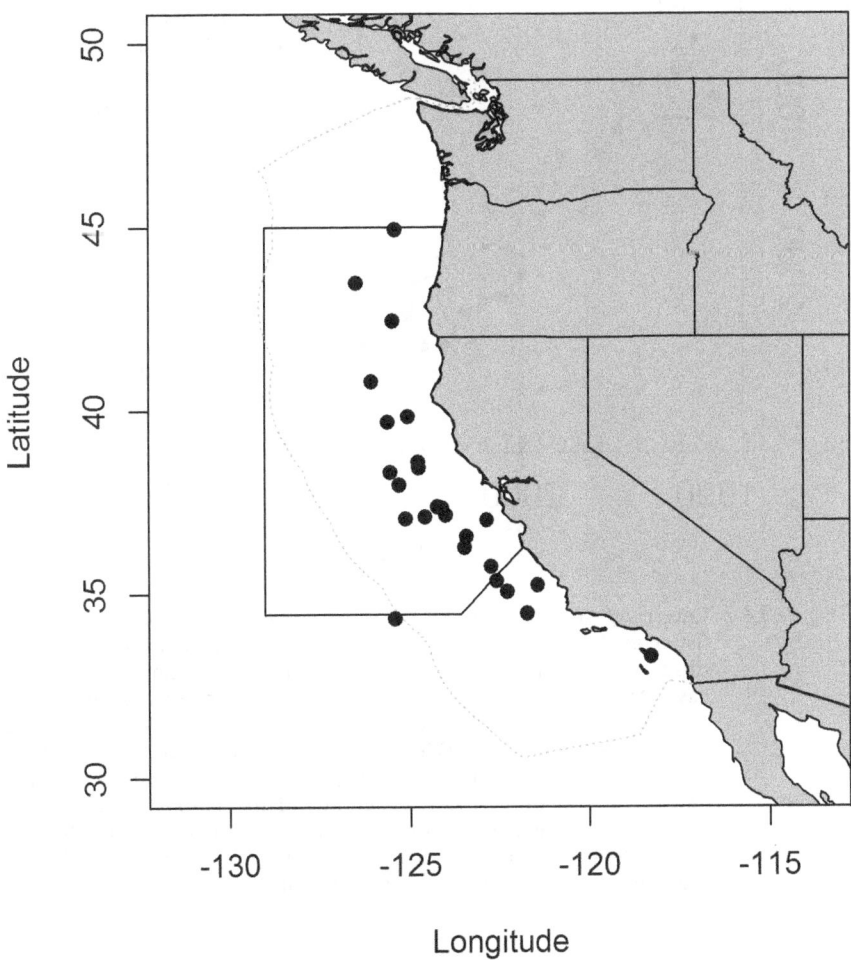

Figure 14.1 Study area (waters off the west coast of the United States) and location of historical observed leatherback entanglements (points) from
1990 to 2013. Dotted line is the US West Coast Exclusive Economic
Zone (EEZ). Solid line demarcates the Pacific Leatherback Conservation Area (PLCA), which has been closed to California drift gillnet
(CDGN) fishing from 15 August to 15 November since 2001.

the CDGN fishery (NMFS 2015). This control rule calls for in-season closure
of the fishery if observed catch reaches a hard cap level. However, as we will
discuss later in the chapter, the rare-event nature of leatherback bycatch makes
a hard cap management scheme problematic. Our analyses (Curtis et al. 2015b;
and this chapter) are intended to help inform this ongoing management process.

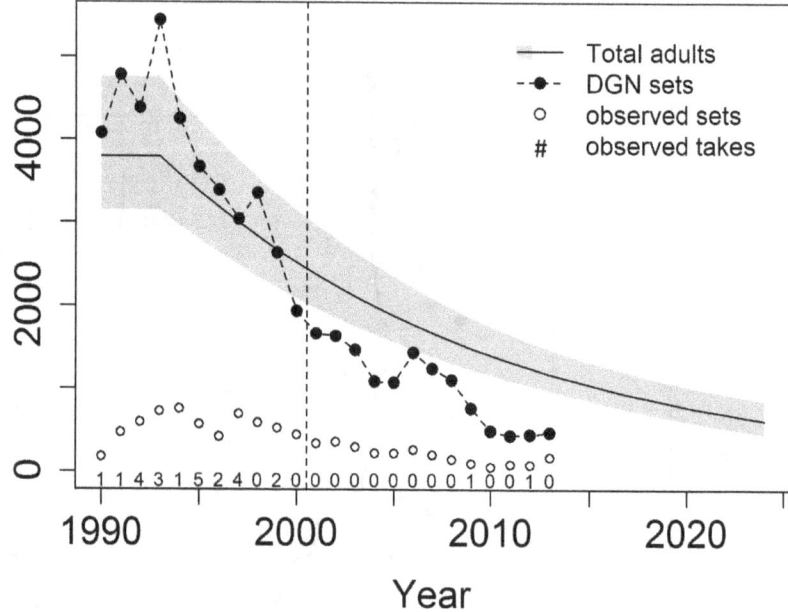

Figure 14.2 Observed and estimated time series used in the current analysis: fisheries data from the NOAA West Coast Regional Office's fishery observer program; and leatherback abundance estimates and forecasts from Curtis et al. (2015b) based on nest count time series reported in Tapilatu et al. (2013). Observer data (1990–2013) include the number of CDGN sets fished, the number observed, and observed leatherback takes (see Figure 14.3 for the number of these known to have died). Leatherback abundance is expressed in terms of the total number of adults (both sexes) in the population. Vertical dashed line indicates establishment of the Pacific Leatherback Conservation Area (PLCA) (between 2000 and 2001).

Think globally, act locally

Western Pacific leatherbacks incur fishing and other impacts from multiple countries and in international waters (Kinan 2005, Wallace et al. 2011), but most management decisions are made by national governments or in some cases international regional fishery management organizations (RFMOs), whose authorities are limited to particular jurisdictions. Appropriate cumulative mortality limits may be estimated for whole populations, but determining limits for individual management jurisdictions is a more complicated matter. A possible way forward (Curtis et al. 2015a, 2015b) is for domestic agencies to implement local limits based on the fraction of the population in the juris-diction, prorated by the fraction of the year spent in the area. This allows for incremental, jurisdiction-wise progress towards limiting mortality for the total population to the overall LRP. The United States uses this approach to estimate PBR for marine mammals whose ranges or migration routes extend beyond US waters (NMFS 2005).

The premise of using local LRPs is that depleted populations will recover to some target level if all countries and sectors do their part. Although many countries do not, we view this as a just, rational, and operationally practical management philosophy. From the standpoint of a country whose impacts to a transboundary population are relatively small, imposing stricter mortality limits than given by the local LRP (e.g. because of strong conservation laws) may unfairly place the conservation burden on a small domestic user group without stemming the problem, whereas allowing more lenient domestic limits (e.g. because "we are not the problem") erodes a country's political and ethi-cal standing to lobby for international conservation efforts. Of course, if the greatest impacts to the population throughout its range are not mitigated, then the population will continue to decline, and local LRPs will concomitantly decrease. This sets up an apparent dilemma: as countries who do not manage their impacts drive a transboundary species toward extinction, should those countries with the strongest protective statutes be forced to shut down their fisheries? Indeed, there is pressure from advocacy groups and politicians in the United States to close the CDGN fishery on the basis of risk to WP leather-backs and other protected species (e.g. Shore 2013, California Assembly Bill 2019, 20 February 2014).

Where local LRPs are small and bycatch events rare, as in the CDGN, man-agement decisions may be aided not just by the local LRP estimates, but also by the likelihood of bycatch exceeding those limits, given characteristics of the fishery. In most cases, as a population and the associated LRPs decline, so too does the likelihood of interacting with the species. The analysis presented in this chapter provides an example of estimating local LRPs and the probability of exceeding them. Evaluating this probability as a function of fishing effort can help inform management, for example by helping to develop input control rules (how much effort to allow) to ensure that the probability of exceeding an LRP remains below a certain threshold, or providing a sense of the extent of monitoring data (e.g. from fishery observer programs) needed to reasonably assess impacts of the fishery.

Development of the management framework

Analytical overview

The objectives of this analysis are to estimate annual leatherback mortality in the CDGN fishery from 1990 to 2013 (the period for which we have appropriate fishery data), evaluate the probability that this exceeded our estimated LRPs over a multiyear time period, and forecast the probability of exceeding LRPs in the near future given projections of turtle abundance and three hypothetical fishing effort levels (current, 100% higher, or 300% higher). The remainder of this section describes the data, how the LRPs were estimated (and the role of computer simulation in doing so), and how we estimated leatherback mortality and probability of this exceeding the LRPs.

Data

Inputs to the current analysis come from two primary sources. The first is from the NOAA West Coast Regional Office's fisheries observer program, which provided information about fishing effort and bycatch in the CDGN fishery from 1990–2013 (Figures 14.1, 14.2). The observer and logbook data describe per-annum fishing effort (number of sets), observer coverage (number of sets observed), observed leatherback entanglements or "takes," and the number of these for which leatherbacks were dead or released alive. The second source of information is an estimated time series of total adult abundance in the boreal summer–nesting portion of the WP leatherback stock from 1993 to 2014, based on nest count data for all but one year from 1993 to 2011 (Tapilatu et al. 2013, Curtis et al. 2015b; Figure 14.2). For this chapter, using the methods of Curtis et al. (2015b), we have hindcasted the abundance-estimate time series to 1990, to fully complement the observer-data timeline, by assuming that abundance in these early years was the same as in 1993, and we have forecasted abundance a decade into the future (to 2024) (Figure 14.2).

Simulation-based development of limit reference point estimators

For sea turtle populations in general, Curtis and Moore (2013) recommended an LRP estimator termed the reproductive value loss limit (RVLL), which expresses the mortality limit in terms of the amount of reproductive potential that can be removed from the population. Sea turtles mature slowly (over many years or decades), but can lay hundreds of eggs in a season, so animals of different age-classes can have vastly different reproductive value (Crouse et al. 1987, Heppell et al. 1996). Moreover, most sea turtles are highly mobile, with life stage–specific habitats in disparate and widely separated regions, such that a given population stressor such as a fishery is likely to interact only with a particular life stage (e.g. Wallace et al. 2008). RVLL expresses the mortality limit across these diverse life stages in a common currency. However, to estimate a local LRP for leatherbacks in the US West Coast EEZ, we take advantage of the fact that leatherbacks using these waters are all large juveniles and adults, which are expected to be equally susceptible to CDGN fishery entanglement. This permits use of the simpler PBR reference point estimator, which expresses the mortality limit in terms of the number of individuals (Curtis et al. 2015a, Curtis et al. 2015b).

We use the general form of the PBR estimator, but with modified notation:

$$LRP = \left[b\left(\lambda_{max} - 1\right)N_{loc}f_a \right]_{min}, \quad\quad (14.1)$$

where $\lambda_{max} - 1$ is the maximum annual net population growth rate in the absence of density-dependent effects (with $\lambda_{max} > 1$); b is the fraction of $\lambda_{max} - 1$ that corresponds to the population's maximum net productivity level (similar to the PBR, this is set to 0.5 based on a simple logistic density-dependent population growth model, which represents a conservative productivity assumption for sea turtles); N_{loc} is a local abundance estimate (prorated for the proportion of the year that animals spend in the local area); and f_a is an adjustment factor that provides a buffer against potential biases in model assumptions or parameters or that addresses other management considerations. Estimates of λ_{max} were based on population growth rates observed in recovering populations elsewhere, adjusted for differences in hatchling production (Curtis et al. 2015b). Local abundance estimates for the US West Coast EEZ (N_{loc}) were calculated by Curtis et al. (2015b), based on: demographic information to derive the abundance of total adults in the population from the nest-count data (Tapilatu et al. 2013); satellite-tag data indicating the proportion of nesters that migrate annually to the US West Coast EEZ and how long they stay there (from Benson et al. 2011); and fishery observer data indicating the proportion of adults among total leatherback turtles caught in the US West Coast EEZ (Jones et al. 2011). Distributions (rather than point estimates) were used for λ_{max} and N_{loc}, resulting in a distribution for the expression within brackets in Equation 14.1. The LRP was given by a lower percentile of this distribution determined through management strategy evaluation (MSE) by Curtis et al. (2015b) to meet performance criteria for population outcomes (hence the outer subscript "min"). The value of f_a was similarly determined through the MSE (see Moore et al. 2013 for a review in the context of bycatch management). In this MSE, the LRP estimator was treated as a harvest control rule based on the assumption that management restricts the bycatch to the LRP.

Curtis et al. (2015b) estimated two LRPs, each corresponding to a different plausible minimum population objective: (1) maintaining a population at or allowing it to rebuild to MNPL (LRP_{upper}) and (2) expediting rebuilding such that time required to rebuild is within 10% of what it would be in the absence of human-caused mortality (LRP_{lower}). For each LRP, a distribution was estimated according to Equation 14.1, given the distributions for N_{loc} and λ_{max}. For the first population objective, our performance criterion required that 95% of simulated populations remain at or above MNPL after 40 years of removals equal to the LRP. A f_a value of 0.6 was chosen because preliminary simulations showed that this value allowed the performance criterion to be met at an intermediate percentile of the LRP distribution, thereby avoiding working with the lower tail. MSE was then used to identify the highest percentile from the LRP distribution at which this performance criterion was met (this was found to be the 15th percentile) and to evaluate sensitivity of population outcomes to important potential sources of bias. For the second population objective, MSE was used to identify a new value for f_a that would allow the objective to be met for the large majority of simulation populations. The lowest value considered, $f_a = 0.1$, allowed more than 75% of populations to meet this objective.

In the biological operating model of the MSE, stochastic age-structured population models with density dependence were used to simulate leatherback populations through time, and it was assumed that all age-classes of the entire population throughout its range incurred the same per-capita removal rate of $0.5(\lambda_{max} - 1)f_a$. To reiterate the simulation results from the previous paragraph, a depleted population would be expected to recover eventually to MNPL if anthropogenic mortality were limited to the 15th percentile of the LRP distribution with $f_a = 0.6$ (LRP$_{upper}$). The lower f_a value of 0.1 provides additional protection for depleted populations to recover to MNPL with *minimal delay* (sensu Wade 1998, Curtis and Moore 2013). We refer the reader to Curtis et al. (2015b) for full details of the MSE.

Estimating bycatch mortality

We took a Bayesian approach to estimating annual bycatch mortality for leatherbacks from fisheries observer data and other inputs. Our approach was similar to that of Martin et al. (2015), who estimated annual leatherback mortality in the CDGN fishery as a function of fishing effort and a bycatch-per-unit-effort (BPUE) parameter estimated for two management eras (before and after implementation of the PLCA). Our analysis differs by virtue of resolving BPUE into the product of population abundance and a catchability parameter, which is essential for forecasting if the population abundance is not stable. Our analysis was implemented in the program OpenBUGS through an interface with R using the R2OpenBUGS package.

Let X_t be the annual observed number of leatherback entanglements in year t. We treated this as a random binomial variable with parameters B_t (true bycatch across all sets in the fishery) and p_t (proportion of sets observed by the scientific observer program):

$$X_t \sim \text{Binomial } (B_t, p_t).$$

B_t was itself treated as Poisson random variable:

$$B_t \sim \text{Poisson } (\mu_t),$$

$$\mu_t = N_t E_t c_t,$$

where the expected number of entanglements, μ_t, is assumed to depend on total adult turtle abundance in the population (N_t) (derived from the nest count data and estimated life-table parameters by Curtis et al. (2015b)), fishing effort in terms of number of sets (E_t) and a scaling parameter referred to as catchability (c_t). The parameters to estimate are B_t and c_t, while X_t, p_t, N_t and E_t are the data. Prior analysis of these data supports use of the Poisson distribution for B_t (Martin et al. 2015).

Note that in our model, c scales bycatch within the area of the CDGN fishery to an abundance index (adult abundance) for the total boreal summer–nesting WP leatherback stock occurring throughout the North Pacific. We are therefore assuming that N_t is related by a constant proportion to the number of animals available to the CDGN fishery in a year. Catchability would have to be reestimated if a different metric were used for N_t (such as a local abundance estimate). We reduced c_t to two values (c_1 and c_2) that represent mean catchability for two time periods: 1990–2000, prior to implementation of the PLCA ($c_t = c_1$); and 2001–2013 ($c_t = c_2$). True catchability may in fact vary annually due to many factors (e.g. oceanographic conditions, turtle and fishing distributions), but we estimated just two mean rates for two reasons – because limited data preclude complex models for c, and because ultimately we are more interested in accurately estimating mean annual bycatch over an extended time period than in the year-to-year values.

Not all entangled animals are killed. Let total mortality for the fishery be

$$M_t = Y_t + Z_t,$$

where Y_t is the number of deaths in the observed sets (including, to be precautionary, injured animals whose fates were recorded as unknown) (Figure 14.3) and Z_t are the deaths in unobserved sets. We estimated a constant mortality rate (m) for entangled animals from the observer data:

$$Y_t \sim \text{Binomial}\ (X_t, m),$$

and then conditional on m,

$$Z_t \sim \text{Binomial}(B_t - X_t, m),$$

where $B_t - X_t$ is the unobserved bycatch. For years without observed bycatch or any observer data (e.g. for forecasting), all bycatch and mortality are unobserved, so total mortality is simply:

$$M_t \sim \text{Binomial}(B_t, m).$$

Additional parameters to estimate are m, Z_t and M_t, while Y_t are data.

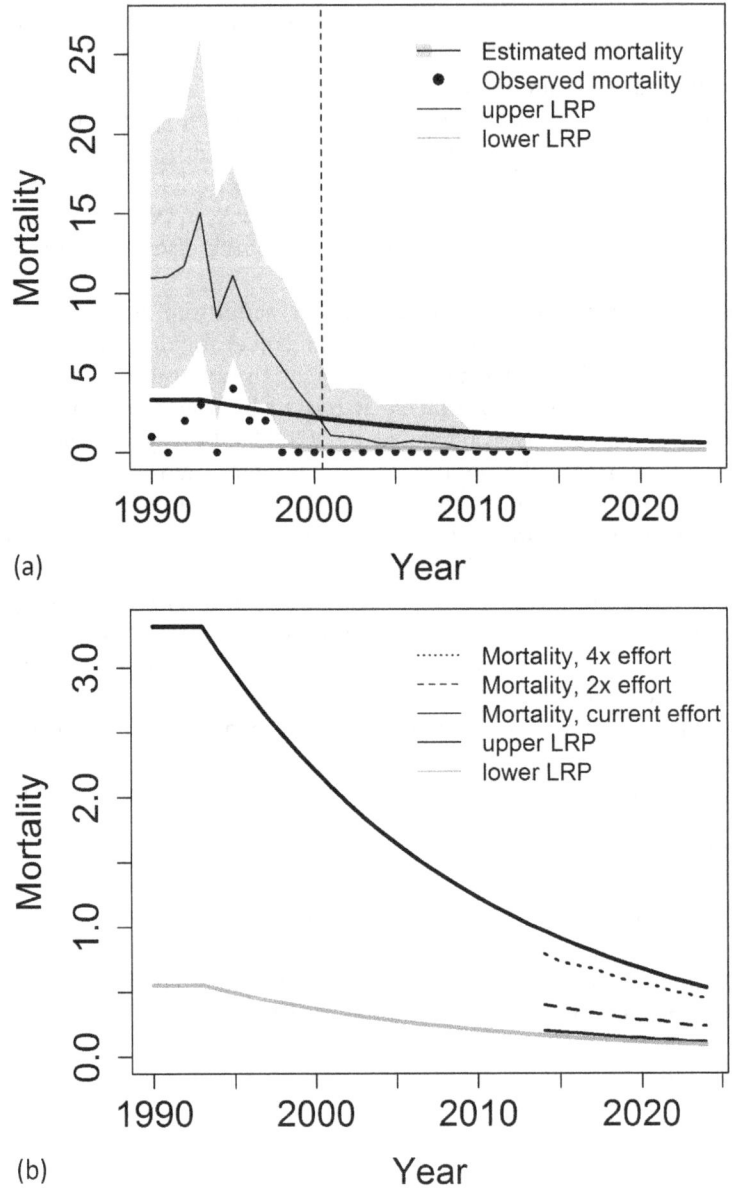

Figure 14.3 (a) Observed and estimated bycatch mortality in terms of the number of individuals (Bayesian posterior mean and 95% Bayesian credible interval), and lower and upper limit reference points (LRP$_{lower}$ and LRP$_{upper}$), based on adjustment factor (f_a) values of 0.1 and 0.6, respectively. (b) The same LRPs as in panel (a), along with forecasts of future bycatch mortality based on three fishing effort levels (current = the mean of recent effort; double this level and quadruple this level). Compared to (a), the vertical axis has been truncated for visibility. (c) Annual probabilities of exceeding the lower and upper LRPs, including for three levels of effort forecast to 2014. Vertical dashed line in panels (a) and (c) indicate establishment of the Pacific Leatherback Conservation Area (PLCA) (between 2000 and 2001).

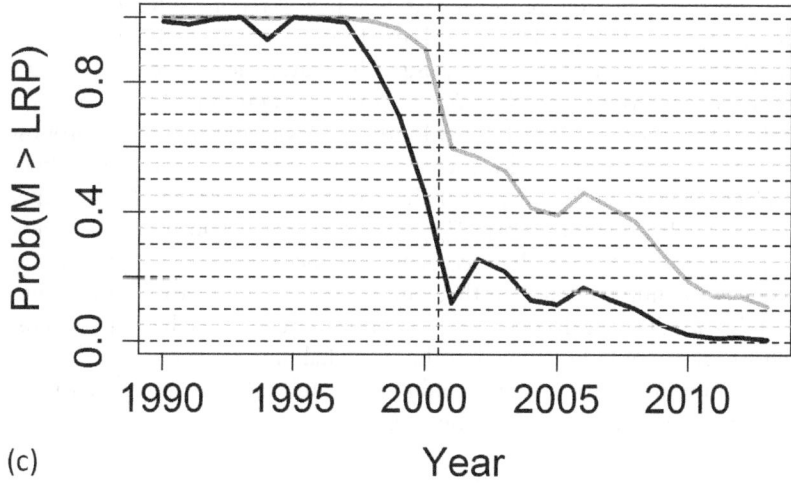

(c)

Figure 14.3 (Continued)

Long-term probability of exceeding reference point thresholds

For each year of the time series, the probability that bycatch mortality exceeds one of the proposed LRPs (LRP$_{upper}$ or LRP$_{lower}$) is simply the proportion of Bayesian MCMC samples for M_t that exceed the LRP. However, we suggest that bycatch in most contexts should be evaluated on a multiyear rather than on an annual basis, especially when bycatch is a rare event. This is in part because annual bycatch estimates in rare-event situations are prone to severe estimation error when observer effort is limited, especially when conventional, annual ratio-type estimators are used (Carretta and Moore 2014, Martin et al. 2015). This can lead to volatile management and fishery instability as managers react to sampling error. The model-based estimation approach described in the previous section greatly improves on ratio-based bycatch estimation, but this approach cannot provide real-time annual estimates, and the assumption of constant catchability (and inability to estimate year or spatially dependent catchability due to limited data) means that individual-year estimates from this method will generally be less accurate than estimates summed over longer time periods. More generally, whereas target species are intentionally fished to set quota levels, bycatch is undesirable, avoided in many fisheries to various extents and for most species occurs at levels that are, on average, within biologically sustainable limits. As long as this is the case, then modest exceedance of the LRP in some years should be of trivial consequence to the bycatch population and thus does not warrant costly management intervention that may harm fisheries without providing any benefit to the species of concern. For all these reasons, we focus on evaluating the probability that total mortality across multiple years exceeds the sum of LRP values for the same time period, in effort to assess whether there is a *persistent* bycatch problem that does in fact require intervention.

We looked at two multiyear periods. First, we evaluated whether total bycatch mortality exceeded the proposed LRPs over the course of 2001–2013 (since implementation of the PLCA). Second, we evaluated whether projected mortality over the next decade (2014–2024) is expected to exceed proposed LRPs given projections of leatherback abundance (Figure 14.1) and three hypothetical levels of fishing effort: (1) maintaining current efforts (E_{future} = mean of $E_{2009-2013}$); (2) doubling effort ($E_{future2x}$ = $2 \times E_{future}$); and doubling it again ($E_{future4x}$ = $4 \times E_{future}$). In all these cases, we assume that the PLCA policies will continue to be implemented, so that catchability in future years is the same as catchability for 2001–2013 (i.e. $c_t = c_2$). Alternatively, the three hypothetical fishing effort levels could be viewed as three catchability levels (c_2, $c_2 \times 2$, $c_2 \times 4$), or some combination of increased effort and catchability, as might be the case if new effort were allowed inside the PLCA.

Application and inferences for managing fishing effort in the CDGN

The estimated number of females nesting annually for the boreal-summer WP leatherback stock declined from about 1,100 in 1993 to about 300 in 2013 (Tapilatu et al. 2013, Curtis et al. 2015b). Given that adult females nest approximately every 2.5 years, and that the proportion of females is estimated at 0.75 (Benson et al. 2011, Tiwari et al. 2013), Curtis et al. estimated the total adult population (N) to have declined from about 3,800 in 1993 to under 1,200 by 2013 (Figure 14.2). If this decline continues at the current pace, there will be just over 600 adults in the population a decade from now (by 2024). To estimate a $N_{loc,t}$ in Equation 14.1, these total adult abundance estimates were multiplied by the fraction of adult females that migrate to this area each year (mean = 0.19; Benson et al. 2011), divided by their length of stay (mean = 3.3 months), and divided again by the proportion of animals in the EEZ that are adults (0.54), which was estimated from size of nesting females measured in boreal summer and CDGN fishery observer program data (Benson et al. 2011, Jones et al. 2011). Although we have provided point estimates for various parameters in this paragraph, we again note that all the aforementioned parameters, with the exception of remigration interval, were expressed in the analysis by distributions, resulting in distributions for $N_{loc,t}$. Similarly, λ_{max} was drawn from a uniform distribution from 1.04 to 1.06. The distributions for $N_{loc,t}$ and λ_{max} provide distributions for the LRPs calculated with the two adjustment factor values of f_d = 0.1 and 0.6. The 15th percentiles of these distributions, which are the LRPs used in the analysis, are plotted through time in Figures 14.3a and 14.3b.

Estimated bycatch mortality from 1990 to present (M_t) is shown in Figure 14.3a. This reflects the product of bycatch (B_t) and estimated 56%

postentanglement mortality rate m (95% Bayesian CRI = 37% to 74%). Bycatch mortality declined dramatically over this time, due to a combination of declining fishing effort, declining turtle abundance (N_t), and reduced catchability (c_2) following implementation of the PLCA. Comparing posterior distributions for c_1 (before 2001) and c_2 (2001 to present) there was a 0.91 probability that c_2 was lower. The mean ratio between these two parameters (c_1/c_2) indicated catchability before 2001 was 3.0 times higher, although there was considerable uncertainty in the true difference, with the 95% CRI for this ratio ranging from 0.7 to 9.7.

The annual probabilities of bycatch mortality exceeding either LRP are plotted in Figure 14.3c. It is clear that prior to 2001, the more conservative LRP$_{lower}$ was almost certainly exceeded in all years (Figures 14.3a and 14.3c), while LRP$_{upper}$ was also probably exceeded in most years. Since 2001, however, the annual probabilities of exceeding the LRPs have been lower and have continued declining as a result of declining fishing effort. For the entire period since the PLCA has been in effect (2001 to most recent data year, 2013), there is a 78% chance that total mortality (posterior mean = 7.8; 95% CRI = 1 to 21) exceeded the sum of LRP$_{lower}$ and a 4% chance that it exceeded the sum of LRP$_{upper}$ (Table 14.1).

For all three future effort levels considered, the expected bycatch mortality falls between the two LRPs (Figure 14.3b). If the fishery continues at current effort levels, then over the next decade (2014–2024) we expect a reasonable chance (50%) that cumulative bycatch mortality (expected value = 1.9; 95% CRI = 0 to 6) will exceed the sum total of the LRP$_{lower}$ (1.3 fatalities), but we estimate only 1% chance that the cumulative LRP$_{upper}$ (eight fatalities) would be exceeded (Table 14.1). Doubling or quadrupling effort results in a high probability (0.76 and 0.86, respectively) that the LRP$_{lower}$ would be exceeded for the whole time period, and fairly low but nontrivial probabilities of exceeding LRP$_{upper}$ (0.11 and 0.25).

Table 14.1 Probabilities that cumulative bycatch mortality (M_{total}) for various time periods exceeds the sum total of limit reference points (LRP$_{total}$), for two different LRP levels corresponding to two different minimum population objectives – maintaining or recovering population size to maximum net productivity level (LRP$_{upper}$), or expediting recovery to that level (LRP$_{lower}$) – and adjustment factors (f_a).

$P(M_{total} > LRP_{total})$ for years:	LRP$_{lower}$ (f_a = 0.1)	LRP$_{upper}$ (f_a = 0.6)
2001–2013	0.50	0.04
2014–2024 | current effort	0.50	0.01
2014–2024 | 2 × current effort	0.76	0.11
2014–2024 | 4 × current effort	0.86	0.25

Discussion

For many commercial fisheries, target species are managed using harvest control rules based on biological reference points. Global imperatives to minimize waste and ecosystem impacts of fishing (e.g. FAO 1995) can be addressed in part by using reference-point based control rules for managing bycatch as well (Moore et al. 2013, Curtis et al. 2015a). The PBR framework used to limit fisheries interactions with marine mammals in the United States serves as a useful precedent for setting LRPs for protected or threatened species (Taylor et al. 2000). Our approach to estimating LRPs for leatherback turtles in US West Coast EEZ waters is based on similar principles, including using an LRP estimator with parameters that can be reasonably estimated or otherwise inferred, and using MSE to inform a precautionary approach in the face of uncertainty by tuning the LRP so that it provides a high likelihood of achieving management objectives for the population if used by all relevant management sectors (sensu Wade 1998, Curtis and Moore 2013).

The case study presented here provides an example of how an LRP-based control rule might be applied to rare species that incur impacts across multiple management jurisdictions. Our MSE work on this problem (Curtis et al. 2015b) suggests that if local LRPs were applied across domestic and international management sectors that interact with the WP leatherback population, the population could likely recover from its currently depleted state. Such multilateral collaboration is presently unlikely, but our approach could nonetheless facilitate population and risk assessments that could lead to more international conservation efforts. For example, LRPs estimated for each relevant country or management region could serve as a means of evaluating management and highlight where the most severe impacts are occurring (e.g. where the LRPs are most severely exceeded), thereby helping to prioritize international conservation efforts. These LRPs could also be used as a metric for evaluating whether international fisheries meet bycatch standards for importing seafood to certain countries (e.g. sensu US MMPA, Magnuson-Stevens Fishery Conservation and Management Act) or for sustainability certifications (e.g. sensu Marine Stewardship Council).

Unless international conservation efforts are rapidly stepped up, the WP leatherback population will continue declining. Our analysis provides guidance for managing an individual domestic fishery under this pessimistic scenario. First intuition might suggest that the rarer a population and hence the lower the LRP becomes, the more important it is to fish less, prompting calls from many nongovernment organizations and legislators to close the California drift gillnet fishery. However, in most cases, as a population becomes rarer, so does the expected rate of interaction. Our analysis quantifies the risk of bycatch mortality exceeding the LRP as a function of population decline to provide useful information for managing fishing effort. Given projected declines in turtle abundance, we estimate that for the period 2014–2024, mortality kept below a cumulative value of 8 (LRP_{upper}) in the US West Coast EEZ would represent the US contribution toward allowing the population to eventually rebuild to MNPL levels. Limiting mortality to below 1.3 (LRP_{lower}) represents the US contribution toward allowing rebuilding to occur at a near-maximum rate. We find it highly unlikely (1% chance) that leatherback mortality over this future time period will exceed 8, but reasonably likely (50% change) that it will exceed 1.3.

To be clear, meeting local LRP goals in the United States alone would not, without international cooperation, recover the species, nor would modestly exceeding local LRPs in the United States appreciably elevate population risk. The potential impact, pro or con, of current US west coast activities on Western Pacific leatherbacks is small compared to other stressors throughout the Pacific. The point of the framework we have outlined is that local LRPs represent a mortality rate for the segment of the population using US West Coast EEZ waters, which, if applied to all segments of the population, should allow for population recovery. In this sense, the local LRPs may be thought of as the US share of responsibility, based on the fraction of the population it interacts with, to conserving the whole population.

Our approach also provides information about the level of monitoring required to ensure proper implementation of a control rule for rare species. For example, one might conclude from our analysis that if current fishing effort levels are maintained over the next decade and LRP_{upper} is the reference point of interest to management, then accurately estimating annual CDGN fishery leatherback mortality is not imperative (though obviously is desirable), since the risk of exceeding LRP_{upper} appears low. If fishing effort were increased substantially, it would become more important to implement high observer coverage levels to provide accurate feedback as to whether the LRPs were being exceeded.

We discuss two caveats of our analysis. First, we have only considered bycatch mortality in this analysis, but there are in fact other mortality sources in the US West Coast EEZ, including boat and ship strikes and interactions with other fisheries (R. LeRoux, NOAA Southwest Fisheries Science Center, USA, personal communication, March 2015); and these are not quantified. It is the cumulative mortality across all these sources that should be compared to the local LRP. Thus, while impacts to the population from the CDGN fishery alone appear to be within LRP_{upper}, we cannot confidently state this to be the case for all impacts within the US West Coast EEZ combined. Second, our estimates of declining expected bycatch mortality assumes a somewhat random spatial process (animals bumping into passive net gear) for which interaction rates vary linearly with the abundance of animals and gear. This is likely reasonable for the leatherback case study, but in generalizing this approach to other systems, if animals were actively pursued or the probability of interaction does not scale with abundance for some other reason, the model may need revision. These issues notwithstanding, this analysis illustrates how Bayesian inference and LRP-based control rules could be used to improve management for bycatch of threatened populations, particularly for highly mobile species or rare-event situations.

Acknowledgements

We thank Lenfest Ocean Program and NOAA Office of Science and Technology for funding to support this work. Stephen Stohs and Summer Martin provided useful comments to improve the chapter.

References

Benson, S.R., Forney, K.A., Harvey, J.T, Carretta, J.V. & Dutton, P.H. 2007. Abundance, distribution, and habitat of leatherback turtles (*Dermochelys coriacea*) off California, 1990–2003. *Fishery Bulletin,* 105, 337–347.

Benson, S.R., Eguchi, T., Foley, D.G., Forney, K.A., Bailey, H., Hitipeuw, C., Samber, B.P., Tapilatu, R.F., Rei,V, Ramohia, P., Pita, J. & Dutton, P.H. 2011. Large-scale movements and high-use areas of western Pacific leatherback turtles, *Dermochelys coriacea. Ecosphere,* 2, art84.

Caddy, J.F. & Mahon, R. 1995. Reference points for fisheries management. *FAO Fisheries Technical Paper* No. 347. FAO, Rome.

Carretta, J.V. & Moore, J.E. 2014. Recommendations for pooling annual bycatch estimates when events are rare. *NOAA Technical Memorandum* NOAA-TM-NMFS-SWFSC-528.

Carretta, J.V., Oleson, E., Weller, D.W., Lang, A.R., Forney, K.A., Baker, J., Hanson, B., Martien, K., Muto, M.M., Orr, A.J., Huber, H., Lowry, M.S., Barlow, J., Lynch, D., Carswell, L., Brownell, Jr., R.L. & Mattila, D.K. 2014. U.S. Pacific marine mammal stock assessments, 2013. U.S. Department of Commerce, NOAA Technical Memorandum NMFS, NOAA-TM-NMFS-SWFSC-532.

Chilvers, B.L. 2008. New Zealand sea lions *Phocarctos hookeri* and squid trawl fisheries: bycatch problems and management options. *Endangered Species Research,* 5, 193–204.

CIRVA. 2014. Report of the Fifth Meeting of the 'Comité Internactional para la Recuperación de la Vaquita' (CIRVA-5). Available at http://vaquita.tv/wp-content/uploads/2014/08/Report-of-the-Fifth-Meeting-of-CIRVA.pdf (last accessed 5 January 2015).

Croxall, J.P., Butchart, S.H.M., Lascelles, B., Stattersfield, A.J., Sullivan, B., Symes, A. & Taylor, P. 2012. Seabird conservation status, threats and priority actions: A global assessment. *Bird Conservation International,* 22, 1–34. doi:10.1017/S0959270912000020

Curtis, K.A. & Moore, J.E. 2013. Calculating reference points for sustainable take of marine turtles. *Aquatic Conservation: Marine and Freshwater Ecosystems,* 23, 441–459.

Curtis, K.A., Moore, J.E. & Benson, S.R. 2015b. Estimating limit reference points for western Pacific leatherback turtles (Dermochelys coriacea) in the U.S. West Coast EEZ. *PLoS ONE,* 10(9), e0136452. doi:10.1371/journal.pone.0136452

Curtis K.A., Moore, J.E., Boyd, C., Dillingham, P.W., Lewison, R.L., Taylor, B.L. & James, K.C. 2015a. Managing catch of marine megafauna: Guidelines for setting limit reference point. *Marine Policy,* 61, 249–263.

Davies, R.W.D., Cripps, S.J., Nickson, A. & Perter, G. 2009. Defining and estimating global marine fisheries bycatch. *Marine Policy,* 33, 661–672.

Dewar, H., Eguchi, T., Hyde, J., Kinzey, D., Kohin, S., Moore, J., Taylor, B.L. & Vetter, R. 2013. Status review of the northeastern Pacific population of white sharks (*Carcharodon carcharias*) under the Endangered Species Act. U.S. Department of Commerce, NOAA Technical Memorandum NMFS, NOAA-TM-NMFS-SWFSC-523.

Diffendorfer, J.E., Beston, J.A., Merrill, M.D., Stanton, J.C., Corum, M.D., Loss, S.R., Thogmartin, W.E., Johnson, D.H., Erickson, R.A. & Heist, K.W. 2015. Preliminary methodology to assess the national and regional impact of U.S. wind energy development on birds and bats. U.S. Geological Survey Scientific Investigations Report 2015–5066.

Dulvy, N.K., Fowler, S.L., Musick, J.A., Cavanagh, R.D., Kyne, P.M., Harrison, L.R., Carlson, J.K., Davidson, L.N.K., Fordham, S.V., Francis, M.P., Pollock, C.M., Simpfendorfer, C.A., Burgess, G.H., Carpenter, K.E., Compagno, L.J.V., Ebert, D.A., Gibson, C., Heupel, M.R., Livingston, S.R., Sanciangco, J.C., Stevens, J.D., Valenti, S. & White, W.T. 2014. Extinction risk and conservation of the world's sharks and rays. *eLife*, 3, e00590.

Dutton, P.H., Hitipeuw, C., Zein, M., Benson, S.R., Petro, G., Pita, J., Rei, V., Ambio, L. & Bakarbessy, J. 2007. Status and Genetic Structure of Nesting Populations of Leatherback Turtles (Dermochelys coriacea) in the Western Pacific. *Chelonian Conservation and Biology*, 6, 47–53.

FAO (Food and Agriculture Organization of the United Nations) 1995. *Code of Conduct for Responsible Fisheries*, Rome, FAO.

Garcia, S.M. & Staples, D.J. 2000. Sustainability reference systems and indicators for responsible marine capture fisheries: A review of concepts and elements for a set of guidelines. *Marine and Freshwater Research*, 51, 385–426.

Gaspar, P., Benson, S.R., Dutton, P.H., Réveillére, A., Jacob, G., Meetoo, C., Dehecq, A. & Fossette, S. 2012. Oceanic dispersal of juvenile leatherback turtles: going beyond passive drift modeling. *Marine Ecology Progress Series*, 457, 265–284.

Gilman, E. 2011. Bycatch governance and best practice mitigation technology in global tuna fisheries. *Marine Policy*, 35, 590–609.

Hall, M.A., Alverson, D.L. & Metuzals, K.I. 2000. By-catch: problems and solutions. *Marine Pollution Bulletin*, 41, 204–219.

Hall, S.J. & Mainprize, B. 2004. Towards ecosystem-based fisheries management. *Fish and Fisheries*, 5, 1–20.

Hanan, D.A., Holts, D.B. & Coan, Jr., A.L. 1993. The California drift gill net fishery for sharks and swordfish, 1981–82 through 1990–91. *Fish Bulletin*, 175, 95.

Heppell, S.S., Crowder, L.B. & Crouse, D.T. 1996. Models to evaluate headstarting as a management tool for long-lived turtles. *Ecological Applications*, 6, 556–565.

Heppell, S.S., Heppell, S.A., Read, A. & Crowder, L.B. 2005. Effects of fishing on long-lived marine organisms. In: Norse, E.A. & Crowder, L.B. (eds.) *Marine Conservation Biology*, pp. 211–231. Washington, DC: Island Press.

Heppell, S.S., Crowder, L.B. & Menzel, T. 1999. Life table analysis of long-lived marine species with implications for management. *American Fisheries Society Symposium*, 23, 137–148.

Hobday, A.J., Smith, A.D.M., Stobutzki, I.C., Bulman, C., Daley, R., Dambacher, J.M., Deng R.A., Dowdney, J., Fuller, M., Furlani, D., Griffiths, S.P., Johnson, D., Kenyon, R., Knuckey, I.A., Ling, S.D., Pitcher, R., Sainsbury, K.J., Sporcic, M., Smith, T., Turnbull, C., Walker, T.I., Wayte, S.E., Webb, H., Williams, A., Wise, B.S. & Zhou, S. 2011. Ecological risk assessment for the effects of fishing. *Fisheries Research*, 108, 372–384.

Jones, T.T., Hastings, M.D., Bostrom, B.L., Pauly, D. & Jones, D.R. 2011. Growth of captive leatherback turtles, Dermochelys coriacea, with inferences on growth in the wild: Implications for population decline and recovery. *Journal of Experimental Marine Biology and Ecology*, 399, 84–92.

Kock, K.-H., Reid, K., Croxall, J. & Nicol, S. 2007. Fisheries in the Southern Ocean: An ecosystem approach. *Philosophical Transactions of the Royal Society B,* 362, 2333–2349.

Lewison, R.L., Crowder, L.B., Wallace, B.P., Moore, J.E., Cox, T.M., Žydelis, R., McDonald, S., DiMatteo, A., Dunn, D., Kot, C.Y., Bjorkland, R., Kelez, S., Soykan, C., Stewart, K.R., Sims, M., Boustany, A., Read, A.J., Halpin, P, Nichols, W.J. & Safina, C. 2014. Global patterns of marine mammal, seabird and sea turtle bycatch reveal taxa-specific and cumulative megafauna hotspots. *Proceedings of the National Academy of Sciences of the United States,* 111, 5271–5276.

Martin, S.L., Stohs, S.M. & Moore, J.E. 2015. Bayesian inference and assessment for rare-event bycatch in marine fisheries: a drift gillnet fishery case study. *Ecological Applications,* 25, 416–429.

Moore, J.E., Wallace, B., Lewison, R., Žydelis, R., Cox, T. & Crowder, L. 2009. A review of marine mammal, sea turtle and seabird bycatch in USA fisheries and the role of policy in shaping management. *Marine Policy,* 33, 435–451.

Moore, J.E., Curtis, K.A., Dillingham, P.W., Cope, J.M., Fordham, S., Heppell, S., Pardo, S.A., Simpfendorfer, C.A., Tuck, G. & Zhou, S. 2013. Evaluating sustainability of fisheries bycatch mortality for marine megafauna: conservation reference points for data-limited populations. *Environmental Conservation,* 40, 329–344.

MRAG Ltd, Poseidon, and Lamans s.a. 2011. Contribution to the preparation of a Plan of Action for Seabirds. European Commission, Final Report. Available at http:// ec.europa.eu/fisheries/documentation/studies/seabirds_2011_en.pdf (last accessed 19 November 2015).

NMFS. 2015. Preliminary draft environmental assessment of drift gillnet hard caps and monitoring alternatives. Prepared by Dept. of Commerce, National Marine Fisheries Service, West Coast Region, Long Beach, California, USA, August 2015. Available at http://www.pcouncil.org/wp-content/uploads/2015/08/G2a_NMFS_Rpt1_DGN_draftEA_and_metrics_SEPT2015BB.pdf (last accessed 12 January 2016).

Read, A.J. 2008. The looming crisis: interactions between marine mammals and fisheries. *Journal of Mammalogy,* 89, 541–548.

Reeves, R.R., Dawson, S.M., Jefferson, T.A., Karczmarski, L., Laidre, K., O'Corry-Crowe, G., Rojas-Bracho, L., Secchi, E.R., Slooten, E., Smith, B.D., Wang, J.Y. & Zhou, K. 2013. *Cephalorhynchus hectori ssp. maui.* The IUCN Red List of Threatened Species. Version 2014.3. www.iucnredlist.org. Downloaded on 3 January 2015.

Reuter, R.F., Conners, M.E., Dicosimo, J., Gaichas, S., Ormseth, O. & Tenbrink, T.T. 2010. Managing non-target, data-poor species using catch limits: lessons from the Alaskan ground fishery. *Fisheries Management and Ecology,* 17, 323–335.

Richard, Y. & Abraham, E.R. 2013. Application of Potential Biological Removal methods to seabird populations. New Zealand Aquatic Environment and Biodiversity Report 108.

Rojas-Bracho, L., Reeves, R.R., Jaramillo-Legorreta, A., Taylor B.L. 2008. *Phocoena sinus.* The IUCN Red List of Threatened Species. Version 2014.3. www.iucnredlist.org. Downloaded on 3 January 2015.

Senko, J., White, E.R., Heppell, S.S. & Gerber L.R. 2013. Comparing bycatch mitigation strategies for vulnerable marine megafauna. *Animal Conservation,* 17, 5–18.

Shore, T. 2013. California's Deadliest Catch: The Drift Gillnet Fishery for Swordfish and Shark, an Exposé and Call for Action. Turtle Island Restoration Network.

Tapilatu, R.F., Dutton, P.H., Tiwari, M., Wibbels, T., Ferdinandus, H.V., Iwanggin, W.G. & Nugroho, B.H. 2013. Long-term decline of the western Pacific leatherback, *Dermochelys coriacea:* a globally important sea turtle population. *Ecosphere,* 4, art25.

Tiwari, M., Wallace, B.P. & Girondot, M. 2013. *Dermochelys coriacea (West Pacific Ocean subpopulation)*. The IUCN Red List of Threatened Species. Version 2014.3. www.iucnredlist. org. Downloaded on 3 January 2015.

Turvey, S.T., Pitman, R.L., Taylor, B.L., Barlow, J., Akamatsu, T., Barrett, L.A., Zhao, X., Reeves, R.R., Stewart, B.S., Wang, K., Wei, Z., Zhang, X., Pusser, L.T., Richlen, M., Brandon, J.R. & Wang, D. 2007. First human-caused extinction of a cetacean species? *Biology Letters*, 3, 537–540.

Wallace, B.P., DiMatteo, A.D., Bolten, A.B., Chaloupka, M.Y., Hutchinson, B.J., Abreu-Grobois, F.A., et al. 2011. Global conservation priorities for marine turtles. *PLoS ONE*, 6(9): e24510.

Wallace, B.P., Heppell, S.S., Lewison, R.L., Kelez, S. & Crowder, L.B. 2008. Impacts of fisheries bycatch on loggerhead turtles worldwide inferred from reproductive value analyses. *Journal of Applied Ecology*, 45, 1076–1085.

Wallace, B.P., Tiwari, M. & Girondot, M. 2013. *Dermochelys coriacea (East Pacific Ocean subpopulation)*. The I. UCN Red List of Threatened Species. Version 2014.3. www.iucnredlist.org. Downloaded on 03 January 2015.

Werner, T., Krauss, S., Read, A. & Zollett, E. 2006. Fishing techniques to reduce the bycatch of threatened marine animals. *Marine Technology Society Journal*, 40, 50–68.

Zhou, S. & Griffiths, S.P. 2008. Sustainability assessment for fishing effects (SAFE): a new quantitative ecological risk assessment method and its application to elasmobranch bycatch in an Australian trawl fishery. *Fisheries Research*, 91, 56–68.

Zhou, S., Smith, A.D.M. & Fuller, M. 2011. Quantitative ecological risk assessment for fishing effects on diverse non-target species in a multi-sector and multi-gear fishery. *Fisheries Research*, 112, 168–178.

Žydelis, R., Bellebaum, J., Österblom, H., Vetemaa, M., Schirmeister, B., Stipniece, A., Dagys, M., van Eerden, M. & Garthe, S. 2009. Bycatch in gillnet fisheries – An overlooked threat to waterbird populations. *Biological Conservation*, 142, 1269–1281.

## Chapter 15

Using simulation evaluation to account for ecosystem considerations in fisheries management

Éva E. Plagányi

Introduction

There is growing recognition of the need to extend more traditional single-species approaches to account for "ecosystem considerations," and hence achieve an ecosystem approach to fisheries (EAF) (Smith et al., 2007). This necessitates taking into account the knowledge and uncertainties about biotic, abiotic and human components of ecosystems and their interactions and applying an integrated approach to fisheries within ecologically meaningful boundaries (FAO 2003).

Simulation models are increasingly being used to evaluate alternative management approaches or harvest control rules, or to identify the potential for trade-offs among fisheries management objectives (Christensen and Walters, 2004, Link, 2010, Gaichas et al., 2012, Levin et al., 2013). Although the goals of EAF can be accomplished in some cases without the need for models, use of appropriate models and management strategy evaluation (MSE) (Smith et al., 2007) are widely accepted as integral components of advancing an EAF approach (Levin et al., 2009, Link, 2010). This is also considered best practice by FAO (FAO, 2008). MSE is an ideal tool because of its ability to account for uncertainty as well as to make the trade-offs between diverse societal objectives explicit.

MSE approaches can serve as formal risk assessment methods, given their focus on the identification and modelling of uncertainties as well as in balancing different representations of resource dynamics (Sainsbury et al., 2000). This includes consideration of the implications, for both the resource and its stakeholders, of alternative combinations of monitoring data, analytical procedures, and decision rules (Sainsbury et al., 2000, Rademeyer et al., 2007, Smith et al., 2007). By identifying and evaluating trade-offs in performance across a range of management objectives, it provides indicators on whether different objectives can be reconciled and whether the outcomes are robust to inherent uncertainties in the inputs and assumptions on which decisions are based (Cooke, 1999).

In this chapter, I summarise MSE approaches that have been developed to include ecosystem considerations as part of advice on fisheries management. I provide examples of the full spectrum of approaches used to advance an EAF, from the most simple coupling of predator-prey population dynamics, to using comprehensive end-to-end models, such as *Atlantis* and *InVitro*, that represent entire systems through their physical, biological and human interactions (Fulton, 2010, Rose et al., 2010). To aid discussion I distinguish between comprehensive models, that are designed to describe the larger contextual setting, and minimalist models that are focused on specific components of the system. This latter group have been termed models of intermediate complexity for ecosystem assessments (MICE) (Plagányi et al., 2014), because they only consider ecosystem components of direct relevance to the management question being addressed. MICE are simpler than whole-of-ecosystem models in terms of the number of components (e.g. species or functional groups) or processes represented. Characteristically they can be fitted to data, apply standard statistical methods for parameter estimation and account for a broad range of uncertainties. This means they can be applied for tactical as well as strategic fisheries decision-making (Danielsson et al., 1997, Plagányi and Butterworth, 2012, Plagányi et al., 2014).

Ecosystem models have been variously classified and compared, both in a local context (e.g. Hollowed et al., 2011b) and globally (e.g. Hollowed et al., 2000, Fulton et al., 2003, Plagányi, 2007, FAO, 2008), but the distinction between tactical and strategic models is most important here. Tactical models are intended to support specific management decisions. For example, they may be used within a management procedure (MP) as part of a system for recommending the total catch (either single-species or for a multispecies complex, or the catch of a target species that takes into account impacts on other species in the ecosystem). Strategic models are focused on a broad-scale assessment of direction and change in the ecosystem. Their uses include improving understanding of the impact of different management alternatives on the structure and functioning of an ecosystem, as well as the social and economic consequences. Typically, comprehensive models are used strategically, whereas MICE can be used for both strategic and tactical purposes.

In support of tactical decision-making, ecosystem models allow for MPs that can explicitly account for both biological and economic factors, as demonstrated by the bioeconomic multispecies model (including technical but not trophic interactions) applied to Australia's Northern Prawn Fishery (NPF) (Dichmont et al., 2010, Dichmont et al., Chapter 10, this volume). However use of an ecosystem model in this way is still rare, and the major developments so far in application of an EAF have constituted the use of ecosystem models to simulation evaluate proposed MPs. This represents a strategic application, and when applied in this way they are referred to as operating models (OMs).

Multispecies and ecosystem operating models

Operating models provide the simulation testing frameworks used in MSE. They are mathematical-statistical models taken to represent the true underlying resource dynamics (Rademeyer et al., 2007) (see also Dankel and Edwards, Chapter 1, this volume). Multispecies OMs are here defined as models that include at least two species but do not explicitly include a trophic link between the species (although they may have a technical interaction, e.g. Dichmont et al., 2010). On the other hand, ecosystem OMs are similarly defined as any model including at least two species, as well as a trophic interaction (Plagányi et al., 2014). OMs (especially single-species models) are typically "conditioned" on available information (including fisheries and ecological data, analogous to fitting an assessment model), by adjusting parameter values to ensure plausibility and consistency with this information (Rademeyer et al., 2007). This becomes increasingly difficult as the OM expands to include more ecosystem components, but there are nevertheless increasing efforts to use ecosystem models as OMs within an MSE context (Gamble and Link, 2012, Link and Bundy, 2012).

In applications where an ecosystem rather than single-species OM is required, there are a number of guidelines as to the number of additional species or species groups to add. These include: (1) not aggregating serially linked groups (predator and prey), or species, age-classes or functional groups with rate constants which differ by more than two- to threefold (Fulton et al., 2003), (2) select model aggregation appropriately based on the question at hand; (3) similarly, select appropriate spatial and temporal scales to simulate ecological processes for the current question (Essington and Plagányi, 2014); (4) include sufficient (but no more) fishing sectors and predators to account for at least 90% of the mortality on each of the key species groups (Punt and Butterworth, 1995, Plagányi et al., 2014); and (5) evaluate whether there are key prey species that need to be included (without including all prey groups; e.g. Smith et al., 2011) in order to address a specific management question.

Ultimately, the OM should be designed to represent the system components that are important to management. For example, fishery harvest policies are increasingly turning to the maximum economic yield (MEY) as the target harvest yield (Dichmont et al., 2010), requiring economic components to be included in the OM. It may also be important to balance not only conflicting conservation needs and economic development, but also societal and cultural backgrounds (Castilla and Defeo, 2005, Andrew et al., 2007, Beddington et al., 2007). This is particularly the case where resource use is a key part of tradition and custom of indigenous communities (Berkes et al., 2000, Plagányi et al., 2013b). Hence underlying successful fisheries management are "triple bottom line" (Elkington, 1994) sustainability objectives involving trade-offs between economic, social and biological performance. Modern fisheries management is therefore challenged by a need to broaden the components included in an OM and modify the overall MSE adaptive management framework accordingly.

In parallel to developments that have recognised the need to account for not only the target species, but also broader interactions in ecosystems, there has been an increased urgency to account for the effects of climate change on fish and fisheries (Hollowed et al., 2011a, Plagányi et al., 2011, Gamble and Link, 2012, Szuwalski and Punt, 2012, Szuwalski and Punt, Chapter 7, this volume). Given the uncertainty associated with ecosystem responses to climate forcing, management models should ideally include a monitoring component that is capable of capturing key changes in the ecosystem and harvest control rules should ideally be evaluated across a range of biological, economic and social indicators to the extent possible. MSE-based risk management frameworks have been developed to simultaneously account for uncertainty in biological understanding as well as projected climate change impacts (Plagányi et al., 2013a). The schematic in Figure 15.1 illustrates how an adaptive management loop can be included as an outer loop around a MSE testing loop, to accommodate regular updates to the OM, as scientific understanding of climate change impacts affecting each individual fishery improves.

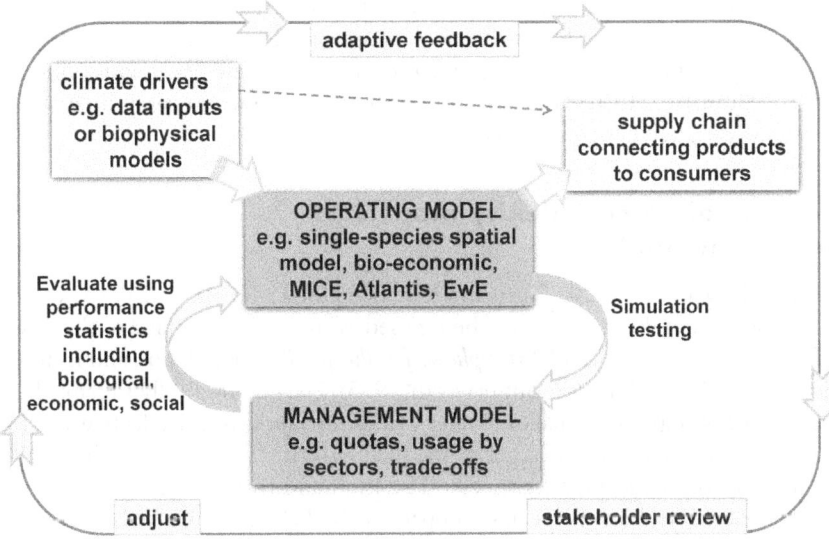

Figure 15.1 Schematic summary of an outer, adaptive feedback loop around a Management Strategy Evaluation (MSE), which consists of an Operating Model and Management Model connected by an inner feedback loop. Allowing adaptive updates to the MSE testing process provides a framework for making fisheries management responsive to a changing climate. Management procedures (with their feedback loops) are predicted to be more successful in terms of detecting and stalling fishery downturns (Plagányi et al., 2011).

Several OM examples have been developed for use within MSE frameworks. For instance, the *Atlantis* modelling framework (Fulton et al., 2005) has been closely aligned to efforts to evaluate ecosystem indicators and alternative regional fisheries management options using an MSE approach. Other work has focused on the use of *Ecosim* models to investigate multispecies management strategies for capture fisheries (Pitcher and Cochrane, 2002). *In Vitro* extends the MSE approach still further to consider multiple use management (McDonald et al., 2008), and *ELFsim* (Mapstone et al., 2008) is an example of how intermediate complexity approaches can be used as an OM to consider fisheries management questions.

There are many practical limitations to including ecosystem effects explicitly in fisheries models and management procedures, but even if we cannot use an ecosystem model directly to make decisions, we can use a model to test whether our decisions are robust to ecosystem effects. In some cases, an interim solution is to implement potential effects in MP evaluation exercises only implicitly (Plagányi et al., 2007). For example, rather than developing complicated multispecies testing models to contribute to its Revised Management Procedure (RMP) development process, the Scientific Committee of the International Whaling Commission adopted a simpler approach of allowing for time-dependence in the intrinsic growth rate and carrying capacity parameters (where the variation is likely due to interspecific interactions) of the single-species operating model for the population under harvest (Punt and Donovan, 2007). Similarly, MP testing procedures for fisheries can use simplified surrogates for climate change impacts.

Intermediate complexity models as operating models

MICE (Plagányi et al., 2014) can and have been used as OMs. One of the earliest examples of a multispecies model being used as an OM was applied to the South African west coast fur seal (*Arctocephalus pusillus pusillus*) population and its interaction with hake (Punt and Butterworth, 1995). The commercially valuable hake consists of two species, a shallow-water (*Merluccius capensis*) and a deep-water (*M. paradoxus*) species, with the larger of the shallow-water species eating the smaller individuals of the deep-water species. The Punt and Butterworth (1995) model was restricted to the two species comprising the hake resource, seals, a grouped category of large predatory fish and the hake fishery. Together these were estimated to account for more than 90% of all mortality of hake. The level of detail taken into account for each component was based on need and data availability. Thus fully age-structured models were used for the two hake species (to capture cannibalism and interspecies predation effects), but the "other" predatory fish components were simply lumped into either a small or large fish category.

The approach of Punt and Butterworth (1995) involved taking explicit account of uncertainty and management issues through the use of a simulation framework that incorporated the feedback control rules actually in place for setting TACs for the hake fishery. The purpose of this approach was to check whether the management system applied to compute TACs was able to take advantage of a potential increase of hake sustainable yields after a seal (predator) reduction.

In the next section, I provide a further example of the use of ecosystem models as OMs in MSE testing of harvest control rules.

Using intermediate complexity approaches: accounting for the needs of dependent predators

A number of multispecies models have been developed in response to requests for scientific advice by the Commission for the Conservation of Antarctic Marine Living Resources (CCAMLR). For example, models have been developed for the Scotia Sea krill (*Euphausia superba*) fishery and its possible impacts on dependent predators (Plagányi and Butterworth, 2012, Watters et al., 2013). Although krill catches (where the total krill catch limit is determined in a separate process) are currently fairly low, it has been recognised that they could nonetheless make an appreciable ecosystem impact if they are concentrated in small localised areas that simultaneously serve as important foraging grounds for dependent predators. Operating models were therefore required to explore alternative scenarios involving subdivision of the precautionary catch limit for krill among 15 small-scale management units (SSMUs) in the Scotia Sea.

An important performance criterion in these examples was the need to assess and compare the current and future potential impacts of fishing on land-based predators, given that krill catches may increase substantially in the future. The OMs were used in the first instance to compare five options for allocating the catch limit among the SSMUs in the Scotia Sea, as presented in Hewitt et al. (2004):

1   proportional to the historical catch within the SSMU;
2   inversely related to estimated predator demand in the SSMU;
3   positively related to estimated standing stock of krill in the SSMU;
4   krill standing stock minus predator demand in the SSMU;
5   dynamic allocation based on land-based predator monitoring conducted just prior to or early in the fishing season (whereas the other options are static options with a proportional allocation that is constant in time).

Krill and predator abundance at the end of a projected 20-year fishing period were compared with equivalent no-fishing trials to compute the probability of being less than 75% of the median no-fishing level (assessing impacts relative to comparable no-fishing trials avoids drawing erroneous conclusions based on population trends; Plagányi and Butterworth 2012). Despite the considerable uncertainties included in the analyses, it was possible to detect differences between the results of catch allocation options 1 (high risk to predators), 4 (better performance) and 2–3, although there is less to discriminate between options 2 and 3 (Plagányi and Butterworth, 2012, Watters et al., 2013).

Based on experience with South African fisheries, Rademeyer et al. (2007) recommend using a reference set in preference to a single reference case when choosing core operating models for MP testing for populations for which there are a number of sources of major uncertainty about the dynamics. This approach was adopted in the analyses of Plagányi and Butterworth (2012), and a reference set comprising 12 alternative combinations of a basic operating model was used to bound the range of uncertainty associated with the krill-predator-fishery system. Appropriate boundaries were selected based on plausible biological ranges for parameters such as natural mortality, as well as accounting for uncertainty regarding the nature and strength of the interaction between predators and prey. In this way, the MP approach provides the appropriate framework for multispecies approaches through its focus on the identification and modelling of uncertainties, as well as through balancing different resource dynamics representations and associated trophic dependencies and interactions (Butterworth and Punt 1999, Sainsbury et al. 2000). More complex ecosystem models can also employ the approach of testing predictions with bounded parameterisations for key parameters, based on insights by the model developer and preliminary sensitivity analyses (Fulton, 2010). There are, however, computational constraints on more thorough explorations of the parameter space for large ecosystem models which can include thousands of parameters (and hence run times are too long to systematically explore the impacts of alternative parameter combinations).

There has been a similar need for modelling to inform management decisions in the South African purse seine fishery. In this system, it is important to ensure adequate escapement of anchovy (*Engraulis encrasicolus*) and sardine (*Sardinops sagax*) to avoid negative impacts on vulnerable marine predator species, such as the African penguin (*Spheniscus demersus*) (Crawford et al., 2006). The anchovy and sardine stocks are jointly managed using an operational management procedure (OMP – analogous to a MP) (de Moor et al., 2011, de Moor and Butterworth, Chapter 11, this volume). Use of an OMP has facilitated responding (without increasing risk) to major changes in resource abundance, as occurred when both species peaked concurrently off South Africa around the turn of the century (Coetzee et al., 2008a, de Moor et al., 2011). Sardine populations on the west coast declined thereafter due to an eastward shift in their distribution, coincidental with cooling of inshore waters that took place on the south coast during the 1980s and 1990s (Roy et al., 2007, Coetzee et al., 2008b, Rouault et al., 2009, Rouault et al., 2010). The decline in the abundance of pelagic fish, the dominant prey of African penguins, was followed by a steep decline in penguin breeding numbers (with a time-lag effect given it takes 3–4 years for birds to mature) at two of the major colonies on the west coast (Cury et al., 2000, Crawford et al., 2006). This motivated development of a penguin population dynamics model that could be linked to the pelagic OM used in developing the OMP. The combined model could take account of the relationship between breeding success of African penguins and the abundance of both fish species (Robinson et al., 2015).

A penguin population dynamics model was fitted to available data on penguin abundance in the form of penguin moult counts and resightings of tagged penguins. The model suggested that penguin adult annual mortality may be a function of local sardine biomass (Robinson, 2013, Robinson et al., 2015) and so it was linked to the pelagic OM. This link meant that penguin population dynamics could be predicted under different future sardine and anchovy abundance trajectories. These abundance trajectories were chosen to match those used in developing the sardine-anchovy OMP. In this way, the performance of alternative MPs for the fishery could be evaluated taking into account potentially negative impacts of the fishery on a dependent predator species. An important component of this research involved the need to separate the role of fishing versus environmental effects in driving changes in a dependent predator, and results suggested that fishing has a relatively small impact on penguins (Robinson, 2013, Robinson et al., 2015). Rather, the spatial shift in the distribution of a key prey species, namely sardine, has likely had a substantial negative impact on the penguins (Robinson, 2013).

Extending intermediate complexity approaches: accounting for social, cultural and economic considerations

Triple bottom line approaches supporting management need to account for social objectives, but these are rarely explicitly included in the definition of fishery targets because of the mismatch between quantitative metrics and more qualitative social information (but see Haapasaari et al., 2012). Plagányi et al. (2013b) developed a framework that substantially expands on previous MSE approaches and introduces new linkages in order to holistically compare social, economic and biological trade-offs anticipated under a range of proposed alternative management strategies and scenarios.

Plagányi and colleagues used the Torres Strait tropical rock lobster (TRL; *Panulirus ornatus*) fishery as an example, which is the most important commercial fishery to Torres Strait Islanders. The fishery is managed by the Protected Zone Joint Authority (PZJA), consisting of representatives from the Australian and Queensland governments under Article 22 of the Torres Strait Treaty (February 1985) between Australia and Papua New Guinea. The Treaty refers not only to the need for optimal utilisation of the resource, but also the need to maximise the opportunities for participation by the traditional inhabitants of both countries. It is therefore important to evaluate the performance of alternative strategies in terms of sociocultural performance indicators for this fishery, in addition to the more conventional biological and economic indicators.

An evaluation of the trade-offs between alternative strategies comprising various derivatives of the introduction of an output system (TAC – total allowable catch) was undertaken. These included an individual quota per indigenous fishing license (TIB – traditional inhabitant fishing boat), a community-based system, or an "Olympic"-type system (involving the setting of a global total quota that is available to be fished by either the whole sector or part thereof). The current status quo (input system) was included as a base case for comparison. The underlying operating model includes 16 spatial areas and is fitted to midyear population survey data available since 1989, as well as catch per unit effort (CPUE) information.

The inclusion of the social dimension was possible because of extensive stakeholder consultation that included a series of dedicated workshops and individual interviews. In order to simulate the outcomes of different management systems, it was necessary to quantify how participation by each fisher subgroup would change. To estimate participation rate changes for the indigenous (TIB) sector, which depends on social as well as economic drivers, a Bayesian network (BN) analysis was used (van Putten et al., 2013). The TIB fleet was divided into subfleets based on a typology of activity and alternative licensing arrangements (as well as technical and economic factors), and this facilitated prediction of the changes in participation under alternative scenarios. The coupled bioeconomic model calculated profit per sector/subfleet but also incorporated a production function and frontier analysis (Pascoe et al., 2013a), a data envelopment analysis (DEA) (Pascoe et al., 2013b) (to estimate which nonindigenous vessels might exit the fishery with lower quota levels) and included estimates of owner-operator returns to labour for the islander traditional boat holder license operators. Finally the framework included the supply chain, which enabled prediction of an additional performance indicator, namely value added.

The MSE outputs of Plagányi et al. (2013b) highlight the complexity of trade-offs between social and economic considerations for islander (indigenous) and nonislander participants, as well as the important role of external agency decision-making. This example was based on extensive data and stakeholder consultation, but is one way forward in terms of quantifying and making explicit the trade-offs in the impact of alternative management strategies on the triple bottom line sustainability objectives.

Comprehensive ecosystem models and MSE

The *Atlantis* modelling framework (Fulton et al., 2011a) is a comprehensive (or end-to-end) model that can be used as an OM for MSE. For example, it has successfully been used to compare alternative fisheries management strategies for a complex multispecies fishery in southeastern Australia (Fulton et al., 2014). *Atlantis* models and agent-based approaches such as *InVitro* represent not only the key biophysical properties of systems, but also extend to link economic, social and psychological modules (e.g. the *InVitro* modelling framework simulates individual operators or demographic actors making decisions based on experience versus expectations as conditioned on attitude profiles), as well as including all relevant industries (such as commercial fishing, tourism, oil and gas exploration, salt production, mariculture, port operations, shipping and road transport), either as fixed inputs or dynamically (Fulton et al., 2015). Representation of the feedback management cycle is then closed by including the management methods and decision-making process. The human side of fisheries is thus captured not only through consideration of more conventional economic drivers, but also through explicit modelling of social, psychological and economic drivers that mediate fishing fleet behavior as well as the behavior of other actors in the system. Fulton et al. (2011b) stressed that it is important to consider human dimensions in ecosystem models, for example to avoid unintended consequences (such as effort displacement) that can result due to management that modifies the incentives of resource users. Fishing fleet models are one way to improve prediction of changes in incentives, and hence in fisher location choice and impact on natural resources (Fulton et al., 2011b).

Given the challenges in parameterising whole ecosystem models and the associated paucity of suitable data, *Atlantis* is not statistically fitted to data. Instead historic trajectories for key species are subjectively compared with historical observations. Calibration of the model means that they generally compare well (Fulton et al., 2011a).

Another example is from the Bering Sea Project (the combined Bering Ecosystem Study, BEST, and the Bering Sea Integrated Ecosystem Research Program, BSIERP), which used comprehensive ecosystem models including climate drivers, low trophic levels, fish dynamics and management processes to support fisheries management in the eastern Bering Sea (Punt et al., 2015). MSE was used to compare alternative management strategies.

A challenge with using comprehensive ecosystem models in MSE is that the range of management objectives considered can increase with complexity of the model. Compared to single-species approaches, a much broader range of performance measures needs to be generated and communicated effectively to a more diverse group of stakeholders; for example the biological status of both the target and bycatch species, biodiversity measures, economic measures and operational indices for an industry as well as social considerations.

Effectively communicating multivariate outputs is a challenge, and various approaches have been proposed and used, such as kite diagrams with each branch representing a different indicator (Shin et al., 2010, Fay et al., 2013). The kite diagrams in Figure 15.2 are from the southeastern Australian multispecies fishery, and compare the status quo strategy with the so-called integrated management strategy, which performed best overall (Figure 15.2), and was subsequently adopted. This strategy involved a mix of quota, gear and spatial management controls and its implementation has improved performance of social, economic and ecological objectives (Fulton et al., 2014).

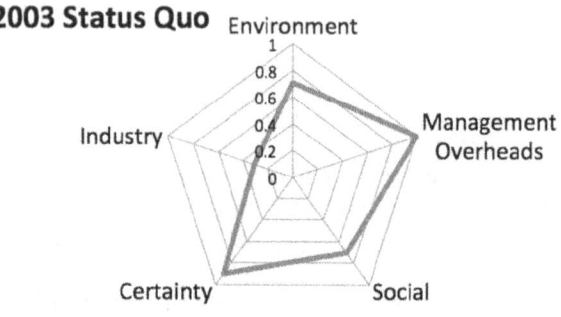

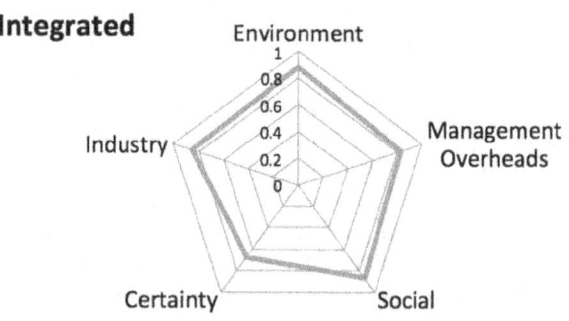

Figure 15.2 Overall performance of the integrated management strategy compared with the status quo strategy for the southeastern Australian multispecies fishery. Integrated (composite) performance measures were obtained using the Atlantis operating model (and normalised so 1.0 = good and 0.0 = poor performance), highlighting the improvement in overall performance of the integrated strategy compared with the status quo. Descriptions of alternative management strategies are given by Fulton et al. (2014).

Source: (Fulton et al., 2014).

Using comprehensive models in MSE: management recommendations for forage fish

Forage fish are small pelagic fish and krill that play critical roles in food webs by transferring energy from small, low trophic level planktonic species to large, valuable species such as large fish, seabirds, cetaceans and pinnipeds (Cury, 2000, Pikitch et al., 2012b). Localised depletion of forage fish is an important consideration for land-based breeding predators which may rely on local availability of prey, and there is increasing recognition that when this occurs in key foraging areas and at critical feeding times it may have a major effect on dependent predators (Hewitt et al., 2004, Plagányi and Butterworth, 2012, Robinson, 2013, Watters et al., 2013). Hence it has been recommended that management objectives for forage species should include consideration of broader ecological effects (Pikitch et al., 2012b). Seafood certification bodies such as the Marine Stewardship Council (MSC) have specified additional certification requirements for these species so that fisheries require an evaluation of whether a species is a "key" forage species (Smith et al., 2011, Essington and Plagányi, 2014, Plagányi and Essington, 2014), and if so, whether biomass targets and limits (i.e. the management strategy) are appropriate to protect dependent predators.

A number of ecosystem OMs have been used to explore how ecosystems respond to various forage fish management strategies. Smith et al. (2011) used a range of multispecies trophic models, namely *Osmose* (Shin and Cury, 1999), *Atlantis* and *Ecopath with Ecosim* (Christensen and Walters, 2004) to critically evaluate alternative harvest control rules for forage fish. Pikitch et al. (2012a) used *Ecopath with Ecosim* with an additional module that enables consideration of observation error and hence facilitates MSE (Christensen and Walters, 2004). Pikitch et al. (2012a) used both deterministic and stochastic models that included perturbations to the fishing mortality at realistic levels of variability. They defined a minimum biomass threshold B_{LIM}, at or below which point there is no fishing, and B_0 was computed as the terminal, unfished biomass of the forage fish. Minimum biomass thresholds of $0.2B_0$ and $0.4B_0$ were tested. A range of harvest control rules were evaluated: constant fishing mortality, constant yield "step" functions (SF, constant fishing mortality rate F until B_{LIM}), and "hockey stick" control rules (Figure 15.3). The hockey stick (HS) control rules, labeled 20/100 HS and 40/100 HS, set a B_{LIM} of 20% and 40% of unfished biomass, respectively, with F constant above B_0 and decreasing linearly for biomass between B_{LIM} and B_0. In the stochastic runs, the fishing mortality rate varied each year with a 30% coefficient of variation for the CF case, as well as for SF and HS cases when the biomass was greater than the lower biomass limit. Harvest control rules were applied separately to each of the 30 forage fish species included in the 10 *Ecopath with Ecosim* models analyzed.

In Pikitch et al. (2012a) the constant yield level strategies performed relatively poorly, with only very low catch levels sustainable by most of the populations modelled. The hockey stick rules performed best, yielding higher forage fish biomasses and lower predator declines for all fishing levels. The superior performance of both step function and hockey stick strategies over the CF strategy was attributed to the implementation of hard lower biomass limits (see also Essington et al., 2015), and the 40/100 HS rule consistently outperformed the 20/100 rule in terms of maintaining target species biomass at reasonably high levels, and avoiding large declines in dependent predators (Figure 15.4).

The simulations suggested that fishing rates corresponding to a fishing mortality F_{MSY} at the maximum sustainable yield (MSY), could negatively impact many dependent predators in ecosystems and incur a high risk of collapse of the forage fish population. Pikitch et al. (2012a) therefore recommended that fishing mortality rates applied to forage fish should not exceed $F = 0.5F_{MSY}$ (or about half the species' natural mortality rate $F = 0.5M$), to ensure with high probability (75 to 95%) that forage fishing will not result in the overly

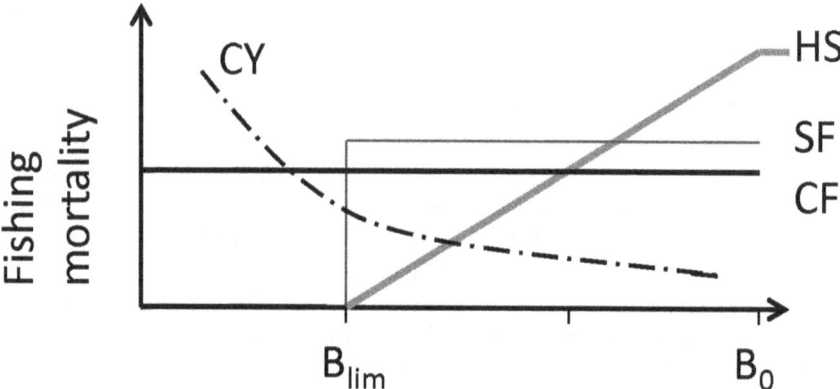

Figure 15.3 Schematic illustration of alternative harvest control rules for forage fish, including constant fishing mortality (CF), constant yield (CY) (this strategy performed poorly and hence wasn't included in the testing using a stochastic model), step functions (SF) (constant fishing mortality rate F until B_{lim}), and hockey stick (HS) control rules, with F decreasing linearly for biomass between B_{lim} and B_0.

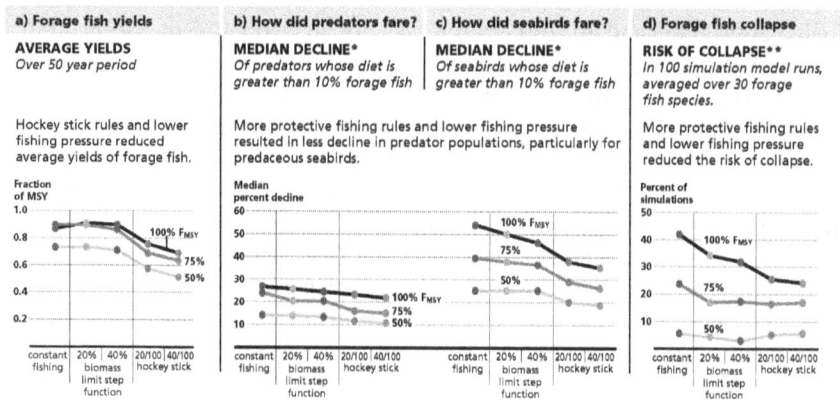

Figure 15.4 The impacts of alternative harvest control rules on both forage fish populations themselves, and their dependent predators, based on stochastic ecosystem models (*Ecopath with Ecosim*) at three different levels of fishing pressure: 100%, 75%, 50% of F_{MSY}. The harvest control rules are: constant fishing at each of the three levels, 20% B_{lim} step function, 40% B_{lim} step function, 20/100 hockey stick and 40/100 hockey stick with upper bounds set at each of the three fishing levels tested. More conservative rules reduced yields (a), but were much better at protecting predators (b and c) and had a lower risk of forage fish stock collapse (d).

Source: Pikitch et al. (2012a), with permission.

concerning declines in dependent predators. Their recommendations that more conservative harvest levels be applied to forage fish to account for their important ecological role were consistent with the recommendations of Smith et al. (2011). These authors showed that when exploitation rates were halved in the models used, it was still possible to achieve 80% of MSY but with substantially lower impacts on marine ecosystems and likely greater economic profitability. These studies demonstrate how ecosystem OMs can be used to evaluate alternative management strategies and harvest control rules in a more holistic manner that takes into account the impacts of harvesting one species on other species in the ecosystem.

Multispecies operating models with climate drivers

Improvements in understanding the functional relationships between climate variability and fish production are increasingly enabling their explicit incorporation in fisheries models (Keyl and Wolff, 2008, Hollowed et al., 2009, Hollowed et al., 2011a, Hollowed et al., 2011b, Ianelli et al., 2011, Stock et al., 2011). For example, (Gamble and Link, 2012) used a simple operating model to represent the northeastern US large marine ecosystem fish community under a range of climate effects scenarios. Their study stresses the need for operating models that are capable of evaluating the effects of a triad of drivers (exploitation, ecological interactions and the physical environment) on fish populations. In concert with these developments, much attention has focused on improving the robustness of projections of climate impacts, for example using IPCC-class climate models (International Panel on Climate Change; Stock et al., 2011).

Hollowed et al. (2011a) have developed new modelling tools to improve predictions as to climate change impacts on the production and distribution of commercial fisheries off Alaska. For example, evaluation of management strategies for walleye pollock (*Theragra chalcogramma*) highlight the potential increase in risk to the fishery when using harvest control rules under a regime with reduced mean recruitment.

Plagányi et al. (2013a) provide an example of using a MSE to integrate across biological and climate uncertainties, and test the performance and risks (biological, multispecies, economic) of alternative management strategies applied to the Torres Strait *bêche-de-mer* (sea cucumber) fishery. A reference set (Rademeyer et al., 2007) of alternative model parameterisations was used to collectively capture some of the key biological uncertainties (e.g. alternative natural mortality estimates and steepness of the stock–recruitment relationship), as well as uncertainty of the likelihood (using high-risk scenarios only versus assuming both high- and medium-risk scenarios occur) and severity (accounting for a doubling of the severity of each postulated effect) of climate change effects. In this way they simultaneously integrated across a range of biological and climate impact uncertainties, and thereby tested a range of alternative harvest strategies to evaluate performance under changing climate. Similarly, Ianelli et al. (2015)

describe a multimodel inference framework that combines information from alternative models to better characterise uncertainty. They present an example applied to three species of groundfish in the Bering Sea, and incorporate both trophic and climate information.

The alternative management scenarios tested by Plagányi et al. (2013a) included options that could be used in data-poor fisheries, as well as those requiring monitoring and spatial management (i.e. adaptive feedback in response to climate change). Performance statistics were also modified so that they were relevant in a climate change context. For example these included statistics such as $B_{2030}^{sp} / \left[B_{2030}^{sp} \left(\text{no fishing or climate change} \right) \right]$, the expected spawning biomass at the end of the projection period, relative to the comparable simulation of no-fishing with no climate change effects, for each species averaged across the entire area. The same set of random numbers was used to generate sets of no-fishing projections for each species and spatial zone, as a baseline for comparisons with the range of projections with fishing and climate change. Finally a multispecies economic performance statistic was computed as average annual profit (US$ million), measured as the landed weight of each species multiplied by current average market prices (but did not account for costs of monitoring and adaptive management).

More recently, Plagányi et al. (2015) have applied a modified form of this model to the multispecies sea cucumber fishery in Australia's Great Barrier Reef Marine Park. They used a spatial and age-structured model that includes nine sea cucumber species with populations distributed across 162 zones. MSE was used to test rotational harvest strategies for sea cucumbers, and demonstrated a substantial reduction in the risk of localised depletion, higher long-term yields and improved economic performance compared with nonrotational strategies.

Conclusions

Management strategy evaluation is increasingly being applied in ecosystem-based fisheries management, because it is an effective tool for evaluating the trade-offs between multiple objectives. Its success can also be attributed to the fact that it is a consultative approach whereby all stakeholders can have input into the candidate operating models and management scenarios (Smith et al., 2007). As the approach demands clear criteria to judge the performance of the alternative strategies, the method forces participants to clarify their objectives and to specify relevant performance indicators. To meet this need, substantial progress has been made developing methods that can quantify diverse societal goals or ecosystem metrics (Murawski, 2000, Marasco et al., 2007).

Consistent with an EAF, MSE frameworks are increasingly being broadened to take into account not only target species, but also broader ecosystem interactions as well as abiotic and human components of ecosystems. Both comprehensive and minimum complexity ecosystem models are being developed and applied, and both the range of models, as well as advantages and disadvantages are reviewed in a number of publications (Hollowed et al., 2000, Plagányi, 2007, FAO, 2008, Fulton, 2010). The best OM to use depends on the questions that need to be addressed as well as the linkages in a system. For example, to test broad management strategies that affect multiple sectors, it may be necessary to use a whole ecosystem model, whereas if the focus is more narrow, such as the need to account for a dependent predator species when fishing is applied to its prey, then a simpler multispecies model may suffice as an OM.

Whole ecosystem OMs that have been used include *Atlantis*, *InVitro* and *Ecopath with Ecosim* and have demonstrated that successful system level management is possible (Fulton et al., 2014, Fulton et al., 2015). Intermediate complexity OMs such as MICE (Plagányi et al., 2014) are tied to observational data, utilise standard statistical tools and have demonstrated that rigorous evaluation and comparison of management strategies is possible in a multispecies context. These models will increasingly include spatial structure, which is seldom accounted for in traditional single-species models, but is an important consideration in models that explicitly represent trophic interactions and fishing activity (see Goethel et al., Chapter 16, this volume). Other important developments will include approaches such as that of Plagányi et al. (2013a) and Ianelli et al. (2015), who provide a biological complement to climate modelling, and account for important sources of uncertainty that are an integral part of effective risk management decision-making.

References

Andrew, N.L., Bene, C., Hall, S.J., Allison, E.H., Heck, S. & Ratner, B.D. 2007. Diagnosis and management of small-scale fisheries in developing countries. *Fish and Fisheries,* 8, 227–240.

Beddington, J.R., Agnew, D.J. & Clark, C.W. 2007. Current problems in the management of marine fisheries. *Science,* 316, 1713–6.

Berkes, F., Colding, J. & Folke, C. 2000. Rediscovery of traditional ecological knowledge as adaptive management. *Ecological Applications,* 10, 1251–1262.

Castilla, J.C. & Defeo, O. 2005. Paradigm shifts needed for world fisheries. *Science,* 309, 1324–1325.

Christensen, V. & Walters, C.J. 2004. Ecopath with Ecosim: methods, capabilities and limitations. *Ecological Modelling,* 172, 109–139.

Coetzee, J.C., Merkle, D., De Moor, C.L., Twatwa, N.M., Barange, M. & Butterworth, D.S. 2008a. Refined estimates of South African pelagic fish biomass from hydro-acoustic surveys: quantifying the effects of target strength, signal attenuation and receiver saturation. *African Journal of Marine Science,* 30, 205–217.

Coetzee, J.C., van der Lingen, C.D., Hutchings, L. & Fairweather, T.P. 2008b. Has the fishery contributed to a major shift in the distribution of South African sardine? *ICES Journal of Marine Science,* 65, 1676–1688.

Cooke, J.G. 1999. Improvement of fishery-management advice through simulation testing of harvest algorithms. *ICES Journal of Marine Science,* 56, 797–810.

Crawford, R.J.M., Barham, P.J., Underhill, L.G., Shannon, L.J., Coetzee, J.C., Dyer, B.M., Leshoro, T.M. & Upfold, L. 2006. The influence of food availability on breeding success of

African penguins Spheniscus demersus at Robben Island, South Africa. *Biological Conservation,* 132, 119–125.

Cury, P. 2000. Small pelagics in upwelling systems: patterns of interaction and structural changes in "wasp-waist" ecosystems. *ICES Journal of Marine Science,* 57, 603–618.

Cury, P., Bakun, A., Crawford, R.J.M., Jarre, A., Quinones, R.A., Shannon, L.J. & Verheye, H.M. 2000. Small pelagics in upwelling systems: patterns of interaction and structural changes in "wasp-waist" ecosystems. *Ices Journal of Marine Science,* 57, 603–618.

Danielsson, A., Stefánsson, G., Baldursson, F.M. & Thorarinsson, K. 1997. Utilization of the Icelandic Cod Stock in a Multispecies Context. *Marine Resource Economics* 12, 329–344.

De Moor, C.L., Butterworth, D.S. & De Oliveira, J.A.A. 2011. Is the management procedure approach equipped to handle short-lived pelagic species with their boom and bust dynamics? The case of the South African fishery for sardine and anchovy. *Ices Journal of Marine Science,* 68, 2075–2085.

Dichmont, C.M., Pascoe, S., Kompas, T., Punt, A.E. & Deng, R. 2010. On implementing maximum economic yield in commercial fisheries. *Proceedings of the National Academy of Sciences of the United States of America,* 107, 16–21.

Elkington, J. 1994. Towards the sustainable corporation – Win-Win-Win business strategies for sustainable development. *California Management Review,* 36, 90–100.

Essington, T.E., Moriarty, P.E., Froehlich, H.E., Hodgson, E.E., Koehn, L.E., Oken, K.L., Siple, M.C. & Stawitz, C.C. 2015. Fishing amplifies forage fish population collapses. *Proceedings of the National Academy of Sciences,* 112, 6648–6652.

Essington, T.E. & Plagányi, É.E. 2014. Pitfalls and guidelines for "recycling" models for ecosystem-based fisheries management: Evaluating model suitability for forage fish fisheries. *Ices Journal of Marine Science,* 71, 118–127.

FAO 2008. Best practices in ecosystem modelling for informing an ecosystem approach to fisheries. *FAO Fisheries Technical Guidelines for Responsible Fisheries.,* No. 4, Suppl. 2, Add. 1, 78.

Fay, G., Large, S.I., Link, J.S. & Gamble, R.J. 2013. Testing systemic fishing responses with ecosystem indicators. *Ecological Modelling,* 265, 45–55.

Fulton, E.A. 2010. Approaches to end-to-end ecosystem models. *Journal of Marine Systems,* 81, 171–183.

Fulton, E.A., Boschetti, F., Sporcic, M., Jones, T., Little, L.R., Dambacher, J.M., Gray, R., Scott, R. & Gorton, R. 2015. A multi-model approach to engaging stakeholder and modellers in complex environmental problems. *Environmental Science & Policy,* 48, 44–56.

Fulton, E.A., Link, J.S., Kaplan, I.C., Savina-Rolland, M., Johnson, P., Ainsworth, C., Horne, P., Gorton, R., Gamble, R.J., Smith, A.D.M. & Smith, D.C. 2011a. Lessons in modelling and management of marine ecosystems: the Atlantis experience. *Fish and Fisheries,* 12, 171–188.

Fulton, E.A., Smith, A.D. & Punt, A.E. 2005. Which ecological indicators can robustly detect effects of fishing? *ICES Journal of Marine Science: Journal du Conseil,* 62, 540–551.

Fulton, E.A., Smith, A.D.M. & Johnson, C.R. 2003. Effect of complexity on marine ecosystem models. *Marine Ecology Progress Series,* 253, 1–16.

Fulton, E.A., Smith, A.D.M., Smith, D.C. & Johnson, P. 2014. An integrated approach is needed for ecosystem based fisheries management: Insights from ecosystem-level management strategy evaluation. *Plos ONE,* 9(1): e84242.

Fulton, E.A., Smith, A.D.M., Smith, D.C. & van Putten, I.E. 2011b. Human behaviour: The key source of uncertainty in fisheries management. *Fish and Fisheries,* 12, 2–17.

Gaichas, S., Gamble, R., Fogarty, M., Benoît, H., Essington, T., Fu, C., Koen-Alonso, M. & Link, J. 2012. Assembly rules for aggregate-species production models: Simulations in support of management strategy evaluation. *Marine Ecology Progress Series*, 459, 275–292.

Gamble, R.J. & Link, J.S. 2012. Using an aggregate production simulation model with ecological interactions to explore effects of fishing and climate on a fish community. *Marine Ecology Progress Series*, 459, 259–274.

Haapasaari, P., Kulmala, S. & Kuikka, S. 2012. Growing into Interdisciplinarity: How to converge biology, economics, and social science in fisheries research? *Ecology and Society*, 17(1): 6.

Hewitt, R.P., Watters, G., Trathan, P.N., Croxall, J.P., Goebel, M.E., Ramm, D., Reid, K., Trivelpiece, W.Z. & Watkins, J.L. 2004. Options for allocating the precautionary catch limit of krill among small-scale management units in the Scotia Sea. *Ccamlr Science*, 11, 81–97.

Hollowed, A., A'mar, T., Barbeaux, S., Bond, N., Ianelli, J., Spencer, P. & Wilderbuer, T. 2011a. Integrating ecosystem aspects and climate change forecasting into stock assessments. *ASFC Quarterly Report Research Feature*, July–August–September, NOAA Alaska Fisheries Science Center.

Hollowed, A.B., Aydin, K.Y., Essington, T.E., Ianelli, J.N., Megrey, B.A., Punt, A.E. & Smith, A.D.M. 2011b. Experience with quantitative ecosystem assessment tools in the northeast Pacific. *Fish and Fisheries*, 12, 189–208.

Hollowed, A.B., Bax, N., Beamish, R., Collie, J., Fogarty, M., Livingston, P., Pope, J. & Rice, J.C. 2000. Are multispecies models an improvement on single-species models for measuring fishing impacts on marine ecosystems? *ICES Journal of Marine Science: Journal du Conseil*, 57, 707.

Hollowed, A.B., Bond, N.A., Wilderbuer, T.K., Stockhausen, W.T., A'mar, Z.T., Beamish, R.J., Overland, J.E. & Schirripa, M.J. 2009. A framework for modelling fish and shellfish responses to future climate change. *Ices Journal of Marine Science*, 66, 1584–1594.

Ianelli, J.N., Hollowed, A.B., Haynie, A.C., Mueter, F.J. & Bond, N.A. 2011. Evaluating management strategies for eastern Bering Sea walleye pollock (Theragra chalcogramma) in a changing environment. *Ices Journal of Marine Science*, 68, 1297–1304.

Ianelli, J., Holsman, K.K., Punt, A.E. & Aydin, K. 2015. Multi-model inference for incorporating trophic and climate uncertainty into stock assessments. *Deep Sea Research Part II: Topical Studies in Oceanography*, doi:10.1016/j.dsr2.2015.04.002

Keyl, F. & Wolff, M. 2008. Environmental variability and fisheries: What can models do? *Reviews in Fish Biology and Fisheries*, 18, 273–299.

Levin, P.S., Fogarty, M.J., Murawski, S.A. & Fluharty, D. 2009. Integrated ecosystem assessments: developing the scientific basis for ecosystem-based management of the ocean. *PLoS Biology*, 7, e1000014.

Levin, P.S., Kelble, C.R., Shuford, R.L., Ainsworth, C., Dereynier, Y., Dunsmore, R., Fogarty, M.J., Holsman, K., Howell, E.A., Monaco, M.E., Oakes, S.A. & Werner, F. 2013. Guidance for implementation of integrated ecosystem assessments: a US perspective. *ICES Journal of Marine Science*. doi:10.1093/icesjms/fst112

Link, J. 2010. *Ecosystem-based Fisheries Management: Confronting Tradeoffs*, Cambridge University Press.

Link, J.S. & Bundy, A. 2012. Ecosystem modeling in the Gulf of Maine region: Towards an ecosystem approach to fisheries. American Fisheries Society Symposium.

Mapstone, B.D., Little, L.R., Punt, A.E., Davies, C.R., Smith, A.D.M., Pantus, F., Mcdonald, A.D., Williams, A.J. & Jones, A. 2008. Management strategy evaluation for line fishing in the Great Barrier Reef: Balancing conservation and multi-sector fishery objectives. *Fisheries Research*, 94, 315–329.

Marasco, R.J., Goodman, D., Grimes, C.B., Lawson, P.W., Punt, A.E., Quinn, II & Terrance, J. 2007. Ecosystem-based fisheries management: some practical suggestions. *Canadian Journal of Fisheries and Aquatic Sciences*, 64, 928–939.

Mcdonald, A., Little, L., Gray, R., Fulton, E., Sainsbury, K. & Lyne, V. 2008. An agent-based modelling approach to evaluation of multiple-use management strategies for coastal marine ecosystems. *Mathematics and Computers in Simulation*, 78, 401–411.

Murawski, S.A. 2000. Definitions of overfishing from an ecosystem perspective. *ICES Journal of Marine Science*, 57, 649.

Pascoe, S., Hutton, T., van Putten, I., Dennis, D., Plagányi-Lloyd, E. & Deng, R. 2013a. Implications of Quota Reallocation in the Torres Strait Tropical Rock Lobster Fishery. *Canadian Journal of Agricultural Economics-Revue Canadienne D Agroeconomie*, 61, 335–352.

Pascoe, S., Hutton, T., van Putten, I., Dennis, D., Skewes, T., Plagányi, E. & Deng, R. 2013b. DEA-based predictors for estimating fleet size changes when modelling the introduction of rights-based management. *European Journal of Operational Research*, 230, 681–687.

Pikitch, E., Boersma, P.D., Boyd, I.L., Conover, D.O., Cury, P.,, Essington, T., Heppell, S.S., Houde, E.D., Mangel, M., Pauly, D., Plagányi, É., Sainsbury, K., & Steneck, R.S. 2012a. Little Fish, Big Impact: Managing a Crucial Link in Ocean Food Webs. *Lenfest Ocean Program. Washington, DC*, 108.

Pikitch, E.K., Rountos, K.J., Essington, T.E., Santora, C., Pauly, D., Watson, R., Sumaila, U.R., Boersma, P.D., Boyd, I.L., Conover, D.O., Cury, P., Heppell, S.S., Houde, E.D., Mangel, M., Plagányi, É., Sainsbury, K., Steneck, R.S., Geers, T.M., Gownaris, N. & Munch, S.B. 2012b. The global contribution of forage fish to marine fisheries and ecosystems. *Fish and Fisheries*, n.p.

Pitcher, T. & Cochrane, K. 2002. The use of ecosystem models to investigate multispecies management strategies for capture fisheries *University of British Columbia Fisheries Centre Research Report*.

Plagányi, É. 2007. Models for an ecosystem approach to fisheries. *FAO Fisheries Technical Paper*, 477.

Plagányi, É.E. & Butterworth, D.S. 2012. The Scotia Sea krill fishery and its possible impacts on dependent predators: modeling localized depletion of prey. *Ecological Applications*, 22, 748–761.

Plagányi, É.E. & Essington, T.E. 2014. When the SURFs up, forage fish are key. *Fisheries Research*, 159, 68–74.

Plagányi, É.E., Punt, A.E., Hillary, R., Morello, E.B., Thebaud, O., Hutton, T., Pillans, R.D., Thorson, J.T., Fulton, E.A., Smith, A.D.M., Smith, F., Bayliss, P., Haywood, M., Lyne, V. & Rothlisberg, P.C. 2014. Multispecies fisheries management and conservation: Tactical applications using models of intermediate complexity. *Fish and Fisheries*, 15, 1–22.

Plagányi, É.E., Rademeyer, R.A., Butterworth, D.S., Cunningham, C.L. & Johnston, S.J. 2007. Making management procedures operational – innovations implemented in South Africa. *Ices Journal of Marine Science*, 64, 626–632.

Plagányi, É.E., Skewes, T., Murphy, N., Pascual, R. & Fischer, M. 2015. Crop rotations in the sea: Increasing returns and reducing risk of collapse in sea cucumber fisheries. *Proceedings of the National Academy of Sciences*, 112(21), 6760-5

Plagányi, É.E., Skewes, T.D., Dowling, N.A. & Haddon, M. 2013a. Risk management tools for sustainable fisheries management under changing climate: a sea cucumber example. *Climatic Change*, 119, 181–197.

Plagányi, É.E., van Putten, I., Hutton, T., Deng, R.A., Dennis, D., Pascoe, S., Skewes, T. & Campbell, R.A. 2013b. Integrating indigenous livelihood and lifestyle objectives in managing a natural resource. *Proceedings of the National Academy of Sciences of the United States of America,* 110, 3639–3644.

Plagányi, É.E., Weeks, S.J., Skewes, T.D., Gibbs, M.T., Poloczanska, E.S., Norman-Lopez, A., Blamey, L.K., Soares, M. & Robinson, W.M.L. 2011. Assessing the adequacy of current fisheries management under changing climate: a southern synopsis. *Ices Journal of Marine Science,* 68, 1305–1317.

Punt, A.E. & Butterworth, D.S. 1995. The effects of future consumption by the Cape fur seal on catches and catch rates of the Cape hakes. 4. Modelling the biological interaction between Cape fur seals Arctocephalus pusillus pusillus and the Cape hakes Merluccius capensis and M. paradoxus. *South African Journal of Marine Science/Suid-Afrikaanse Tydskrif vir Seewetenskap,* 16, 255–285.

Punt, A.E. & Donovan, G.P. 2007. Developing management procedures that are robust to uncertainty: lessons from the International Whaling Commission. *ICES Journal of Marine Science: Journal du Conseil,* 64, 603–612.

Punt, A.E., Ortiz, I., Aydin, K.Y., Hunt, G.L. & Wiese, F.K. 2015. End-to-end modeling as part of an integrated research program in the Bering Sea. *Deep Sea Research Part II: Topical Studies in Oceanography.*

Rademeyer, R.A., Plagányi, E.E. & Butterworth, D.S. 2007. Tips and tricks in designing management procedures. *ICES Journal of Marine Science,* 64, 618–625.

Robinson, W.M.L. 2013. Modelling the impact of the South African small pelagic fishery on African penguin dynamics. PhD thesis, University of Cape Town.

Robinson, W.M.L., Butterworth, D.S. & Plagányi, É.E. 2015. Quantifying the projected impact of the South African sardine fishery on the Robben Island penguin colony. *ICES Journal of Marine Science: Journal du Conseil,* fsv035.

Rose, K.A., Allen, J.I., Artioli, Y., Barange, M., Blackford, J., Carlotti, F., Cropp, R., Daewel, U., Edwards, K., Flynn, K., Hill, S.L., Hillerislambers, R., Huse, G., Mackinson, S., Megrey, B., Moll, A., Rivkin, R., Salihoglu, B., Schrum, C., Shannon, L., Shin, Y.-J., Smith, S.L., Smith, C., Solidoro, C., St. John, M. & Zhou, M. 2010. End-To-End Models for the Analysis of Marine Ecosystems: Challenges, Issues, and Next Steps. *Marine and Coastal Fisheries,* 2, 115–130.

Rouault, M., Penven, P. & Pohl, B. 2009. Warming in the Agulhas Current system since the 1980's. *Geophysical Research Letters,* 36.

Rouault, M., Pohl, B. & Penven, P. 2010. Coastal oceanic climate change and variability from 1982 to 2009 around South Africa. *African Journal of Marine Science,* 32, 237–246.

Roy, C., van der Lingen, C.D., Coetzee, J.C. & Lutjeharms, J.R.E. 2007. Abrupt environmental shift associated with changes in the distribution of Cape anchovy Engraulis encrasicolus spawners in the southern Benguela. *African Journal of Marine Science,* 29, 309–319.

Sainsbury, K.J., Punt, A.E. & Smith, A.D.M. 2000. Design of operational management strategies for achieving fishery ecosystem objectives. *ICES Journal of Marine Science,* 57, 731.

Shin, Y.J., Bundy, A., Shannon, L.J., Simier, M., Coll, M., Fulton, E.A., Link, J.S., Jouffre, D., Ojaveer, H., Mackinson, S., Heymans, J.J. & Raid, T. 2010. Can simple be useful and reliable? Using ecological indicators to represent and compare the states of marine ecosystems. *Ices Journal of Marine Science,* 67, 717–731.

Shin, Y.J. & Cury, P. 1999. OSMOSE: A multispecies individual-based model to explore the functional role of biodiversity in marine ecosystems. *Ecosystem Approaches for Fisheries Management,* 16, 593–607.

Smith, A.D.M., Brown, C.J., Bulman, C.M., Fulton, E.A., Johnson, P., Kaplan, I.C., Lozano-Montes, H., Mackinson, S., Marzloff, M., Shannon, L.J., Shin, Y.-J. & Tam, J. 2011. Impacts of Fishing Low–Trophic Level Species on Marine Ecosystems. *Science*, 333, 1147–1150.

Smith, A.D.M., Fulton, E.J., Hobday, A.J., Smith, D.C. & Shoulder, P. 2007. Scientific tools to support the practical implementation of ecosystem-based fisheries management. *Ices Journal of Marine Science*, 64, 633–639.

Stock, C.A., Alexander, M.A., Bond, N.A., Brander, K.M., Cheung, W.W.L., Curchitser, E.N., Delworth, T.L., Dunne, J.P., Griffies, S.M., Haltuch, M.A., Hare, J.A., Hollowed, A.B., Lehodey, P., Levin, S.A., Link, J.S., Rose, K.A., Rykaczewski, R.R., Sarmiento, J.L., Stouffer, R.J., Schwing, F.B., Vecchi, G.A. & Werner, F.E. 2011. On the use of IPCC-class models to assess the impact of climate on Living Marine Resources. *Progress in Oceanography*, 88, 1–27.

Szuwalski, C.S. & Punt, A.E. 2012. Fisheries management for regime-based ecosystems: A management strategy evaluation for the snow crab fishery in the eastern Bering Sea. *ICES Journal of Marine Science: Journal du Conseil*, 70, 955–967.

van Putten, I., Lalancette, A., Bayliss, P., Dennis, D., Hutton, T., Norman-Lopez, A., Pascoe, S., Plagányi, E. & Skewes, T. 2013. A Bayesian model of factors influencing indigenous participation in the Torres Strait tropical rocklobster fishery. *Marine Policy*, 37, 96–105.

Watters, G.M., Hill, S.L., Hinke, J.T., Matthews, J. & Reid, K. 2013. Decision-making for ecosystem-based management: evaluating options for a krill fishery with an ecosystem dynamics model. *Ecological Applications*, 23, 710–725.

Chapter 16

Incorporating spatial population structure into the assessment-management interface of marine resources

Daniel R. Goethel, Lisa A. Kerr and Steven X. Cadrin

Introduction

Complex spatial structure and population richness are common in marine species (Cadrin and Secor, 2009) and can provide regional stability in the context of localized environmental perturbation, ensuring long-term persistence (Kerr et al., 2010b, 2014, Kritzer and Liu, 2014). By contrast, most fisheries regulations are spatially uniform over broad-scale management units that reflect political, economic, and data collection convenience (Smedbol and Stephenson, 2001, Steneck and Wilson, 2010). Over the last two decades, protection of population structure has been implemented through the use of spatiotemporal fishery closures (Kritzer and Liu, 2014). However, the importance of maintaining spatial structure was realized too late to prevent the depletion of spawning components for some species (e.g. Atlantic cod in Canadian waters; Smedbol and Stevenson, 2001). The importance of complex population heterogeneity has become well documented, but spatial management is often ad hoc with decisions based on the output from broad-scale stock assessment models that ignore fine-scale structure (Cope and Punt, 2011, Berger et al., 2012).

Stock assessment models are generally a central feature of fisheries management science, often providing the most scientifically defensible estimate of resource status. An assessment is applied to a 'stock' unit that usually attempts to match the delineations of true biological populations, but stocks are also defined by management boundaries and do not always accurately reflect biological entities. Therefore, we distinguish 'stock' from 'population' throughout this chapter. When developing model-based management procedures (e.g. management strategy evaluation), assessment models are included within the framework, because assessment model outputs represent the primary information source used to determine resource status in relation to predetermined biological reference points. However, stock assessments can also be used as the basis for operating models in order to simulation test the management procedure itself. Under this paradigm, the operating model is parameterized during the assessment phase. Nevertheless, we distinguish an operating model from an assessment model based on its intended purpose, and the fact that an operating model can be extended to include components not included in the assessment.

As abstractions of reality, stock assessment models require the use of simplifying assumptions, the most common being that of a unit stock with dynamic pool population dynamics. A unit stock implies that immigration and emigration are negligible (i.e. the stock is closed to both adult movement and larval subsidy), while a dynamic pool assumes that vital rates are homogeneous within stocks and fishing pressure is homogenously distributed (Field et al., 2006). Unfortunately, fishery monitoring data is typically inadequate to match the biological structure observed for many marine species (Holland, 2003, Field et al., 2006, Cadrin and Secor, 2009). Due to uncertainty regarding fine-scale stock structure and sample size limitations at spatial scales matching many biological processes, a large majority of global stock assessments utilize the unit stock assumption and are ignorant of the spatial complexities of population and fishery processes or data collection.

Management based on stock assessment models that do not account for population structure ignores fine-scale processes (Frank and Brickman, 2000). For example, if a population falls within more than one stock management boundary, it can lead to the dispersal of management actions across stocks (Hart and Cadrin, 2004, Botsford et al., 2009). The result is that smaller population or subpopulation components within a stock complex can become susceptible to overfishing and serial depletion (Fu and Fanning, 2004, Kerr et al., 2014), while more productive units are underharvested (Tuck and Possingham, 1994, 2000). Declines in population richness (i.e. the number of subpopulations within a stock complex) can lead to reductions in resilience, stability, and productivity of the entire stock complex (Kerr et al., 2010a, 2010b). Mismatching the scale of assessments and management (e.g. enacting fine-scale spatial management based on broad-scale assessments) is common but can be detrimental to stated objectives, because the models are unable to accurately determine population status at the dimensions of fishery regulations (Punt and Methot, 2004, Cope and Punt, 2011).

Spatially explicit modeling approaches have been developed to account for fine-scale dynamics in stock assessment and to loosen the unit stock and dynamic pool assumptions. Approaches to incorporating spatial structure vary, including individual-based models, spatially referenced parameter estimation, modeling substocks as discrete units, and spatial movement models (Goethel et al., 2011). Although the mathematical constructs for implementing stock assessment models that match the scale of spatial management exist, applications remain limited due to increased model complexity and data requirements (Goethel et al., 2015a).

We outline a general framework that can be used to develop spatial simulation models through the combination of quantitative data, best estimates of population parameters, and qualitative knowledge regarding spatial complexity and connectivity. This estimation-simulation framework can be used to validate current assessments, gauge the cost/benefit of collecting spatially referenced data, evaluate the potential impact of future regulations, and develop spatially explicit biological reference points. We draw on our experiences with New England groundfish, particularly yellowtail flounder (*Limanda ferruginea*) and Atlantic cod (*Gadus morhua*), to illustrate the importance of spatial structure in development and implementation of sustainable harvest strategies.

The utility of spatially explicit operating models for evaluation of biological reference points and harvest control rules

It is imperative that biocomplexity be considered in any modeling framework that simulates the impacts of fishery regulations because of the complex spatial interactions among population units. Many simulation studies have been developed to explore how population structure can alter perceptions of productivity and resulting sustainable harvest levels. For instance, Hart and Cadrin (2004) illustrated that when metapopulation structure was simulated for the three yellowtail flounder stocks off of New England, population connectivity resulted in important differences in rebuilding schedules and abundance, compared to simulations where movement was ignored. Similarly, Kerr et al. (2010b, 2014) modeled Gulf of Maine Atlantic cod, and demonstrated that productivity and maximum sustainable yield vary when the substock structure within the management unit is modeled explicitly.

In a simulation of two populations of Atlantic cod off coastal Canada connected by movement, Fu and Fanning (2004) compared the performance of quota-based management that assumed either a single population or two distinct populations. Their results indicated that management of combined populations led to overfishing the smaller stock, and connectivity greatly complicated the effectiveness of management. Similarly, Wilberg et al. (2008) showed that variations in source-sink dynamics within a spatially structured Lake Michigan yellow perch (*Perca flavescens*) population led to differential performance of management measures.

A critical aspect of evaluating harvest strategies that has received little attention is the impact of spatial structure and population connectivity on the calculation of biological reference points (Goethel et al., 2015a, 2015b). Biological reference points (BRPs) are usually simulation-based quantities that define fishing mortality limits or targets as well as threshold stock abundance levels below which a predetermined action should be taken as part of a harvest control rule (HCR; Deroba and Bence, 2008, Botsford et al., 2009). Despite the reliance of productivity on spatial structure (e.g. Kerr et al., 2010b), little research has considered the dependence of reference points on population connectivity and biocomplexity. Many spatial simulations include the critical components necessary for estimating spatially explicit reference points (i.e. dispersal, connectivity, and effort distribution), but calculation has only been pursued in a few case studies.

The effect of spatial structure on interpretation of stock status is not considered in most harvest control rule simulations (e.g. the classification of reference points by Cadrin and Pastoors, 2008). Even all-encompassing generic approaches, which are intended to be holistic tools for evaluating management policies (e.g. ISIS-Fish; Mahévas and Pelletier, 2004, Pelletier and Mahévas, 2005), typically ignore potential changes in stock status indicators when spatial considerations are incorporated into the population dynamics. When spatial population structure is ignored in the calculation of overfishing definitions and rebuilding targets, evaluation of harvest control rules will be inherently biased. Therefore, spatially explicit biological reference points must first be developed before spatial management strategy evaluation can be reliably utilized to appraise harvest control rules. We present the case of marine protected areas as a relatively simple example of how spatial structure can be incorporated into the simulation evaluation of management regulations, which illustrates the multifaceted considerations that are necessary when developing spatial operating models and optimal harvest strategies.

Case study: Evaluating optimal harvest policies for marine protected areas (MPAs) using spatial simulations

Marine protected areas (MPAs) are a type of fishery closure meant to provide relief from fishing pressure for a portion of a stock (along with achieving other conservation goals; Field et al., 2006). The optimal placement and effectiveness of MPAs has been widely discussed, but consensus opinion has not been achieved (Holland, 2003, Field et al., 2006, Fogarty and Botsford, 2007; for a thorough review of MPA modeling approaches and implications see Guénette et al., 1998, Botsford et al., 2003, Pelletier and Mahévas, 2005, Botsford et al., 2009). Much of the disagreement on MPA efficacy stems from the inability of empirical studies to determine why closed areas might have been effective or ineffective (Botsford et al., 2009). Additionally, the wide array of dispersal patterns and spatial structures exhibited by marine species create differential usefulness of spatial management measures, which makes generalizations unfeasible (Pelletier and Mahévas, 2005).

Beverton and Holt (1957) developed spatially explicit modeling approaches to illustrate the impact of marine protected areas on yield-per-recruit. Their results demonstrated that increasing the size of a reserve was synonymous to a delayed age at first capture, but effects on yield-per-recruit were limited (Guénette et al., 1998, Botsford et al., 2003). Per-recruit models highlighted the impacts that marine reserves could have, but did not incorporate the fine-scale processes exhibited in most populations (e.g. larval dispersal mechanisms; Tuck and Possingham, 2000). The development of modeling approaches used for the siting of MPAs was first established for sessile species that had dispersive larval stages (e.g. Quinn et al., 1993). Management by closed area was thought to be more successful for species with a sessile adult stage, because fishing effort could be effectively eliminated for entire patches of the population (as opposed to mobile species that move in and out of closed areas). Modeling efforts and interpretations were also greatly simplified when adult movement could be ignored (Hart, 2003, Botsford et al., 2009).

Models that include larval dispersal indicate that optimal MPA siting occurs when the dispersal kernel is maintained within the reserve or closed areas are separated by a distance that is less than the average dispersal (Quinn et al., 1993, Botsford et al., 2001, Botsford et al., 2009). However, the effectiveness of reserves is debated. Differences in the rate, distance, and form of dispersal lead to varying conclusions on the ability of MPAs to improve population persistence (i.e. the ability to replace itself) or yield (Kaplan, 2006, Fogarty and Botsford, 2007, Botsford et al., 2009). Tuck and Possingham (2000) demonstrated that the source population in a metapopulation with unidirectional dispersal should be harvested to maximize spawning stock biomass, but fishing the sink population would maximize yield. Crowder et al. (2000) concluded that placing reserves to protect only the sink populations can be detrimental to the persistence of the metapopulation. Depending on the goals of management and larval dynamics of the managed populations, the optimal regulations can vary considerably.

When adult movement is included in spatial simulations, the implications for implementation of marine reserves become even more diverse. The combined uncertainties inherent in knowledge of larval and adult dispersal are multiplied when both are considered simultaneously (Botsford et al., 2003, Moffitt et al., 2009). When fish are assumed to exhibit random diffusion, MPAs tend to provide increased resiliency (maximum yield occurring at higher fishing mortality), but lower total yield (Fogarty and Botsford, 2007, Le Quesne and Codling, 2009). Walters (2000) concluded that MPAs need to be much larger than fished areas to conserve spawning stocks of mobile species. Walters et al. (2007) showed that when populations are mobile, management of fisheries in open areas is more important for meeting conservation objectives than MPA size or placement. However, Moffitt et al. (2009) argued that when adult movement is modeled using home ranges (i.e. the spatial extent of food searching and other common activities) as opposed to random diffusion, marine reserves can be effectively implemented even for mobile species. Depending on the larval dispersal distance, MPA effectiveness varied, and nonlinear relationships developed from the combination of adult and larval dispersal mechanisms (Moffitt et al., 2009, Grüss et al., 2011).

Spatial variation in a range of processes can also be relevant to the evaluation of management measures. One of the most important of these is the distribution of fishing effort. With the implementation of MPAs, fishing effort must redistribute to account for the closure of former fishing grounds (Field et al., 2006, Botsford et al., 2009). Redistribution of effort tends to reduce persistence and yield for mobile species (Grüss et al., 2011), while leading to decreased net revenue (Holland, 2003). However, closures can be beneficial for sessile species and associated fisheries by increasing yield- and biomass-per-recruit (e.g.

through the use of rolling closures; Hart, 2003). Ecosystem interactions can also have important spatial implications that impact the effectiveness of management. Multispecies models indicate that MPAs can cause overfishing for some species as a result of redistributed effort (Holland, 2003). Implementation of spatial management measures may lead to unanticipated reorganization of the marine foodweb (Fouzai et al., 2012), while Martell et al. (2005) illustrated that oceanographic variability alone can inhibit the successful implementation of MPAs.

Research on MPAs provides a multidimensional overview of the intricacies faced when evaluating spatial management measures. Depending on the stock dynamics, modeling assumptions, and management goals, different conclusions can be made about the efficacy of a given policy using a spatial operating model (Pelletier and Mahévas, 2005). MPA models demonstrate that spatial biocomplexity must be thoroughly evaluated if the results are to accurately represent the system of interest (Fogarty and Botsford, 2007). The detrimental impact caused by ignoring spatial structure in management actions can be even more pronounced when examined in a wider, regional context with interactions among multiple populations. However, a major limitation of most MPA operating models is the relatively simple population structure typically assumed (i.e. a single population with spatial heterogeneity). Spatial models used to investigate marine protected areas can be adapted to account for more complex spatial interactions, and, in some instances, have been used to develop spatially explicit biological reference points.

The use of spatial simulations to calculate spatially explicit biological reference points

Models that incorporate spatial population structure into calculations of biological reference points can be categorized into four main groupings based on the complexity of the underlying model (Table 16.1). Optimal management of complex interconnected populations (e.g. metapopulations) has received limited attention, which makes it impossible to define appropriate generic target mortalities or biomass for such systems at this time, and these are not considered in our classification.

Equilibrium per-recruit models

Per-recruit models are common practice in fisheries science because of their relative ease of use and limited data requirements (Punt and Cui, 2000). Incorporating spatial complexity by allowing movement between areas within a single population unit (e.g. models of marine protected areas) is relatively straightforward, and was first investigated by Beverton and Holt (1957). Connectivity can be modeled explicitly (e.g. Punt and Cui, 2000) or implicitly (e.g. through spatially varying catchability; Grüss et al., 2014). Hart (2001, 2003) presents an

Table 16.1 Types of spatially explicit biological reference points in order of increasing complexity and data needs.

Model	Model type	Data needs	Equilibrium estimates	Age-based	Types	Examples
Per-recruit	Dynamic pool	Limited	Yes	Yes	Yield-per recruit (YPR) with movement within a single population	Beverton and Holt, 1957
					Individual-based YPR and spawning biomass-per-recruit (SPR)	Hart, 2001
					YPR and SPR with spatially varying parameters	Grüss et al., 2014
					YPR and SPR with connectivity among populations	None*
					Dispersal-per-recruit (DPR)	Kaplan et al., 2006, Grüss et al., 2011
Surplus production	Stock assessment parameter estimates	Moderate	No	No	MSY with connectivity	Ying et al., 2011
Dynamic F_{MSY}	Long-term stochastic simulation	High	No	Yes	Movement among patches	Apostolaki et al., 2002
					Connectivity among populations	Kerr et al., 2014

(Continued)

Table 16.1 (Continued)

Model	Model type	Data needs	Equilibrium estimates	Age-based	Types	Examples
Dynamic E_{OPT}	Long-term stochastic simulation	Very high	No	Yes	Optimal harvest strategy in a metapopulation (no direct model of effort)	Wilberg et al., 2008
					Single population with movement and effort allocation (static effort)	Okamura et al., 2014
					Dynamic effort distribution with movement within a population	Le Quesne and Codling, 2009
					Dynamic effort with connectivity among populations	None

*See text for example with independently estimated YPR from spatially explicit assessment parameters.

alternate approach termed individual-based yield-per-recruit, which includes the exploitation history of a cohort through time-averaged fishing mortality. By including spatial variation in mortality (or other demographic rates) the model avoids the dynamic pool assumption (i.e. that fish are well mixed throughout the stock area), which can lead to biased fishing mortality metrics when the assumption is violated (Hart, 2001). Kaplan et al. (2006) present an alternate metric termed dispersal-per-recruit, which is based on an equilibrium spatial simulation including larval dispersal among areas within a single population. The modeling approach combines spawning-per-recruit and larval dispersal to determine a minimal level of larval settlement, typically chosen as 35% of the equilibrium level in each cell of the model, required for persistence of the population. Extensions to the dispersal-per-recruit method have included both adult movement and fishery effort distribution (Moffitt et al., 2009, Grüss et al., 2011).

Despite spatial per-recruit models allowing for movement among areas within a single population, connectivity among different populations or spawning units has not been investigated. Accounting for alternate population structures (e.g. metapopulations) will probably create unpredictable impacts on per-recruit models, because of the interaction of productivity and movement. One option for implicitly incorporating population connectivity is by using parameter estimates from a spatially explicit stock assessment model to calculate per-local-recruit reference points for each population independently (i.e. expected production from locally spawned recruits). For example, the results of the metapopulation stock assessment model of yellowtail flounder in New England (Goethel et al., 2015a) could be used as inputs for per-local-recruit models. Such an approach would result in a 19% decline in spawning stock biomass rebuilding targets for the smallest stock unit (i.e. Cape Cod–Gulf of Maine) compared to results based on a nonspatial assessment, because of an alternate interpretation of recruitment dynamics (Goethel et al., 2015a). However, movement is not explicitly modeled in these per-recruit calculations.

Simultaneously calculating per-recruit reference points for multiple populations with movement modeled explicitly is a relatively simple extension to current spatial single population per-recruit models. But, a complex stock assessment model and extensive data would be necessary to estimate the required parameters including connectivity, which negate the critical benefits of per-recruit models (i.e. simplicity and limited data requirements). In most cases in which the data and understanding of stock dynamics are available to develop a spatial per-recruit model with population connectivity, it would be more appropriate to pursue a dynamic F_{MSY} approach.

Maximum sustainable yield from surplus production models

The next level of complexity for spatial reference point calculations involves the use of spatially explicit surplus production models to derive estimates of maximum sustainable yield (MSY) and the associated fishing mortality (F_{MSY}) and biomass (B_{MSY}). Both Carruthers et al. (2011) and Ying et al. (2011) developed and applied surplus production models that incorporated multiple stocks with connectivity among units. Nonlinear relationships between spatially explicit and single stock reference point calculations can result within the surplus production framework. For instance, emigration and immigration cause F_{MSY} and B_{MSY} to shift and can lead to a nonsymmetrical MSY curve (Figure 16.1). Because reference points are derived directly from the stock assessment model instead of being based on an operating model, the ability to add complexity is restricted. Although useful tools for initial development of spatially explicit biological reference points and gaining a conceptual understanding of how spatial structure can alter reference point estimates, the simplicity of the modeling approach may limit the reliability of the calculations.

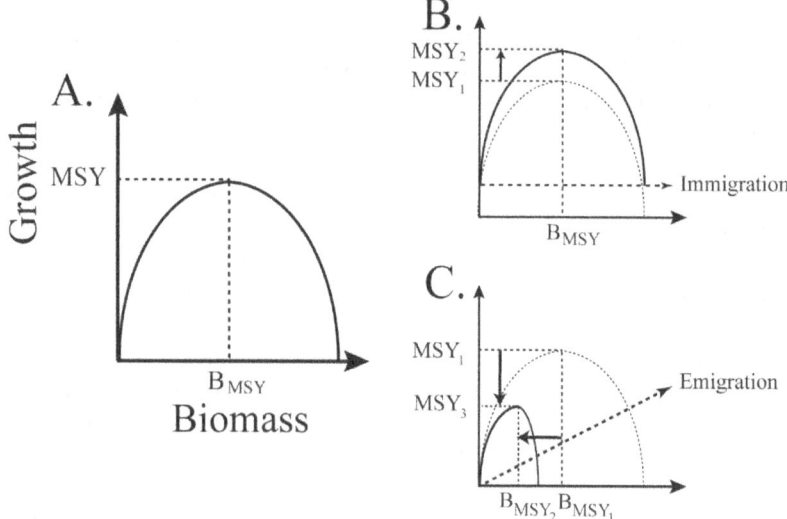

Figure 16.1 Conceptual diagram illustrating the calculation of maximum sustainable yield (MSY) from logistic surplus production models with and without connectivity among stock units. Without mixing among populations (A), MSY occurs at one-half the carrying capacity (located at B_{MSY}). The level of MSY represents the maximum equilibrium growth in the population. When immigration is included (B), the MSY curve is shifted upwards by a constant value, because incoming fish act as a subsidy that increase net growth independent of biomass in the given stock. MSY increases (represented by MSY_2), but B_{MSY} is unaffected. Contrarily, emigration (C) decreases growth as an increasing function of biomass. The shape of the MSY curve is altered (i.e. it may no longer be symmetrical) and both MSY and B_{MSY} decrease (represented by MSY_3 and B_{MSY2}, respectively).

Dynamic F_{MSY}

The role of maximum sustainable yield reference points for fisheries management has changed from considering MSY as a target to considering F_{MSY} as a limit (Mace, 2001). Other developments include application of nonequilibrium approaches to estimating F_{MSY} through stochastic simulation, which avoid equilibrium assumptions associated with per-recruit methods (Deroba and Bence, 2008). Spatial simulations for evaluation of F_{MSY} can take many forms depending on assumed population structure, modeling of larval dispersal, and parameterization of adult movement (see Table 16.2 for a list of considerations necessary for developing a spatially explicit dynamic F_{MSY} operating model).

Table 16.2 Modeling assumptions that must be addressed for spatially explicit dynamic F_{MSY} and E_{OPT} simulations. Many of the subsequent assumptions are dependent on the assumed population structure. The table also illustrates major differences between approaches (i.e. the inclusion of effort dynamics in E_{OPT} models).

Assumption	Potential attributes
Population structure	Unit stock with homogenous distribution
	Unit stock with patchy distribution
	Natal homing with spatial overlap
	Metapopulation
	Other
Recruitment	Time-invariant vs. stochastic
	Stock-specific vs. common pool
Larval dispersal	None
	Advection-diffusion model
	Hydrodynamics model
Connectivity	None
	Time-invariant vs. stochastic
	Overlap or home-range vs. diffusion
	Density-dependent
	Other (e.g. entrainment)
Effort allocation (for E_{OPT} simulations)	None
	Scenario driven
	Gravity model
	Revenue potential
Spatiotemporal scale of reference point determination	Population-specific vs. multiscalar (e.g. metapopulation)
	Long-term vs. short-term/transient

A spatially explicit dynamic F_{MSY} model can assume a single population with spatial patchiness or multiple interconnected reproductive units. From the perspective of a single population it may be of interest for managers to know appropriate reference harvest levels for each patch individually or for the population as a whole. For the former, a model such as that of Cope and Punt (2011) could be used to simulate spatial dynamics with multiple patches being harvested independently, but sharing a common larval pool. The inclusion of movement among patches is relatively straightforward and has been demonstrated in a number of models (e.g. Apostolaki et al., 2002, Holland, 2003, Le Quesne and Codling, 2009). Many of these models were designed for evaluation of marine protected areas and require a slight reformulation to determine F_{MSY} (e.g. calculating the fishing mortality, instead of the fraction of habitat in reserves, that maximizes yield), but the model frameworks and calculations are similar.

Larval dispersal mechanisms also need to be considered for deriving reference points from stochastic simulations. Recruitment can assume a common larval pool or area specific recruitment. Typically, recruitment is based on a stock-recruit relationship for either age-1 recruitment or total egg production, and may include larval dispersal across area boundaries. Other considerations for larval dynamics include whether recruitment and dispersal mechanisms are stochastic or deterministic. Stochastic recruitment can be modeled using a probability distribution, while time-varying dispersal could be modeled using a larval individual-based model linked with a hydrodynamics model (e.g. Heath et al., 2008). Similarly, movement of adults can be modeled assuming time-invariant movement rates, stochastic movement based on a probability function, or some other functional form of movement (e.g. density-dependent movement).

When modeling multiple populations with connectivity, complex interactions between movement and productivity are expected to result in nonintuitive implications for management and reference point estimates (Goethel et al., 2015a, 2015b). Spatial simulations with multiple species (e.g. Holland, 2003) have illustrated that ecosystem effects of fishing can be more complicated than nonspatial counterparts (e.g. Collie and Gislason, 2001). It is expected that simultaneously modeling multiple populations will result in similar complexities, because connectivity among populations leads to transmission of effective fishing mortality across population units throughout the model domain (Heath et al., 2008). Consideration of both regional and local production is needed when developing spatially explicit target reference points for interconnected populations, but has not yet been addressed. Although a dynamic F_{MSY} approach will indicate maximum long-term yield for each population, it does not account for the optimal distribution of fishing effort required to maximize yield from the entire population complex.

Dynamic E_{OPT}

The spatial distribution of fishing effort relative to the distribution of the population can impact optimization of yield (Carruthers et al., 2011). Modeling the optimal distribution of effort can be difficult and requires modeling spatial heterogeneities in production within and among population components (Sanchirico and Wilen, 2005, Botsford et al., 2009, Grüss et al., 2011). Optimal effort distribution (E_{OPT}) takes the same form as spatially explicit dynamic F_{MSY} simulations, but also accounts for the spatial distribution of fishing effort in relation to population structure (Table 16.2). The goal of this type of modeling approach is to optimize the yield from the entire system instead of maximizing harvest on a population by population basis. The same biological implications must be accounted for as in spatially explicit dynamic F_{MSY} simulations (e.g. parameterization of larval dispersal and adult movement). In addition, the movement of fishing effort among areas needs to be considered. Effort distribution is typically modeled assuming the gravity framework in which effort is concentrated in areas of maximum population abundance (e.g. Le Quesne and Codling, 2008, Grüss et al., 2011), but alternate parameterizations can be utilized including basing the movement of fishermen on potential revenue in a given area (e.g. Holland, 2003) or simple time-varying or scenario-driven allocation by region (Apostolaki et al., 2002, Okamura et al., 2014).

Although distribution of effort has been investigated from an economics perspective (e.g. Sanchirico and Wilen, 2005, Mchich et al., 2006), little research has focused on how effort should be spatially allocated to maximize long-term yield. Depending on the assumed spatial population structure, the optimal spatial distribution of the fishery can vary. For example, MacCall's (1990) simulation of a resource with an ideal-free distribution and heterogeneous habitat quality suggested that fishing marginal habitats is optimal for long-term yield. Diverse conclusions have been made for how source-sink populations should be optimally harvested (e.g. whether to focus fishing effort on the source or sink populations) with conclusions often depending on assumed connectivity dynamics (Tuck and Possingham, 2000, Wilberg et al., 2008). Before dynamic optimal effort allocation simulations are used to define spatial reference points, more detailed research is needed to investigate appropriate model frameworks, especially for modeling the movement dynamics of the fishery. However, considering the implications that effort distribution has for the effectiveness of management actions, spatial effort dynamics should be considered when developing operating models for policy evaluation (Holland, 2003, Botsford et al., 2009).

Case study: spatial structure and sustainable yield of Atlantic cod

Current and historical studies of Atlantic cod population structure off the coast of New England have proposed multiple populations with numerous spawning contingents (Ames, 2004, Zemeckis et al., 2014). Recent genetic analysis supports this hypothesis, indicating that there are at least three major, genetically distinct population complexes of cod in US waters (Kovach et al., 2010). These include: (1) a northern spring spawning complex that spawns in inshore Gulf of Maine waters; (2) a southern spawning complex which spawns within the inshore Gulf of Maine in winter and at different offshore locations and seasons within the Gulf of Maine and southern New England waters; and (3) a population that spawns on northeast Georges Bank in the late winter/early spring (Figure 16.2). Despite the spatiotemporal population structure highlighted by genetics, and supported by tagging and demographic studies, the current boundaries for management and assessment of Atlantic cod generally ignore these complexities (Kerr et al., 2014, Zemeckis et al., 2014). Current management assumes two closed unit stocks of cod in New England waters: (1) a Gulf of Maine stock; and (2) a combined Georges Bank and southern New England stock (NEFSC, 2012a; Figure 16.2). The appropriateness of these boundaries has been debated as new stock identification information has become available in recent years. Critical evaluation of these boundaries is particularly relevant due to recent stock assessment results that indicate a pessimistic outlook of cod stock status in New England waters (NEFSC, 2012a).

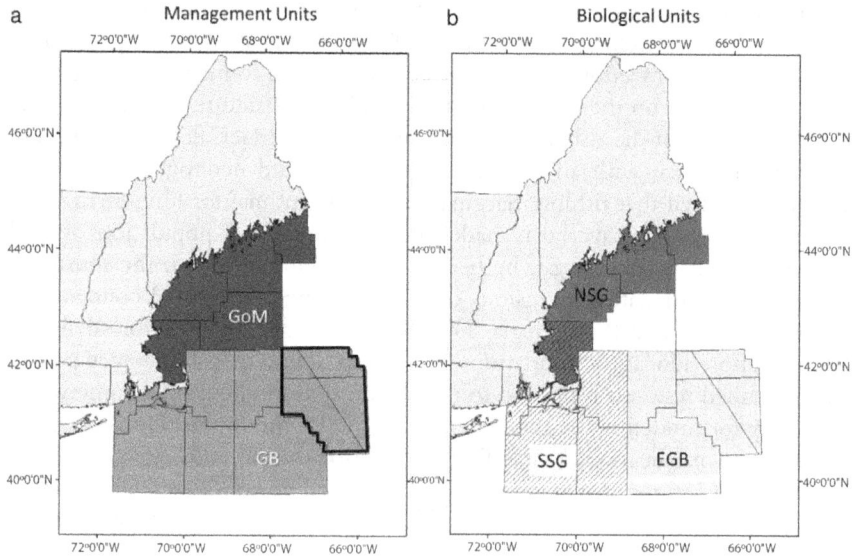

Figure 16.2 Map illustrating stock structure of Atlantic cod in New England. Current management (a) assumes two stocks (GoM: Gulf of Maine, dark grey; GB: Georges Bank, lighter grey). Note that the eastern portion of Georges Bank (outlined in bold) is jointly managed by the United States and Canada as a transboundary resource. Genetic analysis indicates biological population structure (b) with three distinct spawning components (NSG: Northern Spring Spawning Group, dark grey; SSG: Southern Spring Spawning Group, hatched lines; EGB: Eastern Georges Bank Spawning Group, light grey). The simulations of Kerr et al. (2014) used these spatial delineations to define the management unit model (a) and biological unit model (b).

Mismatch between the scale of fisheries management units and biological population structure can potentially result in a misperception of the long-term productivity and sustainable yield of fish stocks. Kerr et al. (2014) used simulation modeling as a tool to compare the expectations of productivity and sustainable yield of Atlantic cod off New England when the resource was modeled according to the two spatially defined US management units (Gulf of Maine and Georges Bank) or as three biologically defined complexes (a northern spawning group, a southern spawning group, and an eastern Georges Bank spawning group; Figure 16.2). The two stochastic age-structured operating models (defined as the management unit or biological unit model) of Atlantic cod were compared over a range of fishing mortality scenarios (extending from $F = 0.0$ to 1.0). The parameters of the management unit model were informed by the most recent stock assessments of Gulf of Maine and Georges Bank cod (NEFSC, 2012a). On the other hand, the biological model parameters were estimated from survey data constrained in space and time to best represent the spawning complexes, while exchange rates between complexes were informed by genetically based estimates of connectivity (proportion of migrants exchanged between populations was derived from F_{st} values and estimates of effective population size).

In the context of the simulations, examining the cod resource based on the current stock boundaries (i.e. the management unit model) presented an overly optimistic view of regional productivity, maximum sustainable yield, and F_{MSY} compared to results derived from the biological unit model (Figure 16.3). The management unit model indicated that the cod complex was more resilient to fishing mortality and not as susceptible to 'collapse' as indicated by the biological unit model. The simulation revealed that ignoring population structure of Atlantic cod could potentially lead to overexploitation of segments of the population, particularly the eastern Georges Bank spawning complex, which appeared to be the minority component in the system. Dynamic F_{MSY} and E_{OPT} models that attempt to capture biocomplexity, such as this cod example, can revise our perception of target harvest levels and expectations regarding stock rebuilding.

Developing spatially explicit operating models

Thorough simulation testing of harvest control rules can be difficult and often requires the use of complex operating models that incorporate multiple modules to emulate alternative plausible scenarios for the fishery, the biological population, the assessment of the resource, and the complete management framework. An increasingly popular tool for analyzing fishery policy is management strategy evaluation (MSE), which is a simulation framework used for determining the performance of management strategies (e.g. harvest policies) across a range of management objectives. Because assessment models are included as part of a simulated set of management procedures (i.e. assessment outputs are

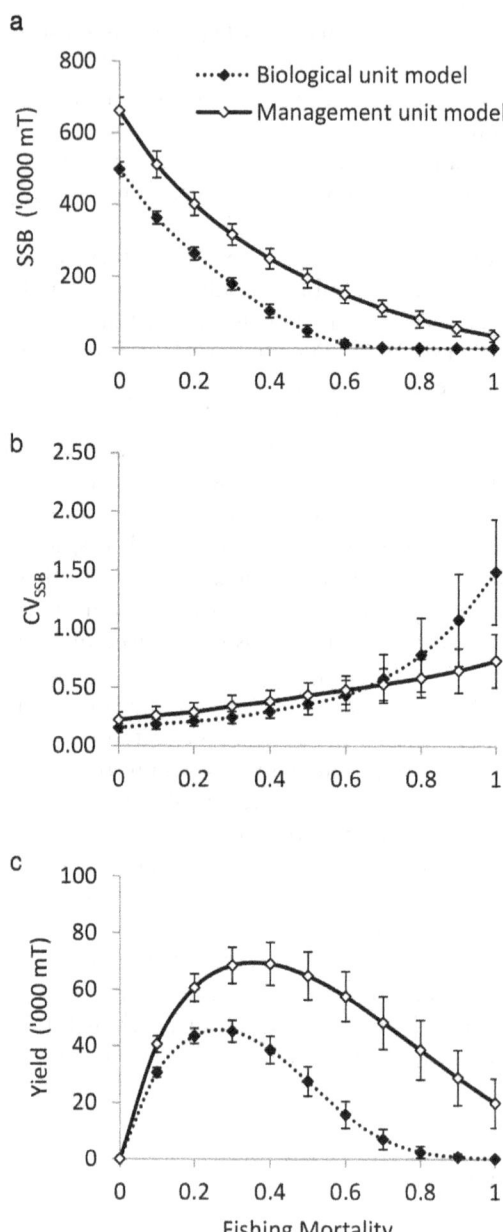

Figure 16.3 Spawning stock biomass (SSB), coefficient of variation in spawning stock biomass (CV$_{SSB}$), and yield for the Atlantic cod simulations of Kerr et al. (2014). Results demonstrate the difference in interpretation of regional productivity and F_{MSY} depending on the spatial population structure assumptions of the simulation. Modeling the stock complex using current management boundaries leads to a more resilient and productive outlook (resulting in a higher F_{MSY}) than the biological model, which assumes a more complex spatial population structure based on genetic analyses.

used as the basis of management decisions), the framework can account for the inherent bias that results from the interplay of assessment and management. However, operational implementation of spatial management strategy evaluation remains difficult, because of the lack of spatially explicit stock assessment models (Goethel et al., 2011, 2015a, 2015b) and the difficulties in parameterizing spatial operating models.

The three-tiered estimation-simulation framework

Constructing a spatial operating model for evaluating management actions requires defining and characterizing the system of interest, designating response variables, verifying model performance, and validating the results (detailed in Kerr and Goethel, 2014). However, determination of accurate values for model parameters and validation of model performance can be extremely difficult, because there is usually no estimation model to verify model outputs with observed data. On the other hand, stock assessment models, which are generally simpler than spatially explicit operating models, are able to directly estimate population parameters from observed data. When appropriate data are available the assessment and simulation frameworks can be combined. A spatially explicit stock assessment model can be used to estimate parameter values, which are then used as the basis of a spatially explicit operating model that can evaluate management procedures. By using a spatially explicit estimation model as the operating model, best-fit parameter values can be input directly to the simulation based on observed data. From a management perspective, this approach ensures that parameters used for developing policy evaluation are based on available data.

The estimation-simulation framework is not appropriate for all situations, because it is data intensive and often requires development of a complex stock assessment model. However, the approach provides an intuitive and consistent method to estimate simulation parameters using direct observations. Unlike decoupled approaches (where operating model parameters are not informed directly from an estimation model), the three-tier framework maintains consistency of assumptions and uncertainty when transitioning from the estimation to simulation models (similar to how integrated analysis is able to do so by analyzing all data within a single framework; see Maunder, 2001). For instance, White (2010) warns that, when using parameter estimates from single stock estimation models, critical processes can be misinformed within spatially explicit models, because spatial processes that are implicit within the estimate are now directly accounted for within the spatial model (e.g. steepness of the stock-recruit curve subsumes many spatial early life history processes). By using a single underlying population dynamics model in the three-tier framework for both the estimation and simulation models, consistency of assumptions is propagated, while uncertainty of estimated parameters can be accounted for in a more appropriate manner.

The basic premise of the estimation-simulation framework is that a spatial assessment is developed and applied to observed data to estimate spatially explicit parameter estimates (Tier 1), these parameters are then used along with various population and spatial assumptions (e.g. see Table 16.2) to develop a spatial operating model (Tier 2), and the operating model is used to simulate and evaluate various management policies or test hypotheses regarding spatial structure (Tier 3; Figure 16.4).

Tier 1: parameter estimation

Spatially explicit stock assessments take many forms depending on the available data and observed population structure (Goethel et al., 2011). Ideally the estimation model would match the scale of the true observed biological structure, but this may not be feasible given the available data. The goal of the estimation

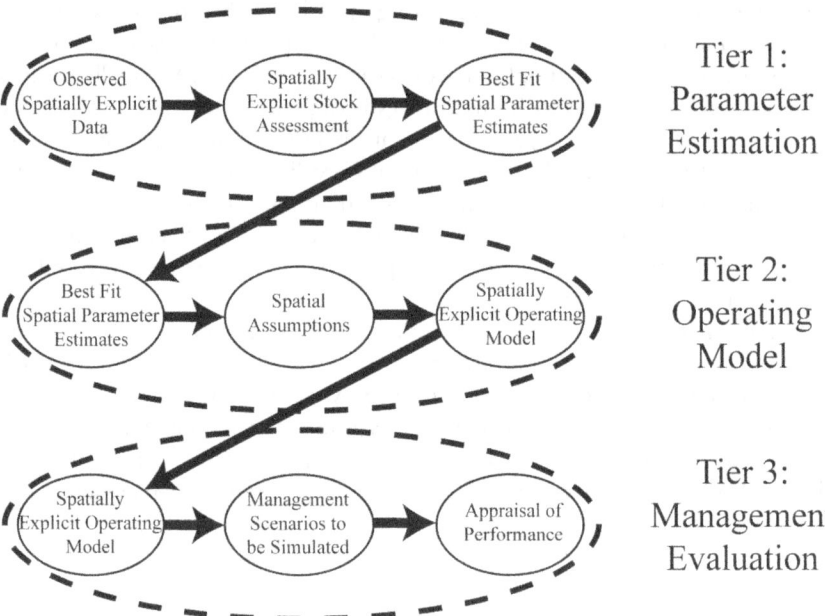

Figure 16.4 Schematic representation of the three-tiered estimation-simulation framework that is proposed for the development of spatially explicit operating models. In Tier 1 the spatial assessment model is fit to observed data to obtain best fit parameter estimates. In Tier 2 the spatially explicit operating model is developed using the parameter estimates from Tier 1 along with any further assumptions that must be addressed (e.g. see Table 16.2). Finally, in Tier 3 the simulation is run to explore various population structure scenarios, while the performance of the different management procedures and reference point calculations is evaluated. The framework can also be used to validate nonspatial models or test an array of hypotheses.

model is to provide best estimates of spatially explicit parameters that can be used to inform the spatial operating model (i.e. narrow down the possible range of values), and it is not expected to be a perfect representation of the spatial population structure.

Tier 2: operating model

The spatially explicit operating model should represent the best available science on population structure and connectivity through a mixture of directly estimated parameters, assumed variables, and stochastic processes (Kerr and Goethel, 2014). The most basic simulation model will use the spatially explicit assessment model, assuming that parameter estimates are the true values. However, when population structure is more complex than the data will support, the simulation model will need to be altered to accommodate these additional intricacies. The estimation model is meant to provide a first approximation of spatial processes and parameter values, which can then be altered or explored within the simulation model through sensitivity analysis.

Considering the high degree of uncertainty in spatial dynamics, allowing for imprecision in our knowledge of a given natural system is critical if the simulation is to properly emulate the array of potential states of nature. Uncertainty can be incorporated by using the posterior distribution of parameter estimates from the estimation model to sample from when simulating the multiple realizations of the system. Typical error sources can also be included by simulating observation error in the various data sources. However, accounting for alternate plausible operating models (e.g. different recruitment or connectivity assumptions; Kerr et al., 2013) is perhaps the most important aspect of uncertainty. In cases when no a priori justification can reduce the plausible dynamics to a single model, multiple operating models should be developed and tested (including rerunning the entire estimation–simulation framework to ensure consistency of population dynamics and resulting parameter values among models in Tiers 1 and 2). Careful consideration of parameterization and uncertainty is warranted, especially when parameters are not estimated directly from a spatial estimation model. Depending on the goals of the simulation, emphasis and importance of modeling assumptions will be placed on different aspects of the operating model. Kerr and Goethel (2014) provide a detailed explanation of how to develop spatial simulations and the various assumptions that should be considered.

Tier 3: management evaluation

Once the operating model is developed, it is relatively easy to address a number of different questions. Spatially explicit biological reference points can be derived, and their performance for meeting policy objectives can be evaluated. It is also possible to use the framework to perform a holistic evaluation of the entire assessment-management system. The complete appraisal involves validating single stock assessments with the spatially explicit assessment model to determine whether the spatial model provides alternate views of population trajectories given the available data. The spatial operating model is then developed and used to perform robustness tests under a variety of assumed population structures and data scenarios with both the single stock and spatially explicit assessments. Simulation testing of the multiple assessment models will indicate how well each performs (e.g. the level of process error that results from simplifying assumptions regarding population structure and dynamics) and when an assessment model is likely to provide highly biased outputs. Finally, alternative simulations or even management strategy evaluation can be performed to evaluate the expected performance of possible management actions given the bias calculated in assessment outputs.

Evaluating harvest control rules using the three-tiered framework

Any of the spatially explicit reference point models discussed earlier can easily be adapted as a submodel within the spatial operating model. If adapted for use within a management strategy evaluation, a variety of scenarios could be explored to evaluate both the reliability of reference point estimates and the performance of harvest control rules. For example, parameter estimates from spatially explicit and closed population assessments could be used to calculate both spatial and nonspatial overfishing definitions. The results could then be compared to explore how close estimated reference points were to the 'true' values (i.e. those calculated from the assumed true population parameters), and how ignoring spatial structure could impact the performance of the simulated harvest control rule.

A comprehensive management strategy evaluation that includes spatial structure in all aspects of the simulation would allow a more informative investigation of the efficacy of a given harvest control rule to attain management objectives. Similarly, it would provide managers a general evaluation of how current assessment-management frameworks, which typically ignore spatial structure, might detrimentally impact resource status or fishery objectives. Spatially explicit operating models provide the ability to explore many questions and hypotheses that are critical to our understanding of population regulation and associated optimal management.

Examples of the three-tiered framework

Using outputs of an estimation model to inform parameters in a simulation is not a new idea, but the use of the three-tiered approach for development of spatially explicit fisheries operating models has only been made possible in the last decade with the expansion of integrated assessment techniques and computing power (Goethel et al., 2011). Kerr et al. (2013) used the three-tier framework to develop a spatially explicit operating model for Atlantic bluefin tuna (*Thynnus thynnus*) based on the results of the spatially explicit Multi-stock Age-Structured Tag-integrated assessment model (MAST; Taylor et al., 2011). The model was developed to explore consequences of stock structure and mixing on stock productivity and rebuilding goals. Even though the operating model was based on an applied assessment, reliability of parameter estimates was still difficult to address and alternate population dynamics assumptions were necessary to reflect observed population structure. For example, exploring alternative maturity scenarios had profound implications for correlated parameters (i.e. seasonal movement coefficients, selectivity, fishing mortality, and recruitment). Despite these uncertainties, the simulation clearly showed that failure to recognize the role of mixing in population and fishery dynamics of bluefin tuna may compromise assessment and management efforts.

Goethel et al. (2015a, 2015b) demonstrated how the three-tiered framework could be applied to yellowtail flounder fisheries management by developing a spatially explicit tag-integrated assessment that simultaneously modeled the three stocks of yellowtail flounder assuming metapopulation structure and connectivity among stocks. The model was used to validate current single stock assessments (Goethel et al., 2015a), and was then used as the basis for a spatial operating model that tested the robustness of both the spatially explicit and single stock assessment models (Goethel et al., 2015b). Although the single stock model was relatively robust, the three-tiered framework highlighted complex interactions between movement and productivity parameters that could cause differential estimates of recruitment trajectories if connectivity was ignored in the assessment.

The examples with bluefin tuna and yellowtail flounder reflect the difficulties in developing spatial operating models, but highlight a number of benefits of the three-tiered approach. By using a spatially explicit estimation model, the three-tiered approach maintains consistency between available data, assessment estimates, and simulation input parameters. Using one model to estimate all parameters within the simulation retains consistency between variables, which is not the case when parameter values are taken from multiple studies (Kerr et al., 2013). We believe that the three-tiered approach can be a useful integrated management tool that explicitly incorporates biocomplexity and population connectivity into the assessment-management system.

The importance of adaptive spatial and real-time management

Even spatially explicit biological reference point derivations assume either equilibrium (e.g. per-recruit simulations) or stationarity in many components of productivity (e.g. most dynamic F_{MSY} simulations). However, some of the most profound implications of spatial stock structure and connectivity result from transient effects (e.g. recruitment episodes in one subpopulation with spillover to adjacent areas). Transient population dynamics represent an important conundrum for evaluating harvest policies, because they represent short-term developments that are essentially unpredictable over expanded time horizons.

For example, Goethel et al. (2015a) demonstrated that emigration from one stock area substantially increased after an extreme recruitment event in the stock, but only for the 3 years immediately following recruitment of the year-class. Such an atypical event would be difficult to incorporate in a spatial simulation, because movement was time-varying, autocorrelated, and density-dependent. Therefore, predicting when that scenario might reoccur in the future is nearly impossible. However, emigration of the recruits would have important implications for the optimal harvesting of the metapopulation, because the stock of origin became a short-term de facto source population. The surplus emigration in the 3 years following the recruitment event would likely be harvested differentially from other years, which would lead to a temporary change in the optimal harvest strategy. A similar scenario may be occurring from extreme recruitment events of haddock on Georges Bank, with subsequent emigration to the Gulf of Maine (Brodziak et al., 2008). Tagging and distribution data suggest low connectivity between the two areas, but some large year-classes were present in both areas (Grosslein and Hennemuth, 1973). However, conventional stock assessment models are not well suited to detect immigration of abundant year-classes, particularly when they immigrate at intermediate ages (e.g. NEFSC, 2012b). Adaptive management and strategic decision-making such as this are likely to become increasingly important as climate change and environmental fluctuations result in shifting spatial distributions and unpredictable oscillations in stock productivity.

Adaptive spatial management can be a useful tool for dealing with transient dynamics, especially issues of bycatch. Hobday et al. (2010) developed a habitat preference model based on temperature gradients from satellite tags of southern bluefin tuna (*Thunnus maccoyii*) and coupled it to an ocean circulation model to create relatively fine-scale and near real-time predictions of abundance distributions. The information has been directly incorporated into management of the species to create dynamic zoning that restricts fishermen's access to areas with high abundance at various times throughout the fishing year. However, some transient spatial processes are best accounted for with tactical rather than strategic management. For example, until the spatial complexity of Atlantic cod populations off New England can be accounted for in stock assessment and management, spawning closures may help to conserve discrete spawning stocks (Zemeckis et al., 2014).

Overarching fishery policy should be based on long-term population dynamics (e.g. dynamic F_{MSY}), but flexibility is needed to maximize utilization on short-term time horizons when transient dynamics are present. Understanding of transient processes is likely to be limited, which makes many adaptive management measures inherently ad hoc, but they should not be overlooked simply because they cannot be explicitly model tested.

Summary and conclusions

The effectiveness of any given management action will be dependent on the spatial complexities of the stock along with potential ecosystem interactions

(Botsford et al., 2003, Pelletier and Mahévas, 2005). This review of the development of spatial operating models for evaluating management policies has illustrated the array of spatial processes that can be modeled, the differential impact that small changes in model assumptions can have, and the challenges faced when attempting to develop spatial simulations.

A predefined set of goals for a given management policy or harvest control rule is essential (Deroba and Bence, 2008). As demonstrated, different policies will have different costs and benefits. Defining how a regulation will be appraised a priori allows objective determinations of the most desirable management action, which is a critical component of the scientific method and a necessary part of any simulation analysis (Kerr and Goethel, 2014). We agree with Deroba and Bence (2008) that the performance of any proposed harvest policy should be tested within a stochastic simulation with a predefined objective function to be maximized that includes uncertainty. However, we contend that spatial population structure needs to be carefully considered within this simulation, and in many cases incorporated explicitly within the evaluation of the policy.

Frameworks already exist for estimating spatial reference points, but minimal research effort has been focused on their application. Large knowledge gaps still exist pertaining to the potential impacts of population connectivity on estimates of target biomass and fishing mortality levels. It is unlikely that there is a linear relationship between the optimal utilization of a single population compared with that of a metapopulation. The potential exists for the development of alternate harvest control rules that take into consideration the multiscalar nature of population processes and account for local and regional productivity (Heath et al., 2008, Steneck and Wilson, 2010, Goethel et al., 2015a), especially when complicated (e.g. source-sink) dynamics are present (Tuck and Possingham, 2000, Wilberg et al., 2008).

Despite the appeal of more biologically realistic and sophisticated simulation methods, ground-truthing model performance remains an important and difficult task (Holland, 2002). The more intricate a model becomes, the more difficult it is to ensure that input parameters are reasonable and model dynamics are representative of the system of interest. A critical limitation to any modeling approach is the scale of data collection in relation to the scale of the framework being used (Holland, 2002). Data to support estimation of all spatially explicit parameters used in spatial operating models is rarely available (Kerr and Goethel, 2014). Therefore, it is imperative that model dynamics are closely validated with what is known about observed spatial dynamics and any relevant information that may be available (Holland, 2002, Pelletier and Mahévas, 2005, Heath et al., 2008). The three-tier estimation-simulation framework eliminates many of the difficulties in determining input variable values for spatial simulation approaches by using observed data to estimate operating model parameters. However, increased collection of fine-scale data will be necessary in the future to reliably implement operating models for evaluating spatial harvest policies

(Fogarty and Botsford, 2007). The spatial dimension of data collection is rapidly achieving finer scales (e.g. through collection of catch locations with vessel monitoring systems and migration pathways using electronic tags; Gerritsen and Lordan, 2011, Sippel et al., 2014). Therefore, we expect that implementation of spatially explicit stock assessments and associated management strategy evaluation will become more widespread.

Evaluating the performance of comparatively simple management procedures using more complex simulation models leads to the challenging question of what level and dimension of complexity should be included in the operating model? Oftentimes, a trade-off exists between the complexity of the model and the reliability of input parameters values. In the context of multispecies processes, Plagányi et al. (2014) suggest that models of intermediate complexity should be used to address specific management questions. Their approach allows estimation of operating model parameters from available data similar to our three-tiered approach. However, the question remains of when is spatial complexity needed considering the many possible dimensions of model complexity (e.g. biological interactions, abiotic factors, and fishing behavior)? We suggest that a review of available information on the significance of each dimension of complexity for the productivity of the fishery and resource, as well as an understanding of the mechanisms of influence, should be considered in the decision to include spatial structure in a given simulation. Unfortunately, a paradox exists such that we cannot know whether spatial structure and connectivity (or other complexities) are important considerations for stock assessment and fishery management until we implement the entire simulation process. For example, there was evidence for connectivity and varying vital rates among yellowtail flounder stocks, but simulations showed that closed population stock assessment models performed relatively well (Goethel et al., 2015a). We would not have made that conclusion a priori.

We conclude that too few spatial operating models have been developed and applied to allow generalizing the situations that necessitate consideration of spatial complexity in management strategy evaluation. In lieu of a meta-analysis basis for determining criteria for including spatial structure and connectivity in operating models, strong evidence for fine-scale (i.e. substock) structure or movement among stocks should justify a spatial operating model. Additionally, when spatial population structure is present, spatially explicit biological reference points should be developed, and multiscalar spatiotemporal management considered. Despite limited understanding of many marine spatial processes, the importance of maintaining spatial biocomplexity through spatial management is well recognized. Although management strategy evaluation provides a powerful tool for analyzing policy performance, most approaches are not spatially explicit and cannot be used to appraise the performance of spatial management measures. In order to ensure that implemented policies are providing the intended benefits for the fish and fishery, spatial operating models need to be more thoroughly investigated, implemented, and continually improved.

Acknowledgements

We thank the editors for inviting us to submit our work for this book and for providing detailed feedback to improve our manuscript. Thanks also to Aaron Berger for providing a review of the manuscript. Our ideas and conclusions are the result of numerous collaborations. We acknowledge the hard work and insight of all those who contributed to these projects especially: Chris Legault, Terry Quinn, and Larry Alade for their involvement with yellowtail flounder; Doug Zemeckis, Adriene Kovach, Jon Loehrke, and Dave Martins for their work on Atlantic cod; and Dave Secor, Doug Butterworth, Clay Porch, and Shannon Cass-Calay for their effort with Atlantic bluefin tuna.

References

Ames, E.P. 2004. Atlantic cod stock structure in the Gulf of Maine. *Fisheries,* 29(1), 10–28.

Apostolaki, P., Millner-Gulland, E.J., McAllister, M.K., & Kirkwood, G.P. 2002. Modelling the effects of establishing a marine reserve for mobile fish species. *Canadian Journal of Fisheries and Aquatic Sciences,* 59, 405–415.

Berger, A.M., Jones, M.L., Zhao, Y., & Bence, J.R. 2012. Accounting for spatial population structure at scales relevant to life history improves stock assessment: the case for Lake Erie walleye *Sander vitreus. Fisheries Research,* 115–116, 44–59.

Beverton, R.J.H., & Holt, S.J. 1957. *On the Dynamics of Exploited Fish Populations.* Fisheries Investment Series 2. 19. UK Ministry of Agriculture and Fisheries, London, Chapman and Hall.

Botsford, L.W., Brumbaugh, D.R., Grimes, C., Kellner, J.B., Largier, J., O'Farrell, M.R., Ralston, S., Soulanille, E., & Wespestad, V. 2009. Connectivity, sustainability, and yield: bridging the gap between conventional fisheries management and marine protected areas. *Reviews in Fish Biology and Fisheries,* 19, 69–95.

Botsford, L.W., Hastings, A.W., & Gaines, S.D. 2001. Dependence of sustainability on the configuration of marine reserves and larval dispersal distance. *Ecology Letters,* 4, 144–150.

Botsford, L.W., Micheli, F., & Hastings, A. 2003. Principles for the design of marine reserves. *Ecological Applications,* 13, 525–531.

Brodziak, J, Traver, M.L., & Col, L.A. 2008. The nascent recovery of the Georges Bank haddock stock. *Fisheries Research,* 94, 123–132.

Cadrin, S.X., & Pastoors, M.A. 2008. Precautionary harvest policies and the uncertainty paradox. *Fisheries Research,* 94, 367–372.

Cadrin, S.X., & Secor, D.H. 2009. Accounting for spatial population structure in stock assessment: past, present, and future. *In:* Beamish, R. & Rothschild, B. (eds.) *The Future of Fisheries Science in North America,* pp. 405–426. Dordrecht, Springer.

Carruthers, T.R., McAllister, M.K., & Taylor, N.G. 2011. Spatial surplus production modeling of Atlantic tunas and billfish. *Ecological Applications,* 21(7), 2734–2755.

Collie, J.S., & Gislason, H. 2001. Biological reference points in a multispecies context. *Canadian Journal of Fisheries and Aquatic Sciences,* 58, 2167–2176.

Cope, J.M., & Punt, A.E. 2011. Reconciling stock assessment and management scales under conditions of spatially varying catch histories. *Fisheries Research,* 107, 22–38.

Crowder, L.B., Lyman, S.J., Figueira, W.F., & Priddy, J. 2000. Source-sink dynamics and the problem of siting marine reserves. *Bulletin of Marine Science,* 66, 799–820.

Deroba, J.J., & Bence, J.R. 2008. A review of harvest policies: understanding relative performance of control rules. *Fisheries Research,* 94, 210–223.

Field, J.C., Punt, A.E., Methot, R.D., & Thomson, C.J. 2006. Does MPA mean 'Major Problem for Assessments'? Considering the consequences of place-based management systems. *Fish and Fisheries,* 7, 284–302.

Frank, K.T., & Brickman, D. 2000. Allele effects and compensatory population dynamics within a stock complex. *Canadian Journal of Fisheries and Aquatic Sciences,* 57, 513–517.

Fogarty, M.J., & Botsford, L.W. 2007. Population connectivity and spatial management of marine fisheries. *Oceanography,* 20, 112–123.

Fouzai, N., Coll, M., Palomera, I., Santojanni, A., Arneri, E., & Christensen, V. 2012. Fishing management scenarios to rebuild exploited resources and ecosystems of the Northern-Central Adriatic (Mediterranean Sea). *Journal of Marine Systems,* 102–104, 39–51.

Fu, C., & Fanning, L.P. 2004. Spatial considerations in the management of Atlantic cod off Nova Scotia, Canada. *North American Journal of Fisheries Management,* 24, 775–784.

Gerritsen, H., & Lordan, C. 2011. Integrating vessel monitoring systems (VMS) data with daily catch from logbooks to explore spatial distribution of catch and effort at high resolution. *ICES Journal of Marine Science,* 68, 245–252.

Goethel, D.R., Legault, C.M., & Cadrin, S.X. (2015a). Demonstration of a spatially explicit, tag-integrated stock assessment model with application to three interconnected stocks of yellowtail flounder off of New England. *ICES Journal of Marine Science,* 72(1), 164–177. doi:10.1093/icesjms/fsu014

Goethel, D.R., Legault, C.M., & Cadrin, S.X. (2015b). Testing the performance of a spatially explicit tag-integrated stock assessment model of yellowtail flounder (*Limanda ferruginea*) through simulation analysis. *Canadian Journal of Fisheries and Aquatic Sciences,* 72(4), 582–601. doi:10.1139/cjfas-2014-0244.

Goethel, D.R., Quinn, T.J., II, & Cadrin, S.X. 2011. Incorporating spatial structure in stock assessment: movement modelling in marine fish population dynamics. *Reviews in Fisheries Science,* 19(2), 119–136.

Grosslein, M.D., & Hennemuth, R.C. 1973. Spawning stock and other factors related to recruitment of haddock on Georges Bank. *Rapports et Proces-verbaux des Réunions/Conseil International pour l'Éxploration de la Mer,* 164, 77–88.

Grüss, A., Kaplan, D.M., & Hart, D.R. 2011. Relative impacts of adult movement, larval dispersal and harvester movement on the effectiveness of reserve networks. *PLoS ONE.* 6(5), e19960.

Grüss, A., Kaplan, D.M., & Robinson, J. 2014. Evaluation of the effectiveness of marine reserves for transient spawning aggregations in data-limited situations. *ICES Journal of Marine Science,* 71, 435–449.

Guénette, S., Lauck, T., & Clark, C. 1998. Marine reserves: from Beverton and Holt to the present. *Reviews in Fish Biology and Fisheries,* 8, 251–272.

Hart, D.R. 2001. Individual-based yield-per-recruit analysis, with an application to the Atlantic sea scallop *Placopecten magellanicus. Canadian Journal of Fisheries and Aquatic Sciences,* 58, 2351–2358.

Hart, D.R. 2003. Yield- and biomass-per-recruit analysis for rotational fisheries, with an application to the Atlantic sea scallop (*Placopecten magellanicus*). *Fishery Bulletin,* 101, 44–57.

Hart, D.R., & Cadrin, S.X. 2004. Yellowtail flounder (*Limanda ferruginea*) off the Northeastern United States: implications of movement among stocks. *In:* Akçakaya, H.R., Burgman, M., Kindvall, O., Wood, C.C., Sjögren-Gulve, V., Hatfield, J.S., & McCarthy, M.A. (eds.) *Applications in RAMAS,* pp. 230–243. New York, Oxford University Press.

Heath, M.R., Kunzlik, P.A., Gallego, A., Holmes, S.J., & Wright, P.J. 2008. A model of meta-population dynamics for North Sea and West of Scotland cod—the dynamic consequences of natal fidelity. *Fisheries Research,* 93, 92–116.

Hobday, A.J., Hartog, J.R., Timmiss, T., & Fielding, J. 2010. Dynamic spatial zoning to manage southern bluefin tuna (*Thunnus maccoyii*) capture in a multi-species longline fishery. *Fisheries Oceanography,* 19(3), 243–253.

Holland, D.S. 2002. Integrating marine protected areas into models for fishery assessment and management. *Natural Resource Modeling,* 18, 369–386.

Holland, D.S. 2003. Integrating spatial management measures into traditional fishery management systems: the case of the Georges Bank multispecies groundfish fishery. *ICES Journal of Marine Science,* 60, 915–929.

Kaplan, D.M. 2006. Alongshore advection and marine reserves: consequences for modelling and management. *Marine Ecology Progress Series,* 309, 11–24.

Kaplan, D.M., Botsford, L.W., & Jorgensen, S. 2006. Dispersal per recruit: an efficient method for assessing sustainability in marine reserve networks. *Ecological Applications,* 16(6), 2248–2263.

Kerr, L.A., Cadrin, S.X. & Kovach, A.I. 2014. Consequences of a mismatch between biological and management units on our perception of Atlantic cod off New England. *ICES Journal of Marine Science,* 71(6), 1366–1381.

Kerr, L.A., Cadrin, S.X., & Secor, D.H. (2010a). Simulation modelling as a tool for examining the consequences of spatial structure and connectivity on local and regional population dynamics. *ICES Journal of Marine Science,* 67, 1631–1639.

Kerr, L.A., Cadrin, S.X., & Secor, D.H. 2010b. The role of spatial dynamics in the stability, resilience, and productivity of an estuarine fish population. *Ecological Applications,* 20(2), 497–507.

Kerr, L.A., Cadrin, S.X., Secor, D.H. & Taylor, N. 2013. A simulation tool to evaluate effects of mixing between Atlantic bluefin tuna stocks. *Collective Volume of Scientifics Papers ICCAT,* 69, 742–759.

Kerr, L.A. & Goethel, D.R. 2014. Simulation modelling as a tool for synthesis of stock identification information. *In* Cadrin, S.X., Kerr, L.A., & Stefano, M. (eds.) *Stock Identification Methods,* pp. 501–533. London, Elsevier.

Kovach, A.I., Breton, T.S., Berlinsky, D.L., Maceda, L., & Wirgin, I. 2010. Fine-scale spatial and temporal genetic structure of Atlantic cod off the Atlantic coast of the USA. *Marine Ecology Progress Series,* 410, 177–195.

Kritzer, J.P., & Liu, O.R. 2014. Fishery management strategies for addressing complex spatial structure in marine fish stocks. *In:* Cadrin, S.X., Kerr, L.A., & Stefano, M. (eds.) *Stock Identification Methods,* pp. 29–57. London, Elsevier.

Le Quesne, W.J.F., & Codling, E.A. 2009. Managing mobile species with MPAs: The effects of mobility, larval dispersal, and fishing mortality on closure size. *ICES Journal of Marine Science,* 66, 122–131.

MacCall, A.D. 1990. *Dynamic Geography of Marine Fish Populations,* Seattle, University of Washington Press.

Mace, P.M. 2001. A new role for MSY in single-species and ecosystem approaches to fisheries stock assessment and management. *Fish Fisheries,* 2, 2–32.

Mahévas, S. & Pelletier, D. 2004. ISIS-Fish, a generic and spatially explicit simulation tool for evaluating the impact of management measures on fisheries dynamics. *Ecological Modelling,* 171, 65–84.

Martell, S.J.D., Essington, T.E., Lessard, B., Kitchell, J.F., Walters, C.J., & Boggs, C.H. 2005. Interactions of productivity, predation risk, and fishing effort in the efficacy of marine protected areas for the central Pacific. *Canadian Journal of Fisheries and Aquatic Sciences,* 62, 1320–1336.

Maunder, M.N. 2001. Integrated tagging and catch-at-age analysis (ITCAAN): Model development and simulation testing. *In:* Kruse, G.H., Bez, N., Booth, A., Dorn, M.W., Hills, S., Lipcius, R.N., Pelletier, D., Roy, C., Smith, S.J., & Witherell, D. (eds.) *Spatial Processes and Management of Marine Populations,* pp. 123–146. Fairbanks, AK, University of Alaska Sea Grant, AK-SG-01–02.

Mchich, R., Charouki, N., Auger, P., Raissi, N., & Ettahiri, O. 2006. Optimal spatial distribution of the fishing effort in a multi fishing zone model. *Ecological Modelling,* 197, 274–280.

Moffitt, E.A., Botsford, L.W., Kaplan, D.M., & O'Farrell, M.R. 2009. Marine reserve networks for species that move within a home range. *Ecological Modelling,* 19(7), 1835–1847.

NEFSC (Northeast Fisheries Science Center). 2012a. 53rd Northeast Regional Stock Assessment Workshop (53rd SAW) Assessment Report. *United States Department of Commerce, Northeast Fisheries Science Center Reference Document* 12–05.

NEFSC (Northeast Fisheries Science Center). 2012b. Assessment or Data Updates of 13 Northeast Groundfish Stocks through 2010. *United States Department of Commerce, Northeast Fisheries Science Center Reference Document* 12–06.

Okamura, H., McAllister, M.K., Ichinokawa, M., Yamanaka, L. & Holt, K. 2014. Evaluation of the sensitivity of biological reference points to the spatio-temporal distribution of fishing effort when seasonal migrations are sex-specific. *Fisheries Research,* 158, 116–123.

Pelletier, D., and Mahévas, S. 2005. Spatially explicit fisheries simulation models for policy evaluation. *Fish and Fisheries,* 6, 307–349.

Plagányi, É.E., Punt, A.E., Hillary, R., Morello, E.B., Thébaud, O., Hutton, T., Pillans, R.D., Thorson, J.T., Fulton, E.A., Smith, A.D.M., Smith, F., Bayliss, P., Haywood, M., Lyne, V., & Rothlisberg, P.C. 2014. Multispecies fisheries management and conservation: tactical applications using models of intermediate complexity. *Fish and Fisheries,* 15, 1–22.

Punt, A.E., & Cui, G. 2000. Including spatial structure when conducting yield-per-recruit analysis. Australian society for fish biology workshop proceedings, pp. 176–182. September, 1999. Bendigo, Victoria, Australia.

Punt, A.E., & Methot, R.D. 2004. Effects of marine protected areas on the assessment of marine fisheries. *American Fisheries Society Symposium,* 42, 133–154.

Quinn, J.F., Wing, S.R., & Botsford, L.W. 1993. Harvest refugia in marine invertebrate fisheries: models and applications to the red sea urchin, *Stronglocentrotus franciscanus. American Zoologist,* 33, 537–550.

Sanchirico, J.N., & Wilen, J.E. 2005. Optimal spatial management of renewable resources: Matching policy scope to ecosystem scale. *Journal of Environmental Economics and Management,* 50, 23–46.

Sippel, T., Evenson, J.P., Galuardi, B., Lam, C., Hoyle, S., Maunder, M.N., Kleiber, P., Carvalho, F., Tsontos, V., Teo, S.L.H., Aires-da-Silva, A. & Nicol, S. 2014. Using movement data from electronic tags in fisheries stock assessment: A review of models, technology, and experimental design. *Fisheries Research.* doi:10.1016/j.fishres.2014.04.006

Smedbol, R.K., & Stephenson, R.L. 2001. The importance of managing within-species diversity in cod and herring fisheries of the North-Western Atlantic. *Journal of Fish Biology,* 59 (Suppl. A), 109–128.

Steneck, R.S., & Wilson, J.A. 2010. A fisheries play in an ecosystem theatre: Challenges of managing ecological and social drivers of marine fisheries at multiple spatial scales. *Bulletin of Marine Science,* 86(2), 387–411.

Taylor, N.G., McAllister, M.K., Lawson, G.L., Carruthers, T. & Block, B.A. 2011. Atlantic bluefin tuna: A novel multistock spatial model for assessing population biomass. *PLoS ONE,* 6(12), e27693.

Tuck, G.N., & Possingham, H.P. 1994. Optimal harvesting strategies for a metapopulation. *Bulletin of Mathematical Biology,* 56, 107–127.

Tuck, G.N., & Possingham, H.P. 2000. Marine protected areas for spatially structured exploited stocks. *Marine Ecology Progress Series,* 192, 89–101.

Walters, C.J. 2000. Impacts of dispersal, ecological interactions, and fishing effort dynamics on efficacy of marine protected areas: how large should protected areas be? *Bulletin of Mathematical Biology,* 66(3), 745–757.

Walters, C.J., Hilborn, R., & Parrish, R. 2007. An equilibrium model for predicting the efficacy of marine protected areas in coastal environments. *Canadian Journal of Fisheries and Aquatic Sciences,* 64, 1009–1018.

White, J.W. 2010. Adapting the steepness parameter from stock-recruit curves for use in spatially explicit models. *Fisheries Research,* 102, 330–334.

Wilberg, M.J., Irwin, B.J., Jones, M.L., & Bence, J.R. 2008. Effects of source-sink dynamics on harvest policy performance for yellow perch in southern Lake Michigan. *Fisheries Research,* 94, 282–289.

Ying, Y., Chen, Y., Lin, L., & Gao, T. 2011. Risks of ignoring fish population spatial structure in fisheries management. *Canadian Journal of Fisheries and Aquatic Sciences,* 68, 2101–2120.

Zemeckis, D.R., Martins, D., Kerr, L.A., & Cadrin, S.X. 2014. Stock identification of Atlantic cod (*Gadus morhua*) in US waters: An interdisciplinary approach. *ICES Journal of Marine Science,* 71(6), 1490–1506.

## Chapter 17

The quantification and presentation of risk

Laurence T. Kell, Polina Levontin,
Campbell R. Davies, Shelton Harley, Dale S. Kolody,
Mark N. Maunder, Iago Mosqueira,
Graham M. Pilling and Rishi Sharma

Introduction

Tuna Regional Fisheries Management Organisations (tRFMOs) are inter-governmental organisations that are charged with data collection, scientific monitoring and the management of tuna and tuna-like species. There are five tRFMOs, namely the Commission for the Conservation of Southern Bluefin Tuna (CCSBT), the Inter-American Tropical Tuna Commission (IATTC), the International Commission for the Conservation of Atlantic Tunas (ICCAT), the Indian Ocean Tuna Commission (IOTC) and the Western and Central Pacific Fisheries Commission (WCPFC). As institutions they are still evolving, as some have been created only recently (i.e. WCPFC) while others are older, dating back to the 1940s (i.e. IATTC). The Commissions are political bodies responsible for making management decisions, which are subsequently adopted by member states. Advice is provided by Scientific Committees, and in some cases stock assessments are performed by working groups comprising representatives from member states (e.g. CCSBT, ICCAT and IOTC), while in others work is either performed in house (IATTC) or contracted out (WCPFC).

As a major step towards harmonisation the tRFMOs have recently agreed on a common management advice framework, known as the Kobe Framework (Anonymous, 2009): an agreement to adopt a common methodology for sharing scientific resources, to facilitate data sharing and to coordinate management, compliance and enforcement approaches. The framework also provides a basis for cooperation on improving how uncertainty is quantified, incorporated into analyses and communicated. A particular emphasis is therefore devoted to risk defined as in Chapter 2 – a chance event with negative consequences. Although there are several actual definitions of risk, most standards define risk as *an uncertainty that, if it occurs, will have an effect on achieving objectives* (Hillson, 2011). The quantification of risk requires estimating the probability of an event occurring and the severity of any consequences.

A discussion of risk needs to begin with an understanding of management objectives. In fisheries objectives are represented by target reference points (TRPs), which are used with limit reference point (LRPs) that indicate the state of a fishery or a stock considered undesirable and which should be avoided with high probability. It is preferable that risk tolerance is made explicit, for instance, stipulating that there should be no more than a 5% chance of breaching an LRP. Reference points can be used as part of a harvest control rule (HCR); a management decision algorithm which specifies in advance what actions need to be taken and when, in order to over time maximise the probability of achieving targets and minimise the risk of breaching limits.

The main management objective when the tRFMOs were established was to achieve the maximum sustainable yield (MSY). This requires maintaining total and adult stock biomass at the levels associated with MSY (i.e. B_{MSY} and SSB_{MSY}) and fishing mortality at a level that would on average achieve MSY (F_{MSY}). Therefore management was based on biomass and fishing mortality TRPs linked to MSY. Subsequently the United Nations Conference on Straddling and Highly Migratory Fish Stocks (United Nations, 1995) redefined F_{MSY} as a least stringent standard for a LRP; that is this implies that now the probability of exceeding (F_{MSY}) must be lower than 50%.

Due to the higher expectations for sustainability of contemporary management agreements (see Brown et al., 1987, Harley et al., 2012), and in order to bring themselves in line with the world mandate for adopting the precautionary approach, tRFMOs are now starting to incorporate reference points (LRPs and TRPs) into advice and evaluating their performance as part of HCR using management strategy evaluation (MSE). MSE provides a formal way to test the robustness of reference points to some of the known sources of uncertainty. When running an MSE control actions from the HCR are fed back into an operating model (OM) that represents the system being managed so that its actions on the simulated stock and hence on future fisheries data is propagated through the stock and fishery dynamics. MSE may be used to simulation test a management procedure (MP), which is the combination of predefined data, together with an algorithm into which such data are input to provide a value for a total allowable catch (TAC) or effort control measure. The MP may include a HCR and a stock assessment estimator, but does not have to; for example CCSBT provides a model-free example of a MP that is based on year-to-year changes and trends in empirical indicators.

MSE can be a part of the Kobe Framework by helping to quantify uncertainties associated with different levels of exploitation or different approaches to adaptive management. When designing elements of a management system using MSE a key benefit is greater conformity to the precautionary principle of resource exploitation, since performance is judged under conditions of (simulated) incomplete knowledge. The degree of this benefit depends on the effort made to elicit and represent important uncertainties within the simulations. The Kobe Framework is not necessarily going to supersede the current institutional arrangements of each Commission when negotiating decisions. However, it may assist in making management measures more robust; that is, ensure that the management objectives are met with high probability despite the presence of uncertainty or stressful environmental conditions (Radatz et al., 1990).

Execution of a MSE can be summarised in the following steps (i.e. described in greater detail by Punt and Donovan, 2007):

1   Identification of management objectives and mapping these into statistical indicators of performance or utility functions;
2   Selection of hypotheses for considering in the OM that represents the simulated versions of reality;
3   Conditioning of the OM based on data and knowledge, and weighting of model hypotheses depending on their plausibility;
4   Identifying candidate management strategies and coding these as MPs;
5   Projecting the OM forward in time using the MPs as a feedback control in order to simulate the long-term impact of management (Ramaprasad, 1983);
6   Identifying the MP that robustly meet management objectives.

Performance of a simulated MP is measured in terms of risk, based on probabilistic measures of the simulated frequencies with which management objectives are met. Statistics from the OM might also be collected to assess additional subjective (according to elicited stakeholder views) consequences, for instance modelled impacts on employment, profits or ecosystem. The involvement of stakeholders is a key point; it is increasingly a normative view within fisheries management that stakeholders should be the foundation of the decision-making process. In order to ensure that stakeholder concerns are included it is essential to communicate to stakeholders how uncertainty is quantified, to solicit their feedback from the very beginning in specifying hypothesis and representing uncertainties within the OM, and respond where possible with amendments to the simulation framework.

The MSE approach is flexible and not restricted to the goals of finding an MP that will run on autopilot (as described in Hillary et al., Chapter 8, this volume); since MSE can be an exploratory tool employed to examine the sensitivity of the system to various beliefs about how it functions (e.g. Kell and Fromentin, 2007). Case-specific MSEs can facilitate decision-making processes within tRFMOs by improving the understanding of risks and reconciling the differences of opinions among stakeholders regarding the implications of outstanding gaps in knowledge.

We first review the application of the Kobe Framework and the development of LRPs and MPs by the tRFMOs. We then discuss how the Kobe Framework can be extended to utilise more fully approaches for identifying, quantifying and communicating risks – in particular, the potential for MSE to make the Kobe Framework more robust and more inclusive.

Scientific advice framework

In traditional stock assessment and management based upon it, it is assumed that the system dynamics are known and expressed in the form of a mathematical model and that a management control (e.g. TAC) can be adjusted based on that knowledge. In engineering this is known as open loop feed-forward control (Velthuis, 2000). Pure feed-forward control is also termed 'ballistic', since once a control has been set it cannot be adjusted (Whitbeck and Wolkovitch, 1982). Corrective adjustment must be done by updating the formal stock assessment on a regular basis and the Commission agreeing to new management measures. An alternative is a closed-loop control system, termed 'cruise control', which automatically adjusts a control in response to feedback from the system (Figure 17.1). Feedback relaxes the requirement of having to have an exact model of the system being managed since the effect of the controls action on the system are monitored and adjusted accordingly, for example as in the case of the southern bluefin MP, where TACs are set based on trends in indices of abundance (see Hillary, Chapter 8, this volume). MSE includes the simulation of closed-loop feedback systems. MSE is widely thought of as a process that is used to create a MP that runs for several years on autopilot, without the need for managers to agree on measures based on an annually updated stock assessment. However, MSE can also be used for strategic proposes to explore robustness of existing or proposed elements of a management regime.

When managing fisheries, decisions have to be made under incomplete knowledge in a stochastic environment. Therefore the precautionary approach (PA, Garcia, 1996) requires that undesirable outcomes should be anticipated and measures taken to reduce the probability of them occurring and/or the

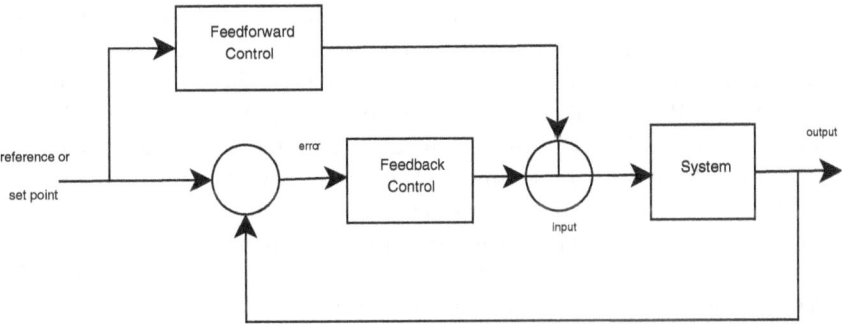

Figure 17.1 A comparison between feed-forward and feedback control systems. The top control shows a feed-forward control system ideally based on an exact model of the system; the bottom control is based on feedback which is reactive and automatically compensates for disturbances (i.e. errors), and in practice can be very simple if the signal used to monitor the system is adequate.

magnitude of their impact. This requires managing the identified causes of uncertainty in order to ensure that management objectives are met, as well as managing the consequences. As such the PA represents a shift towards risk management (see Edwards, Chapter 2, this volume).

The voluntary Code of Conduct on Responsible Fishing (FAO, 1995) and the United Nations Fish Stocks Agreement (United Nations, 1995) provide the formal basis for the application of the PA to fisheries management. The conventions of some tuna Regional Fisheries Management Organisations such as the WCPFC and IATTC refer to these codes directly, whilst others do not explicitly address the PA (De Bruyn et al., 2012) as their conventions were signed before the PA was drafted. Yet all five tRFMOs have agreed to take steps through the Kobe Framework process (Anonymous, 2009), towards a common approach to risk-based management, and hence are moving synchronously towards meeting the needs of the PA.

One of the requirements of the PA is that stocks are assessed regularly with respect to LRPs and TRPs. Given the political complexity of tRFMOs, which include many international partners and stakeholders that deal with a valuable resource about which knowledge is both insufficient and contested, it is unlikely that a system of decision-making that relies on negotiations will be replaced by HCRs run on autopilot. Of the five tRFMOs only CCSBT, which manages a single stock with a limited number of member states, has fully adopted a MP approach. In the other RFMOs a simplified robust decision-making process might be developed based on the current stock assessment processes to provide guidance, comparison and risk-based advice.

The main outcome of the Kobe Framework is the standardisation of the presentation of stock assessment and management advice relative to reference points. The reference points currently used by the tRFMOs (Anonymous, 2015) are summarised in Table 17.1. Management objectives are mainly articulated through MSY-based target reference points, the values of which depend on the productivity of the stock plus the selectivity of the fisheries and their relative effort. Some tRFMOs also derived LRPs from MSY (e.g. ICCAT and IOTC). WCPFC have used $SSB_{F=0}$ (the spawning stock biomass in the absence of fishing derived from a stock assessment); this has an advantage of not depending on the selection pattern of the fleets. Deterministic estimates of F_{MSY} and B_{MSY} may be derived from the parameters of a stock assessment model. In a stochastic framework, MSY-based reference points are usually obtained by simulation. For example, MSY can be calculated as the largest average long-term yield from application of a constant fishing mortality F, that is F_{MSY}, or from a harvest control rule where F varies as a function of stock size. B_{MSY} is then the average biomass that results from fishing at F_{MSY}, where B_{MSY} commonly refers to the spawning stock biomass. In this case how productivity and fishery selectivity varies over time becomes important.

It is an intention of some of the tRFMOs to test reference points within an MSE setting to determine their potential sensitivity to various sources of

Table 17.1 Summary of reference points used by tuna RFMOs.

	CCSBT	IATTC	ICCAT	IOTC	WCPFC
Limits	None	For tropical tunas: $F_{0.5R0}$ and $B_{0.5R0}$ evaluated assuming a steepness of 0.75 (adopted at the 87th Meeting as interim limits). The B limit corresponds to a depletion level of $0.077B_0$. Using the 2014 assessment results, the corresponding F/F_{MSY} values are 2.4 and 1.6 for yellowfin and bigeye.	For N. Atlantic swordfish: $0.4\ B_{MSY}$ (interim limit; Rec. 13–02).	Biomass: Tropical tunas: $0.4B_{MSY}$ ($0.5B_{MSY}$ for BET) $1.4F_{MSY}$ ($1.3F_{MSY}$ for BET and $1.5\ F_{MSY}$ for SKJ)–(interim limits; Res. 12/01 and 13/10).	For tropical tunas and S. Pacific albacore: $0.2SB_{F=0}$ evaluated using recent recruitment levels (adopted at the 2012 annual meeting).
Rebuilding targets	Rebuilding $0.2B_0$ (with 70% probability) in 25 years	None	Western Atlantic bluefin: 20-year program to rebuild to B_{MSY} (Recs. 98–07 and 14–05). Eastern Atl. and Mediterranean bluefin: A 15-year recovery program to reach B_{MSY} with at least 60% probability (Recs 07–05 and 14–04). Past Recommendations: Rec. 06–02 established a 10-year rebuilding program for N. Atlantic swordfish to achieve B_{MSY} with greater than 50% probability. Rec. 09–05 established a rebuilding program for N. Atlantic albacore with the implied rebuilding target of B_{MSY} in 10 years.	None	For BET, reducing F to F_{MSY} by 2017 is an implied rebuilding target under CMM 2014–01.

uncertainty and their ability to achieve management objectives. One possible outcome of MSE is that these reference points might be revised to reflect a better understanding of the risks arising from the scientific, managerial and other sources of uncertainty.

Uncertainty

Characterisation of uncertainty is a requirement of the precautionary approach, but it is a process that can vary widely in scope depending on how representative it is about involving stakeholders. Traditional stock assessment, and advice based upon it, mainly considers measurement and process error despite research showing that uncertainty about the actual dynamics (structural uncertainty) has a larger impact on achieving management objectives (Punt, 2008). Discussions of uncertainty in the context of stock assessment include limitations in our knowledge of system dynamics, the unpredictability of environmental events and their impacts, the lack of precision in our ability to implement management measures and to monitor stocks and fisheries (Kirkwood and Smith, 1995, Leach et al., 2014). Uncertainties in this system are usually classified in the following manner (after Rosenberg and Restrepo, 1994):

> *Process error* due to the underlying stochasticity in population dynamics such as random variability in recruitment or year-to-year changes in distribution.
>
> *Observation error* due to sampling and measurement of quantities such as the catch or average size at age.
>
> *Model error* related to the ability of the model structure to capture system dynamics; including
>
>> *structural uncertainty* due to inadequate models, incomplete or competing conceptual frameworks, or where significant processes or relationships are wrongly specified or not considered. Such situations tend to be underestimated by experts (Morgan and Henrion, 1990);
>>
>> *value uncertainty* model parameters that are treated as fixed inputs because they are difficult to estimate reliably (e.g. stock-recruit relationships).
>
> *Estimation error* arising when estimating parameters of the models used in the assessment procedure; estimation error can result from any of the aforementioned uncertainties, or from limitations of the numerical procedures, and is the inaccuracy and imprecision in the estimated model parameters such as stock abundance or fishing mortality rate.
>
> *Implementation error* where the effects of management actions may differ from those intended, e.g. the inability to achieve a target harvest strategy exactly.

The definitions of Rosenberg and Restrepo (1994) focus on aspects that can be quantified in mathematical models, particularly as stock assessment working groups often focus on technical aspects related to modelling, such as eliciting prior distributions in Bayesian modelling frameworks or ranges for parameter values. The characterisation of uncertainty is ultimately a pragmatic choice depending on the purpose of a particular application. There are other classifications of uncertainty; for example, Edwards (Chapter 2, this volume) defines 'statistical uncertainty', which includes the structural, process and observation uncertainties, and combines 'model error', 'structural uncertainty' and 'value uncertainty' into 'structural uncertainty'; then summarises the different sources based on those that can be reduced and those that are inherent to the system:

Irreducible aleatoric:

- Process
- Implementation

Reducible epistemic:

- Statistical
- Structural

Linguistic sources of uncertainty which play a role in communication and elicitation, or uncertainty related to risk perception or vagueness and the possibly contradictory nature of management goals are also important (Regan et al., 2002). The national discourses about resource use, power dynamics among nations based on cultural, economic and epistemic histories shape the way the problems of internationally shared resources are understood, but these crucial differences are rarely articulated in the language of stock assessments. Uncertainties related to institutional and social norms are also important because they pertain to the very definition of a management problem which is a social construct first and foremost.

The transformation of a management mandate into a modelling problem is often been seen as an unproblematic 'natural' process determined by the available methodology. But with the proliferation of available modelling alternatives (from single-species to ecosystem modelling) it is increasingly evident that uncertainty pertaining to methodology can be pivotal. The scientific advice produced can be radically different depending on the model used. Unless the process of building models is participatory, the modellers have potentially unwarranted control over the means of producing knowledge which influences management decisions. This is why there is an increasing consensus on a definition of risk that covers all known uncertainties, rather than individual elements of it.

Partly for these reasons MSE is increasingly being used to address uncertainties and their effect on management advice. Although MSE cannot address every kind of uncertainty, it can accommodate more sources of uncertainties than stock assessment alone. For example MSE can indicate how improving management through monitoring control and surveillance (MCS) or knowledge through focused scientific data collection and research can ensure that management objectives are met (Fromentin et al., 2014). A strength of MSE is that by agreeing management objectives in advance stability can be added to the management decision process. Particularly as MSE requires a dialogue between scientists, managers and stakeholders on how to evaluate alternative management procedures given uncertainty over what time period, which reference points to use, what are the acceptable levels of risks, what are the possible trade-offs between social and economic objectives and so on (e.g. Röckmann et al., 2012). MSE can lend logical support to a conversation through which outstanding disagreements can be potentially resolved.

Assessment frameworks

Assessment advice within the tRFMOs is increasingly being based on integrated models; for example IATTC uses Stock Synthesis (SS; Methot and Wetzel, 2013), WCPFC uses Multifan-CL (Hampton and Fournier, 2001) and IOTC uses SS alongside a variety of other models. While ICCAT uses integrated models and virtual population analysis (VPA), most advice is based on biomass dynamic models. Management by CCSBT is based not on stock assessment but a management procedure (MP) developed using an integrated age-based OM (Hillary et al., Chapter 8, this volume).

Two main visualisation tools are used as part of the Kobe Framework to present stock assessment advice, namely the Kobe II phase plot (K2PP) and the Kobe II strategy matrix (K2SM). The K2PP presents stock status against fishing mortality relative to TRPs as a two-dimensional phase plot. The K2SM lays out the probability of meeting management objectives under different options, including if necessary ending overfishing or rebuilding overfished stocks. Presenting advice in the K2SM format is intended to facilitate the application of the PA by providing Commissions with a basis to evaluate and adopt management options at various levels of risk (Anonymous, 2009). This enables Commissioners to make management recommendations while taking some sources of uncertainty into account. As an exception the CCSBT does not use the K2SM, since they prefer to consider other performance measures (related to catch levels and catch variability) as well as stock status.

The K2PP identifies quadrants (regions) where the stock is overfished (biomass or SSB is less than B_{MSY}) or overfishing is occurring ($F \geq F_{MSY}$) and a target region (where both SSB $\geq$ SSB$_{MSY}$ and $F \leq F_{MSY}$). In the case of biomass dynamic stock assessment model results biomass may be used instead of SSB. The target region is also called the *green* quadrant, referring to the colour scheme typically used when presenting the K2PP. The plots can be used to indicate for example when management plans to recover the stock to the target region should be implemented. In practice there is a lot of diversity in how status is presented in the K2PPs and a range of examples are shown in Figure 17.2; these are not exhaustive as many variants are used both between and within the tRMFOs.

In some cases results from a single model that was run with different fixed values are presented, as is the case with the IATTC example for the Eastern Pacific bigeye SS assessment (Aires-da Silva and Maunder, 2012) presented in Figure 17.2a. The 26 scenarios represent uncertainty about the values of

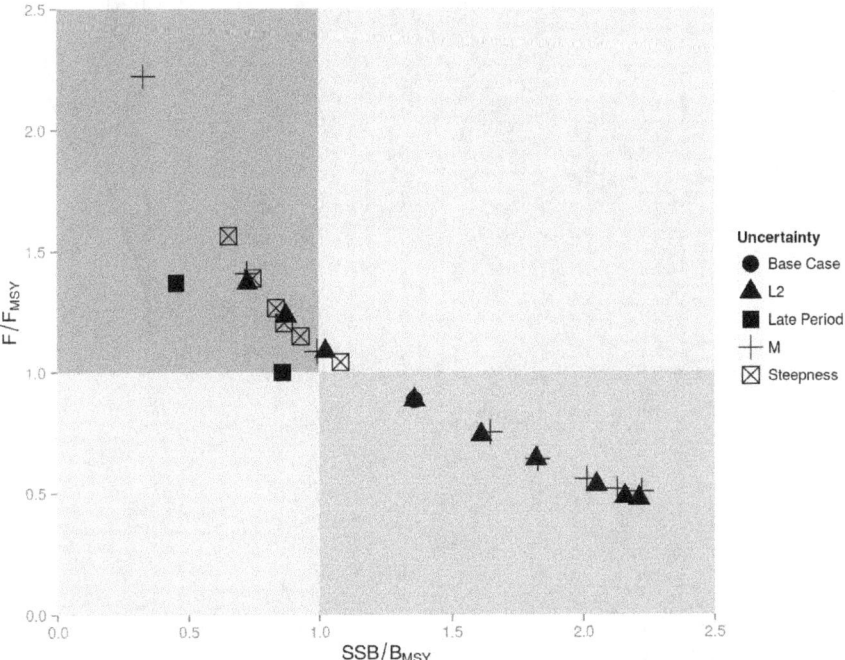

Figure 17.2 Example of Kobe phase plots, quadrants identify where the stock is overfished (biomass or SSB is less than B_{MSY}) or overfishing is occurring ($F \geq F_{MSY}$) and a target region (where both $SSB \geq SSB_{MSY}$ and $F \leq F_{MSY}$). IATTC example is for the Eastern Pacific bigeye Stock Synthesis assessment, with 26 scenarios represent uncertainty about the values of parameters; IOTC example is for the Indian Ocean albacore Stock Synthesis assessment and 36 scenarios results are contoured to generate a probability density. WCPFC shows the silky shark Stock Synthesis assessment for 2,592 scenarios; the size of the circles correspond to plausibility based on expert judgement. The ICCAT example is for South Atlantic albacore assessment using two biomass dynamic models implementations (ASPIC using maximum likelihood and BSP using Bayesian simulation) with two production functions and two catch per unit effort series; the large circles denote the medians from each assessment run and the small dots individual estimates from Monte Carlo simulations, marginal probability distributions are also shown along the x- and y-axes.

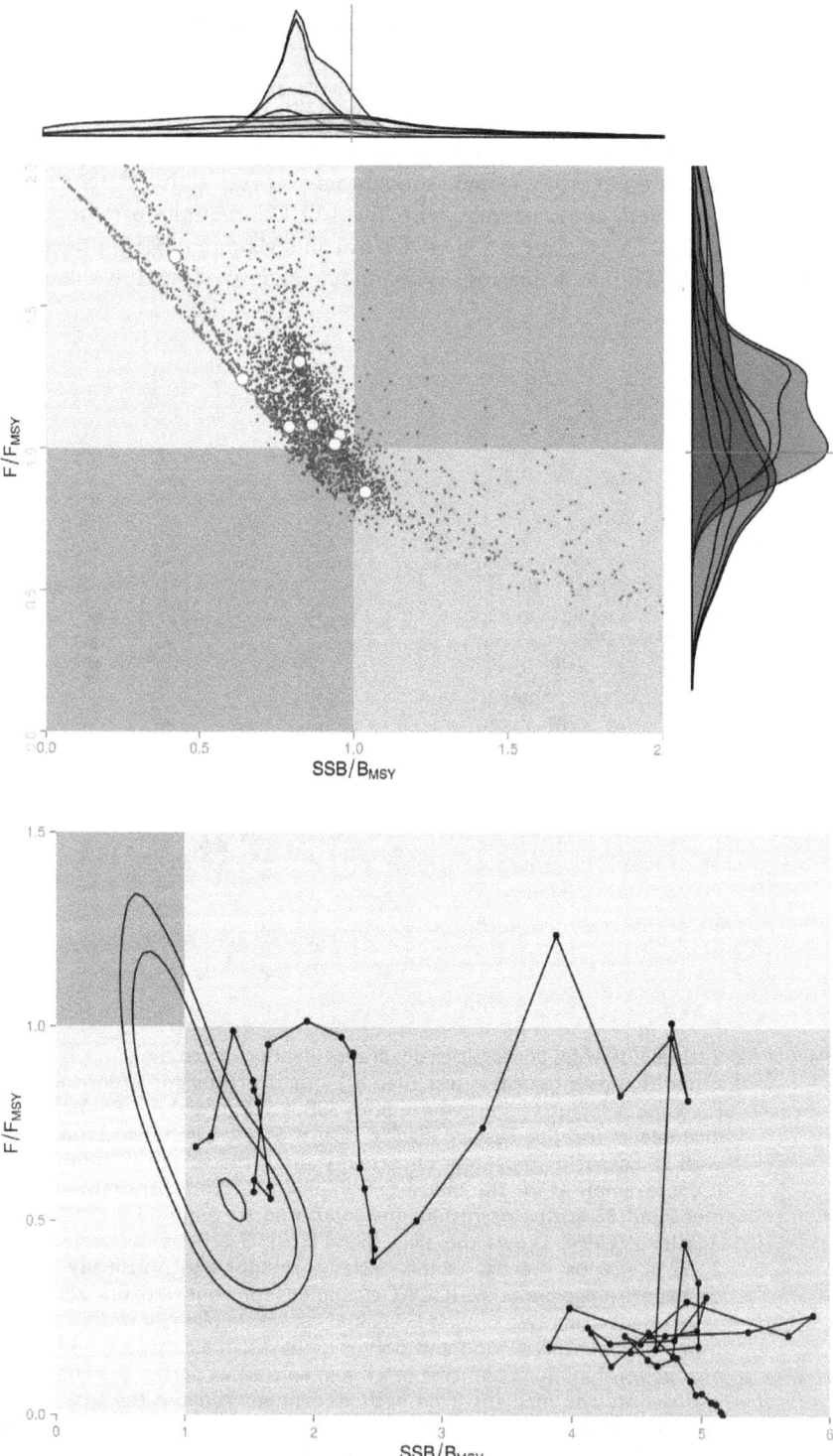

Figure 17.2 (Continued)

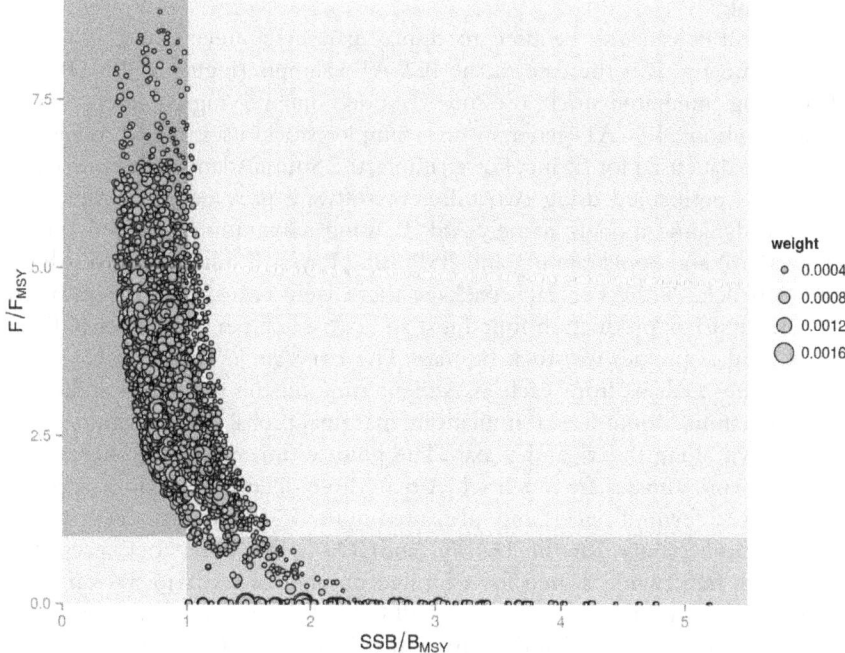

weight
○ 0.0004
○ 0.0008
◐ 0.0012
◯ 0.0016

Figure 17.2 (Continued)

parameters used in the assessment, that is steepness, M and the average length of the oldest individuals, that are fixed structural assumptions in the models. The base case is in the target quadrant (i.e. the lower right quadrant). There is a curvilinear relationship between F/F_{MSY} and SSB/B_{MSY}. Reducing steepness results in the stock becoming overfished (i.e. SSB/B_{MSY} decreases) and overfishing occurring (i.e. F/F_{MSY} increases). While changing M and the length of the oldest individuals results in a decrease/increase in F/F_{MSY} and an increase/decrease in SSB/B_{MSY}. In the IATTC reference points vary by year; B_{MSY} changes as historic recruitment varies and F_{MSY} as selectivity and the mix of gears changes.

The Indian Ocean Tuna Commission adopts a similar approach to IATTC, presenting a range of assessment results based on model assumptions, the Stock Synthesis assessment for Indian Ocean albacore is presented in Figure 17.2c; this time 36 scenarios were evaluated for different values of natural mortality (M) and steepness, assumptions about effort creep and weighting of data. In this example the results were contoured to generate a probability density. The WCPFC also explores value and structural uncertainty, using SS based on an uncertainty grid, where individual combinations of alternative parameter values are run as separate assessments. For the silky shark assessment, Figure 17.2d, alternative stock assessment runs (2,592 scenarios based on steepness, growth,

M, etc.). The sizes of the circles correspond to plausibility based on expert judgement.

The K2PPs can also be used to display parameter uncertainty, as well as point estimates, as is the case in the ICCAT example (Figure 17.2b). Rather than using integrated stock assessment models and varying parameter values or assumptions, ICCAT generally uses simpler stock assessment models and varies the data used for fitting. For example, the South Atlantic albacore assessment was performed using two different software packages that implement biomass dynamic models; namely ASPIC using maximum likelihood (Prager et al., 1996) and bootstrapping and BSP using Bayesian simulation (McAllister and Babcock, 2003). For each package there were two model specifications (logistic and Fox production functions) and two catch per unit effort (CPUE) series used as proxies for stock biomass. The large circles in the ICCAT plot denote the medians from each assessment run and the small dots individual estimates from Monte Carlo simulations; marginal probability distributions are also shown along the *x*- and *y*-axes. The point estimates are all different and the Bayesian estimates are much wider than those obtained by bootstrapping.

The main form of uncertainty presented in the K2PPs was model error due to value uncertainty: for the IATTC and IOTC the same stock assessment modelling framework is used for a limited number of scenarios by varying a single value at a time, whereas the WCPFC used a structural uncertainty grid where all values are varied at the same time. Only in the ICCAT example were uncertainties about the point estimates shown, although other tRFMOs present similar plots. In this case the procedure used to estimate the variance around the point estimates (by either bootstrapping or using Bayesian simulation) give different perceptions of risk (Magnusson et al., 2012). Which in turn changing the probabilities provided by scientists for management advice. Managers are often unaware of these issues, while the uncertainty which concern stock assessment scientists is mainly related to their own personal modelling choices, rather than providing advice related to the management of risk. But this aspect is commonly lost in the process of communicating stock assessment results to decision-makers. Thus, even when some sources of uncertainty were present or accounted for at some stage of the assessment process, this information might not filter through to the stakeholders.

Management

Once the assessment of current stock status is accepted, the next step is to advise on the measures required to achieve management objectives. The K2SM is intended to be a standardised format for presenting advice on measures required to achieve a management target with a certain probability within a given time scale (Anonymous, 2009). For example ICCAT (RES 11–14) requires the Standing Committee for Research and Statistics (SCRS) to provide three Kobe II strategy matrices indicating for different total allowable catches (TACs) the probabilities by year of $B \geq B_{MSY}$, $F \leq F_{MSY}$ and $B \geq B_{MSY}$ and $F \leq F_{MSY}$. An example based on the 2012 ICCAT East Atlantic and Mediterranean bluefin assessment is presented in Table 17.2. The objective is to recover the stock to the target (i.e. bottom right) quadrant of the K2PP. IOTC uses a different format for the K2SM that provides a summary of measures that meet management objectives, for example Table 17.3, which uses the same data as in

Table 17.2 Kobe II strategy matrix, $P(F \leq F_{MSY})$ and $P(SSB \geq B_{MSY})$ based on Eastern Atlantic bluefin.

TAC	2013	2014	2015	2016	2017	2018	2019	2020	2021	2022
0	33	43	51	60	69	79	89	95	99	100
20	33	42	50	59	67	77	87	94	98	100
40	33	41	50	57	66	75	85	93	97	99
60	33	41	48	56	64	73	83	91	96	99
80	33	40	48	55	62	71	81	89	95	98
100	32	40	46	53	61	69	79	87	94	97
120	32	39	45	52	59	67	76	85	91	96
140	32	38	44	50	57	64	73	82	89	94
160	32	38	43	49	55	62	70	78	86	92
180	32	37	42	47	53	59	67	75	82	89
200	31	36	41	46	51	57	63	71	78	84
220	31	35	40	44	48	53	59	67	73	79
240	28	32	36	40	44	49	54	61	67	72
260	25	29	33	36	40	44	49	54	60	65
280	22	25	29	33	36	40	44	49	54	58
300	19	23	26	30	33	37	40	44	49	53

Table 17.3 Strategy matrix in the IOTC format for setting management measures based on Eastern Atlantic bluefin.

Objective	TAC					
	0K	6K	12K	18K	24K	30K
$F_{2022} \leq F_{MSY}$	1.0	1.0	1.0	1.0	0.91	0.72
$SSB_{2022} \geq B_{MSY}$	1.0	0.99	0.96	0.89	0.74	0.59
Green quadrant	1.0	0.99	0.96	0.89	0.72	0.53

Table 17.4 Strategy matrix in the IATTC format integrating over assessment uncertainty and by recruitment, catch and selection pattern scenarios based on
Eastern Atlantic bluefin.

Green by 2022	TAC					
	0K	6K	12K	18K	24K	30K
Combined	0.99	0.98	0.96	0.88	0.72	0.52
Low recruitment	0.99	0.97	0.90	0.79	0.64	0.50
Medium recruitment	1.00	1.00	0.99	0.95	0.74	0.45
High recruitment	1.00	0.99	0.98	0.91	0.77	0.61
Inflated	1.00	0.99	0.99	0.98	0.95	0.85
Reported	0.99	0.98	0.92	0.79	0.48	0.19
Selectivity 2010	0.99	0.98	0.94	0.85	0.69	0.50
Selectivity 2012	0.99	0.99	0.97	0.91	0.74	0.54

Table 17.5 Kobe II strategy matrix for yellowfin tuna in the EPO in 2012.

Proposed reference point	State of nature steepness	Variability	δ required to ensure the following probability of being below target or limit			
			95%	90%	80%	50%
Target $F = F_{MSY}$	Base case	Low	0.972	0.980	0.991	1.010
		High	0.906	0.929	0.957	1.010
	$h = 0.75$	Low	0.604	0.613	0.624	0.644
		High	0.578	0.592	0.610	0.644
Limit $F = 1.4F_{MSY}$	Base case	Low	1.361	1.372	1.381	1.415
		High	1.269	1.301	1.323	1.415
	$h = 0.75$	Low	0.809	0.829	0.854	0.902
		High	0.846	0.846	0.873	0.902

the previous table. This shows the probability of ending overfishing and recovering the stock for different catch levels.

The K2SM presented differs from decision tables; the latter provide performance measures for a set of alternative management actions under different
states of nature (Punt and Hilborn, 1997, Maunder and Aires-da Silva, 2012).
Table 17.5 shows an example from IATTC for yellowfin tuna in the Eastern
Pacific. Assessment scenarios considered were two assumptions for the steepness

of the stock-recruitment relationship and two levels of recruitment variability (Minte-Vera et al., 2013). The Table shows the fraction of the current fishing mortality (δ) required to ensure that fishing mortality is below the target (F_{MSY}) and limit ($1.4F_{MSY}$) with a given probability.

The probabilities presented in the earlier bluefin example were averages derived from different historic assessment and projection scenarios (i.e. uncertainty about historic catch levels, future recruitment and selection pattern) where all were given equal weight. Table 17.4 shows the probabilities of achieving management objectives for each source of uncertainty. A difference between Tables 17.3 and 17.4 is that in the latter the effect on management objectives of the different sources of uncertainty are shown. This is consistent with the definition of risk-based management that requires identifying the consequences of uncertainty in order to manage the impact on management objectives. Mantyniemi et al. (2009) showed there was an economic benefit of resolving uncertainty about the stock-recruitment relationship. If the productivity of the stock is underestimated then yield will be forgone, while if it is overestimated the stock may be overfished. Table 17.3, by averaging over different future recruitment levels the consequences of resolving uncertainty about recruitment is masked while in Table 17.4 the consequences of different recruitment regimes is identified.

MSE

The earlier summaries of the application of the Kobe framework are based on traditional stock assessment. However, MSE is increasingly being used to evaluate the robustness of advice frameworks to the main sources of uncertainty. There is an important difference between conducting an MSE to develop an MP based on optimisation of objectives that will be run on autopilot and comparing alternative assessment and management options, for example for choosing reference points. In the latter case MSE is used to inform management, not to dictate it. The CCSBT has developed a management procedure (MP) using MSE but to date no other tRFMO has implemented MPs based on MSE, while ICCAT for example has used MSE to evaluate the implicit MP of ICCAT (Kell et al., 2003) to develop LRPs. An implicit MP is a set of rules for management of a resource that contains all the elements of an MP but is not run on autopilot (Kell et al., 2005a). According to Rademeyer et al. (2007) an implicit MP is also an MP that has not been simulation tested.

Examples

The CCSBT chose to develop an MP using MSE because they had two plausible assessments that gave contradictory estimates of stock status and productivity, and as a result a TAC could never be agreed. The CCSBT does not have MSY as an objective reflecting the time that the convention was signed (1994) and the improved understanding of why MSY is not very helpful as a specific technical objective (Holt, 2011). The CCSBT does not use the K2SM, but does use the K2PP as part of the agreed reporting to FAO and other tRF-MOs, but is thought of more as a tool for objective elicitation in the context of what is agreed in conventions and the UN Fish Stocks Agreement (United Nations, 1995). When using MSE to develop an MP, performance statistics based as catch, catch variability, CPUE and biomass/recruitment are used, as these ensure that results actually mean something to stakeholders.

A reason for adopting MSE by the CCSBT was to help resolve scientific disputes and embrace uncertainty by developing an OM which included plausible alternative hypotheses about stock and fishery dynamics. This in turn allows the selection of a robust management procedure that meets the CCSBT objectives. To do this the CCSBT used a grid of quantitative uncertainties when designing the OM (Table 17.6). The grid specified values for key parameters where there was little information in the data. This allowed quantitative evaluation of the impact of uncertainty on management objectives. Priors or resampling based on the objective function was also possible (see Table 17.6). During OM development a *reference set* was used in a series of *robustness trials* in order to tune the MP. Subsequently the OM scenarios were refined to narrow uncertainty in order to focus on things that really made a difference to performance measures and management objectives. The main objective was to rebuild the stock by 2035 with a 70% probability, but trade-offs between MPs were also considered. Since if several MPs achieved the rebuilding target there may be other characteristics that made a particular MP more desirable, for example the relative risk of catch limit reductions following previous increases in catches.

In the other tRFMOs, although no MP has been evaluated and implemented, the trend is to use stock assessment models as OMs and then test simpler MPs or alternative reference points for use as part of HCRs. For example the IOTC

Table 17.6 CCSBT reference set of OMs.

	Level	CumulN	Values	Prior	Weighting
h	5	5	0.55, 0.64, 0.93, 0.82, 0.9	uniform	obj. fun.
M_0	4	20	0.3, 0.35, 0.4, 0.45	uniform	obj. fun.
M_{10}	3	60	0.07, 0.1, 0.14	uniform	obj. fun.
Ω	1	60	1	NA	NA
CPUE	2	120	w.5, w.8	uniform	prior
q age-range	2	240	4–18, 8–12	0.67, 0.33	prior
Sample size	1	240	SQRT	NA	NA

has developed an OM for albacore conditioned using SS and a grid of factors and levels based on the stock assessment, while ICCAT's advice, including limit reference points for North Atlantic albacore, is now based on a biomass dynamic model which has been evaluated using an OM based on Multifan-CL, which was previously used to provide stock assessment advice (Kell and De Bruyn, 2013). This makes the transition from stock assessment based on integrated models to advice based on simpler assessments or rules possible for stocks that have been assessed using SS (Maunder, 2014).

Steps

The first step when conducting an MSE is to identify management objectives. The Kobe framework provides a basis for doing this, since it stipulates that the stock should be in the green quadrant of the phase plot, that is it defines a target region based on biomass and F reference points. However, the actual reference points, probabilities and time scales still need to be agreed on a case-specific basis. The Kobe matrix helps to provide a framework in which probabilities and time scales can be discussed. Also the same biological objectives can be achieved with different social and economic consequences. For example F can be reduced through time and/or area closures or capacity reduction as well as TACs. Therefore managers have to consider trade-offs and long/short-term outcomes related to social and economic objectives (ICCAT, 2014). Objectives may include the minimisation of variability in catch and/or effort, as in the case of the CCSBT, since wide annual fluctuations in catch and effort limit the ability of the fishing industry to plan for the future (Kell et al., 2005a, 2005b). In the western Pacific Ocean, the Parties to the Nauru Agreement (PNA) have agreed on a range of management objectives for the skipjack purse seine fishery, related to stability of allocation of fishing effort, resource sustainability, economic goals, limiting impacts on the distribution of skipjack and being risk adverse. There is an overall objective for no increases in catch, maintaining current levels of effort, and limiting impact on other species (McKechnie et al., 2013).

Although MSY may be an important policy goal it is not necessarily a useful technical objective. In an MSE performance measures are based on quantities from the OM, and reference points based upon a model-estimate such as F_{MSY} or B_{MSY} do not need to be used in a MP (or HCR) as long as management objectives related to yield are met. The CCSBT provides a model-free example of a MP that is based on year-to-year changes and trends in empirical indicators (i.e. CPUE and fisheries independent indices); reference levels are then tuned to meet management objectives using MSE, where tuning refers to adjusting the parameters of the MP to try and achieve the stated objectives represented by the OM. Model-based MPs, for example those based on a stock assessment model, may include the estimation of MSY-based reference points, but the values of F, F_{MSY}, B and B_{MSY} from the OM do not need to be equivalent to their proxies in the MP (e.g. if a stock assessment models used in the MP is structurally different from that used to condition the OM).

The choice of OM scenarios is crucial since the MP (or HCR) is tuned to the OM, therefore the best MP is a function of the OM scenarios chosen. Any bias in the OM scenarios will lead to bias in the performance of the MP. An MSE does not have to be based on a complex OM since a relatively simple OM may provide an evaluation of what MPs are likely to perform well (Carruthers et al., 2015). However, to evaluate robustness the choice of scenarios is important, since if an assumption is not modelled in the OM, for example about stationarity (e.g. Szuwalski et al., 2014) or population structure (e.g. Kell et al., 2009), then it will be difficult to say much about its impact on achieving management objectives. There are many ways of constructing OMs (Kell et al., 2006). It is common to use the current stock assessment model as the OM, but alternatively a more flexible model that can represent all available data may be especially constructed for the MSE. However not all relevant data sets may be available and so priors may be required for difficult to estimate parameters and to reflect expert opinion. The use of a stock assessment model implies that the assessment model describes nature as well as possible in a model. However, if a MP cannot perform well for simpler models, it is unlikely to perform adequately for more realistic representations of uncertainty. To test the robustness to alternative hypothesis about the dynamics of the system (e.g. driven by climate and environmental uncertainty) will require hypotheses about how biological parameters may change in the future (Punt et al., 2013) rather than relying on models fitted to historical data sets. Such hypothesis can be weighted in terms of their plausibility qualitatively based on expert opinion. The International Whaling Commission (IWC) also provides an example of combining qualitative judgement of OMs scenarios (high, medium and low likelihood) within MSE (Punt and Donovan, 2007).

Examples based on tRFMOs experiences illustrate the inherent flexibility of MSE methodology, the wide range of situations where it has been applied, as well as the potential for these applications to deepen and expand in the future to better suit the needs of stakeholders within the risk-based management paradigm.

Communicating and assessing risk

As discussed earlier, the management of risk requires the identification of management objectives and an assessment of how uncertainty affects the chances of achieving those objectives. In MSE, once the management objectives are mapped to performance measures, OMs are designed to reflect the main uncertainties about the system. Punt et al. (2014) recommended that when conducting an MSE, the range of uncertainties considered should be sufficiently broad so that new information collected after the management strategy is implemented would generally reduce rather than increase the initial range. A major impact on the risk assessment is how to select, reject and weight alternative OM hypotheses. When specifying external weights or priors in a Bayesian approach (unless noninformative priors are used), expert judgement needs to be applied and consensus amongst experts should be sought. However, agreeing on OM hypotheses and associated weights is potentially problematic. There is a need to avoid weighting choices being influenced by management implications. There-fore, ideally, weights for scenarios should be pre-agreed through informed discussion before any computations are undertaken and results presented. How-ever, after identifying hypotheses that although plausible make little difference in terms of management, it might be acceptable not to consider them further (ACE, 2007).

The Atlantic bluefin risk assessment serves as an example of an attempt to formally include stakeholders when conducting an MSE. First Fromentin et al. (2014) reviewed the historic treatment of uncertainty in the assessment and then Leach et al. (2014) used a risk-based approach with stakeholders to iden-tify and prioritise uncertainties for inclusion in the OM (Figure 17.3). Then Levontin et al. (2014) discuss how to turn a qualitative elicitation exercise into a quantitative procedure for use in MSE.

Leach et al. (2014) describe the elicitation methodology used to compile a prioritised list of uncertainties. Three dimensions of uncertainty were elicited: the first two were importance, being the potential impact on management

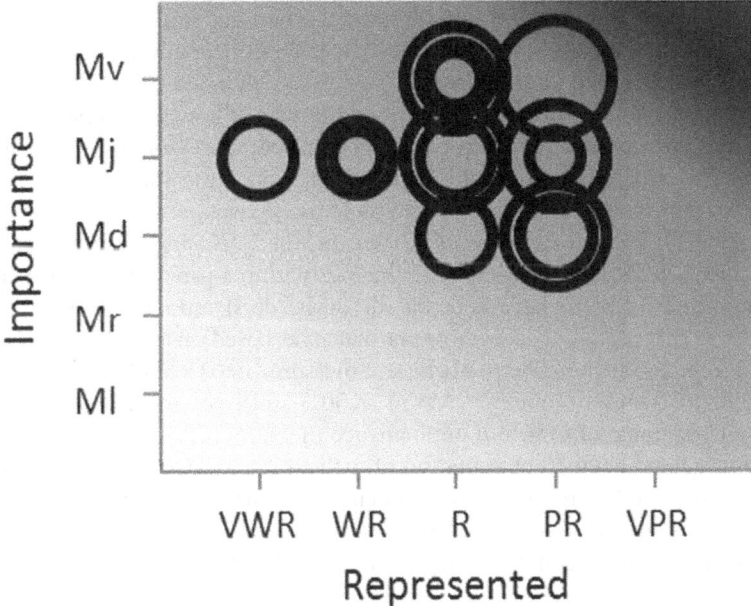

Figure 17.3 Visualisation of stakeholder views.

goals, and knowledge, being the potential to reduce uncertainty through more research (noting that some sources of uncertainty such as natural variability may not be reducible, see Edwards, Chapter 2, this volume). A third component related to the extent to which a given source of uncertainty was already accounted for in the assessment process. Among the stakeholders whose views were solicited were managers, scientists and nongovernmental organisations (NGOs).

Eliciting and representing uncertainties is a necessary step for MSE to ensure that legitimate concerns of stakeholders are part of the testing process for candidate management procedures. The methodology was intended to allow a qualitative prioritisation of uncertainties, while also visualising the degree of consensus among stakeholders on particular issues. An example of the elicitation exercise is shown in Figure 17.3, each hoop shows the views of a single assessor (i.e. stakeholder) and the hoop size the degree of epistemic uncertainty associated with a variable (i.e. small hoops = low uncertainty; large hoops = high uncertainty). Where the variable is for example the assumptions about natural mortality used in the assessment. The vertical axis displays an assessor on beliefs about the importance of that variable (Ml = minimal; Mr = minor; Md = moderate; Mj = major; Mv = massive), and the horizontal axis shows the degree to which the assessor believes it is included in the current assessment.

Perceptions of uncertainty in fisheries often vary widely among scientists, industry and other interest groups, so such tools that can facilitate inclusion and representation of different opinions are useful when decision-making depends on broad agreement and when effective management depends on commitment from stakeholders. The intention is to repeat this analysis after the MSE process has been carried out to provide a quantified measure of how some uncertainties impact the probability of achieving management objectives and how the views of managers and scientists change. This will give us a measure of acceptance among stakeholders of MSE as a valid way to assess risks.

Figure 17.4 shows a decision plot based on the IWC approach; the panels represent the MPs, dots the OMs and the bars within a panel the performance measures. For a MP to be acceptable all OMs (dots) must fall in the lower shaded area that represent acceptable performance. Based on this plot only MP4 and MP5 are acceptable. The performance measures are $P(SSB > B_{MSY}) > 60\%$, $P(F < F_{MSY}) > 60\%$, $P(\text{Yield} > MSY) > 50\%$ and AAV in Yield and Effort $< 30\%$. The choice of OM and performance measures is therefore critical and should be agreed prior to presentation of such a plot; to have a basis for choosing an acceptable MP requires prior agreement on the management objectives, and specifically, quantities, targets, probabilities and time scales over which the values related to management objectives are calculated.

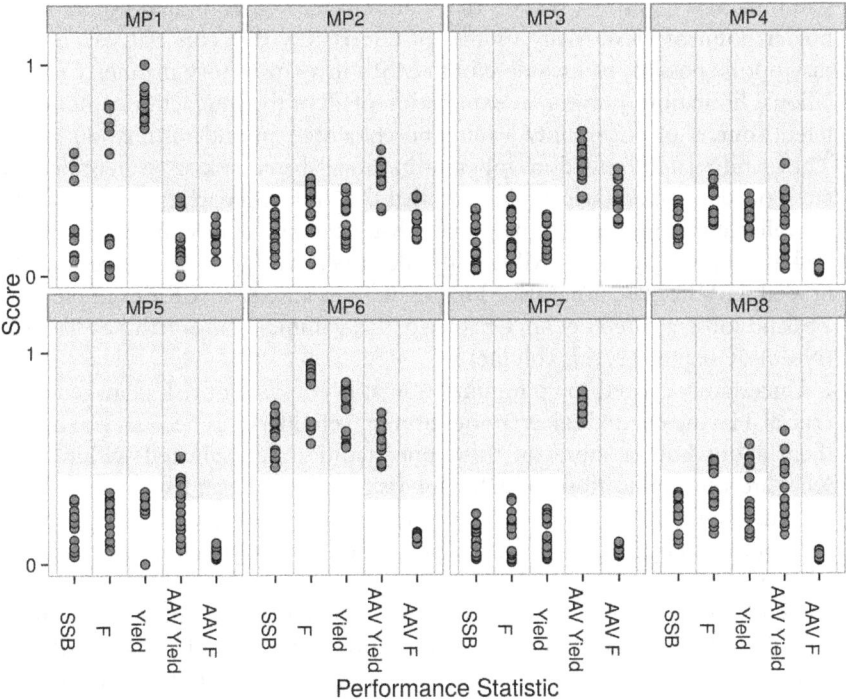

Figure 17.4 Decision plot.

Discussion

The CCSBT, the only tRFMO to have developed an MP through MSE, does not present advice in the Kobe Advice Framework. This is because MSE requires a more proactive approach to uncertainty. The K2SM, while a useful tool for providing a summary of options, by averaging over all the sources of uncertainty fails to help prioritise research, monitoring and enforcement activities to manage risk. Reformatting the K2SM as a decision table (Table 17.3) would be a step towards showing the effect of uncertainty and would help in deciding which uncertainties to include in MSE trials. The K2PP and K2SM are blunt instruments because they only depict a narrow range of objectives that are used to define the green area of the K2PP. By contrast, when conducting an MSE a range of graphical summaries are required to allow decision-makers to understand the results from the MSE and depict a wider range of trade-offs (Punt et al., 2014). Choice plots (Figure 17.4) are another method of visualisation, and when communicating modelling results appropriate tools need to be developed in collaboration with stakeholders.

While a greater range of sources of uncertainty is now commonly considered within MSEs, it is still a process that can be perceived as limited from a stakeholder point of view. Many sources of uncertainty that concern stakeholders may not be possible to include within MSEs in a satisfactory manner. Computational limitations prevent an exploration of all of the interactions among different sources of uncertainty leading potentially to an underestimation of risk. These and related limitations inherent in model-based risk assessment necessitate a need for an empirical validation that the MSE approach to risk assessment is a reliable methodology. Simply put, what is the probability that a MP identified as robust through an MSE will indeed perform safely in the real world? Several MPs have been in place long enough to answer this, for example New Zealand lobster (Breen et al., Chapter 6, this volume) and North Sea haddock (Needle, Chapter 12, this volume).

Uncertainty related to communication at every stage of risk management is crucial. Having elicited uncertainties from stakeholders it is necessary to inform them about how a subset of these uncertainties are evaluated within MSE, what are the implications of MSE for risk perceptions and what is the basis for trusting the MSE process in comparison to other risk assessment methods. Without this, engagement with and buy-in for a MP from all stakeholders will be impossible.

Coding a resource allocation problem in mathematical terms does not make an approach 'objective', meaning it does not in itself resolve the conflicting value systems which may be present and of primary interest to stakeholders. Values attached to resources might depend on gender, age, income, ethnicity, nationality, worldview, culture and language. Similarly, attitudes to risk might vary among stakeholders, complicating MSE ability to assert that risk acceptability criteria have been met. Yet there is a tendency to see modelling as universalist, unaffected by implicit or explicit agendas or politics. The language of programming MSE is anything but that, if only because of its group exclusivity – in most evaluations only a few of the programmers are actually 'fluent' in that language, and the rest of the stakeholders rely on the core group's efforts to communicate results. Many barriers to such communication exist even among the specialists, as it is not uncommon even in a peer-review process to read 'I could not verify the mathematical details, but the authors seem to know what they are doing.' The issue of a lack of trust among stakeholders due to perception that data which supports the modelling is corrupt is well known. Therefore, not just various sources of uncertainties, such as those already mentioned by Edwards (Chapter 2, this volume), but social aspects of values, trust, and communication are also important to consider in order for the MSE process to be successfully inclusive.

Summary

MSE can be used to develop a MP that runs automatically (e.g. Hillary et al., 2013) or to address strategic questions that inform management decisions (e.g.

Kell et al., 2003). One reason for the increased interested in MSE is because LRPs and HCRs are required for certification schemes, for example the Marine Stewardship Council. Scientific Committees of the tRFMOs have made the logical step that LRP make sense as part of a HCR and the best way to evaluate a HCR is to use MSE.

However, there is no consensus within the tRFMOs about moving to MPs. Although the K2SM is an implicit HCR since once objectives and associated probabilities and time scales are agreed then management options such as a TAC can be read from the K2SM. The K2SM could therefore be simulation tested using MSE (see Kell et al., 2003, 2005a, 2005b).

The first step of MSE is the identification of management objectives, which requires a dialogue between managers and stakeholders. The Kobe framework has helped in this respect as have the Kobe phase plot and matrix. However to fully address the effect on uncertainty on achieving management objectives requires a move towards risk management. MSE can help make that step, especially if it helps the tRFMOs to consider social and economic as well as biological objectives. MSE can also be used to indentify what uncertainies matter and therefore provides an objective way to identify research and data collection needs.

Scientists have been buried in abstract concepts, mystifying to even the most informed fisheries policy person for far too long, thereby diverging themselves from matters of direct relevance to stakeholders. The approach suggested in this chapter would help improve the dialogue at all levels.

Acknowledgements

This work was conducted under the Terms of Reference of the Tuna RFMO MSE Working Group.

References

ACE, A. 2007. Report of the study group on risk assessment and management advice (SGRAMA).

Aires-da Silva, A. & Maunder, M.N. 2012. Status of bigeye tuna in the Eastern Pacific Ocean in 2011 and outlook for the future. *Inter-American Tropical Tuna Commission, Stock Assessment Report*, 13, 18–29.

Anonymous 2009. Report of the second joint meeting of tuna regional fisheries management organizations (RFMOs).

Anonymous 2015. Characterizing uncertainty in stock assessment and management advice. *ISSF Technical Report 2015–06*.

Brown, B.J., Hanson, M.E., Liverman, D.M., & Merideth Jr, R.W. 1987. Global sustainability: Toward definition. *Environmental management*, 11(6), 713–719.

Carruthers, T., Laurence Kell, L., Maunder, M.N., Geromont, H., Walters, C., McAllister, M., Hillary, R. an Kitakado, T., Davies, C., Butterworth, D. & P., L. (2015). Performance review of simple management procedures. *ICES Journal of Marine Science: Journal du Conseil*. doi:10.1093/icesjms/fsv212

De Bruyn, P., Murua, H. & Aranda, M. 2012. The precautionary approach to fisheries management: How this is taken into account by tuna regional fisheries management organisations (RFMOs). *Marine Policy*.

Food and Agriculture Organization (FAO) 1995. Code of conduct for responsible fisheries. Food and Agriculture Organization of the United Nations, Rome.

Fromentin, J.-M., Bonhommeau, S., Arrizabalaga, H. & Kell, L. 2014. The spectre of uncertainty in management of exploited fish stocks: The illustrative case of Atlantic bluefin tuna. *Marine Policy*, 47, 8–14.

Garcia, S. 1996. The precautionary approach to fisheries and its implications for fishery research, technology and management: an updated review. *FAO Fisheries Technical Paper*, pp 1–76.

Hampton, J. & Fournier, D.A. 2001. A spatially disaggregated, length-based, age-structured population model of yellowfin tuna (Thunnus albacares) in the western and central Pacific Ocean. *Marine and Freshwater Research*, 52(7), 937–963.

Harley, S., Berger, A., Pilling, G., Davies, N. & Hampton, J. 2012. Evaluation of stock status of south pacific albacore, bigeye, skipjack and yellowfin tunas and southwest pacific striped marlin against potential limit reference points mow1-ip/04 14 Nov 2012. *Evaluation*, 28, 29.

Hillary, R., Ann Preece, A., & Davies, C. 2013. MP estimation performance relative to current input cue and aerial survey data. *CCSBT Extended Scientific Committee held in Canberra*, 1309(19).

Hillson, D. 2011. Risk doctor briefing-the definition debate continued. *Journal of Project, Program & Portfolio Management*, 2(1).

Holt, S. 2011. Maximum Sustainable Yield: The Worst Idea in Fisheries Management. Available at http://breachingtheblue.com/2011/10/03/maximum-sustainable-yield-the-worst-idea-in-fisheries-management (last accessed 10 November 2015).

ICCAT 2014. First meeting of the standing working group to enhance dialogue between fisheries scientists and managers (swgsm).

Kell, L. & De Bruyn, P. 2013. Management strategy evaluation framework for North Atlantic albacore. *ICCAT Collective Volume of Scientifics Papers*, 69(43), 2045–2051.

Kell, L., De Oliveira, J.A., Punt, A.E., McAllister, M.K. & Kuikka, S. 2006. Operational management procedures: an introduction to the use of evaluation frameworks. *Developments in Aquaculture and Fisheries Science*, 36, 379–407.

Kell, L., Dickey-Collas, M., Hintzen, N., Nash, R., Pilling, G. & Roel, B. 2009. Lumpers or splitters? Evaluating recovery and management plans for metapopulations of herring. *ICES Journal of Marine Science: Journal du Conseil*, 66(8), 1776–1783.

Kell, L., Die, D., Restrepo, V., Fromentin, J., Ortiz de Zarate, V., Pallares, P. 2003. An evaluation of management strategies for Atlantic tuna stocks. *Scientia Marina (Barcelona)*, 353–370.

Kell, L. & Fromentin, J. 2007. Evaluation of the robustness of maximum sustainable yield based management strategies to variations in carrying capacity or migration pattern of Atlantic bluefin tuna (thunnus thynnus). *Canadian Journal of Fisheries and Aquatic Sciences*, 64(5), 837–847.

Kell, L., Pastoors, M., Scott, R., Smith, M., Van Beek, F., O'Brien, C. and Pilling, G. (2005a). Evaluation of multiple management objectives for North-east Atlantic flatfish stocks: Sustainability vs. stability of yield. *ICES Journal of Marine Science: Journal du Conseil*, 62(6), 1104–1117.

Kell, L., Pilling, G., Kirkwood, G., Pastoors, M., Mesnil, B., Korsbrekke, K., Abaunza, P., Aps, R., Biseau, A., Kunzlik, P., Needle, C., Roel, B.A., & Ulrich-Rescan, C. (2005b). An evaluation of the implicit management procedure used for some ICES roundfish stocks. *ICES Journal of Marine Science: Journal du Conseil*, 62(4), 750–759.

Kirkwood, G. & Smith, A. 1995. Assessing the precautionary nature of fishery management strategies. *Fisheries and Agriculture Organization. Pre-cautionary approach to fisheries. Part, 2.*

Leach, A., Levontin, P., Holt, J., Kell, L. & Mumford, J. 2014. Identification and prioritization of uncertainties for management of eastern Atlantic bluefin tuna (Thunnus thynnus). *Marine Policy*, 48, 84–92.

Levontin, P., Leach, A., Holt, J., Mumford, J.D. & Kell, L.T. 2014. Specifying and weighting scenarios for MSE robustness trials. *ICCAT Collective Volume of Scientifics Papers*, 66. 1326–1343.

Magnusson, A., Punt, A.E. & Hilborn, R. 2012. Measuring uncertainty in fisheries stock assessment: the delta method, bootstrap, and MCMC. *Fish and Fisheries*, 14(3), 325–342.

Mantyniemi, S., Kuikka, S., Rahikainen, M., Kell, L. & Kaitala, V. 2009. The value of information in fisheries management: North Sea herring as an example. *ICES Journal of Marine Science: Journal du Conseil*, 66(10), 2278–2283.

Maunder, M. N. (2014). Management strategy evaluation (mse) implementation in stock synthesis: 452 application to pacific bluefin tuna. IATTC Stock Assessment Report, 15:100–117.

Maunder, M.N. & Aires-da Silva, A. 2012. Evaluation of the Kobe plot and strategy matrix and their application to tuna in the EPO. *Inter-Amer. Trop. Tuna Comm., Stock Assessment Report*, 12, 191–211.

McAllister, M.K. & Babcock, E.A. 2003. Bayesian surplus production model with the sampling importance resampling algorithm (bsp): a user's guide. Available from www.iccat.es.

Methot, R.D. & Wetzel, C.R. 2013. Stock synthesis: A biological and statistical framework for fish stock assessment and fishery management. *Fisheries Research*, 142, 86–99.

Minte-Vera, C. 454V., Maunder, M. N., and Aires-da Silva, A. (2013). Kobe II Strategy Matrix for the Bigeye and Yellowfin Tuna Stocks of the Eastern Pacific Ocean in 2012.

Morgan, M, G. & Henrion, M. 1990. *Uncertainty: A guide to Dealing with Uncertainty in Quantitative Risk and Policy Analysis.* New York, Cambridge University Press.

Prager, M., Goodyear, C. & Scott, G. 1996. Application of a surplus production model to a swordfish-like simulated stock with time-changing gear selectivity. *Transactions of the American Fisheries Society*, 125(5), 729–740.

Punt, A. 2008. Refocusing stock assessment in support of policy evaluation. *Fisheries for Global Welfare and Environment*, pp. 139–152.

Punt, A. & Donovan, G. 2007. Developing management procedures that are robust to uncertainty: lessons from the International Whaling Commission. *ICES Journal of Marine Science: Journal du Conseil,* 64(4), 603–612.

Punt, A.E., A'mar, T., Bond, N.A., Butterworth, D.S., de Moor, C.L., De Oliveira, J.A., Haltuch, M.A., Hollowed, A.B. & Szuwalski, C. 2013. Fisheries management under climate and environmental uncertainty: control rules and performance simulation. *ICES Journal of Marine Science: Journal du Conseil.*

Punt, A.E., Butterworth, D.S. L M., De Oliveira, J.A. & Haddon, M. 2014. Management strategy evaluation: best practices. *Fish and Fisheries.*

Punt, A.E. & Hilborn, R. 1997. Fisheries stock assessment and decision analysis: the bayesian approach. *Reviews in Fish Biology and Fisheries*, 7(1), 35–63.

Radatz, J., Geraci, A. & Katki, F. 1990. Ieee standard glossary of software engineering terminology. *IEEE Std*, 610121990:121990.

Rademeyer, R.A., Plagányi, É. E. & Butterworth, D.S. 2007. Tips and tricks in designing management procedures. *ICES Journal of Marine Science: Journal du Conseil*, 64(4), 618–625.

Ramaprasad, A. 1983. On the definition of feedback. *Behavioral Science*, 28(1), 4–13.

Regan, H.M., Colyvan, M. & Burgman, M.A. 2002. A taxonomy and treatment of uncertainty for ecology and conservation biology. *Ecological applications*, 12(2), 618–628.

Röckmann, C., Ulrich, C., Dreyer, M., Bell, E., Borodzicz, E., Haapasaari, P., Hauge, K.H., Howell, D., Mäntyniemi, S., Miller, D. & Tserpes, G., 2012. The added value of participatory modelling in fisheries management–what has been learnt? *Marine Policy*, 36(5), 1072–1085.

Rosenberg, A.A. & Restrepo, V.R. 1994. Uncertainty and risk evaluation in stock assessment advice for us marine fisheries. *Canadian Journal of Fisheries and Aquatic Sciences*, 51(12), 2715–2720.

SPC-OFP and PNA 2013. Assessing a candidate target reference point for skipjack tuna consistent with PNA management objectives. Technical report, WCPFC-SC10–2014/MI-WP-09.

Szuwalski, C.S., Vert-Pre, K.A., Punt, A.E., Branch, T.A. & Hilborn, R. 2014. Examining common assumptions about recruitment: a meta-analysis of recruitment dynamics for worldwide marine fisheries. *Fish and Fisheries*, 16(4), 633–648.

United Nations 1995. Agreement for the implementation of the provisions of the United Nations Convention on the Law of the Sea of 10 December 1982 relating to the conservation and management of straddling fish stocks and highly migratory fish stocks. *U.N. Doc. A/Conf. 62/122.*

Velthuis, W.J.R. 2000. *Learning Feed-forward Control-theory, Design and Applications.* Enschede, Netherlands: Universiteit Twente.

Whitbeck, R.F. & Wolkovitch, J. 1982. Optimal terrain-following feedback control for advanced cruise missiles. Technical report, DTIC Document.

## Chapter 18

Introduction to some alternative methods for providing scientific information for management

Verena M. Trenkel, Sarah B. M. Kraak, Jake Rice,
Marie-Joëlle Rochet and Anthony D. M. Smith

Introduction

Fisheries management requires scientific information at most steps of the decision cycle (see Dankel and Edwards, Chapter 1, this volume). The methods for providing this information will vary, and several other methods exist in addition to numerical simulations as demonstrated in previous chapters in this volume. We introduce three categories of such methods for providing information to support management decisions: expert judgment, use of mathematical models not involving simulations, and qualitative approaches. Some of these methods are suitable either as a first step in providing information on which to base advice for fisheries management, to be augmented subsequently by more targeted numerical simulations, or else when numerical simulations are impossible due to lack of knowledge, data, or resources to build appropriate models and carry out extensive simulations, or when the complexity of the systems makes modelling simplifications intractable.

It is challenging to identify the crucial information for decision-making. Even if the information needs are identified, it might be difficult or even impossible to model the stock, fishery or socioeconomic system with sufficient accuracy and precision to inform management decision-making within relevant timeframes. Nevertheless, some of the alternative methods introduced in this chapter might provide sufficient information to inform appropriate management decisions.

We begin with case studies to illustrate the shortcomings of solely simulation-based management advice, before describing alternative approaches that can help to safeguard the fishery, or augment the information on which management is based. Finally we introduce a comprehensive framework designed to ensure that management strategy evaluation studies are conducted in a manner that helps to ensure both their relevance and reliability for management.

The limitations of simulation studies

The simulation-based evaluation of fisheries management measures is often considered a central component of sustainable management. However, this does not mean that it is sufficient, or by itself capable of delineating a complete solution to any management problem. We outline three case studies here illustrating some of these shortcomings.

European cod recovery plan: accounting for fisher behaviour

In December 2008 a new management plan for exploitation of four European cod (*Gadus morhua*) stocks [Council Regulation (EC) No 1342/2008] replaced the previous EU cod management plan. Cod is caught in mixed fisheries, often as bycatch, in all four management regions. A well-known problem of mixed fisheries is that the various stock quotas cannot be filled at the same pace: when the quotas of some stocks have been fully caught, portions of quotas remain to be caught for other stocks. Under landings quotas prevalent in the EU until 2015, this was interpreted as allowing the fishery to continue fishing until all quotas are caught. This resulted in exceeding the total allowable catches (TACs) of the stocks whose quotas are taken up first and discarding of the over-quota catches because landing of them is not allowed. The cod plan set up effort restrictions, as had the previous plan, intended to restrict fishing effort to that needed to catch the allowed cod quota. The intended benefits to cod of these effort restrictions did not materialize, as cod catches continued to be in excess of the quotas and cod fishing mortality (F) did not decline proportionally with effort. Moreover, the fishers were discontented because these effort restrictions limited their ability to fully uptake the quotas for the other species they were targeting (Kraak et al. 2013).

In 2008–2009, the International Council for the Exploration of the Sea (ICES) had carried out impact assessments of the cod management plans, partly after the plans were already in force (ICES 2008, 2009a). However, the results consisted only of the output of simulation scenarios of the harvest control rule (HCR) that specified the annual F targets. Various scenarios were tested, but in its final advice ICES reported only on the consequences for future yield and spawning stock biomass (SSB) assuming that the F specified in the HCR would be fully achieved. ICES (2009b, p. 20) considered the North Sea cod plan to be "in accordance with the precautionary approach *if it is implemented and enforced adequately*" [our italics] and stated that "discarding *in excess of the assumptions* [our italics] under the management plan will affect the effectiveness of the plan," and noted that "the evaluation is most sensitive to assumptions about *implementation error* [our italics] (i.e. TAC overshoot and the consequent increase in discards)." Scenarios exploring the consequences to the cod stocks of nonachievement of the intended F owing to the known mixed-fisheries issues were not presented. It would have been helpful if the request for advice from ICES had included a request for a (qualitative) exploration of the likelihood of adequate implementation of the HCR. Unfortunately, the known issues of stakeholder support, compliance, fleet behaviour, discards – all being derived from the mixed-fisheries problem – were not addressed. These impact assessments did not comment on how likely the plan was to achieve its objectives in the mixed-fisheries context; they assumed that the intended F would be achieved, without discussing how.

In actual fact the F implied by the HCR could only be achieved by changes in fisher behaviour, for which the plan had provisions, but these provisions had not been accounted for in simulations. The simulations had not actually addressed the adequate implementation of the plan as a whole, but only of the HCR. The plan did incorporate two innovative, bottom–up provisions that aimed to provide incentives for changes in the fishermen's behaviour. Under one of these provisions (Article 11) fleet segments were encouraged to take up selective gear or fish outside of the distribution area of cod or otherwise demonstrate that they catch very little cod; these fleets were rewarded with exemption of the restrictive effort regime. The other provision (Article 13) alleviated the effort restrictions for fleet segments that committed to a plan for the reduction of partial cod mortality, for example by using selective gear or implementing real-time closures. Because these provisions were novel, it was difficult to predict how well they would achieve the intended changes in fisher behaviour. With no precedent for the mix of measures, there were no tested formulations for how fishers would respond to them, nor exactly relevant data to parameterize the formulations. Thus, these measures were not very amenable to quantitative simulations. Nevertheless, using expert judgment, scientists could have explored the likeliness of these incentives to change fisher behaviour in various ways. For example, previous fleet-dynamics studies have looked into fleet responses such as effort displacement as a consequence of changed regulations. Other studies have explored the selectivity of alternative gears and spatial differences in catchability. All such studies could have, at least qualitatively, informed the impact assessments (cf. Hamon and Poos, Chapter 3, this volume).

The conclusion from the 2011 joint ICES-STECF (Scientific, Technical and Economic Committee on Fisheries) evaluation of the plan (Kraak et al. 2013) was that the cod management plan had failed to be adequately implemented and had not achieved its objective of meeting the specified target F or even of reducing F in line with the HCR.

We conclude that in this case, evaluations with simulations of the fully implemented HCRs only were not particularly relevant in predicting the chances of success of the plan. It would have been more salient to implement a management strategy evaluation that included qualitative as well as quantitative knowledge of fishermen responses to (changes in) regulations and incentives for behavioural change in addition to a worst-case scenario of status quo fisher behaviour resulting in nonachievement of the intended F.

Northern cod: large and unpredictable changes in stock productivity

Simulation testing is designed to evaluate many things, including changes in productivity of stocks. However the value of any simulation is only as good as the will of decision-makers to act on the results. Whether rules are simple or complex, following them when the news is good and questioning them and delaying action when the news is bad is a path of high risk. The infamous collapse of Northern cod (Newfoundland, Canada) illustrates these points well.

When Canada assumed jurisdiction to 200 nautical miles in 1978, most of the cod stocks on the east coast of Canada had been depleted by years of overfishing by foreign fleets. Canada considered the extended jurisdiction to be an opportunity to both rebuild these depleted stocks through better management and build an economically prosperous domestic fishing industry with both an inshore and offshore component (Munro 1980). At the planning step in the decision cycle (see Figure 1.2 in Dankel and Edwards, Chapter 1, this volume) Canada and North Atlantic Fisheries Organisation (NAFO) adopted fishing at the target level of $F_{0.1}$ as the overall harvest strategy, based on expectations that $F_{0.1}$ would allow stock growth, while providing nearly full fishing opportunity with greater economic return (compared to the increasingly discredited F_{max}) (Deriso 1987). In the case of the flagship Northern cod stock (NAFO Div 2J3KL), NAFO set the target fishing mortality at 80% of $F_{0.1}$, to provide faster recovery of the stock. At the objective-setting step these target F were not based on complex simulations, but the reasoning supporting $F_{0.1}$ had evolved in the fisheries assessment expert meetings in ICES and the Food and Agriculture Organization (FAO; see Deriso 1987 for more background on fishing at $F_{0.1}$ and alternatives like F_{max} as harvest strategies). NAFO also adopted a recovery target for Northern cod spawning biomass, based on assessment estimates of the size of SSB in the early years of the post–World War II international fishery, again lacking the support of complex simulations of stock and ecosystem dynamics due to the limited simulation power of computers of the day.

For all of the 1980s, Canada and NAFO conducted annual assessments and benchmarks of domestic stocks using sequential population analyses tuned to annual trawl survey and offshore mobile fleet catch per unit effort (CPUE). Until 1987, annual harvest advice was for a TAC corresponding to 80% of $F_{0.1}$ applied to the point estimate for fishable biomass, and TACs were set consistent with this scientific advice. Management was successful in keeping quota overruns infrequent and small, and by the mid-1980s an observer program and a system of enterprise allocations (similar to an individual quota) to the companies operating the offshore fleet ensured that size-based discarding and misreporting were rare in those fleets.

The assessment methods were simplistic by contemporary standards (although improving continually though the decade) and the harvest strategy had been chosen based on logic and some simple projections of future population dynamics. Nevertheless, as long as the science advice stuck with the $F_{0.1}$ formulation, TACs followed the science advice, and catches were kept below or at least near the TACs, all the stocks grew (see Rice 2006 for analysis of annual patterns). The rebuilding target for Northern cod was reached by 1986 (when a computational error in setting the target was corrected; Rice and Evans 1988).

Catches increased consistently as the TACs increased, and industry returns were generally good (Parsons 1993).

By 1988, however, the assessment for Northern cod concluded that stock growth had plateaued, and harvest expectations based on projections assuming population growth rates that had been observed in the first half of the 1980s were overly optimistic. Sometime in the mid-1980s stock productivity had declined, initially from an increase in total mortality that had not been captured in the assessment and quota advice. The expectations of continually increasing quotas were unrealistic, and in fact quotas by the mid 1980s were probably already above 80% $F_{0.1}$. When the retrospective pattern in the assessments was found in the mid- to late 1980s, the causes were not known. With a major source of inaccuracy in the annual assessments lacking a clear explanation, and entrenched expectation of continual rebuilding of the stock under Canadian management, the political process initially would not allow reductions in catch at all, despite the advice that stock growth had stopped a couple years earlier, and later quota reductions lagged well behind science advice. The total quotas were harvested annually, so the initial decline in stock growth resulting from a decline in productivity quickly became a decline in stock by fishing well above the 80% $F_{0.1}$ rule. Stock depletion was rapid (Rice 2006).

Two key lessons can be taken from this experience. First, simple harvest control rules applied to results of simple (by today's standards) assessments can work well, and produce the expected benefits to both stock and industry, as long as they are applied and followed rigorously. Second, even if a harvest control rule is working, it will not necessarily always work. Simulation studies can help scientists and managers to build control rules that may be resilient to a wider range of circumstances, but the breadth of scenarios investigated will be guided by ranges of conditions considered plausible by the practitioners. If unforeseen circumstances do occur, a contingency plan should be in place. In the case described here, large unforeseen changes in productivity undermined faith in the assessment, causing decision-makers to deviate from the prescribed control rule in an ad hoc manner. Following an HCR when it advises increases in catches and deviating optimistically from it when it advises decreases in catches is a path to difficulties, whatever the basis for the rule. Robustness to changes in productivity was built into management by selecting a F somewhat lower than the target $F_{0.1}$. This precaution could also have accommodated some error in the assessment. However, it became ineffective when quota setting deviated from the harvest control rule. Had the control rule been followed, this might have provided time for the stock assessment to pick up the productivity changes reliably and for management to react while maintaining more stable catches. However, at the time the drop in productivity observed after the mid-1980s ran counter to anyone's experience or expectations. Had simulations been done they would have had to assume productivities never recorded for the stock to protect against the changes that occurred. This situation exemplifies an important limitation of simulation studies, in that they often rely on historical records to bound possible future outcomes, and in this case would have most probably been inadequate even if conducted to modern standards. Contemporary management science can sometimes draw on sufficient expert knowledge and experience from cases worldwide to select a sufficiently conservative harvest rule and provide a buffer for productivity changes. However such practices still require decision-makers to accept simulations assuming stock productivity parameters that might be far outside those in the historical data sets, and providing rationales for such pessimistic assumptions can be difficult to justify. To guard against unforeseen events, well-considered management plans will include a structured plan and criteria for deviations from the agreed control rule.

Pacific herring: changing economic objectives

Unforeseen circumstances of importance to fisheries management but outside the usual scope of simulation studies may be environmentally derived or represent a phase shift in the ecosystem (as in the previous example). But another reason can also include the economic drivers that determine industry objectives for the fishery. In the case of spring spawning North Pacific herring (*Clupea harengus pallasi*), changing economic incentives changed the quantity of fishery-dependent data, with downstream repercussions for the quality of its management.

North Pacific herring was managed in Canadian waters as six and later five separate stocks, although each stock was itself a combination of a number of local spawning aggregations in the bays and fjords along the British Columbia coast (Schweigert 1991). Recruitment variation is very large and driven much more strongly by oceanographic conditions than by spawning biomass (Hay 1985, Stocker et al. 1985). For decades before the mid-1980s, these stocks had been fished largely for reduction markets. In the early 1980s a new and lucrative market for herring roe was developed, providing an economic incentive for fisheries that took much smaller volumes, but required high product quality and consistent supply.

Hence, in the mid-1980s the Canadian Department of Fisheries and Oceans Science and Management Sectors worked with the fishing industry to develop a new harvest strategy for Pacific herring in British Columbia. The objective setting and strategy selection dialogue was informed by quantitative analyses of stock dynamics under various management regimes. The analyses considered the historical evidence of what sizes of spawning biomasses had been associated with consistently poor recruitment under all environmental conditions, and took into account that if unfavourable environmental conditions occurred, even very large SSBs would be unlikely to produce strong recruitment (Hall et al. 1988). All government, industry and First Nations participants in the discussions converged on a management strategy based around setting a biologically based escapement goal for each stock, estimated from historical stock–recruit data and habitat surveys (Schweigert and Stocker 1988). Although historical data were analyzed thoroughly, most analyses were graphical, simple linear regressions, or ordinations. Simulations of population trajectories by today's standards were never run.

As new technologies have become available, nearly all aspects of the stock assessment and in-season management of the roe fishery improved over the approximately 25 years since the management strategy was adopted. Despite this, the basic components comprising the strategy persisted. Three of the five stocks have fallen below their escapement goal at least once over the 25 years, due to poor recruitment, but the low stock sizes lasted more than two years in only one case. In general, the stocks were considered healthy, and when declines occurred, they were linked to oceanographic conditions leading to a reduction in productivity (Schweigert et al. 2010). In aggregate across the stocks, catches were sufficient to satisfy the market with both quality and volume, and both harvesters and processors considered the fishery lucrative. The industry sector invested heavily in data collection to improve the accuracy and precision of the assessments, so the quotas set could be as reliable as possible and the value of the fishery protected.

However, the roe market is a luxury market and suffered badly during the economic downturn around 2008. The fishery is still recovering from this change because the market for roe has been slow to recover. Loss of markets meant that harvest objectives for the industry changed in one year, from high-value catches to monopolize one luxury market to catches to compete on the much lower value human consumption market. This in turn has meant the industry can no longer support the data collection streams on which the assessments depended, affecting the quality of the whole assessment and management cycle. No amount of testing could have predicted the global economic crisis that ever so affected the Pacific herring roe fishery. But in retrospect, uncertainties, whatever their origins, can and should be accounted for in MSE simulation testing. These uncertainties include information not only about the supply (the resource), but also about the demand. When quantitative information is lacking, qualitative information and relevant expert judgment is needed (in this case economic experts were very relevant) to anticipate and prepare for changes in the fishery.

Providing information to augment or substitute a simulation-based approach

Smith et al. (2007) proposed to structure the scientific tools for fisheries management along two axes: scope and method type. Scope ranges from populations to ecosystems to socioeconomics. Method types are expert judgment (an informal or formal, qualitative method that reflects the predominant opinion within a group of well-informed people), empirical (based on statistical methods using experimental or observational data), and modelling (methods using mathematical models), with several options for each type. The previous chapters in this book provide examples for how mathematical models are used for simulation testing. Here we describe three categories of analysis that can complement such simulations: (1) use of expert judgment, (2) formal analysis of mathematical models used for deriving generic harvesting rules, and (3) qualitative MSE approaches.

Incorporation of expert judgment

Expert judgment is based on knowledge acquired in various ways (scientific studies, fishing, living in the study area, etc.), and this knowledge can concern different aspects of the management decision cycle (Figure 1.2 in Dankel and Edwards Chapter 1, this volume). An important class of knowledge enabling expert judgment is the traditional knowledge held by users of the sea. Traditional knowledge, also called local (ecological) knowledge, relates to the ecology of the system, its exploitation, and management (Johannes 1981, Berkes and Turner 2006). Since the end of the 20th century, much effort has been devoted to develop methods to gather stakeholder knowledge (e.g. Haggan et al. 2007) and to use it in conjunction with scientific knowledge for management (e.g. Motos and Wilson 2006). For traditional knowledge to be representative and robust to scrutiny, the experts (e.g. users of the sea) must be selected rigorously. For example, in a survey of local communities each respondent is asked who he/she thought were the knowledgeable persons (e.g. Davis and Wagner 2003). Using this peer recommendation approach, a short list of people whose detailed knowledge should be collected is then identified. Ranges of methods are then used to collect knowledge, from highly structured frameworks to open conversations through semistructured interviews; graphical tools such as maps or diagrams are often used.

How to use expert knowledge in the management decision cycle is an active area of research. Even in formalized frameworks, expert input is known to depend on the background, level of information, and level of detail required in the assessment, and is prone to subjectivity and value judgment, (e.g. Rochet and Rice 2005). The following practices might limit these drawbacks to the extent possible (ICES 2011):

1   The group of experts contributing the "expert knowledge" should be identified by a rigorous process and include experts with a variety of perspectives, not just disciplines.
2   The expert group should be encouraged to lay out the framework for its analysis, detailing the different steps and elements considered in making their evaluations and criteria used in assembling conclusions.
3   The experts should seek out consensus relative to this framework, but they should not seek forced compromises or make conclusions vaguer as the strategy to get broader agreement.

4   In addition, the experts should provide at least a qualitative measure of uncertainty in their conclusions (Halpern et al. 2007).
5   The expert group should annotate, comment, and provide explanation for its conclusions.

Some expert knowledge, for example on the biology and ecology of the target or bycatch species or on the fleet dynamics and its social and economic drivers, may not be amenable to being incorporated directly into models because it is qualitative or anecdotal knowledge. As a consequence, even if acknowledged by the participants in an MSE process, it often has little or no influence on the conclusions of the work. In the case of the evaluation of the European cod plan described earlier, a considerable amount of knowledge existed among the scientists involved about the complex nature of the harvest composition in the mixed fishery and the spatiotemporal fleet dynamics as well as socioeconomic aspects. Part of this knowledge was qualitative and acquired directly through interaction with industry representatives. We argue that this type of knowledge, although not incorporated in the quantitative models, should not be ignored in the preparation of scientific advice. Methods for its inclusion are affected by the advisory processes. However, in all cases the experts need to ensure that scientific standards are maintained, for example by rigorous peer review that might require an extended peer community including various categories of users in addition to scientists. One possible format for such a review process could consist of scenario discussions in a multistage iterative process between various scientists, managers, and other stakeholders (Kraak et al. 2010), as illustrated by the qualitative MSE for southeastern Australia below.

In other cases, such as the management of fish and marine mammal fisheries in the Arctic, there is "scientific information" on some population parameters of the target species, but the knowledge of many aspects of their life history and behaviour is held primarily by the indigenous people of the region (e.g. Knopp 2010). Only the former has been amenable to modern modelling methods. Moreover, many holders of traditional knowledge presume ownership of it and respectful use applies frameworks in which they can participate fully, rather than handing it over to mathematical modellers. However successful management of many fisheries can only be achieved by an integration of the best of both types of knowledge (see Armitage et al. 2011 for some examples).

Analytical derivation of generic principles for harvesting

Control theory is a branch of mathematics that has been developed for engineering purposes; it provides rules to control a dynamical system, so that its output follows a desired signal. Control theory overlaps with decision theory when it uses stochastic models, but it is less concerned with values and utility. Control theory has been widely applied to fisheries to derive general results about management strategies. During the 1960s and 1970s a large part of "fisheries theory" was devoted to optimal exploitation of fish populations, of which Hilborn and Walters (1992) provide a crisp summary. The approach consists in building a dynamic model of the fish population and fishery as a system of a few differential or difference equations. The long-term properties of the model are then examined by analyzing its equilibria. Then general solutions (harvest strategies) to optimize specific properties (often but not always yield) of the generic models are identified. The information requirements differ among instruments and need to be accounted for when choosing a given strategy.

Harvest strategies usually explored with these methods include constant catch, constant escapement, and constant exploitation rate. If the objective is to maximize yield, generally the constant stock size strategy is optimal under a number of assumptions, including a single homogeneous stock unit and perfectly known model parameters. Optimization has also been used extensively with bioeconomic models, often involving a wider range of objectives addressing conservation, economic efficiency, or equity (Clark 1985). Optimizing economic efficiency favours managing input control (limited entry and transferable quotas, or catch taxes) over or on the top of output controls (catch quotas). Equilibrium analysis has also been applied to simple models for multispecies fisheries (e.g. May et al. 1979).

Stochastic models have been developed to take account of variability and uncertainty. Instead of a single fixed rule needed for optimizing with a deterministic model, stochastic dynamic programming optimizes a sequence of decisions when each decision affects the subsequent system state – thus, any future decision. This approach has, for example, allowed analysts to assess the trade-off between maximizing and stabilizing yield when knowledge is imperfect (Walters 1975). The strategies that aim at minimizing variance typically result in higher and more predictable stock sizes. On the other hand, fixed harvest rate strategies remain optimal in the context of long-term unpredictable environmental fluctuations affecting carrying capacity for a stock. This suggests that sometimes when environmental effects on stock productivity are large it may be wiser to work out efficient implementation tools for fixed harvest control rules rather than investing in predicting climate effects (Walters and Parma 1996).

More recently, viability theory (De Lara and Doyen 2008) has been used to address multiple objectives; this family of tools does not strive to find "best" solutions, but to identify the controls that will enable a system to persist indefinitely within a given set of constraints (or objectives). For example, this approach has been used to demonstrate that harvest control rules based on spawning stock biomass (SSB) and fishing mortality are sufficient to ensure sustainability only when recruits make up a significant part of SSB; otherwise, additional information about the reproductive potential of the stock is required to ensure sustainability (De Lara et al. 2007).

The models that are mathematically tractable generally rely on simplified assumptions which are not met in the real world. However, the examples given earlier illustrate that their analysis can provide useful guidance for designing and evaluating a class of management strategies, because they help to understand the conditions necessary for a given strategy to meet a given objective – even if these conditions are not sufficient. Moreover, mathematically derived strategies have the advantage that they do not depend on any specific parameter value; they are generic and can be used even in systems with little quantitative data and/or large uncertainty, in contrast with simulation results, which are only valid for the set of parameters used in the simulations.

Simulation-based MSE analyses have also provided generic results that can usefully inform management decisions even in fisheries where there is little analytical capability or limited data. For example, Smith (1994) summarizes the general trade-offs inherent in broad harvest strategies such as fixed catch, fixed effort, and fixed escapement strategies. Similarly, Punt (1995) discusses generic findings from harvest strategies that adopt surplus production models to assess fishery state and provides useful advice on the situations where such strategies could work well – and just as importantly, where they may fail.

Qualitative evaluation methods

Including large parts of the decision cycle in a single numerical model is often difficult. In such cases a more qualitative approach can often be used. Here we explain how such a qualitative management evaluation can be organized into five steps. The necessity and extent of each step will of course depend on the case at hand. A case history from southeastern Australia illustrates the insights that can be gained from a qualitative evaluation of the decision cycle.

A five-step qualitative MSE framework

Based on Trenkel et al. (2015), we propose a five-step evaluation framework for a rapid appraisal of management plans. The outcome of applying the framework would be an assessment of:

1   the consistency of the management plan objectives with sustainability;
2   the qualitative outcome of the selected tactics (catch or effort control, gear restrictions, etc. i.e. likely to fail or possibly could succeed) predicted from the structural analysis of the bioeconomic system;
3   the expected overall outcome of the plan based on past performance of the complementary measures in the plan;
4   quantitative results for specific issues;
5   the expected suitability of institutional and management settings to ensure wide support.

The objective of the first three steps is to systematically screen out management plans that have little chance of success before resorting to time-consuming quantitative simulations in step 4 and comparably demanding analyses of institutional capacities.

Step 1: Are the objectives of the management plan consistent with ecological sustainability? The aim of this step is to evaluate whether the objectives of the management plan are consistent with sustainable exploitation. It consists in considering the elements described in the management plan (e.g. biomass and fishing mortality limits and targets) in combination with relevant stock information (life history traits) and fisheries (economic) information.

Step 2: Are the proposed management tactics likely to achieve the objectives of the management plan for the target stock(s)? The adequacy of the planned management tactic (TAC, effort control, etc.) or combination of tactics for achieving the objectives are evaluated qualitatively, for example using qualitative modelling with a bioeconomic model of stock-fishery dynamics to evaluate the proposed management tactics. The principle of qualitative modelling, also called loop analysis, is to consider the expected directions of change from the current to the next (theoretical) equilibrium when a permanent change occurs in one or several of the model state variables (Puccia and Levins 1985, Dambacher et al. 2009). These expected changes are the combined results of direct effects and indirect effects created by feedback loops when perturbations propagate through the links between system state variables. In fisheries applications they correspond to the expected direction of change of stock biomass or fisheries revenues or other state variables, that is predictions of likely increases or decreases of these variables in case of implementing the planned management tactics.

Step 3: Are the proposed socioeconomic, ecosystem, and monitoring measures likely to enhance the success of the management plan? Here all complementary socioeconomic, ecosystem, and monitoring measures included in the management plan are evaluated regarding their contribution to the overall success of the plan. For this Trenkel et al. (2015) derived five criteria from retrospective empirical evaluations of management plans found in the scientific literature: industry support, large initial mortality reduction (if the stock is overfished), multispecies provisions, performance indicators, and monitoring programs.

Step 4: Which harvest control rules might perform best? In this step, quantitative simulations are carried out for specific implementation issues. Typically different bioeconomic models are used to investigate particular aspects of the expected effects of parts of the management (cf. Dichmont, et al., Chapter 10, this volume).

Step 5: How can the institutional and management arrangements be modified to enhance effective implementation of the management plan, including buy-in and compliance from the industry? Here the competency (legal and skill sets) and capacities (of institutions and distributed governance arrangements) are assessed relative to the likelihood of successful implementation of the management plan. This may include research on how fishers' behaviour – in this context particularly compliance with provisions of management plans – is affected by the provisions themselves, as well as stock and ecosystem, economic and social, and governance factors. This is new territory in fisheries management, but under development (Kraak 2011, Kraak et al. 2015). At this time the fine-tuning of management may have to be ad hoc and interactive between the industry and managers, but is nevertheless invaluable in achieving a fishery that actually unfolds as expected (Rice and Richards 1996).

Qualitative MSE for southeastern Australia

The southeastern Australian MSE which focused on management options for Australia's Southern and Eastern Scalefish and Shark Fishery (SESSF), a multispecies, multigear fishery (Smith and Smith 2001), was conducted at a whole ecosystem level. These fisheries were languishing both biologically and economically in the early 2000s. While the quantitative (simulation-based) MSE undertaken using the Atlantis modelling framework was influential in setting a new direction for the management of the fishery (Fulton et al. 2014), it was actually the second stage of a larger study. It was preceded by a highly influential qualitative MSE undertaken prior to the quantitative analysis.

By the mid-2000s, the SESSF had been under quota management for a decade and a half, with over 30 individual stocks allocated in an ITQ scheme. This management system was intended to have improved both the biological and economic performance of the fishery, but it had failed to do so. Many of the stocks were overfished and the net economic returns from the fishery had been negative for most of the early 2000s.

The Alternative Management Strategies (AMS) project, as it came to be known in southeastern Australia, was always conceived of as an MSE study. However, since no one had undertaken such an approach at a whole-fishery level, it was not even clear if such an approach was feasible. A project team comprising one fishery manager and six fishery scientists (both government based and independent, and including one economist) had over 150 years of collective involvement in and knowledge of the fishery. This group was overseen by a steering committee comprising fishers from each of the main gear sectors (trawl, long-line, and gillnet), a senior fishery manager, and one environmental NGO with an active interest in the fishery.

The project team engaged actively through a series of workshops with a much larger group of fishers and other stakeholders in the fishery during the course of the 9-month stage 1 analysis. The method adopted focused on the first three steps listed earlier, and did not include quantitative modelling. It involved identifying objectives for the status of target species, broader ecological impacts of fishing, economic performance, and management efficiencies, and then turning these into more than 20 quantitative indicators, identifying alternative management strategies, and then evaluating their likely performance, based on the expert judgment of the research team, while taking into account uncertainties.

The choice of objectives, indicators, and strategies was undertaken with the diverse set of stakeholders as well as the steering group during the first workshop. The alternative strategies included broad classes such as a principal reliance on quota management, more emphasis on input controls, and mixed strategies including spatial management. The strategies were developed in considerable detail so that all stakeholders in the process could consider how each strategy would impact on the broad range of indicators and objectives. The alternative strategies were evaluated by the project team by qualitatively projecting forward each of the indicators 20 years, using the expert judgment of the team. Uncertainty was taken into account by considering both "optimistic" and "pessimistic" scenarios in the projections. These results were then presented to the broad set of stakeholders in a second workshop, where the results were examined critically. The stakeholders, having developed an appreciation of what such a qualitative MSE analysis looked like and how it might be used, then enthusiastically developed an extended suite of strategies for further analysis and consideration. By the end of stage 1 of the project, 10 broad classes of management strategies had been evaluated using this qualitative approach.

In hindsight, the impact of this stakeholder-driven process was enormous. An important part of the process involved asking stakeholders themselves to design strategies to manage this complex fishery. This resulted in a much broader set of classes of strategies being identified and tested than is normally the case in such MSE exercises. It also completely changed the thinking about what might be possible; it forced explicit consideration of trade-offs across multiple objectives; and by the end of this relatively short exercise, the strategy originally put forward as "blue skies" (because it was so different to the status quo strategy) was seen as a viable and even desirable option. A subset of the strategies was selected for full quantitative analysis, as described in Fulton et al. (2014).

Based solely on the qualitative stage 1 MSE, significant changes had been made to the management of the fishery, in the direction of the "blue skies" option. Importantly, there turned out to be a large correspondence in predictions of the qualitative analyses with the later quantitative, simulation-based study. Predictions were not identical for all indicators, but the rank order of strategies across broad management objectives matched very closely (Fulton et al. 2014).

Conclusions

Information needs for management decisions are diverse and variable over time. The knowledge, time, and/or resources available to meet these needs do not always allow a simulation approach to be adopted, and such approaches do not always address all of the important factors in the problem. We illustrated this with three case studies. In the case of the European cod plan, some knowledge predicting fisher reactions was available but not included in the simulation-based assessment of the HCR and was not taken into account; moreover, the non-HCR provisions of the plan had not been addressed at all, probably because fisher reactions to them could not be quantitatively predicted. The Northern cod case illustrated how simple management rules established without extensive simulations can be successful under stable environmental conditions and when outputs of the rule meet expectations of policy-makers and industry, but the decision-making system could not respond when changes in environmental conditions changed the stock productivity so rule outputs deviated from expectations. In the case of Pacific herring, a change in economic conditions meant the objectives for the fishery changed rapidly and the assessment and management systems performing well for the earlier fishery could not be afforded when the objectives of the fishery changed. Based on our knowledge and experiences from these case studies, we put forward three methods for the wider MSE toolbox: expert judgment based on knowledge, use of mathematical models not involving simulations, and qualitative approaches including loop analysis. All these methods have been applied successfully in practice, with qualitative approaches being the most recent ones. As illustrated by the southeastern Australia case study, the insights gained from qualitative MSE go far beyond selecting optimal harvest control rules. We suggest that students and practitioners of management science for fisheries take these alternative approaches into account through deliberation in an extended peer community (including relevant stakeholders) before tailoring their own MSE toolbox for their specific case.

References

Armitage, D., Berkes, F., Dale, A., Kocho-Schellenberg, E. & Patton, E. 2011. Co-management and the co-production of knowledge: Learning to adapt in Canada's Arctic. *Global Environmental Change-Human and Policy Dimensions*, 21(3), 995–1004.

Berkes, F. & Turner, N.J. 2006. Knowledge, learning and the evolution of conservation practice for social-ecological system resilience. *Human Ecology*, 34, 479–494.

Clark, C.W. 1985. *Bioeconomic Modelling and Fisheries Management*. New York, John Wiley & Sons.

Dambacher, J.M., Gaughan, D.J., Rochet, M.-J., Rossignol, P.A. & Trenkel, V.M., 2009. Qualitative modelling and indicators of exploited ecosystems. *Fish and Fisheries*, 10, 305–322.

Davis, A. & Wagner, J.R. 2003. Who knows? On the importance of identifying "Experts" when researching local ecological knowledge. *Human Ecology*, 31(3), 463–489.

de Lara, M., & Doyen, L. 2008. Sustainable management of natural resources. Mathematical models and methods. *In: Environmental Science and Engineering,* Berlin, Springer-Verlag.

De Lara, M., Doyen, L., Guilbaud, T. & Rochet, M.-J. 2007. Is a management framework based on spawning stock biomass indicator sustainable? A viability approach. *ICES Journal of Marine Science*, 64, 761–767.

Deriso, R.B. 1987. Optimal F0.1 criteria and their relationship to maximum sustainable-yield. *Canadian Journal of Fisheries and Aquatic Sciences*, 44, 339–348.

Fulton, E.A., Smith, A.D.M., Smith, D.C. & Johnson, P. 2014. An integrated approach is needed for ecosystem based fisheries management: Insights from ecosystem-level management strategy evaluation. *PLoS ONE*, 9, e84242.

Haggan, N., Neis, B. & Baird, I.G. Eds. 2007. *Fishers' Knowledge in Fisheries Science and Management.* Paris, UNESCO.

Hall, D.L., Hilborn, R., Stocker, M. & Walters, C.J. 1988. Alternative harvest strategies for Pacific herring (*Clupea harengus pallasi*). *Canadian Journal of Fisheries and Aquatic Sciences*, 45(5), 888–897.

Halpern, B.S., Selkoe, K.A., Micheli, F. & Kappel, C.V. 2007. Evaluating and ranking the vulnerability of global marine ecosystems to anthropogenic threats. *Conservation Biology*, 21, 1301–1315.

Hay, D.E. 1985. Reproductive biology of Pacific herring (*Clupea harengus pallasi*). *Canadian Journal of Fisheries and Aquatic Sciences*, 42, 111–126.

Hilborn, R. & Walters, C.J. 1992. *Quantitative Fisheries Stock Assessment: Choice, Dynamics and Uncertainty.* New York, Chapman and Hall,.

ICES 2008a. Report of the Ad hoc Group on Cod Recovery Management Plan (AG-CREMP).

ICES 2009. Report of the Working Group on the Celtic Seas Region (WGCSE).

ICES 2009b. Report of the ICES Advisory Committee 2009. ICES Advice, 2009. Book 6. pp. 236

ICES 2011. Report of the Working Group on the Ecosystem Effects of Fishing Activities (WGECO).

Johannes, R.E. 1981. *Words of the Lagoon: Fishing and Marine Lore in the Palau District of Micronesia.* Berkeley, University California Press.

Knopp, J.A. 2010. Investigating the effects of environmental change on Arctic char (*Salvelinus alpinus*) growth using scientific and inuit traditional knowledge. *Arctic*, 63(4), 493–497.

Kraak, S.B.M. 2011. Exploring the 'public goods game' model to overcome the Tragedy of the Commons in fisheries management. *Fish and Fisheries*, 12(1), 18–33.

Kraak, S.B.M., Kelly, C. J., Codling, E. A. & Rogan, E. 2010. "On scientists' discomfort in fisheries advisory science: the example of simulation-based fisheries management-strategy evaluations." *Fish and Fisheries*, 11(2), 119–132.

Kraak, S.B.M., Bailey, N., Cardinale, M., Darby, C., de oliveira, J.A.A., Eero, M., Graham, N., Holmes, S., Jakobsen, T., Kempf, A., Kirkegaard, E., Powell, J., Scott, R.D., Simmonds, E.J., Ulrich, C., Vanhee, W., Vinther, M. 2013. Lessons for fisheries management from the EU cod recovery plan. *Marine Policy*, 37, 200–213.

Kraak, S.B.M., Kelly, C., Anderson, C., Dankel, D., Galizzi, M., Gibson, M., Gonçalves, D., Hamon, K., Hart, P., Maguire, J.-J., Nieboer, J., van Putten, I., Reeson, A., Reid, D., Richter, A. & Tavoni, A. 2015. ICES Science Fund 2014 project report: Insights from Behavioural Economics to improve Fisheries Management. Report of the Workshop jointly funded by ICES and FSBI held 21–23 October 2014 at ICES HQ, Copenhagen, Denmark, pp. 46. http://ices.dk/community/icessciencefund/Documents/06_Report%20of%20the%20 BehavFish%20Workshop_final.pdf.

May, R.M., Beddington, J.R., Clark, C.W., Holt, S.J., Laws, R.M. 1979. Management of multispecies fisheries. *Science*, 205, 267–277.

Motos, L. & Wilson, D.C. Eds. 2006. *The Knowledge Base for Fisheries Management.* Developments in Aquaculture and Fisheries Science. Amsterdam, Elsevier.

Munro, G.R. 1980. *A Promise of Abundance: Extended Fisheries Jurisdiction and the Newfoundland Economy.* Economic Council of Canada, Ottawa.

Parsons, L.S. 1993. *Management of Marine Fisheries in Canada.* Ottawa, NRC Press.

Puccia, C.J. & Levins, R. 1985. *Qualitative Modeling of Complex Systems: An Introduction to Loop Analysis and Time Averaging.* Cambridge, MA, Harvard University Press.

Punt, A.E. 1995. The performance of a production model management procedure. *Fisheries Research*, 21(3–4), 349–374.

Rice, J.C. 2006. Every which way but up: The sad story of Atlantic groundfish, featuring Northern cod and North Sea cod. *Bulletin of Marine Science*, 78, 429–465.

Rice, J.C. & Evans, G.T. 1988. Tools for embracing uncertainty in the management of the cod fishery of NAFO Divisions 2J+3KL. *Journal Conseil International pour l'Exploration de la Mer*, 45(1), 73–81.

Rice, J. & Richards, L.J. 1996. A framework for reducing implementation uncertainty in fisheries management. *North American Journal of Fisheries Management*, 16, 488–494.

Rochet, M.J. & Rice, J. 2005. Do explicit criteria help selecting indicators for Ecosystem-based fisheries management? An experimental test. *ICES Journal of Marine Science*, 62, 528–539.

Schweigert, J.F. 1991. Multivariate description of Pacific herring (*Clupea harengus pallasi*) stocks from size and age information. *Canadian Journal of Fisheries and Aquatic Sciences*, 48(12), 2365–2376.

Schweigert, J.F., Boldt, J.L., Flostrand, L., Cleary, J.S. 2010. A review of factors limiting recovery of Pacific herring stocks in Canada. *ICES Journal of Marine Science*, 67(9), 1903–1913.

Schweigert, J.F. & Stocker, M. 1988. Escapement Model for estimating Pacific Herring stock size from spawn survey data and its management implications. *North American Journal of Fisheries Management*, 8, 63–74.

Smith, A.D.M. 1994. Management strategy evaluation – the light on the hill. *In:* Hancock, D.A. (ed.) *Population Dynamics for Fisheries Management. Australian Society for Fish Biology Workshop Proceedings, Perth, 24–25 August 1993*. Perth, Australian Society for Fish Biology, pp. 249–253.

Smith, A.D.M., Fulton, E.J., Hobday, A.J., Smith, D.C. & Shoulder, P. 2007. Scientific tools to support practical implementation of ecosystem-based fisheries management. *ICES Journal of Marine Science*, 64, 633–639.

Smith, A.D.M. & Smith, D.C. 2001. A complex quota-managed fishery: Science and management in Australia's South-East Fishery. Introduction and overview. *Marine and Freshwater Research*, 52(4), 353–359.

Stocker, M., Haist, V. & Fournier, D. 1985. Environemental variation and recruitment of Pacific herring (Clupea harengus pallasi) in the strait of Georgia. *Canadian Journal of Fisheries and Aquatic Sciences*, 42, 174–180.

Trenkel, V.M., Rochet, M.J., Rice, J.C. 2015. A framework for evaluating management plans comprehensively. *Fish and Fisheries*, 16, 310–328.

Walters, C. & Parma, A.M. 1996. Fixed exploitation rate strategies for coping with effects of climate change. *Canadian Journal of Fisheries and Aquatic Sciences*, 53, 148–158.

Walters, C.J. 1975. Optimal harvest strategies for salmon in relation to environmental variability and uncertain production parameters. *Journal of the Fisheries Research Board of Canada*, 32, 1777–1784.

Part IV

The practice of fisheries management

Fisheries management science is intended to lead to informed management actions, and ultimately, it is up to the stakeholders to judge the efficacy of fisheries science for management. Part IV of this volume takes a user perspective and reflects upon the scientific processes and their relevance for fisheries management.

The herring in the North Sea is an important stock for several European nations. It is thus not surprising that one of the first multinational management plans in Europe was developed by scientists and stakeholders of this historically significant and dynamic resource. Chapter 19 provides a review and an insider's view of the development of the management plan for North Sea herring (*Claupea harengus*), also focusing on the complexities of balancing international agreements and directives for healthy ecosystems with other stakeholder priorities.

Stakeholder engagement in management procedures, plans and evaluations is an impetus for good science for policy. The author of Chapter 20 has been active on two sides of the science-policy interface: as an assessment scientist and former chair of the International Council for the Exploration of the Sea (ICES) Advisory Committee and presently as a chief science officer of the Pelagic Freezer-trawler Association (PFA). There are few better individuals to review the status of stakeholder engagement in fisheries science and the complex backdrop of European governance, and Chapter 20 illustrates this with four contrasting case studies: western horse mackerel, deep sea red shrimp, western Baltic spring spawning (WBSS) herring and North Sea *Nephrops*.

There are few places where the credibility crisis among stakeholders and government scientists is so volatile and vocal as in the Northeast United States. Chapter 21 reviews a recent fisheries management groundfish case study (Framework 50) from two seemingly opposing stakeholder views: industry and environmental conservation. However, their conclusions on the process of Framework 50 are in agreement; stakeholders from opposite ends of the table are still trying to successfully engage with scientists, which has proven to be an uphill battle in the highly bureaucratized, technocratic fisheries management system in New England.

The practice of fisheries management brings together the best available science to inform policy needs. This book has focused on the theory, design and application of simulation-based methods. The final Chapter 22 takes a bold look at the qualities of fisheries management science and its readiness and fitness-for-purpose for the science-policy interface. The path forward for sustainable fisheries will no doubt have to further focus on issues of credibility and legitimacy of the science, which can be achieved by refining stakeholder engagement processes in simulation studies and applying innovative ways of underscoring the inevitable role of uncertainties, known and unknown, in responsible decision-making.

## Chapter 19

North Sea herring

Longer term perspective on management science behind the boom, collapse and recovery of the North Sea herring fishery

Mark Dickey-Collas

Introduction

There is a long history of fisheries on North Sea herring and these fisheries have been socially and politically very important to Northern Europe for the last 400 years (Poulsen, 2008). The exploitation of North Sea herring is also seen as a classic example of the boom and bust consequences of industrialisation of the fleet and poor fisheries management, as characterised by the 1970s stock collapse caused by recruitment overfishing and management inaction when the productivity declined (Nichols, 2001, Dickey-Collas et al., 2010). The collapse in the fish stock and its economic consequences (5-year closure of the fishery and loss of markets) led to an awareness of the need for better fisheries management to ensure sustainable exploitation (Simmonds, 2007). In 1995 and 1996, fisheries scientists warned that the stock was being dangerously overexploited again and was "outside safe biological limits." This led to a recommendation of within-year reductions in total allowable catch (TAC) of 50% (Simmonds, 2007). Perhaps as the previous collapse was well within living memory, both the managers and the industry worked with the shared objective to maintain sustainability of the fishery. This led to the first version of the EU-Norway North Sea herring management plan[1] being initiated in 1997. The management plan was an agreement between Norway and the EU to limit catches following a pre-agreed harvest control rule (HCR). This was one of the first applications of HCRs in the EU/Norway arena. Since then the plan has been revised in 2004, 2008 and 2014. Each revision involved fresh evaluations and has kept the core objective, namely to keep spawning stock biomass (SSB) above 800,000 tonnes, but each revision has reflected a new set priorities which developed in response to changes in fisheries management, the quality of the assessment and changes to the biological productivity of North Sea herring. This series of revisions is documented in this chapter and the 2014 revision is described in more detail. As the management of North Sea herring is dynamic and iterative, it is very likely that as this chapter goes to press, the details on the current situation will be outdated and further developments will have occurred.

The core objective of the entire series of management plans was to keep the SSB of North Sea herring above 800,000 tonnes. This figure was initially based on an estimate representing the minimum biologically acceptable level of biomass (MBAL; Figure 19.1a). It was seen as the breakpoint beyond which recruitment would rapidly decline commensurate with declining SSB (Nash et al., 2009). Despite the development in the fisheries management framework, many new stock assessments and new analyses, this MBAL figure has remained the cornerstone of the management plan, although it pragmatically morphed into the reference point B_{lim} under the precautionary approach (ICES, 1998).

The revisions to the management plan were responses to new challenges to the sustainable and optimum exploitation of North Sea herring fisheries. Initially one primary concern was overcatching of the TAC, what was euphemistically called "implementation error" by the scientists that evaluated the plan. The initial 1997 plan set a target SSB (1.3 million tonnes) and a target fishing mortality once the fish stock had reached that target (Table 19.1). The SSB of 1.3 million tonnes was chosen through simulations and assumed to ensure that fishing would have a low risk of impacting recruitment. A key concept within the plan was separate fishing mortality targets for the fishery that exploited

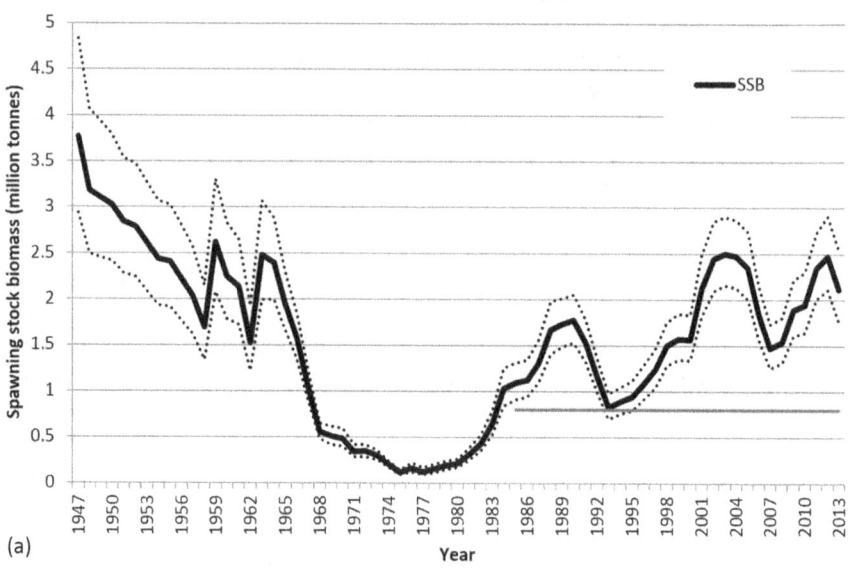

(a)

Figure 19.1 Time series of estimates from 1947 to 2013 of spawning stock biomass (a) and fishing mortality (b) from the 2014 stock assessment for North Sea herring (ICES, 2014). (a) shows the SSB and the primary object of the management plan, avoiding an SSB of 800,000 tonnes or below. (b) shows mean fishing mortality for adults (aged 2–6) and juveniles (aged 0–1) and the approximate advice for the fishery on adults stemming from the management plan. Solid lines show the mean estimate of the stock assessment; the dotted lines indicate the 95% confidence interval of the stock assessment model.

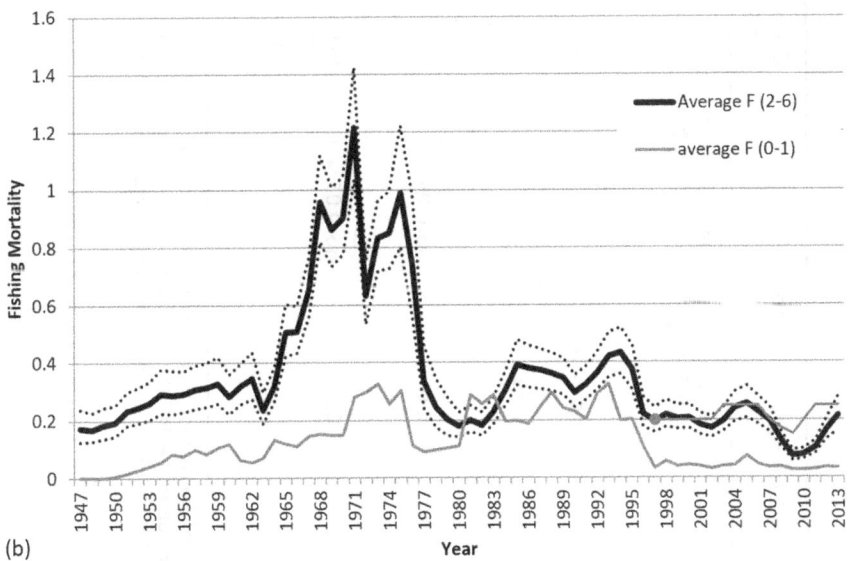

(b)

Figure 19.1 (Continued)

adult herring for human consumption and for the industrial fishery for fish-meal, where juvenile herring are a major bycatch. With the stock biomass less than the MBAL at this time, the scientists were asked by Norway and the EU to provide advice on the adult and juvenile fishing mortalities required to reach the target SSB. In 2004, the first revision of the plan resulted in a more elaborate mechanism to maintain sustainable and stable catches, including a formalised rule for fishing at lower SSBs (< 1.3 million tonnes, see Table 19.1). The revision to the plan in 2008 was mostly driven by unprecedented changes in stock productivity, which resulted in a decline in the SSB despite otherwise effective fisheries management (Payne et al., 2009). The revised plan had a higher $B_{trigger}$ reference point of 1.5 million tonnes, meaning that the stipulated fishing mortality was lowered for biomass values less than this new target. The ability of the plan to deliver sustainable and yet high catches was then challenged again in 2011 and 2012, when uncertainty in the assessment and the catch stability mechanisms resulted in the secondary objective of the plan to "provide for stable and high yields" not being met. This issue was raised by both managers and the industry. The catch stability mechanisms meant that the catch could not be increased fast enough to keep fishing mortality at the desired level as the biomass increased. This resulted in the fishing industry being tied to lower fishing mortalities than one would expect at the estimated biomass of fish (Figure 19.1b), if the plan was followed without the catch stability mechanisms. In 2008, when the plan was being developed, this magnitude of change of the stock was not anticipated. Thus the need for this last revision was caused by a change in the perception of expected stock variability (see Simmonds, 2009).

Table 19.1 Evolution of EU-Norway management plan for North Sea herring.

	Avoid biomass	Trigger biomass	Target F – Adults (2 and older)			Target F – Juveniles 0 and 1)			Catch stability
			Above trigger	Below MBAL/B_{lim}	Slope between trigger and B_{lim}	Above trigger	Below MBAL/B_{lim}	Slope between trigger and B_{lim}	
	million t	million t							
1997	0.8 (MBAL)	1.3	0.25	science advice	no	0.12	science advice	no	no
2004	0.8 (B_{lim})	1.3	0.25	0.1	yes	0.12	0.04	yes	15% max TAC change
2008	0.8(B_{lim})	1.5	0.25	0.1	yes	0.05	0.04	no 0.05	15% max TAC change
2014	0.8(B_{lim})	1.5	0.26	0.1	yes	0.05	0.04	no 0.05	15% max TAC change, near 10% F

Importantly the advised target F for a management year must be based on projections of biomass for that management year that account for some of the fishing prior to spawning. This requires a method of iterative projection to provide advice. MBAL= minimum biologically acceptable level of biomass. B_{lim} = limit reference point for spawning stock biomass. F = fishing mortality. TAC = total allowable catch. Trigger biomass = the point that additional management action is initiated, usually through reductions in F.

Despite these revisions, the plan is still aimed at the original objective of preventing recruitment overfishing of the stock and maintaining a viable fishery. The fishery has become much more efficient, with fewer boats exploiting the TAC. This is probably not an effect of the plan. Throughout the process the plan has been developed in partnership with fisheries managers and scientists and in the last 10 years directly with the fishing industry and the regional advisory councils (RACs), now reformed to advisory councils (ACs).

The decision rule

The initial construction of the decision rule was designed by scientists to perform best in terms of the trade-off between yield and risk and aimed to achieve both catch stability and stock recovery. The creators of the plan cited May et al. (1978) that "what seems really needed is not further mathematical refinement, but rather robustly self-correcting strategies that can operate with only fuzzy knowledge about stock levels and recruitment curves" (Patterson et al., 1997, p. 4). To some extent this has happened via the series of revisions to the management plan since 1997. Although no industry members took part in the evaluation of the initial plan, the creators stated that the plan must be robust, simple and acceptable on social and economic grounds. At the time there were great concerns about overcatching of the TAC, and the general milieu was one of a need for stronger control and enforcement mechanisms, not of managers/ industry partnerships. The rule was similar to others being evaluated in the International Council for the Exploration of the Sea (ICES) and the North Atlantic Fisheries Organisation (NAFO) (ICES, 1997, Serchuk et al., 1997). Over nine different approaches were evaluated and the conclusion was "management using the $F_{0.1}$ reference point achieved a higher yield for low risk of stock depletion than the other scenarios modelled" (Patterson et al., 1997, p. 5). They also stated that "attempting to exploit a stock at constant fishing mortality could lead to high risks of stock depletion" (Patterson et al., 1997, p. 5), thus the rule to reduce F at biomasses that approach MBAL/B_{lim} was devised (ICES, 1997, 1998). The evaluations were projected forward over the forthcoming decades.

The core decision rule provides a target fishing mortality for that management year for each of the two fisheries (adult and juvenile) that is intended to result in a catch equal to MSY over the long term. This target fishing mortality is equivalent to F_{MSY}, but is reduced as the SSB approaches the biomass limit reference point, to allow the stock to rebuild to more productive levels (Figure 19.2a). The rule is designed to maintain a minimum spawning biomass by reducing F as the biomass decreases, thereby ensuring adequate escapement, whilst maintaining a high and sustainable catch. The initial rule was developed using MSY concepts by Patterson et al. (1997) (see ICES, 1997), was adapted to conform to the precautionary approach (ICES, 1998), and has now been assimilated into the new MSY framework for European fisheries management

(ICES, 2014; see later). The rule is similar in structure to the generic MSY rule used by ICES as part of this framework (ICES, 2014).

The allocation of proportions of the TAC for Norway and the EU is also written into the management plan. Within the EU catches, small trade-offs between fleets and between adult and juvenile fisheries are incorporated into the catch advice from ICES, but are not directly stipulated in the plan. This trade-off is required because the advised target fishing mortality per management year must be based on projections of biomass for that management year and must account for some of the fishing prior to spawning.

The SSB reference points for the decision rule are:

- $B_{trigger}$ – where below this trigger the target F is reduced in a linear fashion dependent on the size of the SSB within that management year, and above this use an appropriate F_{MSY} as a target.
- B_{lim} – below which recruitment becomes impaired. It must be avoided with a certainty of > 95% over the whole times series.
- B_{pa} – which is a precautionary buffer on B_{lim} accounting for stock assessment uncertainty.

Most evaluations of the rule use a fixed B_{lim} and an exploration of $B_{trigger}$ over a range of potential F_{MSY} targets. B_{pa} is independent of the management rule. B_{pa} is set at a level that the risk of being below B_{lim} due to assessment uncertainty

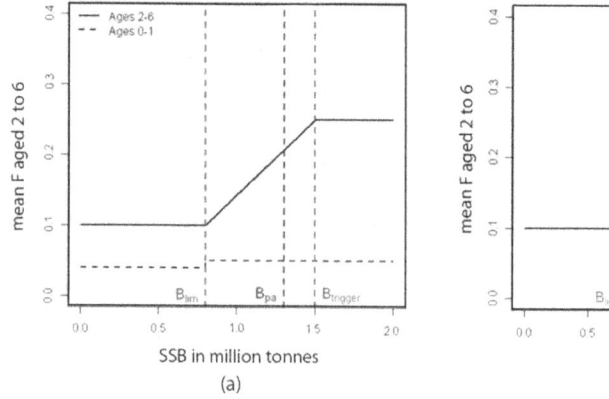

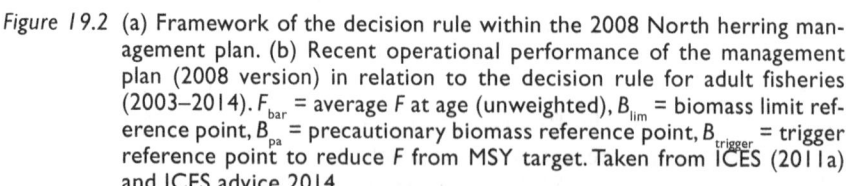

Figure 19.2 (a) Framework of the decision rule within the 2008 North herring management plan. (b) Recent operational performance of the management plan (2008 version) in relation to the decision rule for adult fisheries (2003–2014). F_{bar} = average F at age (unweighted), B_{lim} = biomass limit reference point, B_{pa} = precautionary biomass reference point, $B_{trigger}$ = trigger reference point to reduce F from MSY target. Taken from ICES (2011a) and ICES advice 2014.

is small (< 5%). Although B_{pa} is no longer part of the rule, it still plans a role in accreditation audits for sustainability labelling for marketing and now plays a role within the flexibility mechanisms of the new Common Fisheries Policy (CFP) landing obligation.

The estimates of F_{MSY} are derived from stochastic simulations, based on an analysis with three stock-recruit relationships: Ricker, Beverton and Holt and Hockey Stick, approximated by a continuous function as suggested by Mesnil and Rochet (2010). This assumes that the knowledge base on the recruitment dynamics is uncertain and thus runs a range of simulations accounting for different stock to recruitment relationships.

The recent evaluation process

The evaluation of the decision rule has been carried out by an evolving group of scientists, usually under the auspices of ICES and sometimes through ad hoc EU/Norway workshops (Table 19.2). The scientists form a group that is associated with the ICES herring assessment working group (HAWG) and have tended to come from fisheries institutes in Scotland, England, Norway, the Netherlands and Denmark. Fishing industry representatives have participated in all the evaluations since 2008 and the Pelagic Regional Advisory Council (RAC, now called the Pelagic Advisory Council) has participated since 2011 (Table 19.2).

The evaluations were initiated by formal requests for advice to ICES or member countries. In most cases, prior to the evaluation, the performance diagnostics (evaluation criteria) were agreed with managers and stakeholders. These tended to be related to probability of the spawning biomass falling to 800,000 tonnes or below (an interpretation consistent with the precautionary approach), stability of the catches close to MSY catches and the trade-off between the fisheries on juveniles and adults.

The operating model

The operating model used was originally evaluated using medium-term stochastic simulations for different levels of fishing mortality on adults and juveniles (ICES, 2007). The performance of the harvest control rules were explored by stochastic simulations of population numbers and recruitment, and with assessment uncertainty modelled as a simple random multiplier on the catches. The scientists estimated distributions of equilibrium stock size and of yield by fleet by extending the medium-term projection for 100 years and taking the terminal distributions as approximating equilibrium.

With each revision a slightly different method was used for the evaluation, but all were based on stochastic medium-term projections (looking forward over two decades). The last full exploration of the operating model was carried out in 2011. It was instigated by the EU and Norway and was paid for by

Table 19.2 Time line of the evaluations of the versions of the management plan (MP) for North Sea herring fisheries. Objectives should be considered cumulative.

Year	1997	2004	2008	2011	2012	2014
MP	Version 1	Version 2	Version 3	Explore within year revision	Preparation for Version 4	Version 4
Objectives	Prevent collapse/ overexploitation	Management plan proper (described as an MP) and stable and high yields	Cope with reduced productivity	Failure to optimise yield	Revise MP and reference points	All of the previous
Workshops	ICES answer EU/ Norway request	EU-Norway ad hoc WK Multiannual management plans for shared stocks	SGMAS, WKHMP	WKHERMP and WKHIAMP	WKHELP	
Participation	Scientists	Scientists	Scientists, industry	Scientists, industry, RAC	Scientists, industry, RAC	
References	ICES, 1997, Patterson et al., 1997.	EC, 2004.	ICES, 2007, 2008	ICES, 2011a, 2011b	ICES, 2012	

national governments and as a special request to ICES through the EU/ICES Memorandum of Understanding (see Table 19.2). Here the medium term simulation model simulated the biological herring population and the behaviour of the fishing fleets and surveys, while the stock assessment was mimicked to estimate the stock status. Finally, the management advice and implementation were based on the adjusted management plan scenarios. In turn, management fed back into the biological population and the fishery the year after. The simulations were run with 100 Monte Carlo realisations to obtain a broad range of possible outcomes given the variability in the input data. Stochasticity (randomness) was added to variables and parameters to ensure that biological variation, and the uncertainty in the historic perception of the stock was thus reflected. The amount of stochasticity could still be considered the minimum likely, as since 1997 is has been clear that the uncertainty in the whole fisheries management system has been underestimated, leading to the revisions of the plan. This was predicted by the original creators of the plan (Patterson et al., 2007).

Performance diagnostics

The performance diagnostics have changed throughout the revisions of the management plan as the objectives of the plan have evolved. In 2012, the EU and Norway requested for eight potential harvest control rules to be evaluated. For this 2012 evaluation of the revised management plan, 16 potential criteria were discussed. The workshop with participants from industry and national fisheries research institutes highlighted that evaluation criteria were needed for risk, stock performance (development of the biomass of the stock), fishing mortality, yield (catch) and catch stability (ICES, 2012). The following five evaluation criteria were chosen for the evaluation:

1   Risk: percentage of iterations in which SSB falls below B_{lim} at least once during the simulation period.
2   Stock performance: SSB in 2022 (median of all iterations).
3   Fishing mortality: Mean fishing mortality of herring aged 2–6 over the simulation period.
4   Yield: Mean catch of the dominant human consumption fleet (called the A-fleet) over the simulation period.
5   Stability in TAC: Mean percentage absolute TAC change of the A fleet between consecutive years over the simulation period.

Selection of a harvest control rule

Ninety-nine scenarios were tested during the 2012 evaluation of the eight potential harvest control rules. The results and detailed analysis of these evaluations are given in Section 5 and Annex 3 of ICES (2012). The evaluation workshop (ICES, 2012) found that all eight potential harvest control rules had options that could be considered precautionary, that is low risk of biomass fall-ing below the precautionary reference point. Five of the eight performed best against the five evaluation criteria. By reviewing the performance criteria, the workshop then advised that two out of the eight were operationally best and had the ability to quickly adapt to changes in recruitment and stock produc-tivity (ICES, 2012). The decision as to which of the two should be used was passed back to the managers and their discussions with stakeholders. This illus-trates that, depending on the framework created, there are bounds to the tested criteria that scientists can explore. However well designed the evaluation is, the decision might be based on preferences (policy, operational, cultural) which have not been tested in an existing framework by scientists.

Implementation

Since 1997, the results of the evaluations were fed into the management process through formalised ICES advice. This provided the evidence base, which was supplemented by the same scientists working for their national institutions, into the EU/Norway negotiations. The required revisions of the plan from 1997 to 2014 highlight that the assumptions of each evaluation did not foresee emerg-ing challenges or shifting management priorities. In the 1990s, the likelihood of fisheries sticking to the TAC was overestimated. In the early 2000s stability in the stock's productivity was overestimated, and in the late 2000s the scientists were overconfident in the certainty of the time series coming from the stock assessments (Simmonds, 2009). The catch stability mechanisms initiated in the 2004 revision resulted in the fishery harvesting substantially below MSY from 2008 to 2012, despite being above the trigger biomass (Figure 19.2b). This effect was increased by an upwards revision of the estimate of SSB in 2011 caused by new input data. Further, a reworking of the entire stock assessment series, resulting from a new modelling approach in 2012, increased the esti-mates of SSB by 30%. These kind of retrospective changes in the stock assess-ment time series are often seen in fisheries science, and a failure to consider the appropriate amount of stock assessment uncertainty in any evaluation of a management plan might lead to a problem as shown here. Conversely, it could also lead to an unsustainable exploitation of the fish stock if the estimates of biomass are reduced by new input data or a change in stock assessment method.

It is clear that the development of the plan is an iterative process. No single evaluation group has correctly characterised the uncertainty associated with the whole process. As with all simulations, the scientists have necessarily simpli-fied the assumptions. These simplifications have consequences in terms of the objectives of the management plan. To some extend the time-limited nature of each plan (each revision has a defined deadline and a need for a new evalua-tion) shows that most of the players accept the limitations of scientific foresight.

The SSB of North Sea herring has not fallen below 800,000 tonnes since 1995 (with 95% confidence in the estimate; Figure 19.1a) and catches have been approximately 400,000 tonnes per year (1995–2013, with CV of 35%). This suggests that according to the management plan objectives, the management plan has been broadly a success (considering the initial objectives in 1997). The estimates of recent SSB (between 1.5 to 2.5 million tonnes; Figure 19.1a) are similar to those in the late 1950s, which considering the recent declines in recruits per spawner (Figure 19.3) should also be considered a success. The tension between the fisheries that target adults and the industrial fisheries that catch juveniles as bycatch is now reduced compared to the 1970s and 1980s. This was caused by a reduction in the importance of the industrial fishery, although this might return as a result of the new EU landings obligation. In terms of environmental impact, following both the EU Common Fisheries Policy (CFP) and the Marine Strategy Framework Directive (MSFD), the fisheries have been fishing below F_{MSY} since 1995, with no discernable negative impact on the genetic makeup of the stock, the age profile, fish size or the sex ratio. It is also likely that the stock is repopulating vacant spawning grounds (Payne, 2010).

The way that the plan has iteratively developed through a growing partnership between scientists and stakeholders should be viewed as a success. The

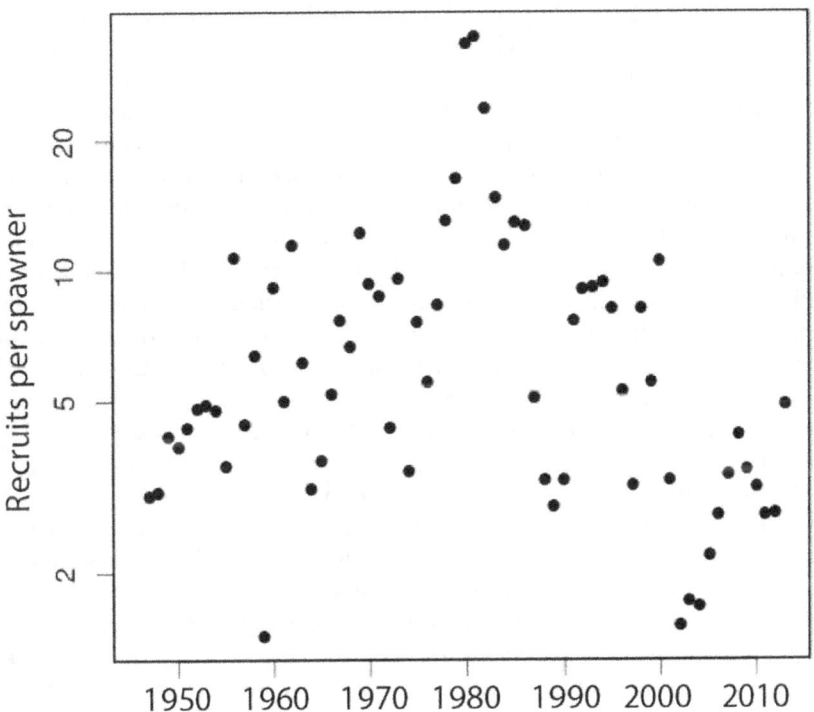

Figure 19.3 North Sea herring recruits per spawner from 1947 to 2012, showing the low productivity of the stock since 2002.

evaluation workshops have been successfully used to build trust and share knowledge and priorities between scientists and industry representatives. It could be argued that the success of the North Sea herring management plan was the impetus for the horse mackerel management plan proposed by the Pelagic RAC, the management plans for herring stocks around Ireland, and the efforts to create a plan for western Baltic spring spawning herring. Herring management plans are a key case study in the GAP2 project (Connecting Science, Stakeholders and Policy, http://gap2.eu/) due to their dynamic, cross-cutting and cross-sector collaborations.

Still, the management of North Sea herring fisheries faces upcoming challenges. The fishery targets a "forage fish" (Dickey-Collas et al., 2014). The stock assessment includes annually variable estimates of natural mortality, with inbuilt provision for seals, birds and cetaceans (Dickey-Collas et al., 2014). However, the experience from New England herring suggests that some campaigners do not see this as enough and are lobbying for further adaptations to harvest control rules to further account for top predators. This has been stimulated by the Lenfest Forage Fish task force (www.oceanconservationscience.org/foragefish/), which did not consider the EU context (information and analysis rich), did not consider the approach taken for North Sea herring (building in dynamic predation into the stock assessment) and did not work with stakeholders in exploring the consequences of the suggested approaches.

Under the MSFD, the impact of the fishery on biodiversity and foodweb (structure and function) needs to be shown to be consistent with achieving Good Environmental Status (GES). We hope that the inclusion of herring removals by predators within the stock assessment will prioritise the ecosystem demands over those of the fishery, but this assumption has not been evaluated through simulations (Dickey-Collas et al., 2014). Likewise, the complex interactions of fish in the marine system have led some scientists to suggest that the fisheries on North Sea cod cannot be managed in isolation of the fisheries on North Sea herring (Speirs et al., 2010, van Denderen and van Kooten, 2013). Cod eat juvenile herring, and herring eat cod eggs and larvae. If managers want to explore this interaction, it would be a challenge to design management plans for fisheries that target demersal and pelagic fish and account for the trophic and behavioural interactions. There are examples of explorations of management plans that consider single-species fisheries objectives and environmental objectives (e.g. Rijnsdorp et al., 2012), but this area of multispecies and ecosystem-wide research is in its infancy (see de Moor and Butterworth, Chapter 11, and Goethel et al., Chapter 16, this volume). One major challenge is that the current management plan operates using a tool that is region-wide (or stock-wide), namely the fishing mortality F. This is assessed for the whole stock. However many of the ecosystem management objectives (including ensuring that impacts on biodiversity and the foodweb are consistent with GES) require more spatially resolved assessments. An example being the breeding success of seabirds in specific bird colonies may reflect local processes rather than region-wide (or stock-wide) processes. How can management plans be set up and evaluated in these instances?

The North Sea herring management plan can be seen as a vehicle for consistency and adaptive management. The two are not inconsistent. The plan has not changed that much since 1997 and the overall framework appears robust. Each revision to the plan has been caused by each contemporary version being challenged by changing circumstances or switching management priorities. This has led to further evaluations and adjustments in response to these shifts. Continual improvements to the plan are what make it adaptive. Over the time period of the plan, stakeholders have been increasingly engaged in the priority setting and choosing of evaluation criteria. Industry, regional advisory council representatives and scientists have used the mechanism of regular revisions and evaluations of the plan to build a true dialogue. And at the moment, North Sea herring is fished sustainably, with the biomass of the stock well above limit reference points.

Note

1 The term "management plan" is usually specific to a European context and should not be confused with "management strategy" or "management procedure" as used elsewhere in this volume. The concept of a management plan is only loosely defined, subject to change, and may be most usefully understood as a "management arrangement" (see Pastoors, Chapter 20, this volume).

References

Dickey-Collas, M., Engelhard, G.H., Rindorf, A., Raab, K., Smout, S., Aarts, G., van Deurs, M., Brunel, T., Hoff, A., Lauerburg, R.A.M., Garthe, S., Haste Andersen, K., Scott, F., van Kooten, T., Beare, D. & Peck, M.A. Ecosystem-based management objectives for the North Sea: riding the forage fish rollercoaster. *ICES Journal of Marine Science*, 71, 128–142.

Dickey-Collas, M., Nash, R.D.M., Brunel, T., Damme, C.J.G. van, Marshall, C.T., Payne, M.R., Corten, A., Geffen, A.J., Peck, M.A., Hatfield, E.M.C., Hintzen, N.T., Enberg, K., Kell, L.T. & Simmonds, E.J. 2010. Lessons learned from stock collapse and recovery of North Sea herring: A review. *ICES Journal of Marine Science*, 67, 1875–1886.

EC 2004. Report of the EU-Norway Ad Hoc Scientific Working Group on multi-annual management plans for stocks shared by EU and Norway. Commission staff working paper. SEC2004 1209.

ICES 1997. Report of the comprehensive fishery evaluation working group. *ICES CM 1997/Assess: 15.*

ICES 1998. Report of the Study Group on the Precautionary Approach to Fisheries Management. Feb 1998. *ICES CM 1998/ACFM:10.*

ICES 2007. Report of the study group on management strategies [SGMAS]. *ICES CM 2007/ACFM:04.*

ICES 2008. Report of the Workshop on Herring Management Plans [WKHMP], 4–8 February, ICES Headquarters Copenhagen. *ICES CM 2008/ACOM:27.*

ICES 2011a. Report of the Workshop on the evaluation of the long-term management plan for North Sea herring [WKHERMP], 14–15 March 2011, ICES Headquarters, Copenhagen. *ICES CM 2011/ACOM.*

ICES 2011b. Report of the Workshop on Herring Interim Advice on the Management Plan [WKHIAMP], 24 October 2011, ICES Headquarters, Copenhagen, Denmark. *ICES CM 2011/ACOM:62.*

ICES 2012. Report of the Workshop for Revision for the North Sea Herring Long Term Management Plan [WKHELP], 1–2 October 2012, ICES HQ, Copenhagen, Denmark. *ICES ACOM:72.*

ICES 2014. General context of ICES advice. Available at www.ices.dk/sites/pub/Publication%20Reports/Advice/2014/2014/1.2_Advice_basis_2014.pdf, 20pp (last accessed 19 November 2015).

May, R.M., Beddington, J.R., Horwood, J.W. & Shepherd, J.G. 1978. Exploiting natural populations in an uncertain world. *Mathematical Biosciences,* 42, 219–252.

Mesnil, B. & Rochet, M-J. 2010. A continuous hockey stick stock–recruit model for estimating MSY reference points. *ICES Journal of Marine Science,* 67, 1780–1784.

Nash, R.D.M., Dickey-Collas, M. & Kell, L.T. 2009. Stock and recruitment in North Sea herring (Clupeaharengus); compensation and depensation in the population dynamics. *Fisheries Research,* 95, 88–97.

Nichols, J.H. 2001. Management of North Sea herring and prospects for the new millennium. *In:* Funk, F., Blackburn, J., Hay, D., Paul, A.J., Stephenson, R., Toresen, R., & Witherell, D. (eds.) *Herring 2000: Expectations for a New Millennium,* pp. 645–665. 18th Lowell Wakefield Fisheries Symposium, Anchorage, Alaska, 23–26 February 2000. Fairbanks, University of Alaska Sea Grant, AK-SG-01-04.

Patterson, K.R., Skagen, D., Pastoors, M. & Lassen, H. Harvest control rules for North Sea herring. WD to ACFM 1997.

Payne, M.R. 2010. Mind the gaps: A state-space model for analysing the dynamics of North Sea herring spawning components. *ICES Journal of Marine Science,* 67, 1939–1947.

Payne, M.R., Hatfield, E.M.C., Dickey-Collas, M., Falkenhaug, T., Gallego, A., Gröger, J., Licandro, P., Llope, M., Munk, P., Röckmann, C., Schmidt, J.O. & Nash, R.D.M. 2009. Recruitment in a changing environment: the 2000s North Sea herring recruitment failure. *ICES Journal of Marine Science,* 66, 272–277.

Poulsen, B. 2008. *Dutch Herring. An Environmental History c. 1600–1860,* Amsterdam, Aksant.

Rijnsdorp, A.D., van Overzee, H.M.J. & Poos, J.J. 2012. Ecological and economic trade-offs in the management of mixed fisheries: A case study of spawning closures in flatfish fisheries. *Marine Ecology progress Series,* 447, 179–194.

Serchuk, F., Rivard, D., Casey, J. & Mayo, R. 1997. Report of the Ad Hoc Working Group of the NAFO scientific council on the precautionary approach. NAFO SCR Working Paper 97/30.

Simmonds, E.J. 2007. Comparison of two periods of North Sea herring stock management: Success, failure, and monetary value. *ICES Journal of Marine Science,* 64, 686–692.

Simmonds, E.J. 2009. Evaluation of the quality of the North Sea herring assessment. *ICES Journal of Marine Science,* 66, 1814–1822.

Speirs, D.C., Guirey, E.J., Gurney, W.S. C. & Heath, M.R. 2010. A length-structured partial ecosystem model for cod in the North Sea. *Fisheries Research,* 106, 474–494.

van Denderen, P.D. & van Kooten, T. 2013. Size-based species interactions shape herring and cod population dynamics in the face of exploitation. *Ecosphere,* 4(10): art130. doi:10.1890/ES13–00164.1

## Chapter 20

Stakeholder participation in the development of management strategies

A European perspective

Martin A. Pastoors

Background

Fishery management in the Northeast Atlantic is largely at a level of multiple coastal states attempting to agree on common rules for shared stocks, like mackerel, herring, blue whiting and cod. The European Union (EU) counts as one coastal state, which negotiates with other coastal states like Norway, the Faroe Islands, Iceland and Russia.

The focus of this chapter is the role of stakeholder participation in the EU. Because of the multinational environment of the EU, I will illuminate the stakeholder-science links with the decision-making at the coastal states level.

Fishery management in the EU is organized in a layered approach. The European Commission (EC) is the executive branch of the EU. The EC is responsible for generating the legislative proposals and for implementing and controlling the adopted measures. Until recently (2007) the political responsibility for the European policy was with the Fisheries Council of Ministers, consisting of the fisheries ministers of all EU member states. The Fisheries Council would decide on the management measures proposed by the EC. Since the signing of the Lisbon Treaty (Council of the European Union 2007), the arrangements have changed because the European Parliament (EP) obtained a much stronger role in the fishery policy. While previously, the EP was mostly informed about the fishery policy decisions, since 2007 the EP has become the co-legislator with the Council of Ministers.

European fisheries are managed under a shared policy: the Common Fisheries Policy (CFP). The first CFP was agreed in 1983 (Council of the European Communities 1983) and was renewed roughly every 10 years in 1992 (Council of the European Communities 1992), 2002 (Council of the European Union 2002) and 2013 (European Parliament and the Council of the European Union 2013). The common policy means that the key decisions are taken at the European level even though the implementation is largely still at the member state level. For example, total allowable catch (TAC) setting arrangements, management plans and technical measures are agreed at the European level, but how the quota are distributed among the fishermen, how the control is organized and how the fleet size is managed are decided by the different member states.

The 2002 reform of the CFP expanded the role of stakeholder organizations in the European fishery policy through the establishment of regional advisory councils (RACs). Prior to 2002, stakeholder participation had mainly been channeled through the Advisory Committee on Fisheries and Aquaculture (ACFA) which addressed fisheries and aquaculture issues for all of Europe. Over the years 2004–2007, five regional seas RACs were installed (Baltic Sea, North Sea, Western Waters, South Western Waters, Mediterranean Sea) and two for specific fisheries (pelagic and long-distance fisheries; Commission of the European Communities 2008). According to the Council Decision, the Executive Committees of each of the RACs consisted of two-thirds of fisheries organizations and one-third other organizations affected by the CFP (Council of the European Union 2004). The RACs have become important platforms where stakeholders organize their inputs into the European decision-making framework.[1] Several of the RACs have been proactive in the development of management plans for specific stocks and fisheries. I discuss some examples here.

Fisheries science plays an important role in fishery management in the Northeast Atlantic, both for the decision-making between coastal states and for the decision-making in the EU. The two main mechanisms for fisheries science to deliver their outputs to the management system are via the International Council for the Exploration of the Sea (ICES) and via the Scientific, Technical and Economic Committee on Fisheries (STECF). Whereas the fisheries advice produced by ICES is developed as part of an international council of member states on both sides of the Atlantic Ocean, STECF is the advisory body that is formally part of the EC. ICES hosts most of the expert groups that deal with stock assessment and management advice. Those groups bring together the basic information and apply the methods to assess the size of stocks and the level of exploitation. STECF has a formal role in reviewing the ICES advice on behalf of the EC but also to specifically assess economic and technical implications of the advice. In the context of the topic of this chapter (stakeholder participation in management strategies in Europe), it is important to note that both ICES and STECF have been used as suppliers of scientific advice on proposed management plans.

Management strategies? Management plans? Harvest control rules?

The title of this chapter refers to management strategies, but in the European context it is not straightforward to select a single core concept that could frame this chapter. Many different concepts have been used by different actors to describe elements of pre-agreed management actions for fisheries management, for example harvest control rules (EU-Norway 2002, Patterson 2002), multiannual management plans (EC 2002, NSRAC 2009), multiannual rebuilding plans (EC 2002) and management arrangements (EU–Norway 1997, 1999). Despite attempts to clarify the terminology (e.g. ICES 2006), in practice different concepts are used interchangeably without very clear definitions.

The first application of a "management arrangement" resulted from the EU–Norway negotiations in 1997 (EU–Norway 1997). The year before, ICES had identified an imminent stock collapse for North Sea herring and advised a drastic reduction in the catches. To avoid these rapid fluctuations from year to year, fisheries managers from EU and Norway were looking for a TAC setting mechanism that would "ensure a rational exploitation pattern and provide for stable and high yields" and that relied the state of the stock (biomass) and the exploitation level (fishing mortality) (EU–Norway 1997). The management arrangement had many of the attributes of a harvest control rule and was loosely coupled to the scientific exploration and advice using simulations tools (Patterson et al. 1997). The linkages between science and management in the domain of harvest control rules that started in 1997 have come to be an important feature of the European management system in the 2000s and beyond. With the initiation of the RACs in the mid-2000s, a stronger role for stakeholders in the practice of management plan development and evaluation has started to develop.

The role of science in the evaluation harvest control rules has become more and more institutionalized. The rules for evaluations have been discussed and agreed during many scientific expert groups in ICES (e.g. ICES 2005, 2006, 2007b, 2009b, 2013) and STECF (e.g. STECF 2007, 2008, 2009, 2010). During the process of institutionalization, the prime focus has been on the judgement of whether or not a "plan" would comply with the precautionary approach, interpreted as a less than 5% chance of the stock going below the biomass limit reference point B_{lim}. In most cases, only the harvest control rule embedded within a management plan would be evaluated according to those standards. The other elements of the plan (e.g. effort management, control, stowage) would not be taken into account (Kraak et al. 2010).

Case studies on stakeholder participation

Let us delve a bit further in the stakeholder participation. The four case studies described here demonstrate different types of stakeholder participation in the development of long-term management approaches for specific fisheries in Europe:

- The western horse mackerel management plan developed by the Pelagic RAC
- The deep sea red shrimp management plan in Palamós, Spain
- The efforts to develop a management plan for ICES Area IIIa herring
- The initial efforts to develop a management plan for North Sea *Nephrops*.

The four cases were chosen to represent different institutional settings, different initiating actors and different spatial scope. However, the cases were also chosen because of some personal involvement in the cases that which allowed for relatively quick assessments.

The western horse mackerel plan developed by the Pelagic RAC

The development of a long-term management plan for Western horse mackerel was initiated by the Pelagic RAC in 2006 in combination with scientists from different EU member states (Clarke et al. 2007, Hegland and Wilson 2009). The stock of western horse mackerel had been characterized as a stock with limited information and the advice produced by ICES in 2004 was essentially based on a fixed, low TAC. Since there was no accepted assessment of the stock, there was also no analytical advice supplied by ICES. Fishery stakeholders in the Pelagic RAC were dissatisfied by this situation and initiated a development towards a management plan by engaging horse mackerel scientists from different member states (Coers et al. 2012). By that time, two scientists had already carried out work on potential harvest control rules that were less data dependent than the traditional harvest control rules based on full analytical assessments (Roel and De Oliveira 2007). It was also clear that the EC, not ICES, would take initiative in developing a management plan.

The first meeting between the Pelagic RAC and scientists took place in November 2006, and through a sequence of meetings during the first half of 2007 a management plan was proposed and accepted by the Pelagic RAC in July 2007 (PRAC 2007). Initially the scientists tried to get input from the fisheries stakeholders in the form of a questionnaire, but the response to the questionnaire was limited. Hegland and Wilson, both observers and action researchers on the management plan development, noted that

> the hesitance of the industry might be related to two issues: 1) the nature of the questionnaire as a communication tool and 2) the nature of the questions posed. [. . .] [I]t seems that they were more comfortable discussing freely within their mandate compared to having to consult their members to be able to provide a fixed answer to a question. [. . .] Most importantly, simply, may be that the industry actors are culturally accustomed to meetings, not to questionnaires.
>
> (Hegland and Wilson 2009, p. 84)

After the questionnaires were replaced by regular meetings, the close cooperation between industry representatives and individual scientists resulted in an innovative approach to the management of a data-poor fish stock. The results were submitted to ICES for evaluation and ICES advised that the plan was in conformity with the precautionary approach for a period of 3 years only (ICES

2007a). This demonstrates the double role of science in the development of management plans: a role in developing new approaches for specific fisheries and situations and a role in overall evaluation of the results. Hegland and Wilson hypothesized that the evaluation by ICES may have acted as a way of constraining the fisheries stakeholders from "pushing the limits of the precautionary approach" within the science-industry collaboration because the stakeholders were aware of the external review and evaluation that was planned to take place (Hegland and Wilson 2009).

Participation in this case was effectively limited to fisheries representatives and scientists. The environmental nongovernmental organizations (NGOs) who are members of the RAC did not participate because of budget constraints and possibly also different priorities of operating in the RAC. This is an area of concern because the legitimacy of the output of such participatory modelling in the policy process could be lower if some stakeholder groups are "prevented" from participating (Hegland and Wilson 2009).

An important issue that came up after the development of a proposed management plan by the Pelagic RAC was the delineation what the plan was actually about. In the evaluation process, the scientists and the RAC where working on the hypothesis that the plan would cover the whole area where catches of western horse mackerel are taken, including catches taken by Norwegian vessels. Although the management plan was never formally accepted as European legislation,[2] it was applied as a basis for setting TACs. However, the management rule was applied to the EU catches only, thereby departing from the assumptions underlying the scientific evaluation. This is where the political aspects of fisheries management clash with the modelling reality in fisheries science. And it is an important aspect of how the results of those evaluations are communicated and used as discursive resources in decision-making.

Deep sea red shrimp management plan developed in Palamós, Spain

The red shrimp (*Aristeus antennatus*) is an important and valuable deep sea species that is fished in trawl fisheries in the western and central Mediterranean (Gorelli et al. 2014). The city of Palamós (located 100 km north of Barcelona) is one of the main fishing ports of the Catalan coast for red shrimp, where they make up to 10% of the total landings but 50% of the total income. In recent years, a collaboration between scientists of the Institute of Marine Sciences of Barcelona, Fishermen's Association of Palamós and the Autonomous Government of Catalonia has been established to seek the sustainable exploitation of this resource. The collaboration really started from the grassroots level of fishermen and scientists talking about the knowledge base for red shrimp and potential measures to safeguard the stock in the Palamós area. This led to frequent observer trips on board of the vessels and a mutual interpretation of the results. From this basis a management plan was developed by the fishermen and scientists. The plan includes several measures to reduce fishing effort and preserve the juvenile population. The technical measures established by the plan are a fishery closure during two months in winter, the use of a more selective mesh size and the reduction of the number of trawlers in the fleet. The plan was already implemented by the whole fleet from the beginning of 2012, more than a year before the publication of the Management Plan by the Spanish Government (Boletín Oficial del Estado 2013).

The overall characteristic of this case, in the light of participatory research, is that the focus has been strongly on the development of a common knowledge base between fisheries and science. On the basis of that common knowledge, measures could be discussed and agreed, because they referred back to the common knowledge base. It should be noted that the management plan did not involve a formal model-based evaluation (e.g. MSE) but instead focused more on measures and monitoring. It should also be noted that this case did not involve potentially difficult sharing arrangements as the scale of the case was limited to a relatively small area within the exclusive economic zone (EEZ) of one single EU member state.

Efforts to develop a management plan for western Baltic spring spawning (WBSS) herring

The ambition to develop a management plan for herring in the Skagerrak and Kattegat dates back to 2008, when the EC requested that ICES identify options for multiannual management of pelagic stocks in the Baltic Sea. The western Baltic spring spawning (WBSS) herring was one of the stocks to be considered. This stock is a complex of several herring populations all spawning in the western Baltic Sea. The stock migrates annually towards the Kattegat and Skagerrak and up to the eastern North Sea (Ulrich et al. 2010), where mixing occurs with the North Sea autumn spawning (NSAS) herring. The stock spans different management zones and is shared between EU and Norway. However, the request for the development of a management plan initially came from the EU only (ICES 2009c).

The initial approach to the development of a management plan was handled through a direct interaction between science (ICES) and management (EC), first via a group of experts who applied the "standard" management strategy evaluation (MSE) framework to the WBSS herring stock (ICES 2008: Workshop on Herring Management Plans [WKHMP]). Several stakeholder representatives were present in the meeting of that group and contributed to the discussions. During a follow-up meeting in 2009, the results were further elaborated but without real changes in the results obtained (ICES 2009a: Workshop on Multi-annual Management of Pelagic Fish Stocks in the Baltic [WKMAMPEL]). The management rule suggested by WKHMP and WKMAMPEL was very simple: in the absence of a trigger biomass below which the target fishing mortality would start decreasing, the only biomass reference point that remained was B_{lim}. If the stock were above B_{lim}, fishing mortality should be set at $F = 0.25$; if the stock were below B_{lim}, $F = 0$ (ICES 2008). This type of binary rule (either open or closed) was certainly not preferred by stakeholders, and they initiated a process together with scientists to look for alternative harvest rules (Ulrich et al. 2010)

Three European funded research projects have facilitated the development of an alternative management plan for WBSS herring: GAP1 (Mackinson et al. 2011), JAKFISH (Judgement and Knowledge in Fisheries Management) and GAP2 (www.gap2.eu), in which participatory research has been a key delivery mechanism. In the first stage of the participatory process, the explicit aims were to improve learning about assessment and evaluation techniques and to improve decision quality by joint scenario selection. This resulted in an agreement and commitment on a preferred harvest control rule in 2010 (BSRAC 2010, JAKFISH 2010, PRAC 2010). Scientists and stakeholders jointly selected scenarios and evaluation criteria, which ensured a broad scope and high relevance (salience) of the evaluation process. A key feature of the proposed harvest control rule was that the binary rule suggested by ICES would be replaced by a sliding rule, and a stability clause on TAC changes was added.

Afterwards, it became clear that there were still unresolved political issues around the sharing of the TAC across areas, fleets and coastal states. The WBSS case exemplified the need to discuss all potentially conflicting issues, and especially the politically sensitive ones, early in the process. Mutual understanding of motivations, concerns, wishes and expectations were identified as essential to a successful participatory modelling process (Coers et al. 2012). If taken on appropriately, this could lead to common perspectives on future management.

An important challenge in the participatory stakeholder-science work around WBSS herring has been to get representatives from EU and Norway both around the table. As part of research projects like GAP, this has not yet been resolved. However, a change occurred in 2013 when at the political level between EU and Norway an agreement was reached to jointly develop TAC setting arrangements for this stock in the Skagerrak and Kattegat (EU-Norway 2013). This joint expert group of fisheries managers and scientists did not focus on the management plan as a whole, but instead focused on the political sensitive issues of TAC sharing. The expert group did not include stakeholder representatives.

The WBSS case shows some of the sensitivities of stakeholder participation in the development of management plans that straddle the EEZs of multiple coastal states. Stakeholders and scientists have constructively worked together but only within their own coastal states. Management plans are essentially political compromises on the sharing of access to resources. Because of that political constellation, stakeholder participation may be more difficult to include from the start.

The initial efforts to develop a management plan for North Sea **Nephrops**

The example of the development of a management for North Sea *Nephrops* is quite distinct from the examples given earlier. The North Sea Regional Advisory Council (NSRAC) has been working on the development of management plan for *Nephrops* for several years already (Stange et al. 2015), but here the focus is explicitly on the role of participatory modelling vis-à-vis the development of a management plan.

North Sea *Nephrops* is managed as an EU resource under a single TAC for the whole area. *Nephrops* are known to aggregate in relatively fixed aggregations with little exchange between them. In 2009, the NSRAC seemed to be well underway with the development of a management plan for the stock(s) (NSRAC 2009, Stange et al. 2015). The JAKFISH research project attempted to contribute to that development by offering participatory modelling approaches for *Nephrops* (Pastoors et al. 2012, Röckmann et al. 2012).

During the initiation of the JAKFISH *Nephrops* case, stakeholders and scientists did not really know what they could expect from each other. The objectives and processes were not clearly formulated. In the absence of a joint framing of the research question, scientists assumed that the problem they had in mind would also be relevant for the stakeholder of the NSRAC. Scientists saw the major challenge in the computer simulation programming (Fisheries Library in R, FLR) (Kell et al. 2007) to solve the "age and length" modelling dilemma for this species. Stakeholders in the NSRAC were unaware of the potential contributions to be expected from the scientists but were apparently not very interested in the modelling approach proposed by the scientists.

The *Nephrops* case is an example of lack of communication and mutual understanding between scientists and stakeholders. The JAKFISH scientists experienced a case with significant delays and problems which negatively affected outcomes. From the stakeholders' perspective almost all the fishers believed that it was right to protect the stocks via long-term management plans, and fishers felt they had been listened to (Park 2011). Mutual problem framing in an open, transparent and flexible way is an essential first step in a participatory modelling process. This allows the identification of stakes, problems, possibilities and needs. The actual modelling should only start after the need for modelling has been jointly stated and the goal of modelling has been identified.

In this case, the historical relationship between fisheries and science has left some legacies of mistrust among parties. The ability to overcome these is crucial to the success of mutual problem framing. Mutual learning is often necessary to create a common knowledge basis of what is required and possible. One-way education – also known as the "deficit model" (Irwin and Wynne 1996, Davies 2008, Besley and Nisbet 2013) – where scientists are "teaching" the stakeholders – should be avoided (Röckmann et al. 2012).

The JAKFISH project finished in 2012. At this time the development and implementation of long-term management plans in Europe has been in a stalemate position between the European Parliament and the Council of Ministers for a number of years already. No new management plans were being agreed and existing plans were not being renewed. In that situation, no real progress could be made with the development or implementation of a management plan for North Sea *Nephrops*. So despite the long efforts invested by the NSRAC, at the time of this writing a management plan has yet to be agreed.

Discussion

The fisheries management system in Europe and the Northeast Atlantic is a complex and layered system with regulations at the level of Regional Fisheries Management Organizations (RFMOs), coastal states and EU member states. The EU together with the other coastal states (Norway, Russia, Iceland, Greenland, and the Faroe Islands) frame the overall approach to management. This somewhat complex governance system also creates specific challenges for stakeholder participation in fisheries management and for participatory research involving science and fisheries.

Long-term management plans have mostly been agreed at a relatively high political level (e.g. during coastal state negotiations) and have largely been limited to harvest control rules (HCRs), that is specified rules on TAC setting given the perceived stock abundance and fishing mortality. The introduction of the precautionary approach into the ICES advice in 1998–1999 has strongly promoted the development of HCRs on the one hand, and has also lead to science-dominated evaluation processes using mathematical models. While this has reinforced the links between science and policy, the participation of stakeholders was initially rather limited.

With the introduction of RACs in Europe (2004 and beyond), the stakeholder "voice" became more strongly embedded into the decision-making processes and in the relationships with fisheries science. But how were the stakeholders going to contribute? How were they going to develop their knowledge base to be able to sit at the table and speak the same language? Several examples were discussed earlier on how participatory stakeholder-science processes have attempted to contribute to future decision-making. Some generic lessons from these examples are:

* When management plans are expected to operate in a complex political environment, it is very challenging to prepare new management plans through a participatory science-stakeholder process. On the one hand it is difficult to engage all relevant stakeholders, and on the other hand the political agreement may be dependent on issues other than the content of the management plan itself.
* When applying quantitative evaluation tools like management strategy evaluation (MSE) in a participatory process, it is important to be able to discuss the rationale behind the evaluations and the way uncertainties have

been identified and addressed. The examples of the WBSS herring and western horse mackerel have shown how such discussions have helped to develop a shared understanding.

- In the European context, management plans have become closely linked to the concept of harvest control rules. The main element of management plans is often the rule that determines how the quota will be set depending on the development in a stock or a number of stocks. This has also lead to a strong reliance on quantitative models for ex ante evaluations and on scientific judgement on whether a plan could be expected to comply with the agreed management objectives. It could also lead to an overemphasis on the quantitative evaluation procedures that could sometimes prevent finding common-sense solutions instead of complicated rules (Rochet and Rice 2009, Kraak et al. 2010).

- Developing a management plan without the "burden" of requiring a full ex ante evaluation of the plan creates an environment where it is relatively straightforward for fishing industry and science to collaborate. The example of the red shrimp indicates that in such a situation the debate is focused on shared knowledge generation, monitoring and measures that could help to remedy unwanted situations.

Stakeholder participation in EU fisheries management has changed substantially over the past 10 years. The establishment of RACs has offered an institutionalized role for stakeholder organizations in an advisory capacity to policy-makers (Mackinson et al. 2011). As part of that new role, RACs have been proactive in the development of management plans for fisheries management. This chapter has shown some successes and some drawbacks of those processes. However, it is important to realize that the development of management plans plays out on different levels involving policy, stakeholders and science. The governance challenge of developing appropriate strategies for managing fish resources is obviously much wider than just developing a common knowledge base (Linke et al. 2011). And the roles performed by different actors may change depending on the specific context in which they operate (Röckmann et al. 2015).

The future of long-term management plans and harvest control rules in Europe remains somewhat unclear at the time of this writing. Although the deadlock between the European Parliament and the Council of Ministers seems to have been resolved (EU 2014), concrete implementations of the new generation of management plans have not yet materialized. What the contents of those co-decided management plans will entail cannot therefore be assessed at this stage, although there is a general expectation that they will mostly specify the overall objectives of management instead of the operational choices that need to be made (EU 2014).

At the same time the development of specific harvest control rules agreed between different coastal states seems to continue like before (with new HCRs being proposed for North Sea herring, western Baltic spring spawning herring, northeast Atlantic mackerel and blue whiting, for example). The specific roles of stakeholder participation in those developments are less clearly defined because there are no specific stakeholder organizations that are encompassing the decision-making frameworks of multiple coastal states. In those situations we can observe that the scientific evaluations processes within ICES provide the most direct platforms for stakeholder participation in the development of those HCRs.

Fisheries governance in the Northeast Atlantic takes place against the backdrop of complex governance arrangements. This obviously has implications for the way harvest control rules or long-term management plans are developed and evaluated, and it has major implications for the inclusion of stakeholders in those processes. If we want to deliver management plans that have a strong resonance with all stakeholders involved, I would argue that we should invest in creating environments with effective communication on the objectives for management plans, on the methods to be applied, and on the co-production of the relevant knowledge base.

Acknowledgements

I would like to thank Kari Stange, Clara Ulrich, Lotte Worsøe Clausen, Troels Jacob Hegland and the editors of this book for their helpful comments on earlier versions of this manuscript.

Notes

1 In the most recent reform of the CFP (2013), the Regional Advisory Councils (RACs) have been changed into Advisory Councils (AC). Because in this chapter I am mostly looking at the period prior to 2013, I will keep referring to RACs.
2 The development of management plans came to a halt after the Lisbon Treaty that called for co-decision between the European Parliament and the Council of Ministers. For many years a deadlock existed between parliament and council on the role of long-term management plans and whether or not they should fall under the co-decision procedure. This resulted in a stop on the development or renewal of management plans in the period 2008–2014.

References

Besley, J.C. & Nisbet, M. 2013. "How scientists view the public, the media and the political process." *Public Understanding of Science,* 22(6), 644–659.
BSRAC 2010. BS RAC recommendations for a multi-annual plan for pelagic stocks in the Baltic Sea, Baltic Sea RAC.
Clarke, M., van Balsfoort, G., Coers, A., Campbell, A., Dickey-Collas, M., Egan, A., Ghiglia, M., Harkes, I., Kelly, C. & O'Donoghue, S. 2007. *A new scientific initiative with the Pelagic RAC to develop a management plan for western horse mackerel. ICES C.M. 2007 / O:20.*
Coers, A., Raakjær, J. & Olesen, C. 2012. Stakeholder participation in the management of North East Atlantic pelagic fish stocks: The future role of the Pelagic Regional Advisory Council in a reformed CFP. *Marine Policy,* 36(3), 689–695.

Commission of the European Communities 2008. Communication from the Commission to the Council and the European Parliament: Review of the functioning of the Regional Advisory Councils. COM 2008 364.

Council of the European Communities 1983. Council Regulation (EC) No 170/83 of 25 January 1983 establishing a Community system for the conservation and management of fishery resources. OJ L 24, 27.1.1983.

Council of the European Communities 1992. Council Regulation (EC) No 3760/92 of 20 December 1992 establishing a Community system for fisheries and aquaculture. OJ L 389, 31.12.1992.

Council of the European Union 2002. Council Regulation (EC) No 2371/2002 of 20 December 2002 on the conservation and sustainable exploitation of fisheries resources under the Common Fisheries Policy, Official Journal of the European Communities. OJ L 358, 31.12.2002: 59–80.

Council of the European Union 2004. Council Decision of 19 July 2004 establishing Regional Advisory Councils under the Common Fisheries Policy (2004/585/EC). Council of the European Union. OJ L 256/17, 3.8.2004.

Council of the European Union 2007. Treaty of Lisbon amending the Treaty on European Union and the Treaty establishing the European Community, signed at Lisbon, 13 December 2007. OJ C 306, 17.12.2007

Davies, S.R. 2008. "Constructing communication: Talking to scientists about talking to the public." Science Communication, 29(4), 413–434.

EC 2002. Council Regulation (EC) No 2371/2002 of 20 December 2002 on the conservation and sustainable exploitation of fisheries resources under the Common Fisheries Policy.

EU-Norway 1997. Agreed record of conclusions of fisheries consultations between the European Community and Norway for 1998, Brussels, 2 December 1997. Available at www.regjeringen.no/no/dokumenter/stmeld-nr-47-1997-98-/id191835/?q=&ch=9 (last accessed 19 November 2015).

EU-Norway 1999. Agreed record of conclusions of fisheries consultations between the European Community and Norway for 2000, Brussels, 2 December 1999.

EU-Norway 2002. Report of a two-day meeting of scientists from Norway and the Community on the evaluation of Harvest Control Rules for North Sea cod (Brussels, 18–19 March 2002).

EU-Norway 2013. Agreed record of fisheries consultations between the European Union and Norway on the regulation of fisheries in Skagerrak and Kattegat for 2013, Clonakilty, 18 January 2013.

EU 2014. Task Force on multiannual plans. Final report April 2014., EU. Available at www.europarl.europa.eu/meetdocs/2009_2014/documents/pech/dv/taskfor/taskforce.pdf (last accessed 19 November 2015).

European Parliament and the Council of the European Union 2013. Regulation No 1380/2013 of the European Parliament and of the Council of 11 December 2013 on the Common Fisheries Policy, Official Journal of the European Communities. OJ L 354/22, 28.12.2013.

Gorelli, G., J.B. Company and F. Sarda 2014. "Management strategies for the fishery of the red shrimp Aristeus antennatus in Catalonia (NE Spain)." Marine Stewardship Council Science Series May 2014.

Hegland, T.J. & Wilson, D.C. 2009. "Participatory modelling in eu fisheries management: Western Horse Mackerel and the Pelagic RAC." MAST, 8(1): 75–96.

ICES 2005. Report of the Study Group on Management Strategies (SGMAS), Copenhagen 31 January – 4 February 2005. *ICES C.M. 2005 / ACFM: 09, ref. D, G.*

ICES 2006. Report of the Study Group on Management Strategies (SGMAS), Copenhagen 23–27 January 2006. *ICES C.M. 2006 / ACFM: 15, ref. RMC, LRC.*

ICES 2007a. EC Request on evaluation of management plan for Western horse mackerel. In: *Report of the ICES Advisory Committee, 2007.* Book 9, Section 9.3.2.9.

ICES 2007b. Report of the Study Group on Management Strategies (SGMAS), Copenhagen 22–26 January 2007. *ICES C.M. 2007 / ACFM: 04.*

ICES 2008. Report of the Workshop on Herring Management Plans (WKHMP), 4–8 February 2008. *ICES C.M. 2008 / ACOM:27.*

ICES 2009a. Report of the Workshop on Multi-annual management of Pelagic Fish Stocks in the Baltic (WKMAMPEL), 23–27 February 2009. *ICES C.M. 2009 / ACOM:38.*

ICES 2009b. Report of the ICES-STECF Workshop on Fishery Management Plan Development and Evaluation (WKOMSE), 28–30 January 2009, EEA, Copenhagen, Denmark. *ICES C.M. 2009 / ACOM:27.*

ICES 2009c. Report of the ICES Advisory Committee, 2009, ICES.

ICES 2013. Report of the Workshop on Guidelines for Management Strategy Evaluations (WKGMSE), 23–23 January 2013, Copenhagen. *ICES C.M. 2013 / ACOM:39.*

Irwin, A. & B. Wynne 1996. *Misunderstanding Science?: The Public Reconstruction of Science and Technology*, Cambridge, Cambridge University Press.

JAKFISH 2010. Conclusions from Focus Group meeting on Western Baltic Spring Spawning herring, 9th March 2010, JAKFISH.

Kell, L.T., Mosqueira, I., Grosjean, P., Fromentin, J. M., Garcia, D., Hillary, R., Jardim, E., Mardle, S., Pastoors, M.A., Poos, J.J., Scott, F., and Scott, R. D. 2007. "FLR: an open-source framework for the evaluation and development of management strategies." *ICES Journal of Marine Science*, 64(4), 640–646.

Kraak, S.B.M., Kelly, C.J., Codling, E.A. & Rogan, E. 2010. "On scientists' discomfort in fisheries advisory science: the example of simulation-based fisheries management-strategy evaluations." *Fish and Fisheries*, 11(2), 119–132.

Linke, S., Dreyer, M. & Sellke, P. 2011. "The Regional Advisory Councils: What is their potential to incorporate stakeholder knowledge into fisheries governance?" *Ambio*, 40(2), 133–143.

Mackinson, S., Wilson, D.C., Galiay, P. & Deas, B. 2011. Engaging stakeholders in fisheries and marine research. *Marine Policy*, 35(1), 18–24.

NSRAC 2009. Meeting of the Nephrops Long Term Management Plan Development Group. Tuesday 14th April 2009. Pentland House, Edinburgh.

Park, M. 2011. Long Term Management Plan for North Sea Nephrops – Incorporating Stakeholders into the decision making process. Presentation at the JAKFISH final symposium, Brussels, 9 March 2011.

Pastoors, M.A., Ulrich, C., Wilson, D.C., Röckmann, C., Goldsborough, D., Degnbol, D., Berner, L., Johnson, T., Haapasaari, P., Dreyer, M., Bell, E., Borodzicz, E., Hiis Hauge, K., Howell, K., Mäntyniemi, S., Miller, D.S., Aps, R., Tserpes, G., Kuikka, S. & Casey, J. 2012. JAKFISH D1.5 Final Report.

Patterson, K.R. 2002. *Possible Outcomes of Harvest Control Rules concerning Recovery Stocks.* Working Document to a Two-day meeting of scientists from Norway and the Community on the evaluation of Harvest Control Rules for North Sea cod, Brussels, 18–19 March 2002.

Patterson, K.R., Skagen, D., Pastoors, M.A. & Lassen, H. 1997. Harvest control laws for North Sea herring. Working Document to ACFM 1997.

PRAC 2007. Draft Management Plan for Western Horse Mackerel.

PRAC 2010. A LTM plan for WBSS herring, Pelagic RAC.

Rochet, M.J. & Rice, J.C. 2009. "Simulation-based management strategy evaluation: ignorance disguised as mathematics?" *ICES Journal of Marine Science,* 66(4), 754–762.

Röckmann, C., Ulrich, C., Dreyer, M., Bell, E., Borodzicz, E.P., Haapasaari, P., Hauge, K.H., Howell, D., Mäntyniemi, S., Miller, D.S., Tserpes, G. & Pastoors, M.A. 2012. The added value of participatory modelling in fisheries management – what has been learnt? *Marine Policy,* 36(5), 1072–1085.

Röckmann, C., van Leeuwen, J., Goldsborough, D., Kraan, M. & Piet, G. 2015. The interaction triangle as a tool for understanding stakeholder interactions in marine ecosystem based management. *Marine Policy,* 52, 155–162.

Roel, B.A. & De Oliveira, J.A.A. 2007. Harvest control rules for the Western horse mackerel (Trachurus trachurus) stock given paucity of fishery-independent data. *ICES Journal of Marine Science,* 64(4), 661–670.

Stange, K., van Tatenhove, J. & van Leeuwen, J. 2015. Stakeholder-led knowledge production: Development of a long-term management plan for North Sea Nephrops fisheries. *Science and Public Policy,* 42 (4): 501–513. doi:10.1093/scipol/scu068

STECF 2007. Report of the SGMOS-07-01 Working group on the Evaluation of "Policy statement harvest rules". A. Sinclair, H. Dörner and F. Hölker.

STECF 2008. Report of the Working Group on Harvest Control Rules (SGRST 08-02). M. Haddon and F. Hoelker.

STECF 2009. Report of the STECF Study Group on the Evaluation of Fishery Multi-annual Plans (SGMOS 09-02). E.J. Simmonds.

STECF 2010. Report of the Sub Group on Management Objectives and Strategies (SGMOS 10-06, Part b). Impact assessment of North Sea plaice and sole multi-annual plan. E.J. Simmonds, D. Miller, H. Bartelings and W. Vanhee.

Ulrich, C., Coers, A., Hauge, K.H., Clausen, L.W., Olesen, C., Fisher, L., Johansson, R. & Payne, M.R. 2010. Improving complex governance schemes around Western Baltic Herring, through the development of a Long-Term Management Plan in an iterative process between stakeholders and scientists. *ICES C.M. 2010 / P:07.*

Chapter 21

Stakeholder involvement in New England fisheries

A case study

Jackie Odell and Sarah Lindley Smith

Background

Fisheries management in the United States is designed to offer numerous opportunities for stakeholder participation throughout the management process, and venues for providing input are both formal and informal. However, stakeholder engagement is generally aimed at the policy-making stage, with input into the science of fisheries management often more limited to traditional science roles.

Within the United States, the Magnuson-Stevens Fishery Conservation and Management Act (usually referred to as the Magnuson-Stevens Act, or MSA), established in 1976, is the principal law governing the management of most fisheries extending from 3 nautical miles through the extent of the Exclusive Economic Zone, or EEZ (200 nautical miles). The MSA established the creation of eight regional fishery management Councils throughout the United States and its territories, each of which is responsible for the establishment of fishery management plans within its given jurisdiction.

Each of the regional Councils is intended to consist of fishery stakeholders who encompass the diversity of fisheries expertise within each region. In addition to state fisheries directors and the Regional Administrator of the National Marine Fisheries Service (NMFS), the Council is made up of stakeholders with some interest or involvement in commercial or recreational fisheries. Council members in some instances may also include academic researchers, representatives of fisheries-related industries (e.g. seafood processing), or representatives of environmental nongovernmental organizations (ENGOs). Representation on the Councils is supposed to be balanced and not overly weighted toward one particular fishery or interest group.

The National Marine Fisheries Service (NMFS) is the federal agency responsible for implementing and administering the policies voted on by the Councils. NMFS also must approve any fishery management plan or amendment based on whether the action taken by a Council is consistent with the MSA and other federal regulations. The NMFS staff includes fishery managers and policy and legal experts, in addition to a large number of scientists who develop most of the research and analyses driving management decisions. NMFS predominantly collects fisheries data through surveys (e.g. trawl surveys), fisheries observers, port sampling, and fisheries landings records. These data are all used to develop assessments of managed stocks within each region, conducted by NMFS science centers.

In the Northeast United States, stock assessments conducted by the NMFS Northeast Fishery Science Center follows a thoughtful and often arduous process. To provide a simplified overview, there are typically two processes established for evaluating stocks: benchmark assessments conducted through the Northeast Regional Stock Assessment Review Workshop (SAW/SARC); and operational assessments, which typically follow the benchmark outcomes. The Northeast Fishery Science Center also engages the Transboundary Resource Assessment Committee, which conducts stock assessments for those stocks that straddle the border between the United States and Canada. The Northeast Regional Stock Assessment Review Workshop and the TRAC processes are both formal scientific peer-reviewed processes. The final results of these NMFS-generated assessments are utilized by the Councils' technical teams (e.g. Plan Development Team) and Scientific and Statistical Committees who establish overfishing limits and acceptable biological catch limits for full Council deliberation, consideration, and approval.

The regional Councils represent the most significant avenue for stakeholder engagement in the United States fisheries management process. All meetings are required to be open and advertised to the public, and the public may provide written and/or oral testimony prior to or during Council meetings. During meetings of the New England Fishery Management Council, the Council often provides the public an opportunity to comment on specific motions to be voted on by the full Council as well as during Committee meetings of the Council. These meetings are also broadcast to the public by webinar. In addition, NMFS facilitates a formal written comment period for proposed management before determining whether to approve, reject, or modify an action.

The stock assessment process and the resulting science, on the other hand, are more technical in nature and not as conducive to effective stakeholder participation. Although the meetings are fully open and participatory, it is often very difficult for a less informed individual to participate effectively without extensive knowledge of stock assessment methods and the science that drives them. However, stakeholders may be able to offer insight into catch or abundance trends seen in data sets being utilized by an assessment.

Case study: Framework 50

The New England Fishery Management Council (NEFMC) is responsible for managing fisheries off the coasts of Connecticut, Rhode Island, Massachusetts, New Hampshire, and Maine in the United States. New England has a long history of commercial fishing, dating back at least as far as the first European settlers in the early 1600s, and fishing remains an important part of the region's cultural identity. Many argue that these centuries of fishing efforts have led to the decline of many of the region's important groundfish stocks. Legal mandates to rebuild these fish stocks and end overfishing have held federally managed fisheries to a high standard over the past 20 years. Adding complexity to these legal mandates has been fluctuating scientific assessment reports. In hindsight, revised stock assessment reports have often found that the biomass for a given stock was either overestimated or underestimated. This means the management measures implemented following the prior assessment either did not reduce mortality enough or may have restricted catch more than was necessary. The result has been that fisheries science and management in New England has been exceedingly contentious for decades and, in many cases, stakeholders have become strongly polarized.

In January 2013, the NEFMC voted on a series of highly controversial decisions on Framework 50, an adjustment to the Northeast Multispecies Fishery Management Plan. This fishery management plan regulates 20 stocks of 13 species of groundfish: Atlantic cod, haddock, pollock, yellowtail flounder, winter flounder, witch flounder, windowpane flounder, American plaice, white hake, Atlantic halibut, Acadian redfish, Atlantic wolffish, and ocean pout. The management plan, which was first implemented in 1986, has been amended numerous times through amendments and framework adjustments to establish or revise management measures as necessary to reduce fishing mortality.

Within Framework 50, the Council was tasked with revising the overfishing limits, acceptable biological catches, and annual catch limits for several species for the fishing years 2013–2015. These changes were based on outcomes of benchmark stock assessments for Gulf of Maine and Georges Bank cod as well as operational assessments for other groundfish stocks. For many stocks under consideration, the advice was to significantly reduce catch based on the determination that these stocks were either overfished ($B < B_{THRESHOLD}$), or undergoing overfishing ($F > F_{MSY}$). Under the MSA, the Council and the NMFS Regional Administrator are required to use the best available science to set specifications for a managed fish stock, end overfishing, and in most cases where a stock is determined to be overfished, the Council must establish a plan to rebuild the stock within 10 years.

The decision to establish a new Gulf of Maine cod acceptable biological catch and annual catch limit was particularly contentious during this framework. A previous assessment conducted in 2008 had reported Gulf of Maine (GOM) cod to be rebuilding rapidly and the NEFMC had supported catch levels that were based on these results. However, the stock assessment held 3 years later for GOM cod completely reversed the prior results. The updated assessment reported the stock biomass had been significantly overestimated in 2008 and GOM cod had not made significant progress towards its rebuilding deadline scheduled for 2014. This dramatic shift in perceived stock status came as a shock to the fishing industry, which had believed the efforts to reduce fishing mortality over the course of many years of the rebuilding plan were successfully rebuilding cod. At the same time, those in the environmental community expressed frustration that a more precautionary approach to the positive assessment reports in 2008 had not been taken.

While the peer-reviewed stock assessments conducted by NMFS are deemed to be the best available science, the Gulf of Maine cod assessment was laden with questions and uncertainties, arising partly out of the poor performance of the previous assessment. The result was a lack of confidence from some in the results, particularly among members of the fishing industry, who claimed what they see and experience on the water is not reflected in the assessments that are heavily influenced by NMFS trawl survey indices. Some scientists as well as stakeholders questioned some of the parameters of the Gulf of Maine cod stock assessment, including whether the time series used in the assessment was appropriate, and whether the level of natural mortality (M) and the biological reference points used in the models were adequate.

A year earlier, as a result of the 2011 assessment and prior to the vote for Framework 50, the NEFMC had recommended and NMFS adopted a 1-year interim catch level for GOM cod. Under provisions in the MSA, NMFS has the authority to develop interim catch levels for stocks meeting specific criteria while the NEFMC develops longer-term management measures. GOM cod met the criteria. The interim catch level implemented allowed NMFS to reduce but not end overfishing on GOM cod and allowed additional time for uncertainties in the assessment to be more closely examined. Framework 50 was the groundfish management action that was intended to provide longer-term management measures for GOM cod.

Some industry advocates recommended that a second year of interim measures be implemented for GOM cod consistent with the MSA, which indicates the US Secretary of Commerce can implement interim measures for up to two years, allowing the Council to develop management actions accordingly. However, NMFS, which has explicit discretion over whether to permit interim measures, did not support a second year of interim measures for GOM cod. For their part, the environmental community did not support extending the interim measures for a second year. Implementing the interim catch level through the emergency measures that preceded Framework 50 was seen by many in the environmental community as a highly risky decision, given that the stock assessment pointed to the much larger cuts that were eventually implemented the following year.

When voting on catch limits for GOM cod under Framework 50, NEFMC was limited in its latitude to do something other than what was recommended by the final assessment report and the correlating acceptable catch limits recommended by the Scientific and Statistical Committee. Ultimately, based on the results of the updated assessments, the NEFMC voted under Framework 50 to decrease the acceptable biological catch by 77% for Gulf of Maine cod, by 71% for Gulf of Maine haddock, and by more than 50% each for Georges Bank cod, plaice, witch flounder, and Gulf of Maine yellowtail. The cumulative reduction in GOM cod ABC was closer to 83%, as the ABC had been reduced from 9,012 mt for 2011 to 6,700 mt for 2012, before being reduced again to 1,550 mt. All of these stocks are mainstays for the New England groundfish fishery. The only significant increases were to the Southern New England winter flounder stock, which was increased by more than 150%, and to redfish.

Stakeholder perspectives on Framework 50: introduction

At the time of the Framework 50 debate, numerous stakeholders weighed in through public comment at the NEFMC meetings, through written testimony, and by verbal communications with managers. Many members of the fishing industry were alarmed that the NEFMC would vote to implement such large cuts, and cause significant harm to the fishing industry based on the results of new assessments which many questioned. Some stakeholders were calling for a second year of interim catch levels for GOM cod and possibly GOM haddock, which some argued also met guidelines under MSA. Some questioned whether the drastic reductions in catch limits would actually rebuild fish stocks within the timeline prescribed or whether stocks could even rebuild to such levels because of poor recruitment or a possible regime shift. The environmental community, on the other hand, was of the opinion that these cuts were necessary to rebuild groundfish stocks, many of which are thought to be in poor shape regardless of the uncertainty.

The Northeast Seafood Coalition and Environmental Defense Fund are two stakeholder organizations involved in fisheries management within New England, representing the groundfish industry and the environmental community, respectively. Our organizations work to shape the fisheries management process in New England and its outcomes through direct participation and engagement in the process. As stakeholders, we attend public meetings of the NEFMC and its various Committees to publicly comment on management actions being developed, in some cases offering alternative recommendations, and submit written comments to the NEFMC and NMFS regarding Council decisions. We also engage in the decision-making process in less formal means; our organizations develop relationships with Council members and speak with them ahead of scheduled meetings. We also submit opinion or news pieces in regional newspapers or otherwise use media to publically convey our perspectives.

Stakeholder perspectives on Framework 50: Environmental Defense Fund

Environmental Defense Fund (EDF) aims to promote the long-term sustainability of fisheries, including both ecological and economic sustainability, in

New England and elsewhere. We collaborate with fishermen, fisheries scientists and managers, and other environmental groups toward achieving this goal. We actively engage in the NEFMC process, and also work with the stock assessment process to promote the best available science and improved scientific outcomes for fisheries management.

An important component of EDF's involvement in fisheries science is having scientists on staff who can understand and contribute to fisheries science, including collaborating with other scientists in the region. This scientific capacity gives EDF the knowledge and resources to participate in stock assessments, Science and Statistical Committee meetings, and other scientific processes in a way that is unique for a stakeholder organization. During the stock assessment process for Framework 50, for example, EDF's senior scientist, along with another scientist in the region who also serves on the Science and Statistical Committee, submitted a discussion paper on the spatial ecology of Gulf of Maine cod in the interests of informing the stock assessment debate and broadening the discussion. EDF's scientists and other staff also regularly participate in scientific workshops and other events where our organization can contribute to fisheries science in the region.

EDF's position on the Framework 50 decision was to recommend caution in setting annual catch limits, following the recommendations of the majority of scientists on the Science and Statistical Committee and the Stock Assessment Workshop. In cases where more than one option for the annual catch limit had been provided, EDF recommended following a precautionary approach and selecting the lower number. At the time of this decision, EDF submitted a letter to the NEFMC expressing support for the revised annual catch limits, as well as commenting on other actions the Council was considering under the framework action. It was EDF's opinion that adopting the annual catch limit cuts was necessary to meet the requirements of the MSA. EDF supported the NEFMC's decision to accept the annual catch limits recommended by the Science and Statistical Committee.

On the whole, EDF agreed the stock assessments conducted by NMFS could be improved by taking into account environmental variability and other ecosystem considerations, and recognized that the assessments that produced the Framework 50 specifications contain a great deal of uncertainty. To this end, EDF is working with scientists in the region to develop new tools for both data collection and stock assessment that can improve outcomes of fisheries science. However, the Framework 50 assessments also represented the best available science, and it was EDF's position that they needed to be the basis for this decision. Additionally, many fishermen were not catching their allocations of certain stocks in that year, despite high ex-vessel prices. This led EDF to conclude that existing annual catch limits were in some cases derived from stock assessments that overestimated the stock abundance, strengthening the argument for making the cuts. As painful as they were for the industry, we believe these substantial annual catch limit cuts were necessary to rebuild and promote the long-term sustainability and productivity of fish stocks, and that future prosperity of the industry would be compromised if these cuts were not made. EDF believes that strict adherence to scientific advice gives the best chance of rebuilding groundfish stocks.

From EDF's perspective, the organization's engagement in the process for this particular issue was successful. EDF had participated in, understood, and agreed with the science behind the decisions. The interest of the organization was to support the stock assessment results rather than challenge them, and as those results were accepted and implemented under Framework 50, EDF's participation in this issue was effective. This is not the case for every management decision. While the organization typically has the influence and legitimacy to successfully engage in most management issues, EDF's preferred outcome is not always chosen. At times EDF's positions are in line with segments of the fishing industry, but at other times the environmental community and the fishing industry may have opposing viewpoints on particular issues, frequently including setting catch limits. Management decisions at the Council sometimes pit these two groups of stakeholders against each other, with each lobbying Council members for its preferred outcome. Some NEFMC members, many of whom are commercial fishermen or have close ties to the commercial fishing industry, will often sympathize with and choose to support the fishing industry on contentious issues.

Stakeholder perspectives on Framework 50: Northeast Seafood Coalition

The Northeast Seafood Coalition (NSC) founded in 2002 is a membership organization that represents commercial fishing businesses in the Northeast United States on political and policy issues affecting their interests as participants in the multispecies (groundfish) fishery. NSC's work is geared toward crafting creative and realistic management solutions to complex fishery problems. The organization strives to find solutions which rebuild fish stocks and preserve family-owned fishing businesses that support a diverse groundfish fishing fleet that includes different vessel sizes, gear types and ports.

On behalf of our membership and at the direction of the NSC Board of Directors, NSC directly engages in the NEFMC process with the aid of three people: an executive director, a volunteer policy advisor who is an NSC Board member and fellow fisherman, and a fisheries consultant employed by NSC to provide technical assistance on matters relating to federal legislation. Quite often NSC collaborates with other stakeholders to find common ground, and it often engages scientists on a range of issues that have direct relevance to the groundfish fishery. Furthermore, NSC follows a lengthy internal process to arrive at decisions regarding NSC positions. The NSC Board of Directors, which includes individuals with a diverse set of fisheries expertise, meets regularly to digest management issues and advise NSC leadership on potential solutions. In some instances, NSC will also seek input from its members in order to assist the NSC Board of Directors in its decision-making.

In the Northeast multispecies fishery, stock assessments have proven to be volatile. This is most likely due to the nature of the Gulf of Maine and Georges Bank ecosystems, the unknown and often unpredictable interactions between stocks within the groundfish complex and with other fish stocks, the effects of water temperature fluctuations, and a long list of other factors that contribute to stock recruitment, natural mortality, and growth. It may also be the result of limited data sources used to estimate relative abundance, placing a greater weight upon trawl survey indices. Such complexity has led to large fluctuations in perceived stock status from one assessment to the next, in turn causing extreme fluctuations in allowable catches, further exaggerated by attempts to get some stocks back onto a trajectory to meet their rebuilding target. The result has been a highly unstable business environment for small, family-owned fishing businesses that are the mainstay of the Northeast groundfish fishery.

The updated stock assessment reports released in 2012, which included an operational update conducted for groundfish stocks as well as a more thorough benchmark assessment completed for Gulf of Maine and Georges Bank cod, provide a perfect example of drastic changes that can occur in scientific reports. These findings significantly revised previous reports of stock status for a number of stocks. For Gulf of Maine cod alone, the new benchmark assessment lowered the previously reported spawning stock biomass by approximately 68%. A stock that was previously nearing its rebuilding target in 2008 was now reportedly years away from being considered rebuilt.

The catch reductions that were put in place for Gulf of Maine cod and several flatfish species under Framework 50 in 2013 have devastated the commercial fishing fleet. Despite years of stringent management measures, where the commercial fishery adhered to catch limits prescribed by the science at the time, the reductions in the annual catch limits implemented in the groundfish fishery were so severe that the fishery received a fishery disaster declaration, as prescribed under federal law, by the US Secretary of Commerce on September 12, 2012. This has furthered NSC's belief that fisheries science simply cannot predict nature nor should science be required to do so. Predictions of recruitment, natural mortality, growth, and oceanographic dynamics are uncertain and unpredictable, and current rebuilding mandates under US law simply demand more of the science than science is able to deliver in terms of predicting how nature will respond in future years.

Prior to the development of Framework 50, NSC recommended that NMFS utilize authority set forth in Section 304(e) of the MSA to implement an interim catch level for Gulf of Maine cod rather than stricter catch levels in that year. This authority and action was later recommended by the NEFMC to NMFS, who adopted an interim catch limit for fishing year 2012.

During the Framework 50 discussion, NSC again advised the NEFMC to recommend NMFS implement interim catch limits that reduced overfishing for Gulf of Maine cod and Gulf of Maine haddock, but did not go so far as the cuts being recommended by the Scientific and Statistical Committee. It was NSC's belief that interim catch limits on these stocks would help to mitigate the economic impacts associated with catch reductions on other stocks slated under Framework 50. NMFS did not grant the NEFMC's second request and instead implemented Framework 50 measures with the draconian catch reductions on May 1, 2013.

NSC strives to be a positive and constructive participant in the management decision-making process. Solutions put forth by the NSC have a clear and sound legal basis and, despite our deep concerns about scientific advice often provided, are realistic and based upon the best scientific information available at the time.

However, taking the groundfish management experiences to heart, NSC has become a staunch advocate for putting more tools in the toolbox for fisheries managers. NSC has strongly advocated for changes to rebuilding requirements under the MSA to be adopted. Such changes would place a higher weight upon current scientific reports rather than trying to forecast or predict the future. NSC's recommendations are consistent with a key finding concluded in a report released in 2013 by a Committee established under the National Research Council:

> Rebuilding plans that focus more on meeting selected fishing mortality targets than on exact schedules for attaining biomass targets may be more robust to assessment uncertainties, natural variability and ecosystem considerations, and have lower social and economic impact.[1]
>
> (p. 2)

NSC has also greatly encouraged fisheries guidelines and policies that that would allow for smoothing management responses and multiyear evaluations of overfishing. The NEFMC needs to be able to adapt management responses to fluctuations in stock assessments in a manner that is timely yet least disruptive to the fishery.

From NSC's perspective, the management process is anything but fisherman-friendly. The process is laden with technical, scientific, management, and often political meetings which take time and direct attention that working fishermen are simply not able to provide. NSC leadership therefore works within the process established for fisheries management to represent the interests and well-being of our members – to craft management solutions that work for the good of the fishery – while keeping conservation and the requirements of the law at the forefront.

Conclusions

The Framework 50 example is a case study demonstrating how two organizations engaged in the fisheries management process. Both organizations participate in the fisheries management process primarily through the NEFMC, and largely using direct and focused engagement with fishery managers. Both of our organizations have historically had success affecting fisheries management outcomes at this level. However, what has proven more difficult from the perspective of stakeholder engagement is engaging in the fisheries science that drives these decisions.

Aside from recent experiences surrounding the development and implementation of Framework 50, fisheries science and management have been suffering a crisis of confidence for quite some time in New England. For many fishermen there exists an inherent distrust of government as well as the Council process, which many believe to be less than transparent. Additionally, many fishermen believe the science directing management decisions is poor, the assessments are faulty, and the data is outdated and or skewed. While attention is usually focused on negative stock assessment reports, fishermen often share these same concerns about positive stock assessment reports, which at times have proven during a subsequent assessment to have been overly optimistic.

Many fishermen, along with other stakeholders, are also critical of the NMFS trawl survey and random sampling process, which is heavily weighted in the estimates of relative abundance for groundfish stocks. Fishermen often point to their daily on-the-water experiences as differing greatly from scientific reports. In order to achieve a more comprehensive understanding of stock abundance, alternative data sources should be utilized to supplement estimates of relative abundance offered by trawl survey data.

Concerns about stock assessments extend beyond the fishing industry. Environmental groups and external scientists have also at times criticized the data, methods, and institutions used in fisheries science. Many environmentalists feel that managers too often do not give scientific findings and uncertainties sufficient weight, and therefore make more risk-prone decisions that compromise ecological and economic outcomes. Most stakeholders have a vested interest in improving the fisheries science process. While many of those engaged in fisheries science and management at the NEFMC and at NMFS agree that the process could be improved, institutional inertia sometimes presents a barrier to change current practices.

A successfully managed fishery will require addressing doubts on the part of stakeholders. One way to do this is to make the science of fisheries management more open and transparent to stakeholders. This would require going beyond making scientific meetings open for stakeholder participation.

Instead, this could include holding separate meetings or other facilitated discussions with stakeholders to review data and methods being considered before assessments occur. NMFS has started to organize some focused meetings in recent years and should continue to improve this step in the assessment process.

Additionally, direct engagement by nongovernmental scientists in the assessment process on behalf of stakeholder groups can also provide an important avenue to strengthen the reliability and credibility of stock assessment reports. Due to the level of expertise required, the most effective way for stakeholder groups to engage in the assessment process is by working with nongovernmental experts who can participate directly in scientific discussions and decision-making.

Finally, as mentioned earlier, alternative data sources should be utilized to supplement estimates of relative abundance offered by trawl survey data. Over the years significant resources have gone toward collaborative research initiatives in the groundfish fishery. These projects have been aimed at scientists and fishermen working cooperatively to gather scientific information and data. Both scientists and fishermen have reported an increased appreciation for the other's work after participating in such projects, and there have been calls from many stakeholders to increase the number and scope of collaborative research projects within the region. However, these research projects are most typically directed at collecting biological data (e.g. tagging studies), and less frequently are used as a direct data source in abundance estimates that drive fisheries management. This has led to criticism of and sometimes disillusionment with the cooperative research program on the part of some industry participants. There are exceptions, and NMFS has taken steps to utilize cooperative research data in assessments for some fisheries. For example, NMFS has used video survey data collected through cooperative research to inform abundance estimates for stock assessments used in New England's sea scallop fishery. Further efforts to work with stakeholders to either utilize existing data sources or develop new data sources, particularly in the groundfish fishery, would be one way to improve stakeholder participation in the fisheries science process.

To conclude, stakeholders must navigate a complex, technocratic system in order to be effective participants that provide meaningful input to the process. EDF's and NSC's participation in the Framework 50 management process and our ongoing engagement in the scientific assessment process which drives management decisions requires a significant investment in staff time and capacity that many stakeholders are simply unable to provide. The question remains how to democratize a highly technocratic scientific process for the benefit of improved science and management for the groundfish fishery. We remain open to constructive dialogue with scientists and managers and other stakeholders to improve the scientific management process to meet our common goals of sustainability.

Acknowledgements

The authors would like to thank Tom Nies (New England Fishery Management Council), Jake Kritzer (Environmental Defense Fund), and Matt Mullin (Environmental Defense Fund) for their comments.

Note

1 *Evaluating the Effectiveness of Fish Stock Rebuilding Plans in the United States* (2013). Committee on Evaluating the Effectiveness of Stock Rebuilding Plans of the 2006 Fishery Conservation and Management Reauthorization Act; Ocean Studies Board; Division on Earth and Life Studies; National Research Council.

## Chapter 22

Defining a responsible path forward for simulation-based methods for sustainable fisheries

Dorothy J. Dankel

Introduction

Looking from the outside in, fisheries management can seem riddled with "gloom and doom" stories from the popular media fed by hard-core conservationists. Looking from the inside out, one may infer optimism in global fisheries management given news from one highly successful fishery. The truth is that fisheries are a very mixed bag: there are both successes and failures in each sea, around each bend. Astute students of fisheries will be able to see the forest for the trees and to carefully make statements about global fisheries in a balanced manner. But the question remains how to successfully manage for sustainable fisheries. The truth is, we are only beginning to find this out. What is more well known is what *not* to do, because we can learn from past failures (Daw and Gray, 2005, Rice, 2006, Beddington et al., 2007, Nielsen and Holm, 2007, Simmonds, 2007, Dankel et al., 2008, Pahl-Wostl, 2009).

The theme session "Development and Use of Management Strategy Evaluation (MSE) and Procedures for Robust Fisheries Management" at the World Fisheries Congress in Edinburgh in 2012, co-chaired by Ray Hilborn (University of Washington) and myself, included presentations and discussions by many of the case study authors in Part 2 of this volume. The "MSE Session," as it became known, was in part an attempt to clarify some of the background to the framework as applied in the Southern Hemisphere (Smith et al., 1999a, Butterworth, 2008, Butterworth et al., 2010), some critique of MSEs from application in Europe (Kraak et al., 2010) and issues surrounding a more detailed mathematical critique and responses (Rochet and Rice, 2009, Butterworth et al., 2010, Rochet and Rice, 2010), and to discuss definitions of what makes up an MSE compared to a management procedure approach, based on the terminology given by Rademeyer et al. (2007).

One upshot of the MSE session, as well as preceding and following discussions, was that there are fundamental differences in how the MSE framework is applied by different practitioners, which has attracted criticism of the approach (Rochet and Rice, 2009), but should not be interpreted as flaws in MSE per se (Butterworth et al., 2010). In fact, when done in appropriate ways, MSE ticks many of the boxes of "best practices" in management science for fisheries – including stakeholder involvement – and should be recognized as an important methodology in fisheries science (Rochet and Rice, 2009). But the real-life practice of MSE in different contexts varies in quality and adherence to accepted norms (as it always will), and this warrants ongoing critical examination.

Keeping this in mind, this concluding chapter complements the preceding chapters by examining what I call "meta-issues" in simulation-based methods in fisheries. As discussed by Dankel and Edwards (Chapter 1, this volume) and exemplified by Dickey-Collas (Chapter 19, this volume) and Pastoors (Chapter 20, this volume), effective management science in fisheries needs solid and well-defined frameworks of objective setting, simulation testing and dissemination of scenarios. Management strategy evaluation is designed as such a framework (see Figure 1.2 in Dankel and Edwards, Chapter 1, this volume) for a sketch including the MSE-like "planning phase") but is still a component of the management system (within the broader fishery system, Figure 1.1). A more holistic approach to the problem of management within the fishery system should also include meta-issues of management science for fisheries. These relate to paradigms of science, the science-policy interface, dealing with uncertainty and the need to extend the peer community for increased knowledge quality in decision-making.

Caution: contexts ahead!

If fisheries scientists (be they biologists, economists or social scientists), policy-makers or decision-makers (be they managers, politicians or bureaucrats) are not aware of the whole of the fishery system and that other parts of the system will influence what they are doing (Dankel and Edwards, Chapter 1, this volume), they may be accused of having myopic tunnel vision and become irrelevant to the task at hand: providing knowledge and other support for sustainable fisheries. I argue that part of the problem is that despite all the developments of MSEs over the last two decades, there are still some useful tools the fishery science community could add to their common repertoire that may support effective decision-making in the wider fisheries system. To do this, I examine MSE and other simulation-based techniques in their wider domain, the science-policy interface, and argue in this chapter that some additional tools, that of the extended peer community (a concept introduced in Dankel and Edwards, Chapter 1, this volume, and in more detail here) and knowledge quality assessment, are needed in order to move towards the common goal of sustainable fisheries. But first, I place fisheries science into a new context classification, that of "post-normal science."

Fish are normal, but fisheries science is post-normal

The concept of "normal science" was introduced by Thomas S. Kuhn in his seminal book *The Structure of Scientific Revolutions* (1962). Kuhn describes normal science as "puzzle-solving" within a distinct scientific paradigm. A fisheries example of normal science is a fisheries biologist sampling a fish stock in order to determine an index of relative abundance. The purpose of the relative abundance index is to populate a stock assessment model financed by an independent research funding authority. In this example, the paradigm in which this scientist is working is recognized to be independent from policy or management. It is a purely "normal" science paradigm in the sense that it fulfills the basic requirements of the scientific method: experimental setup, data collection and the publishing of results, without concern for the implications.

This simple example suddenly becomes not so innocent in the following scenario. The scientist, in her annual report to the research funders, is asked to describe how this new model can be used to support management decisions for that stock. The scientist describes the state-of-the-art data collection and statistical processing, which increases the knowledge of the size of the stock compared to status quo methods. Impressed with this achievement, governance authorities commission more work from this scientist and her model to strengthen the scientific base for the annual quota setting for this stock. Stakeholders and managers ask the scientist for a harvest control rule (HCR) to be designed to fulfill their (perhaps opaque) objectives. The scientist can then achieve this aim through simulation (i.e. MSE). However, the difference between understanding how an HCR performs in a stochastic model in silico and how it performs in real life is large. HCR application in vivo yields a learning curve dictated by ecological, logistic and financial limitations. Although adaptive management will allow learning to take place in response (see Dankel and Edwards, Chapter 1, and Edwards, Chapter 2, this volume), large-scale management experiments designed to facilitate learning are rare (Walters, 2007) and expensive trial-and-error experiences predominate (e.g. Eikeset et al., 2011).

At this moment, the scientist (perhaps unknowingly) is forced from a normal science paradigm into another domain. Some may label this domain "science for policy" or science under uncertainty, mission-driven science or another name that describes the application of "science" outside the lab or experimental arena that is context-driven, problem-focused and interdisciplinary. This trajectory has been fully described and applied in the scientific literature under the term "post-normal science" coined by Silvio Funtowicz and Jerome Ravetz (1990) and further framed and applied (Funtowicz and Ravetz, 1993, Funtowicz and Ravetz, 1997, van der Sluijs, 2005, Funtowicz and Ravetz, 2008, Petersen et al., 2010, Hauge, 2011, Dankel et al., 2012, Gluckman, 2014). Table 22.1 points to some defining characteristics of post-normal science.

Funtowicz and Ravetz (1994) outline elements of a post-normal science as the following: the appropriate management of uncertainty and quality, plurality of commitments and perspectives, intellectual structures and social structures. For post-normal science to work, scientists should use fundamental understandings of their discipline to guide them in their cross-disciplinary dialogues.

Table 22.1 Symptoms of post-normal science.

- Research is issue driven
- External pressure is exercised on the research groups involved, because associated policy decisions are urgent, decision stakes are high and values are in dispute
- No single paradigm dominates
- Complications within scientific venture are confronted and not skirted
- Research is focused on a whole web of extensive problems
- Research concerns many large (partly irreducible) uncertainties
- Conflicting certainties coexist
- Scientists are confronted with incomplete control and unpredictability of the analyzed system
- A multitude of legitimate scientific and ethical perspectives coexist
- Basic research is transplanted into strategic research, with a view to long-term application
- Established boundaries between the political and the scientific arena are subject to continual renegotiation

Source: From Nolin (1995) as modified by van der Sluijs (1997).

One common misconception of post-normal science is that it throws the baby (normal science) out with the bathwater (murky waters of hotly contested issues). This is not the case. "Normal" science (intersubjective, testable, significant facts and perceptions) is absolutely fundamental for the foundations of knowledge. And it is precisely this knowledge foundation that science-based advice for management draws upon. However, lifting "normal" science up to policy requires a (sometimes surprising) revelation of the intricacies of the science-policy interface, causing a reevaluation of the science as fit-for-purpose. Indeed, Garcia and Charles (2007) outline several arguments on why fisheries science should take a "soft watch" approach instead of a hard "clockworks" approach, the latter being anchored in the false belief that stock forecasts are reliable, and which is the basis of the widely used quota management regulatory control. Garcia and Charles argue that analytical processes should be complemented with participatory processes, including problem-oriented multidisciplinary knowledge production (also known as Mode 2 science; Nowotny et al., 2001) and post-normal science approaches (Funtowicz and Ravetz, 1995). Furthermore, writing from a European perspective, Hauge (2011) outlines how frameworks of management advice in ICES (International Council for the Exploration of the Sea, Europe's marine science and advice body) are built on a narrow framing of risks, management units and uncertainties. These framings are based on the philosophical foundation that there is a strict division between management and science. Instead, multidisciplinary and cross-disciplinary settings are needed, with explicit stakeholder involvement, to discuss problem framing, value-ladenness, and the management of uncertainty (Funtowicz and Ravetz, 1990, Wynne, 1992, Funtowicz and Strand, 2007, Hauge, 2011, Dankel et al., 2012).

Meeting the needs for the science-policy interface in fisheries: some ingredients of the post-normal science approach

Extending the peer community

The science-policy interface is where scientists, their models and managers meet. This meeting has its own contexts: social, economic, political and scientific. In the past decades, literature examining cases on the science-policy interface points to concepts of participatory processes (van Asselt and Rijkens-Klomp, 2002, Kasemir et al., 2003), stakeholder engagement (Hartley and Robertson, 2006, Nutters and Pinto da Silva, 2012), integration and interdisciplinary knowledge quality assessment (Gunnarsdottir et al., 2015) and the extended peer community (Funtowicz and Ravetz, 1996, Kloprogge and van der Sluijs, 2006). These concepts help guide nontrivial human processes related to the development and dissemination of simulation exercises in management science for fisheries.

There is undoubtedly an increasing trend of techno-science in fisheries, where scientists derive more complex models as the amount of data and the horizon of technology ever expand. The technicalities and complexities in fisheries models pose serious threats to stakeholder inclusion and participation. Most MSE and manangement procedure evaluations are set within a stakeholder process that involves many opinions and much expertise besides just modellers (Cox and Kronlund, Chapter 5, Dichmont et al., Chapter 10, and de Moor and Butterworth, Chapter 11, this volume, Jones et al., 2015). This participatory setting is an example of an extended peer community, which is part of the post-normal toolbox. As briefly discussed in Dankel and Edwards (Chapter 1, this volume), an extended peer community may be considered as both a group of people and a process. The first ingredients are the people: scientists from various relevant disciplines (environmental science, fisheries economics, sociology, etc.) in addition to nonscientists (e.g. fishermen, local businessmen, community members). This new community's action is a process: collecting more relevant information about the fishery system, applying alternative problem-framings, discussing the communication of risk and so forth

Funtowicz and Ravetz (1997) and Kloprogge and van der Sluijs (2006) describe a scientific setting where the extended peer community is needed:

> When problems lack neat solutions, when environmental and ethical aspects of the issues are prominent, when the phenomena themselves are ambiguous, and when all research techniques are open to methodological

criticism, then the debates on quality are not enhanced by the exclusion of all but the specialist researchers and official experts. The extension of the peer community is not merely an ethical or political act; it can positively enrich the process of scientific investigation.

(Funtowicz and Ravetz, 1997, p. 174)

But in order to meet the participatory requirements of science for policy, skills and exercises like trust building, mutual learning, transparency, setup and design of inclusive processes and meetings need to be applied. These skills may not be appreciated in fisheries science, and students may be short-changed by not learning and appreciating these skills early on in their career. My advice is the following: if your goal is to produce something meaningful for policy-makers and/or stakeholders, include them early and often. Listen to their concerns and objectives and aid them in formulating research questions that are testable (i.e. What is the null hypothesis? What is the time horizon?). Revise these research questions with them. Report on preliminary results often. All these tips can help early career scientists form credibility, trust and legitimacy with their audience (see WGMARS [Working Group on Maritime Systems of the International Council for the Exploration of the Sea] section 3.8 (ICES, 2012)).

Assessing the knowledge quality

Different framings of uncertainty can be applied by decision-makers to complex models in fisheries science, which has an affect on the perception of the quality of knowledge used for policy making (Wynne, 1992). Some may take a "deficit view" of uncertainty related to complex systems: the idea that more research and more data are needed in order to arrive at the correct and final scientific "answer." MSE and similar approaches lead you in the opposite direction: the issue is not about precision in the science, but in finding adequate performance to meet management objectives (although more and better data could help prediction). Fisheries scientists and managers are not alone in this boat and can learn a lot from the discourse and literature in climate science; scientists at the International Panel on Climate Change (IPCC) have not only won a Nobel Peace Prize for their high-impact science, but also criticism on how they have framed and communicated uncertainty in IPCC reports (Curry, 2011).

In post-normal science, the traditional search for robust scientific findings, ideally based on scientific consensus, is replaced by a search for robust policy strategies, which are useful regardless of which of the diverging scientific interpretations of the knowledge is correct (van der Sluijs et al., 2010, Dankel et al., 2012). "Advice under uncertainty in the marine system" provides an introduction to post-normal science, the benefits of reframing fisheries science as a post-normal science. A more detailed application of post-normal science to specific cases in Europe is given by Hauge (2011). Here, Hauge points to "a sustained practice of neglecting uncertainty" (p. 180) in the ICES framework for fisheries quota setting. A tool developed to specifically address uncertainties arising from science at the policy interface is called NUSAP (Numeral Unit Spread Assessment Pedigree; Funtowicz and Ravetz, 1990, Risbey et al., 2005). NUSAP is an analytical and notational system which extends the classic notational system for quantitative scientific information (usually provided as a number, a unit and a standard deviation) with two additional qualifiers: expert judgement of the reliability (the assessment) and a multicriteria characterization reflecting the origin and status of the information (the pedigree). The classical notational system using a number of significant digits does not reveal the distinction between nearly perfect information (such as the speed of light in a given medium) and highly imperfect information (Ravetz, 2015), such as the size of a fish stock. Assessment scientists struggle to quantify precision when preparing quota advice for fish stocks, and often break the commonsense rule of significant digits (Hauge, 2011) with what can be called hyperprecision or pseudoprecision (Ravetz, 2015). For example, for a given data set scientists can estimate natural mortality, M, with an estimate of the precision as well. But estimating the uncertainty associated with depletion of the stock is much more difficult because it depends on the structural assumptions of that particular model. Communicating this type of uncertainty to the policy-maker is very difficult using the classic notational system. Drawing further on this example, if the natural mortality is assumed to be 0.2 (the common assumption for many stocks where lack of informative data prevents direct estimation) and the quota derived from the model under this assumption has five significant digits, then the classical notational system has been misused. In this case it will have failed to communicate the underlying uncertainty in the quota advice.

So, how to distinguish between what can (and should) be quantified versus what cannot be in complex fisheries systems? How to better assess, represent and communicate imperfect scientific information? One solution is two additional qualifiers to supplement the number, the unit, and the standard deviation: assessment and pedigree, which attempt to remedy the problem. The pedigree analysis is a qualitative structural process to clarify the knowledge base on which scientists and stakeholders frame their perceptions of a problem, by appraising the information underpinning the numbers and theories that form the basis of scientific advice, often model derived. Mapping components of the knowledge base in a diagnostic diagram therefore reveals the weakest elements of an assessment, helps in the setting of priorities for improvement, and assists in the choice of adequate policy strategies to cope with uncertainty. Furthermore, the uncertainty matrix (Walker et al., 2003, and the improved version by Knol et al., 2009) allows to typify and characterize the various sources of uncertainty for a given case following an application of quantitative methodology. The uncertainty matrix has been used in other places, for example by the Netherlands Environmental Assessment Agency (van der Sluijs et al., 2008b, Wardekker et al., 2008, Petersen et al., 2010), introduced in a fisheries context

(Dankel et al., 2012) and applied to a European fisheries case (Ulrich et al., 2010). However relevant, a mainstream application of the NUSAP system to management science for fisheries is currently lacking, and would add a new dimension to "fitness for purpose" within MSEs.

One can hypothesize to this last observation: why wouldn't scientists and managers want uncertainties, in their different quantiative and qualitiative forms, to come to the center of the table? The answer to this naïve but pertinent question undoubtedly varies, and here I list some of the possibilities: (1) genuine ignorance of the uncertainties, (2) a severe dysfunction of the scientific system (Pereira and Funtowicz, 2015, preface, Ravetz, 2015) or (3) convenient ignorance of these uncertainties. Hauge (2011) points out the importance of problem framing in fisheries science, and examples from ICES assessment working groups where uncertainties have not been presented to clients because of perceived irrevelence to management decisions. Charles (1995) reviews the infamous Canadian cod fishery collapse in the early 1990s in light of the lack of public and community buy-in to the management science processes, and the role of the burden of proof for balancing uncertainty and risks in the fishery. Charles concludes that sustainable fishing needs a change of human attitudes together with appropriate management, since attitudes drive decision-making.

In a normal science paradigm, uncertainties can, and should, be reduced by whatever means are achieveable. But due to the inherent complex issues of the fishery system, "the only certainty may be just more uncertainty" (Garcia and Charles, 2007, p. 585). This is precisely why MSE and similar tools are so precious now: they allow you to investigate the consequences of the almost certain assumption that you are wrong, and on the flip side, help you figure out how much information is *enough* to manage the fishery.

It can be quite convenient to not be explicit about uncertainties. In fact it can be quite empowering to be perceived as having *the answer*. But modern society is also coming to terms with the need to act on mitigation policies despite inherent uncertainties in the knowledge base, which is partly why Al Gore also received the Nobel Peace Prize for his work in climate science communication in his documentary movie *An Inconvenient Truth* (2006). The consensus is that the climate, as well as the marine ecosystem, is changing faster because of negative anthropogenic effects. But there remains a lot of consensus-building to be done as to how we should mitigate these affects. Dialogue is a good place to start.

Steps to meet the human system in fisheries

Dialogue has proved to be an important part of trust building between scientists and stakeholders for different reasons. First, through active dialogue, communication of uncertainty is improved; scientists are able to inform stakeholders of the many uncertainties involved with annual stock assessment that use scientific survey and fisheries dependent data. Second, stakeholders can communicate

their experiences at sea from which scientists may derive important informa
tion. Dialogue between stakeholders and scientists is crucial in the development of management regulations (Smith, 1993, FAO, 1995, Smith et al., 1999b,
Charles, 2001, Caddy and Seijo, 2005, Cunningham and Bostock, 2005, ICES,
2006, Wilson and Pascoe, 2006, Commission and Affairs, 2007, ICES, 2007,
Schwach et al., 2007, ICES, 2008b, Brady and Waldo, 2009, Mäntyniemi et al.,
2009, Ulrich et al., 2010, Mackinson et al., 2011) for a number of reasons,
including elucidating objectives to build management strategies for simulation testing, communicate the results of these evaluations, problem framing
and including other types of relevant knowledge (i.e. tacit knowledge) to the
knowledge base (Hauge, 2011).

The human subsystem in the fishery system (see Figure 1.1 in Dankel and
Edwards, Chapter 1, this volume) has traditionally been the system least connected to fisheries science and, ironically, it is the human system that fisheries
management strives to control. Stakeholder inclusion and dialogue in management science for fisheries has been commonplace and part of management
strategy evaluation and management procedure protocols for a long time. However, in Europe stakeholder processes did not concurrently gain prominent recognition in fisheries management.

In many cases in Europe, for example, fisheries management is a top-down
bureaucratic exercise with centralized control (Gray and Hatchard, 2003, Prince,
2003, Daw and Gray, 2005). Top-down control has several convenient advantages (such as clear lines of control and the ability to enforce tough decisions)
but has a tendency of disconnecting the human system in fisheries management
by not explicitly including the human dimension and its user groups. The lack
of stakeholder integration in management decisions was recognized as a major
impediment to sustainable fisheries in the European Union and led to a reform
of the European Common Fisheries Policy in 2002. One of the results of this
reform was the creation of seven regional advisory councils (RACs) intended
to be a forum for stakeholders to provide consensus advice to the European
Union regarding fisheries management. The birth of the RACs was an attempt
to include stakeholder voice in top-down management. At first, however, the
EU was logistically unable to handle this newfound consensus voice to the
detriment of some of the key RACs (Gray and Hatchard, 2003).

Reflections on the reform of the EU Common Fisheries Policy (Sissenwine and Symes, 2007) give promise to closer stakeholder collaboration in the
realm of management strategy development (or multiannual plans; see European Commission, 2013) in Europe, although this rhetoric of integration and
dialogue in the previous reform has been disputed (Gray and Hatchard, 2003).
The scientific advisory council organization of the International Council for
the Exploration of the Sea (ICES) is one way the European scientific community is striving to give advice in a post-normal world (ICES, 2008b) and there
are examples of how ICES is starting to put these ideas into practice (Aps et al.,
2007, ICES, 2007, 2008a, 2008b, Ulrich et al., 2010). In management strategy

development cases, it is scientists, or scientific organizations like ICES, who use their status as an objective entity of the concerned resource to initiate a dialogue regarding management issues like objectives and appropriate strategies (Roel and De Oliveira, 2007).

In this book, there are a lot of case studies in the chapters that reflect an alternative approach to incorporation of the human system in fisheries, that of "properly" conducted MSE (Butterworth et al., 2010). Indeed stakeholder participation has been an important quality of MSEs in South Africa (Butterworth, 2008, de Moor and Butterworth, Chapter 11, this volume), Australia (Smith et al., 1999b, Dichmont et al., Chapter 10, this volume), New Zealand (Breen et al., Chapter 6, this volume) and North America (Cox and Kronlund, 2008, Cox and Kronlund, Chapter 5, and Hicks et al., Chapter 4, this volume). Stakeholder inclusion and integration varies in form and function from case to case, but the fact that MSEs are specifically designed to incorporate stakeholder involvement (Smith et al., 1999a) shows MSE's readiness for credibility, legitimacy and saliency among stakeholders.

After the stakeholder landscape is recognized and objectives for each interest group are collaboratively discussed and identified, it is possible to proceed with the next building block of successful fisheries management: integration of objectives. But stakeholder objectives often seem conflicting. According to American social anthropologist and management philosopher Mary Parker Follett (1868–1933), there exist three ways to resolve conflict: domination, compromise and integration. Domination is by far the most common method due to its familiarity and ease. In a compromise, some stakeholders have to sacrifice some of their desires to achieve consensus. Integration of a common solution starts with each individual, or stakeholder, reevaluating their desires. An integrated solution is conceived if this reevaluation produces a reasonable homogeneity of objectives that consensus may arise.

> Integration involves invention, and the clever thing is to recognize this and not let one's thinking stay within the boundaries of two alternatives which are mutually exclusive. ... Compromise does not create, it deals with what already exists; integration creates something new.
>
> (Follett, 1955, pp. 33–34)

Since integration produces a new, collaborative view as a solution to a conflict, the conflict is settled and not likely to come about in the future. Compromise, on the other hand, is only a temporary give-and-take scenario that is likely to be faced again. Integration is therefore the preferred method to resolve conflict (Follett, 1955). Follett (1955) summarized steps towards integration including uncovering the real conflict and taking all stakeholder groups' objectives and breaking them up into their constituent parts.

Integration can sometimes be achived by recognizing that some objectives may not be as conflicting as previously thought (Hilborn, 2007). Ecosystem preservation and fishers' profit are an example; in most situations, both objectives are fulfilled at a fishing level lower than that which gives maximum sustainable yield (MSY). In this case, fishers concerned about maximizing profit and stakeholders interested in conserving the resource are likely to come to a consensus that total fishing levels should be below MSY.

Some practitioners in fisheries management are sure to dismiss Follett's description of an integrated solution as wishful thinking. But do not let me make the impression that an integrated solution is easy. Before the integration steps are put into place, it is important to identify obstacles and devise strategies to avoid possible pitfalls. Follett (1955) outlines some typical obstacles that can block successful integration. The first one is a requirement of a "higher order of intelligence"; recognizing the need for integration requires a superior level of consciousness since domination and compromise are much easier and more common alternatives. Another obstacle, and in my opinion the most relevant, is the lack of training in integration and the need for courses in the art of cooperative thinking (Follett, 1955). It is possible that many fisheries crises can be avoided if scientists and managers recognize the need for cooperative thinking techniques and put them to use.[1] Domination in fisheries tends to favor the strongest stakeholders, which are usually the ones with highest economic interests (and most likely the largest and most risky capital investments). Cooperative thinking towards an integrated solution is also a natural component of democracy contrary to domination. In a democratic society, all stakeholders should be represented in appropriate ways when management objectives are discussed and drafted.

In summary, although management strategy and management plan evaluation frameworks are a methodological step forward towards the goal of sustainable fisheries, these processes can further be improved by integrating different forms of knowledge and deliberation from an extended peer community and integrating a knowledge quality assessment framework. As such, management science for fisheries would begin to develop best practices (Saltelli and Funtowicz, 2014) from different fields and cases where this has previously been applied (van der Sluijs, 2002, van der Sluijs et al., 2003, Walker et al., 2003, van der Sluijs, 2005, van der Sluijs et al., 2005, van der Sluijs, 2007, van der Sluijs et al., 2008, Wardekker et al., 2008, Knol et al., 2009,, Petersen et al., 2010, Ulrich et al., 2010, van der Sluijs et al., 2010, Maxim and van der Sluijs, 2011).

The path forward for management science for fisheries

This volume describes the state-of-the-art in simulation-based techniques for fisheries management science and attempts to place it in the context of the wider fisheries system. Some of the experiences and insight in the previous chapters can help the student and practitioner of fisheries management design dialogues, methods, models and processes towards the common goal of sustainable fisheries. I challenge you to consider this question: "How could a post-normal paradigm and related methodologies help in my work?" I believe reflexive questions like this one can inspire and guide you to develop suites and processes of best practice in model building and simulation exercises tailored for your specific fishery system contexts, data and objectives. This will provide fodder for the new narratives of management science and sustainability for fisheries for future volumes.

In review, what is the goal of management science for fisheries? To routinely and reliably collect high quality data to feed management strategy evaluations? To deliberate with diverse stakeholders about objectives, outcomes and regulations? To engineer complex and predictive models? To design robust processes that include a capable, extended peer community? Relating back to Figures 1.1 and 1.2 in Chapter 1 of this volume, we argue that all of these are visions and components of the same goal: sustainable fisheries. The path to an integrated solution to this grand challenge will be inclusive, will be technical, will be iterative and will not be neat, and actually a bit messy. But this is something to aspire to.

Acknowledgements

I thank Anthony D.M. Smith (CSIRO, Hobart, Australia), Silvio O. Funtowicz and Jeroen P. van der Sluijs (Centre for the Study of the Sciences and the Humanities, University of Bergen), and Charles T. T. Edwards (NIWA Ltd., Wellington, New Zealand) for critical reviews and insights that improved this chapter. I also recognize two influential interdisciplinary milieus full of inspiring colleagues that have inspired my work: the Working Group on Maritime Systems of the International Council for the Exploration of the Sea (WGMARS) and the Centre for the Study of the Sciences and the Humanities, University of Bergen.

Note

1 Anthony Charles (Foreword, this volume) discusses his experiences with the Canadian cod collapse and the need for humility in fisheries science.

References

Aps, R., Kell, L.T., Lassen, H. & Liiv, I. 2007. Negotiation framework for Baltic fisheries management: striking the balance of interest. *ICES Journal of Marine Science,* 64, 858–861.

Beddington, J.R., Agnew, D.J. & Clark, C.W. 2007. Current problems in the management of Marine fisheries. *Science,* 316, 1713–1716.

Brady, M. & Waldo, S. 2009. Fixing problems in fisheries—integrating ITQs, CBM and MPAs in management. *Marine Policy,* 33, 258–263.

Butterworth, D.S. 2008. *Some Lessons from Implementing Management Procedures,* Tokyo, TERRAPUB.

Butterworth, D.S., Bentley, N., De Oliveira, J.A.A., Donovan, G.P., Kell, L.T., Parma, A., Punt, A.E., Sainsbury, K., Smith, A.D.M. & Stokes, K. 2010. Purported flaws in management strategy evaluation: basic problems or misinterpretations? *ICES Journal of Marine Science,* 67, 567–574.

Caddy, J.F. & Seijo, J.C. 2005 This is more difficult than we thought! The responsibility of scientists, managers and stakeholders to mitigate the unsustainability of marine fisheries. *Philosophical Transactions of the Royal Society B,* 360, 59–75.

Charles, A.T. 1995. The Atlantic Canadian groundfishery: Roots of a collapse. *Dalhousie Law Journal,* 18, 65–83.

Charles, A.T. 2001. *Sustainable Fishery Systems,* Blackwell Science, Oxford.

Commission, E. & Affairs, D.G.F.F.A.M. 2007. Resources and relations with stakeholders, Interinstitutional relations and dialogue with the sector, programming and evaluation. Fifth RACs Coordination Meeting, Brussels, 11 December 2007.

Cox, S.P. & Kronlund, A.R. 2008. Practical stakeholder-driven harvest policies for ground-fish fisheries in British Columbia, Canada. *Fisheries Research,* 94, 224–237.

Cunningham, S. & Bostock, T. (eds.) 2005. *Successful Fisheries Management: Issues, Case Studies and Perspectives.* Oxford, Eburon Academic Publishers.

Curry, J. 2011. Reasoning about climate uncertainty. *Climatic Change,* 108, 723–732.

Dankel, D.J., Aps, R., Padda, G., Röckmann, C., van der Sluijs, J.P., Wilson, D.C. & Degnbol, P. 2012. Advice under uncertainty in the marine system. *ICES Journal of Marine Science,* 69, 3–7.

Dankel, D.J., Skagen, D.W. & Ulltang, Ø. 2008. Fisheries management in practice: A review of 13 commercially-important fish stocks. *Reviews in Fish Biology and Fisheries,* 18, 201–233.

Daw, T. & Gray, T. 2005. Fisheries science and sustainability in international policy: A study of failure in the European Union's Common Fisheries Policy. *Marine Policy,* 29, 189–197.

Eikeset, A.M., Richter, A.P., Diekert, F.K., Dankel, D.J. & Stenseth, N.C. 2011. Unintended consequences sneak in the back door: making wise use of regulations in fisheries management. *In:* Cunningham, S. & Bostock, T. (ed.) *Ecosystem Based Management for Marine Fisheries,* pp. 183–217. Cambridge: Cambridge University Press.

European Commission 2013. Regulation (EU) No 1380/2013 of the European Parliament and of the Council of 11 December 2013, on the Common Fisheries Policy, amending Council Regulations (EC) No 1954/2003 and (EC), No 1224/2009 and repealing Council Regulations (EC) No 2371/2002 and (EC) No 639/2004 and Council Decision 2004/585/EC. *In:* Parliment, E. (ed.). Brussels, Belgium: Official Journal of the European Union.

FAO 1995. *Code of Conduct for Responsible Fisheries. Fisheries and Aquaculture Department of the United Nations,* Rome, FAO.

Follett, M.P. 1955. *Dynamic Administration: the collected papers of Mary Parker Follett,* New York, Harper & Row Publishers.

Funtowicz, S. & Ravetz, J. 1997. Environmental problems, post-normal science, and extended peer communities. *Etudes et Recherches sur les Systèmes Agraires et le Développement,* 30, 169–175.

Funtowicz, S. & Strand, R. 2007. *Models of Science and Policy,* Trondheim, Tapir Academic Press.

Funtowicz, S.O. & Ravetz, J. Risk management, uncertainty and post-normal science. *In:* Haaland, H.R.A.H. (ed.) *Proceedings of the First International Symposium on Sustainable Fish Farming,* 28–31 August 1995, Oslo. Rotterdam, A.A. Balkema, 261–270.

Funtowicz, S.O. & Ravetz, J., Eds. 1996. Risk Management, post-normal science, and extended peer communities. *Accident and Design, Contemporary Debates in Risk Management,* UCL Press, London, pp. 172–182.

Funtowicz, S.O. & Ravetz, J.R. 1990. *Uncertainty and Quality in Science for Policy*, Dordrecht, Kluwer.

Funtowicz, S.O. & Ravetz, J.R. 1993. Science for the post-normal age. *Futures,* 25, 735–755.

Funtowicz, S.O. & Ravetz, J.R. 1994. The Worth of a Songbird – Ecological Economics as a Post-normal Science. *Ecological Economics,* 10, 197–207.

Funtowicz, S.O. & Ravetz, J.R. 2008. Post-normal science. *In:* Cleveland, C.J. (ed.) *Encyclopedia of Earth.* Washington, DC: Environmental Information Coalition, National Council for Science and the Environment.

Garcia, S.M. & Charles, A.T. 2007. Fishery Systems and Linkages: From Clockwork to Soft Watches. *ICES Journal of Marine Science* 57, 580–587.

Gluckman, P. 2014. The art of science advice to government. *Nature,* 507, 163–165.

Gray, T. & Hatchard, J. 2003. The 2002 reform of the Common Fisheries Policy's system of governance – rhetoric or reality? *Marine Policy* 27, 545–554.

Gunnarsdottir, K., Dijk, N., Wynne, B., Gutwirth, S. & Hildebrandt, M. 2015. *Observations and Reflexivity: Responsibilising Interdisciplinarity and Integration.* Lancaster, Lancaster University.

Hartley, T. & Robertson, R.A. 2006. Stakeholder Engagement, Cooperative Fisheries Research and Democratic Science: The Case of the Northeast Consortium. *Human Ecology Review,* 13, 161–171.

Hauge, K.H. 2011. Uncertainty and hyper-precision in fisheries science and policy. *Futures,* 43, 173–181.

Hilborn, R. 2007. Defining success in fisheries and conflicts in objectives. *Marine Policy* 31, 153–158.

ICES 2006. Report of the Study Group on Management Strategies (SGMAS).

ICES 2007. Report of the Study Group on Management Strategies (SGMAS). *ICES SGMAS Report 2007.*

ICES 2008a. Report of the Study Group on Management Strategies (SGMAS).

ICES 2008b. Report of the Working Group on Fishery Systems (WGFS). *In:* ICES (ed.). Copenhagen, DK: ICES.

ICES 2012. Report of the Working Group on Maritime Systems (WGMARS), 31 October – 3 November 2011, Bergen, Norway. *ICES CM 2011/SSGSUE:13.*

Jones, M.L., Catalano, M.J., Peterson, L.K. & Berger, A.M. (eds.) 2015. *Stakeholder-centered Development of a Harvest Control Rule for Lake Erie Walleye,* Oxford: Routledge (Earthscan).

Kasemir, B., Jäger, J. & Jaeger, C.C. 2003. *Citizen Participation in Sustainability Assessments,* Cambridge, Cambridge University Press.

Kloprogge, P. & van der Sluijs, J.P. 2006. The inclusion of stakeholder knowledge and perspectives in integrated assessment of climate change. *Climatic Change,* 75, 359–389.

Knol, A., Petersen, A., van der Sluijs, J. & Lebret, E. 2009. Dealing with uncertainties in environmental burden of disease assessment. *Environmental Health,* 8, 21.

Kraak, S.B.M., Kelly, C.J., Codling, E.A. & Rogan, E. 2010. On scientists' discomfort in fisheries advisory science: the example of simulation-based fisheries management-strategy evaluations. *Fish and Fisheries,* 11, 119–132.

Kuhn, T.S. 1962. *The Structure of Scientific Revolutions,* Chicago, University of Chicago Press.

Mackinson, S., Wilson, D.C., Galiay, P. & Deas, B. 2011. Engaging stakeholders in fisheries and marine research. *Marine Policy,* 35, 18–24.

Mäntyniemi, S., Haapasaari, P. & Kuikka, S. 2009. Incorporating stakeholders' knowledge to stock assessment: how? *In:* ICES (ed.) *ICES CM 2009/O:12.* Copenhagen, ICES.

Maxim, L. & van der Sluijs, J.P. 2011. Quality in environmental science for policy: Assessing uncertainty as a component of policy analysis. *Environmental Science & Policy,* 14, 482–492.

Nielsen, K.N. & Holm, P. 2007. A brief catalogue of failures: Framing evaluation and learning in fisheries resource management. *Marine Policy*, 31, 669–680.

Nolin, J. 1995. *Stratospheric Ozone and Science: A Study of Post Normal Science (in Swedish, with English summary)*, Ph.D. Thesis. Ph.D., Göteborg University.

Nowotny, W., Scott, P. & Gibbons, M. 2001. *Re-thinking Science. Knowledge and the Public in an Age of Uncertainty*, London, Polity Press.

Nutters, H.M. & Pinto Da Silva, P. 2012. Fishery stakeholder engagement and marine spatial planning: Lessons from the Rhode Island Ocean SAMP and the Massachusetts Ocean Management Plan. *Ocean & Coastal Management*, 67, 9–18.

Pahl-Wostl, C. 2009. A conceptual framework for analysing adaptive capacity and multi-level learning processes in resource governance regimes. *Global Environmental Change*, 19, 354–365.

Pereira, Â. G. & Funtowicz, S. (eds.) 2015. Preface by Ravetz. In *The end of the Cartesian dream*, London, Routledge.

Petersen, A., Cath, A., Hage, M., Kunseler, E. & van der Sluijs, J.P. 2010. Post-Normal Science in Practice at the Netherlands Environmental Assessment Agency, Science, Technology, & Human Values. *Science Technology Human Values*. doi:10.1177/0162243910385797

Rademeyer, R.A., Plagányi, É.E. & Butterworth, D.S. 2007. Tips and tricks in designing management procedures. *ICES Journal of Marine Science*, 64, 618–625.

Ravetz, J. 2015. How quantitative evidence can be used irresponsibly by responsible scientists. *Significant Digits: Responsible Use of Quantitative Information*. Fondation Universitaire, Rue d'Egmont, 11, 9–10 June 2015, Brussels, Belgium: organized by the Joint Research Centre of the European Commission.

Rice, J.C. 2006. Every Which Way but Up: The Story of Cod in the North Atlantic. *Bulletin of Marine Science*, 78, 429–465.

Risbey, J.S., van der Sluijs, J.P., Kloprogge, P., Ravetz, J.R., Funtowicz, S.O. & Corral Quintana, S. 2005. Application of a Checklist for Quality Assistance in Environmental Modelling to an Energy Model. *Environmental Modeling & Assessment*, 10, 63–79.

Rochet, M.-J. & Rice, J.C. 2009. Simulation-based management strategy evaluation: ignorance disguised as mathematics? *ICES Journal of Marine Science*, 66, 754–762.

Rochet, M.J. & Rice, J. 2010. Comment on "Purported flaws in management strategy evaluation: basic problems or misinterpretation?" by Butterworth et al. *ICES Journal of Marine Science*, 67, 575–576.

Saltelli, A. & Funtowicz, S. 2014. When all models are wrong. *Issues in Science and Technology*, Winter, 79-95.

Schwach, V., Bailly, D., Christensen, A.-S., Delaney, A.E., Degnbol, P., van Densen, W.L.T., Holm, P., Mclay, H.A., Nielsen, K.N., Pastoors, M.A., Reeves, S.A. & Wilson, D.C. 2007. Policy and knowledge in fisheries management: a policy brief. *ICES Journal of Marine Science*, 64, 798-803.

Simmonds, E.J. 2007. Comparison of two periods of North Sea herring stock management: success, failure, and monetary value. *ICES Journal of Marine Science.*, 64, 686–692.

Sissenwine, M. & Symes, D. 2007. Relections on the Common Fisheries Policy. *Report to the General Directorate for Fisheries and Maritime Affairs of the European Commission.*

Smith, A.D.M. 1993. Risk assessment or management strategy evaluation: what do managers need and want? *ICES CM*, D:18, 1–6.

Smith, A.D.M., Sainsbury, K.J. & Stevens, R.A. 1999a. Implementing effective fisheries-management systems – management strategy evaluation and the Australian partnership approach. *ICES Journal of Marine Science*, 56, 967–979.

Smith, A.D.M., Sainsbury, K.J. & Stevens, R.A. 1999b. Implementing effective fisheries-management systems – management strategy evaluation and the Australian partnership approach. *ICES Journal of Marine Science*, 56, 967–979.

Ulrich, C., Coers, A., Hauge, K.H., Clausen, L.W., Olesen, C., Fisher, L., Johansson, R. & Payne, M.R. 2010. Improving complex governance schemes around Western Baltic Herring, through the development of a Long-Term Management Plan in an iterative process between stakeholders and scientists. *In:* ICES (ed.) *ICES CM 2010 / P:07.* Copenhagen, ICES.

van Asselt, M.B.A. & Rijkens-Klomp, N. 2002. A look in the mirror: Reflection on participation in integrated assessment from a methodological perspective'. *Global Environ. Change,* 12, 67–184.

van der Sluijs, J.P. 1997. Anchoring amid uncertainty: On the management of uncertainties in risk assessment of anthropogenic climate change. Ph.D., Utrecht University.

van der Sluijs, J.P. 2002. A way out of the credibility crisis of models used in integrated environmental assessment. *Futures,* 34, 133–146.

van der Sluijs, J.P. 2005. Uncertainty as a monster in the science policy interface: four coping strategies. *Water Science and Technology,* 52, 87–92.

van der Sluijs, J.P. 2007. Uncertainty and precaution in environmental management: Insights from the UPEM conference. *Environmental Modelling & Software,* 22, 590–598.

van der Sluijs, J.P., Craye, M., Funtowicz, S.O., Kloprogge, P., Ravetz, J.R. & Risbey, J.S. 2005. Combining quantitative and qualitative measures of uncertainties in model based environmental assessment: The NUSAP system. *Risk Analysis,* 25, 481–492.

van der Sluijs, J.P., Petersen, A.C., Anssen, P.H.M., Risbey, J.S. & Ravetz, J.R. 2008. Exploring the quality of evidence for complex and contested policy decisions. *Environmental Research Letters,* 3.

van der Sluijs, J. P., Risbey, J. S., Kloprogge, P., Ravetz, J. R., Funtowicz, S. O., Corral Quintana, S., Guimarães Pereira., De Marchi, B., Petersen, A. C., Janssen, P.H.M., Hoppe, R. & W F Huijs, S. 2003. RIVM/MNP Guidance for Uncertainty Assessment and Communication: Detailed Guidance (Volume 3), Utrecht University, 2003.

van der Sluijs, J.P., van Est, R. & Riphagen, M. 2010. Beyond consensus: reflections from a democratic perspective on the interaction between climate politics and science. *Current Opinion in Environmental Sustainability.* http://dx.doi.org/10.1016/j.cosust.2010.10.003.

Walker, W.E., Harremoës, P., van der Sluijs, J.P., van Asselt, M.B.A., Janssen, P. & Krayer von Krauss, M.P. 2003. Defining Uncertainty: A Conceptual Basis for Uncertainty Management in Model-Based Decision Support. *Integrated Assessment,* 4, 5–7.

Walters, C.J. 2007. Is Adaptive Management Helping to Solve Fisheries Problems? *AMBIO: A Journal of the Human Environment,* 36, 304–307.

Wardekker, J.A., Van der Sluijs, J.P., Janssen, P.H.M., Kloprogge, P. & Petersen, A.C. 2008. Uncertainty communication in environmental assessments: views from the Dutch science-policy interface. *Environmental Science & Policy,* 11, 627–641.

Wilson, D.C.& Pascoe, S. 2006. Delivering complex scientific advice to multiple stakeholders. In: L. Motos, D.C. Wilson (eds.), The Knowledge Base for Fisheries Management (Vol. 36 of Developments in Aquaculture and Fisheries Science), Amsterdam et al.: Elsevier, pp. 329–353.

Wynne, B. 1992. Uncertainty and environmental learning: Reconceiving science and policy in the preventive paradigm. *Global Environmental Change,* 2, 111–127.

Index

Page numbers in italic format indicate figures and tables.